Corrosion Mitigation Coatings

Also of interest

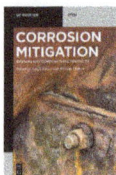

Corrosion Mitigation.
Biomass and Other Natural Products
Editors: Ashish Kumar and Abhinay Thakur, 2022
ISBN 9783110760576, e-ISBN (PDF) 9783110760583

Surface Characterization Techniques.
From Theory to Research
Rawesh Kumar, 2022
ISBN 9783110655995, e-ISBN (PDF) 9783110656480

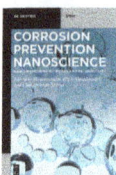

Corrosion Prevention Nanoscience.
Nanoengineering Materials and Technologies
Editors: Berdimurodov Elyor Tu and Chandrabhan Verma, 2023
ISBN 9783111070094, e-ISBN (PDF) 9783111071756

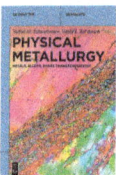

Physical Metallurgy.
Metals, Alloys, Phase Transformations
Vadim M. Schastlivtsev and Vitaly I. Zel'dovich, 2022
ISBN 9783110758016, e-ISBN (PDF) 9783110758023

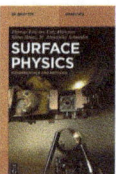

Surface Physics.
Fundamentals and Methods
Thomas Fauster, Lutz Hammer, Klaus Heinz
and M. Alexander Schneider, 2020
ISBN 9783110636680, e-ISBN (PDF) 9783110636697

Corrosion Mitigation Coatings

—

Functionalized Thin Film Fundamentals and Applications

Edited by
Ashish Kumar, Chandrabhan Verma and Abhinay Thakur

DE GRUYTER

Editors
Prof. Ashish Kumar
Department of Science and Technology
Nalanda College of Engineering (NCE)
Chandi 803108, Bihar, India
drashishchemlpu@gmail.com

Dr. Chandrabhan Verma
Department of Chemical Engineering
Khalifa University of Science and Technology
Abu Dhabi, United Arab Emirates
Chandraverma.rs.apc@itbhu.ac.in

Abhinay Thakur
Department of Chemistry
Faculty of Technology and Science
Lovely Professional University
GT Road
Phagwara 144411, Punjab, India
thakurabhinay96@gmail.com

ISBN 978-3-11-101530-9
e-ISBN (PDF) 978-3-11-101616-0
e-ISBN (EPUB) 978-3-11-101698-6

Library of Congress Control Number: 2023941015

Bibliographic information published by the Deutsche Nationalbibliothek
The Deutsche Nationalbibliothek lists this publication in the Deutsche Nationalbibliografie;
detailed bibliographic data are available on the internet at http://dnb.dnb.de.

© 2024 Walter de Gruyter GmbH, Berlin/Boston
Cover image: MR.SUTIN YUUKRUNG/iStock/Getty Images Plus
Typesetting: Integra Software Services Pvt. Ltd.
Printing and binding: CPI books GmbH, Leck

www.degruyter.com

Preface

Corrosion is a persistent and costly problem affecting a wide range of industries, from infrastructure and automotive to biomedical and aerospace. The detrimental effects of corrosion not only lead to significant economic losses but also pose safety and environmental concerns. In recent years, the development of functionalized thin film coatings has emerged as a promising solution to mitigate corrosion and enhance the durability of various materials and structures. This book delves into the comprehensive study of functionalized thin film coatings and their applications in corrosion protection. It aims to provide a comprehensive overview of the fundamental principles, fabrication strategies, synthesis techniques, and characterization methods related to functionalized thin film coatings. Furthermore, it explores the properties, applications, and performance of various types of functionalized thin film coatings in diverse fields. Chapter 1 serves as an introduction, presenting an overview of corrosion and functionalized thin film coatings. It discusses the significance of corrosion mitigation and introduces the concept of functionalized thin film coatings as a potential solution. Chapter 2 delves into the fabrication strategies, synthesis techniques, and characterization methods employed in the development of functionalized thin film coatings. It provides an in-depth understanding of the processes involved in creating these coatings and the tools used to analyze their properties.

Chapter 3 delves into the basics, properties, and applications of functionalized thin film coatings. It explores the diverse range of materials and technologies employed in creating these coatings and discusses their potential applications in corrosion protection. Chapter 4 focuses specifically on functionalized superhydrophobic surfaces and thin film coatings for corrosion protection. It delves into the unique properties of superhydrophobic coatings and their ability to repel water, thereby preventing corrosion. Chapter 5 explores the use of porous materials-based functionalized thin film coatings as corrosion inhibitors. It examines how the porosity of materials can be harnessed to enhance corrosion resistance and protect against various environmental factors. Chapter 6 shifts the focus to functionalized conductive polymer-based thin film coatings as corrosion inhibitors. It highlights the unique electrical properties of conductive polymers and their potential for corrosion protection. Chapter 7 delves into the use of functionalized nanomaterials and nanocomposites-based thin film coatings as corrosion inhibitors. It explores the applications of nanotechnology in developing advanced corrosion-resistant coatings. Chapter 8 explores functionalized carbon allotropes-based thin film coatings as corrosion inhibitors. It delves into the unique properties of different carbon allotropes and their potential in mitigating corrosion. Chapter 9 discusses the application of metal-organic frameworks (MOFs) based on functionalized thin film coatings as corrosion inhibitors. It explores the unique characteristics of MOFs and their potential in corrosion protection. Chapter 10 shifts the focus to functionalized quantum dots and carbonaceous quantum

https://doi.org/10.1515/9783111016160-202

dots-based thin film coatings as corrosion inhibitors. It delves into the unique properties of quantum dots and their potential for corrosion mitigation.

Chapter 11 explores the use of functionalized hybrid sol–gel thin film coatings for corrosion inhibition. It discusses the benefits of combining different materials in sol–gel coatings and their potential applications in corrosion protection. Chapter 12 delves into the incorporation of microcapsules-embedded corrosion inhibitors in functionalized thin film coatings. It explores the concept of self-healing coatings and their potential in enhancing the durability of coatings. Chapter 13 focuses on surface functionalized bioceramics coatings for anti-corrosion performance. It explores the use of bioceramic materials and their potential applications in corrosion protection. Chapter 14 shifts the focus to functionalized thin film coatings for reinforced concrete engineering. It discusses the challenges associated with corrosion in concrete structures and explores the potential of thin film coatings in enhancing the durability of concrete. Chapter 15 explores the applications of functionalized thin film coatings in the biomedical field. It discusses their biocompatibility, antibacterial properties, and potential for preventing fouling in biomedical applications. Chapter 16 delves into the applications of functionalized thin films for automotive coatings. It explores the unique challenges faced by automotive coatings and the potential of functionalized thin film coatings in enhancing their performance. Chapter 17 examines the influence of inorganic/organic additives on the mechanical and electrochemical properties of functionalized thin film coatings. It discusses the role of additives in improving the performance of coatings. Chapter 18 explores the role of surface functionalization on corrosion resistance and thermal stability of functionalized thin film coatings. It discusses how surface modifications can enhance the properties of coatings and improve their resistance to corrosion and high temperatures. Lastly, Chapter 19 reflects on the challenges and future outlooks in the field of functionalized thin film coatings for corrosion mitigation. It highlights the current limitations and areas for future research, paving the way for continued advancements in this important field.

This book serves as a comprehensive reference for researchers, scientists, engineers, and professionals involved in corrosion mitigation, materials science, surface engineering, and related fields. It brings together the latest research and insights into the fascinating world of functionalized thin film coatings and their applications in corrosion protection. It is our hope that this book will inspire further research and innovation in this critical area, ultimately leading to the development of more effective and sustainable solutions for corrosion mitigation.

Ashish Kumar, Chandrabhan Verma, Abhinay Thakur

Contents

List of contributing authors

Jasdeep Kaur
Department of Chemistry, Chandigarh
University, Gharuan, Mohali, Punjab, India
Jasdeep.e9012@cumail.in
Chapter 1

Akhil Saxena
Department of Chemistry, Chandigarh
University, Gharuan, Mohali, Punjab, India
akhil.uis@cumail.in
Chapter 1

Saumya Jaithlia
Department of Chemical Engineering, University
of Petroleum and Energy Studies, Dehradun
248007, Uttarakhand, India
saumyajaithlia8@gmail.com
Chapter 2

Manash Protim Mudoi
Department of Chemical Engineering, Indian
Institute of Technology, Roorkee 247667,
Uttarakhand, India
mp_mudoi@ch.iitr.ac.in,
mpmudoi@ddn.upes.ac.in
Chapter 2

Sheerin Masroor
Department of Chemistry, A. N. College,
Patliputra University, Patna 800013, Bihar, India
masroor.sheerin@gmail.com
Chapter 3

Sanjukta Zamindar
Surface Engineering and Tribology Group, CSIR-
Central Mechanical Engineering Research
Institute, Mahatma Gandhi Avenue, Durgapur
713209, West Bengal, India
and
Academy of Scientific and Innovative Research
(AcSIR), CSIR-HRDC Campus, Sector 19, Kamla
Nehru Nagar, Ghaziabad 201002, Uttar Pradesh,
India
sanjuktazamindar2016@gmail.com
Chapter 4, 12

Surya Sarkar
Surface Engineering and Tribology Group, CSIR-
Central Mechanical Engineering Research
Institute, Mahatma Gandhi Avenue, Durgapur
713209, West Bengal, India
and
Academy of Scientific and Innovative Research
(AcSIR), CSIR-HRDC Campus, Sector 19, Kamla
Nehru Nagar, Ghaziabad 201002, Uttar Pradesh,
India
and
Department of Physics, Durgapur Women's
College, Durgapur 713209, West Bengal, India
suryasarkar52@gmail.com
Chapter 4

Manilal Murmu
Surface Engineering and Tribology Group, CSIR-
Central Mechanical Engineering Research
Institute, Mahatma Gandhi Avenue, Durgapur
713209, West Bengal, India
and
Academy of Scientific and Innovative Research
(AcSIR), CSIR-HRDC Campus, Sector 19, Kamla
Nehru Nagar, Ghaziabad, 201002, India
and
Present Affiliation: Department of Nuclear and
Quantum Engineering, Korea Advanced Institute
of Science and Technology (KAIST), 291, Daehak-
ro, Yuseong-gu, Daejeon 34,141, Republic of
Korea
mani.murmu@gmail.com
Chapter 4, 5, 10, 12

Priyabrata Banerjee
Surface Engineering and Tribology Group, CSIR-
Central Mechanical Engineering Research
Institute, Mahatma Gandhi Avenue, Durgapur
713209, West Bengal, India
and
Academy of Scientific and Innovative Research
(AcSIR), CSIR-HRDC Campus, Sector 19, Kamla
Nehru Nagar, Ghaziabad 201002, Uttar Pradesh,
India
pr_banerjee@cmeri.res.in
Chapter 4, 5, 10, 12

https://doi.org/10.1515/9783111016160-204

Sukdeb Mandal
Surface Engineering and Tribology Group, CSIR-Central Mechanical Engineering Research Institute, Mahatma Gandhi Avenue, Durgapur 713209, West Bengal, India
and
Academy of Scientific and Innovative Research (AcSIR), CSIR-HRDC Campus, Sector 19, Kamla Nehru Nagar, Ghaziabad 201002, Uttar Pradesh, India
sukdebmandal1995@gmail.com
Chapter 5, 12

Naresh Chandra Murmu
Surface Engineering and Tribology Group, CSIR-Central Mechanical Engineering Research Institute, Mahatma Gandhi Avenue, Durgapur 713209, West Bengal, India
and
Academy of Scientific and Innovative Research (AcSIR), CSIR-HRDC Campus, Sector 19, Kamla Nehru Nagar, Ghaziabad 201002, Uttar Pradesh, India
murmu@cmeri.res.in
Chapter 5, 10

Pragnesh N. Dave
Department of Chemistry, Sardar Patel University, Vallabh Vidyangar 388 120, Gujarat, India
pragnesh7@yahoo.com
Chapter 6

Pradip M. Macwan
B. N. Patel Institute of Paramedical and Science (Science Division), Sardar Patel Education Trust, Bhalej Road, Anand 388001, Gujarat, India
pradipspu@gmail.com
Chapter 6

Elyor Berdimurodov
Faculty of Chemistry, National University of Uzbekistan, Tashkent 100034, Uzbekistan
elyor170690@gmail.com
Chapter 7, 17

Ilyos Eliboev
Faculty of Chemistry, National University of Uzbekistan, Tashkent 100034, Uzbekistan
ilyoseliboev883@mail.com
Chapter 7, 17

Abduvali Kholikov
Faculty of Chemistry, National University of Uzbekistan, Tashkent 100034, Uzbekistan
abduvali0079@gmail.com
Chapter 7, 17

Khamdam Akbarov
Faculty of Chemistry, National University of Uzbekistan, Tashkent 100034, Uzbekistan
akbarov_Kh@rambler.ru
Chapter 7, 17

Dakeshwar Kumar Verma
Department of Chemistry, Government Digvijay Autonomous Postgraduate College, Rajnandgaon 491441, Chhattisgarh, India
dakeshwarverma@gmail.com
Chapter 7

Mohamed Rbaa
Laboratory of Organic Chemistry, Catalysis and Environment, Faculty of Sciences, Ibn Tofail University, PO Box 133, 14,000, Kenitra, Morocco
mohamed.rbaa10@gmail.com
Chapter 7

Omar Dagdag
Centre for Materials Science, College of Science, Engineering and Technology, University of South Africa, Johannesburg 1710, South Africa
omar.dagdag@uit.ac.ma
Chapter 7

Khasan Berdimuradov
Faculty of Industrial Viticulture and Food Production Technology, Shahrisabz Branch of Tashkent Institute of Chemical Technology, Shahrisabz 181306, Uzbekistan
khasanberdimuradov@gmail.com
Chapter 7, 17

Deepak Sharma
Department of Chemistry, Deenbandhu Chhotu Ram University of Science and Technology, Murthal 131039, Haryana, India
vatsdeepak2014@gmail.com
Chapter 8

Hari om
Department of Chemistry, Deenbandhu Chhotu
Ram University of Science and Technology,
Murthal 131039, Haryana, India
hariom.chem@dcrustm.org
Chapter 8

Abhinay Thakur
Department of Chemistry, School of Chemical
Engineering and Physical Sciences, Lovely
Professional University, Phagwara 144411,
Punjab, India
thakurabhinay96@gmail.com
Chapter 8, 13, 14, 18, 19

Ashish Kumar
Department of Science and Technology, NCE,
Bihar Engineering University, Chandi 803108,
Bihar, India
drashishchemlpu@gmail.com
Chapter 8, 13, 14, 18, 19

Rajimol P. R.
Material Science and Technology Division CSIR-
NIIST, Trivandrum, Kerala, India
rajipr19@gmail.com
Chapter 9

Sarah Bill Ulaeto
Rhema University, Aba Abia, Nigeria
sarahbillmails@yahoo.com
Chapter 9

Ben John
Material Science and Technology Division CSIR-
NIIST, Trivandrum, Kerala, India
benjohncherpu67@gmail.com
Chapter 9

T. P. D. Rajan
Material Science and Technology Division CSIR-
NIIST, Trivandrum, Kerala, India
tpdrajan@gmail.com, tpdrajan@niist.res.in
Chapter 9

Khalid Bouiti
Laboratory of Molecular Spectroscopy
Modelling, Materials, Nanomaterials, Water and

Environment, CERNE2D, ENSAM, Mohammed V
University, Rabat, Morocco
khalid.bouiti.khb@gmail.com
Chapter 11

Nabil Lahrache
Laboratory of Molecular Spectroscopy
Modelling, Materials, Nanomaterials, Water and
Environment, CERNE2D, ENSAM, Mohammed V
University, Rabat, Morocco
lahrachenabil2007@yahoo.fr
Chapter 11

Ichraq Bouhouche
Laboratory of Molecular Spectroscopy
Modelling, Materials, Nanomaterials, Water and
Environment, CERNE2D, ENSAM, Mohammed V
University, Rabat, Morocco
bouhouche.ichraq@gmail.com
Chapter 11

Najoua Labjar
Laboratory of Molecular Spectroscopy
Modelling, Materials, Nanomaterials, Water and
Environment, CERNE2D, ENSAM, Mohammed V
University, Rabat, Morocco
najoua.labjar@ensam.um5.ac.ma
Chapter 11

Meriem Bensmlali
Laboratoire de chimie organique, bio-organique
et environnement, Faculty of Sciences, Chouaib
Doukkali University, El Jadida, Morocco
bensemlali.meryem@gmail.com
Chapter 11

Souad El Hajjaji
Laboratory of Molecular Spectroscopy
Modelling, Materials, Nanomaterials, Water and
Environment, CERNE2D, Faculty of Science,
Mohammed V University, Rabat, Morocco
s.elhajjaji@um5r.ac.ma
Chapter 11

Richika Ganjoo
Department of Chemistry, School of Chemical
Engineering and Physical Sciences, Lovely
Professional University, Phagwara 144411,

Punjab, India
richikaganjoo@gmail.com
Chapter 13, 18

Shveta Sharma
Department of Chemistry, School of Chemical
Engineering and Physical Sciences, Lovely
Professional University, Phagwara 144411,
Punjab, India
shveta1chem@gmail.com
Chapter 13, 18

Praveen K. Sharma
Department of Chemistry, School of Chemical
Engineering and Physical Sciences, Lovely
Professional University, Phagwara 144411,
Punjab, India
praveen.14155@lpu.co.in
Chapter 13

Nancy George
Department of Chemistry, School of Chemical
Engineering and Physical Sciences, Lovely
Professional University, Phagwara 144411,
Punjab, India
nancygeorge679@gmail.com
Chapter 13

Humira Assad
Department of Chemistry, School of Chemical
Engineering and Physical Sciences, Lovely
Professional University, Phagwara 144411,
Punjab, India
humiraassad888@gmail.com
Chapter 14

Ishrat Fatma
Department of Chemistry, School of Chemical
Engineering and Physical Sciences, Lovely
Professional University, Phagwara 144411,
Punjab, India
ishratfatma0120@gmail.com
Chapter 14

Sonam Singh
Indian Institute of Technology (Indian School of
Mines), Dhanbad, Dhanbad 826004, Jharkhand,
India
sonamamism@gmail.com
Chapter 15

Sayantan Guha
Institute of Technical Education and Research
(ITER), Siksha 'O' Anusandhan (Deemed to be
University), Bhubaneswar 751030, Odisha, India
sayantanguha.maths@gmail.com
Chapter 15

Amita Somya
Department of Chemistry, School of Engineering,
Presidency University, Bengaluru, Karnataka,
India
amitasomya@presidencyuniversity.in
Chapter 16

Amit Prakash Varshney
Tech Lead, Tata Elxsi India, Bengaluru,
Karnataka, India
amit.varshney.kec@gmail.com
Chapter 16

Anvar Khamidov
Faculty of Chemistry, National University of
Uzbekistan, Tashkent 100034, Uzbekistan
a.xamidov8997@gmail.com
Chapter 17

Oybek Mikhliev
Karshi Engineering Economics Institute, Karshi
City 130100, Uzbekistan
elyor150697@gmail.com
Chapter 17

Harpreet Kaur
Department of Chemistry, School of Chemical
Engineering and Physical Sciences, Lovely
Professional University, Phagwara 144411,
Punjab, India
kaurhhh@gmail.com
Chapter 19

Ramesh Chand Thakur
Department of Chemistry, Himachal Pradesh
University, Summer Hill 171005, Himachal
Pradesh, India
drthakurchem@gmail.com
Chapter 19

Jasdeep Kaur, Akhil Saxena

1 An overview of corrosion and functionalized thin film coating

Abstract: Corrosion poses a significant challenge as it makes the metallic component susceptible to damage, leading to both economic losses and environmental hazards for Mother Nature. It is an electrochemical reaction that occurs due to interaction of metal with the environment when affected by the pH, temperature, pressure, etc. Corrosion is a destructive and continuous problem and is a challenging problem for the entire world. Complete eradication would be impractical and impossible but prevention is more practical and feasible. To overcome this problem, thin film coating is used. A thin film coating known as an anti-corrosive thin film coating is applied to the surface of metal to prevent corrosion. Coating of the surface, also known as a passivation layer, is a traditional technique for preventing corrosion. This book chapter provides an overview of corrosion and various types of thin film coating.

Keywords: Corrosion, Functionalization, Thin Film, Mitigation, Metals

1.1 Introduction

1.1.1 Corrosion and its types

Corrosion is a degradation of a substance, usually a metal, or its characteristics because of a material's reaction with the surrounding [1]. Corrosion is defined scientifically by the International Standard Organization as the "physio-chemical reaction between a metal and its environment, resulting in alteration to the metal's properties and degradation of the metal's function [2]." Wet corrosion and dry corrosion are the two categories into which this degradation can be divided. Dry corrosion occurs when there a liquid phase is absent or when the environment's dew point is exceeded. In this situation a metal usually reacts with vapors and gases at high temperatures. On metal surfaces, these reactions may result in a scaling and pitting attack. In wet corrosion, a metal often reacts with electrolytes or aqueous solutions [3].

Basically, corrosion occurs due to the creation of electrochemical cells as shown in Figure 1.1 and for it to occur, five components must be there. If any of these components is eliminated, probably corrosion will not start. The following elements must exist for corrosion to occur [4]:

https://doi.org/10.1515/9783111016160-001

Anode is a part from where loss of electrons (oxidation) occurs. This part is of more negative potential as compared to a cathode.

Cathode is a part where gain of electrons (reduction) occurs. This part is of more positive potential as compared to an anode.

Electrolyte is a medium containing ion. An electrolyte may be a salt, acid, or base dissolved in the polar solvent.

Electrical connection is the connection of a cathode and an anode.

Potential difference must be there between a cathode and an anode to start reaction.

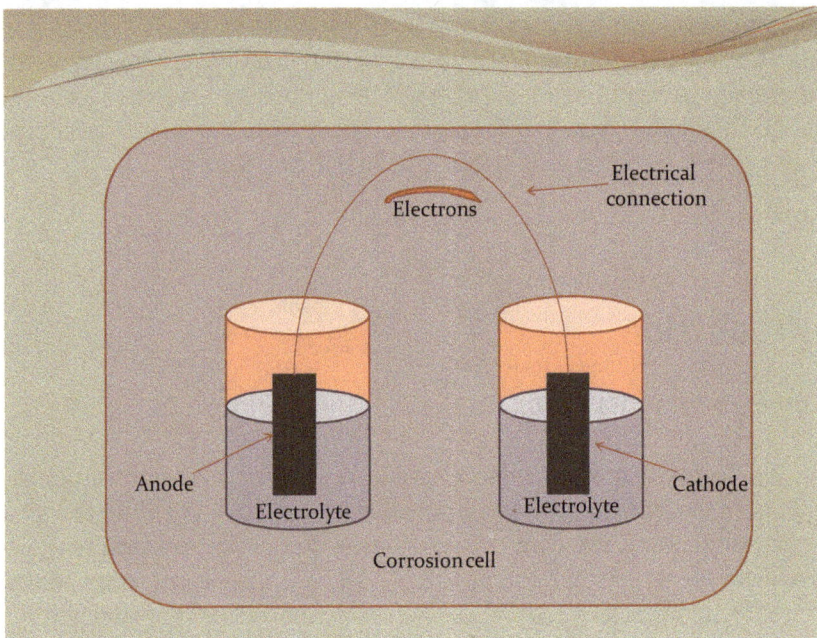

Figure 1.1: Electrochemical cell structure.

The various types of corrosion of low-carbon steel, which depends upon the surrounding are shown in Figure 1.2 and briefly discussed in Table 1.1.

1.1.2 Loss due to corrosion

It causes major economic losses for the entire world. For example, one-fourth of the world's steel production is thought to be lost to corrosion. In addition to having negative

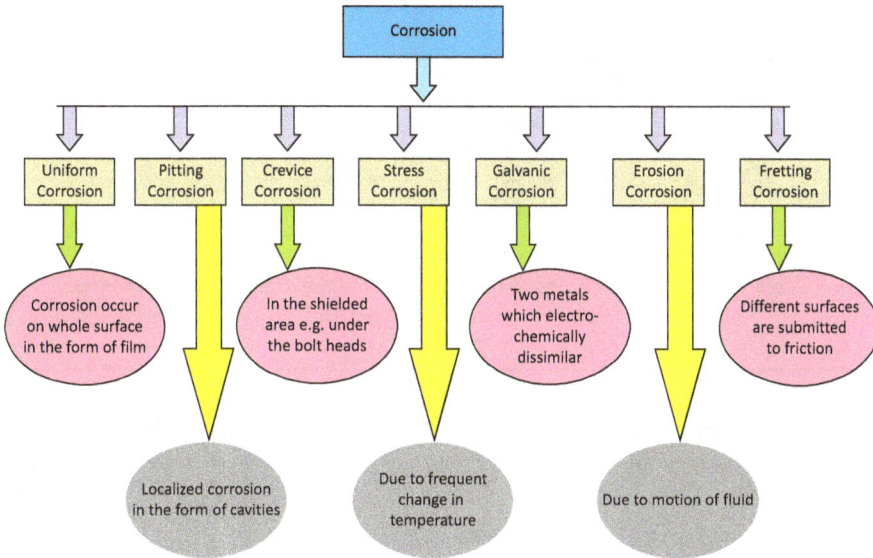

Figure 1.2: Various types of corrosion.

Table 1.1: Types of corrosion.

Types of corrosion	Description
Uniform corrosion	Degradation happens over the whole surface of metal, which makes the surface very thin. When the outer layer is exposed to air, soil, or water, there is a rust coating on the surface.
Pitting corrosion	Localized type of corrosion created in the material by cavities or "holes." Pitting is considered more dangerous than uniform corrosion because corrosion materials cover the holes.
Crevice corrosion	Localized type of corrosion occurs predominantly under the bolt heads in the shielded region, due to a decrease in oxygen content and change in pH.
Stress corrosion cracking	Occurs due to external stress due to frequent change in temperature. Stresses may be generated due to welding, heat treatment, machining, and grinding.
Galvanic corrosion	Three conditions must be present for galvanic corrosion to happen: 1. There must be electrochemically dissimilar metals. 2. These metals have to be in electrical contact. 3. It is necessary to expose the metals to an electrolyte.

Table 1.1 (continued)

Types of corrosion	Description
Erosion corrosion	In the gas and oil sectors, this is more popular. In this type of corrosion, fluid motion removes the corrosion coating that typically consists of metal oxides and metal hydroxides, thereby re-exposing the metal surface to corrosion. Erosion corrosion occurs frequently where fluid moves with high velocity.
Fretting corrosion	When two separate substances or surfaces are exposed to friction, this type of corrosion occurs. The metallic surface's proactive coating is eliminated by rubbing activity and the active surface is re-exposed. This corrosion happens because most equipment is prone to vibration.

economic effects, corrosion has caused a number of structural failures that have had a negative impact on both the environment as well on human health as shown in Figure 1.3.

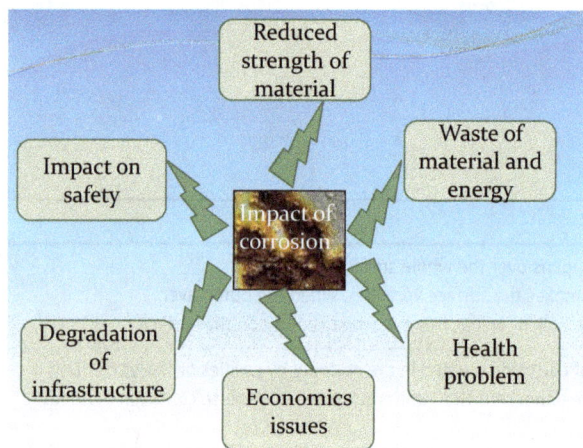

Figure 1.3: Impact of corrosion.

Corrosion is also harmful for human life. For example, in the West Virginia bridge collapse in 1967, which lost 46 lives, stress corrosion cracking was identified as the cause of the collapse.

In another instance, the combined effects of stress and atmospheric corrosion caused the fuselage of an airliner to rip open in Hawaii.

As a result, corrosion issues must be resolved to ensure safety and to reduce environmental contamination. Metal deterioration occurs when the surface comes into contact with a gas or liquid and exposure to warm temperatures, acids, and salts accelerates the process. In chemical industries, oil, and refineries, mainly where mild steel containers are used for various syntheses, acid pickling harms those containers at a very high rate. In accordance with one report, $7.2 billion is spent on corrosion inhibitors globally, which is expected to increase $9.6 billion by 2026 with a growth rate of 3.6% [5]. The major consequences of corrosion are economical loss in a chemical industry and damage of equipment, which overall decrease the efficiency of valuable products. Every year, it is estimated that 10% of total global metal is destroyed due to corrosion, affecting the economic growth of the country [6].

According to NACE International's 2016 estimate, there is a $2.5-trillion economic loss worldwide as a result of corrosion [7, 8]. The direct cost of corrosion in India in the years 1984–1985 was estimated to be INR 40.76 billion, of which INR 18.04 billion was thought to be avoidable [9].

Another study on the cost of corrosion in India, conducted in 1997, estimated a loss of INR 250 billion annually [10]. According to the most recent global assessment by NACE International, corrosion is predicted to cost India's economy 4.2% of its GDP.

1.2 Corrosion thermodynamics

As we have seen, electron transport is a necessary part of corrosion reactions; because of this, corrosion can be considered as an electrochemical reaction. The energy changes connected to the corrosion reaction can be supported by thermodynamics. In general, it can detect when corrosion might occur.

We can use the Gibb's free energy, which is given by the following equation, to determine whether a corrosion reaction is thermodynamically feasible or not.

$$\Delta G = -nFE$$

Here, ΔG = Gibbs free energy
n = number of electrons
F = Faraday's constant
E = Standard potential

The corrosion reaction will not occur when ΔG is positive and is possible only when ΔG is negative.

1.2.1 Factors affecting rate of corrosion

Nearly all metals are chemically unstable in the presence of air. When metal comes in contact with air it forms metal oxide.

Various factors that are responsible for corrosion are shown in Figure 1.4.

Electrochemical reactions may start when materials interact with the environment. The material corrodes as a result of this process. Corrosion is a serious concern because corrosion degrades and eventually destroys materials.

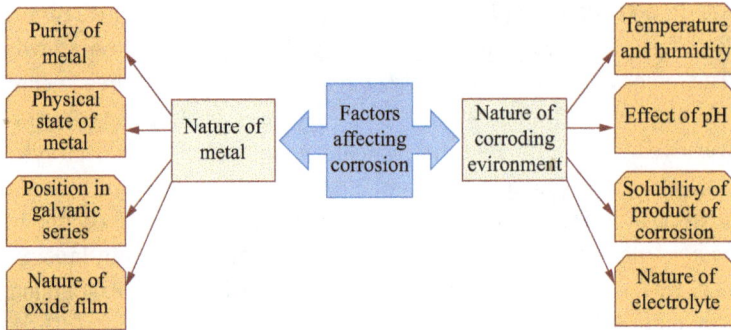

Figure 1.4: Factors affecting corrosion.

The factors that are responsible for corrosion are discussed in the sections that follow:

1.2.1.1 Nature of metal

Metals having the lowest electrode potentials are more susceptible to corrosion; however, higher electrode potential metals resist corrosion better as shown in Figure 1.5.

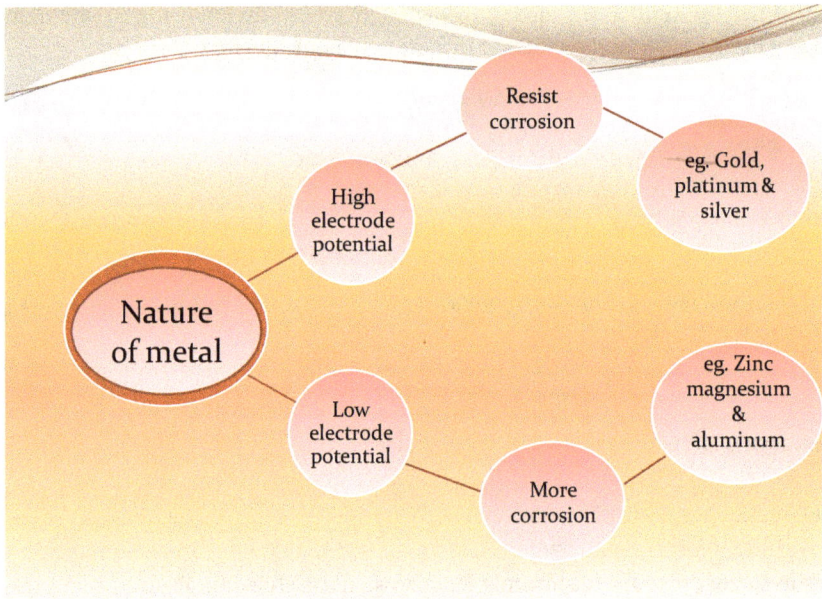

Figure 1.5: Corrosion behavior of metal on the basis of electrode potential.

Higher the electrode potential difference between two metals when they are in contact, more the corrosion that will occur. For instance, there is a greater potential difference between iron and copper (0.78 V) than between iron and tin (0.3 V). As a result, when in touch with copper rather than tin, iron corrodes more quickly. On account of this, it is best to avoid using different metals wherever possible. Other factors are:

1.2.1.2 Purity of metal

The purity of a metal refers to the absence of any other element or impurities in the metal. A pure metal is made up of only one type of atom, and this makes it less likely to corrode or react with other substances. Corrosion is the gradual destruction of a metal by chemical or electrochemical reaction with its environment. When a metal is exposed to environment containing water, air, and other chemicals, it can corrode, forming a layer of rust, oxide, or other unwanted substances on the surface of the metal. Impurities in a metal can create tiny electrochemical cells that can initiate corrosion. These cells can create a voltage gradient across the metal surface, leading to the flow of electrons and the formation of corrosion products. The presence of impurities can also weaken the metal and make it more susceptible to mechanical damage. For example, iron is a metal that easily reacts with oxygen to form iron oxide (rust). However, if the iron is pure, it will not corrode easily. But if the iron contains impurities such as carbon, it can form different types of corrosion, including galvanic corrosion, crevice corrosion, or pitting corrosion, which can significantly reduce the lifespan of the metal.

1.2.1.3 State of metal

The physical state of a metal plays a significant role in its susceptibility to corrosion. Corrosion is the gradual destruction of a metal by chemical or electrochemical reaction with its environment. When a metal is exposed to environment containing water, air, and other chemicals, it can corrode, forming a layer of rust, oxide, or other unwanted substances on the surface of the metal. In general, corrosion occurs more slowly in metals that are under less stress, and the rate of corrosion will be greater if the metal is subjected to excessive stress. Stress can come from various sources, including mechanical forces, temperature changes, and chemical exposure.

When a metal is stressed, it becomes more vulnerable to corrosion because the stresses can create microcracks or defects on the metal surface. These cracks and defects provide entry points for water, air, and other chemicals, which can accelerate the corrosion process. In contrast, metals that are under less stress tend to have fewer defects on their surface, making it harder for corrosive agents to penetrate and cause damage. For example, consider a metal bridge subjected to constant vibrations from heavy traffic. The constant vibration causes stress on the metal, leading to microcracks and defects on its surface. The exposure to air, water, and other chemicals accelerates the corrosion process, leading to a reduction in the strength and durability of the bridge.

Similarly, the physical state of a metal can also affect its corrosion resistance at different temperatures. At high temperatures, metals are more susceptible to corrosion because the heat can cause changes in the structure of the metal, making it easier for corrosive agents to penetrate and cause damage.

1.2.1.4 Position of metal in the galvanic Series

The galvanic series is a ranking of metals and alloys based on their tendency to corrode in certain environments. Metals that are situated higher in the galvanic series have higher oxidation powers and a greater tendency to act anodic, which results in higher rates of corrosion. The galvanic series is based on the electrochemical potential of metals in seawater or other reference electrolytes. The more negative the potential, the more anodic the metal is, and the more likely it is to corrode. Metals with a more positive potential are less anodic and less likely to corrode.

When two different metals are in contact in the presence of an electrolyte, such as seawater or humid air, an electrochemical reaction can occur, leading to corrosion of the more anodic metal. This process is known as galvanic corrosion or bimetallic corrosion. For example, consider a steel pipe that is connected to a copper fitting. If water flows through the pipe, the steel will be more anodic than the copper and will corrode more quickly. The position of the two metals in the galvanic series determines the severity of the corrosion.

In addition to galvanic corrosion, the position of a metal in the galvanic series can also affect its susceptibility to other types of corrosion, such as pitting corrosion and crevice corrosion. Metals that are more anodic tend to be more susceptible to these forms of corrosion because they can initiate the corrosion process.

1.2.1.5 Characteristics of the corrosion product

The characteristics of the corrosion product, as well as the environment in which the metal is placed, are both important factors that can affect the rate of corrosion. When a metal corrodes, a layer of corrosive product is formed on the surface of the metal. If the metal oxide is stable, insoluble, and adherent, it may serve as a protective coating that acts as a barrier between the metal surface and the corrosive medium, stopping further corrosion. In contrast, if the oxide layer is soluble, porous, and unstable, corrosion will be accelerated. For example, aluminum has a natural oxide layer that forms on its surface, which is stable and prevents further corrosion. This makes aluminum an excellent choice for outdoor applications where it is exposed to the elements. In contrast, iron forms a porous oxide layer, which does not protect the underlying metal, and leads to rapid corrosion.

The corrosive environment also affects the rate of corrosion. Temperature, humidity, and the presence of corrosive gases in the atmosphere are all factors that can affect the rate of corrosion. Corrosion is directly related to temperature. With rise in temperature, there is rise in the corrosion rate. This happens due to the increase in the diffusion rate of ions with an increase in temperature. Higher temperatures can also cause changes in the chemical reactions that occur during corrosion, leading to more rapid degradation of the metal. Humidity in the air also directly affects the corrosion rate because moisture in the air behaves as a solvent for the gases present in the atmosphere, resulting in the formation of an electrolyte. This electrolyte can facilitate the corrosion process by allowing for the transfer of electrons and ions between the metal surface and the environment. Higher humidity levels can lead to more rapid corrosion. In addition, the presence of corrosive gases in the atmosphere can also accelerate corrosion. For example, sulfur dioxide and nitrogen oxides can react with water vapor to form acids that can corrode metals. Similarly, chlorides in the air can promote the formation of corrosion products that are more soluble and less protective, leading to accelerated corrosion.

In summary, the characteristics of the corrosion product and the environment in which the metal is placed are both important factors that can affect the rate of corrosion. A stable and insoluble oxide layer can protect the metal, while an unstable and porous oxide layer can accelerate corrosion. The temperature, humidity, and presence of corrosive gases in the atmosphere can all affect the rate of corrosion and should be taken into account when selecting materials for various applications. [11–14].

1.2.1.6 Effect of pH

The pH of the environment surrounding a metal can have a significant impact on the rate of corrosion. Generally, an increase in pH can slow the pace of corrosion, while a decrease in pH can accelerate it. The reason for this is related to the presence of H^+ and OH^- ions in the solution, which can influence the anodic and cathodic reactions that result in corrosion. In an acidic solution with a lower pH, the cathodic process is accelerated by the presence of more H^+ ions. This can result in an increase in corrosion rates. On the other hand, at neutral or basic pH levels, more OH^- ions are produced by the cathodic reaction. This can slow down the corrosion reaction at the anodic site.

To protect metals from corrosion, various methods can be used, including the use of thin film coatings. Thin film coatings are applied to the metal surface and act as a barrier, preventing the metal from coming into contact with the surrounding environment. This can prevent the corrosion reaction from occurring. There are many types of thin film coatings available, including organic coatings, inorganic coatings, and hybrid coatings. Organic coatings are composed of polymers that are applied to the metal surface as a liquid and then cured to form a solid film. These coatings are often used for decorative purposes and to protect against environmental factors such as UV radiation. Inorganic coatings, on the other hand, are composed of inorganic materials such as ceramics, metals, or metal oxides. These coatings are often used in harsh environments where organic coatings may not provide adequate protection. Hybrid coatings are a combination of organic and inorganic materials and offer the benefits of both types of coatings. They can provide good adhesion to the metal surface and excellent protection against corrosion.

1.3 Functionalized thin film coating

Corrosion is the process of degradation and destruction of metals due to their interaction with the environment. Corrosion occurs due to the formation of electrochemical cells on the metal surface, which leads to the oxidation of the metal and the release of electrons. Corrosion can result in the loss of mechanical properties, reduced lifespan, and failure of the metal component. Therefore, corrosion inhibition is critical to ensure the reliability and longevity of metal structures and devices. Functionalized thin film coatings are a promising solution to prevent and inhibit corrosion in various applications. Thin film coatings can be designed to provide a range of properties, such as improved adhesion, durability, hardness, and resistance to wear, corrosion, and chemical attack. They can also be tailored to have specific surface properties, such as hydrophobicity or hydrophilicity, which can be useful in applications where surface interactions are important, such as in biomedical devices, sensors, and electronics. Functionalized thin film coatings can inhibit corrosion by various mechanisms. The most common mechanism is the creation of a physical barrier between the metal surface and the corrosive environment. This barrier can prevent the diffusion of corrosive species, such as oxygen, water, and ions, to the metal surface, thereby reducing the corrosion rate. The thickness and porosity of the coating can significantly affect its barrier properties. Thicker and more compact coatings provide better protection against corrosion than thinner and more porous coatings.

Another mechanism of corrosion inhibition is the formation of a passivating layer on the metal surface. Passivation occurs when a thin layer of a corrosion-resistant material is formed on the metal surface, which prevents further corrosion. Passivation can occur naturally, as in the case of stainless steel, or can be induced by the application of a passivating agent, such as a functionalized thin film coating.

Functionalized thin film coatings can also inhibit corrosion by modifying the electrochemical properties of the metal surface. For example, some coatings can act as a cathodic or anodic inhibitor, which alters the current flow in the electrochemical cell and reduces the corrosion rate. The composition and morphology of the coating can significantly affect its electrochemical properties. Due to the increase in the durability of the metal structures and components in acidic environments, effective anti-corrosion pre-treatments for metallic substrates have drawn a lot of attention in these days. Thin films have developed over several decades and more than a century ago. For example, the first film of diamond-like polycrystalline carbon was formed in 1911, and amorphous carbon thin films were actively researched and developed in the 1970s [15]. Unique characteristics, such as anti-corrosion, self-cleaning, chemical resistance, and scratch resistance, are produced by a variety of thin film coatings [16]. Because of the effectiveness in preventing corrosion, conventional chromate conversion coatings (CCC) were initially seen as good choices. The industry adopted chromate coatings early because of their high electric conductivity, simplicity of application, and capacity for self-healing [17–19]. However, it has been demonstrated that exposure to chromates increases the risk of lung cancer and damages human DNA due to their potent oxidation properties. Due to rigorous environmental legislation and stricter rules on the usage of hexavalent chromate, the need for more ecologically friendly coatings increased.

1.3.1 Types of functionalized thin film coatings

There are several different types of functionalized thin film coatings, each of which is designed to provide specific properties and functions. We will discuss some of the most common types of functionalized thin film coatings and their properties.

1.3.1.1 Corrosion-resistant coatings

Corrosion-resistant coatings are designed to protect metal surfaces from the effects of corrosion, which can cause damage and degradation of the material over time. These coatings are typically composed of materials such as oxides, nitrides, and carbides, which are deposited onto the surface of the metal using techniques such as physical vapour deposition (PVD) or chemical vapor deposition (CVD). Corrosion-resistant functionalized coatings are specialized coatings that provide corrosion protection and additional functionalities. These coatings are composed of a base material that provides corrosion resistance, as well as other materials that provide self-healing, self-cleaning, anti-microbial, or smart properties. Self-healing coatings contain microcapsules that release a healing agent when the coating is damaged. Self-cleaning coatings repel dirt and other contaminants, while antimicrobial coatings inhibit the growth of bacteria and other microorganisms. Nanocomposite coatings combine a base material with nanoparticles to provide additional functionalities. Smart coatings can sense changes in the environment and respond accordingly. One of the most commonly used corrosion-resistant coatings is titanium nitride (TiN), which is deposited onto the surface of metal components to improve their resistance to corrosion, wear, and abrasion. TiN coatings are highly durable and can withstand exposure to harsh environments, making them ideal for use in a wide range of industrial applications.

1.3.1.2 Wear-resistant coatings

Wear-resistant coatings are designed to protect surfaces from the effects of friction and wear, which can cause damage and reduce the lifespan of components. These coatings are typically composed of materials such as diamond-like carbon (DLC), which are deposited onto the surface of the material using techniques such as PVD. DLC coatings are highly durable and can withstand exposure to harsh environments, making them ideal for use in a wide range of industrial applications. They are commonly used to protect the surfaces of cutting tools, molds, and dies, as well as in the aerospace and automotive industries.

1.3.1.3 Anti-fouling coatings

Anti-fouling coatings are designed to prevent the accumulation of fouling agents such as bacteria, algae, and other microorganisms on the surface of the material. These coatings are typically composed of materials such as polymers, which are deposited onto the surface of the material using techniques such as ethyl cellulose dispersion (ECD) or spray coating. Anti-fouling coatings are commonly used in marine environments, where the accumulation of fouling agents can cause damage to ships and other vessels. They are also used in medical applications, where the accumulation of bacteria and other microorganisms can cause infections.

1.3.1.4 Optical coatings

Optical coatings are designed to modify the optical properties of materials, such as their reflectivity, transmission, and absorption of light. These coatings are typically composed of materials such as metals, oxides, and nitrides, which are deposited onto the surface of the material using techniques such as PVD or CVD. Optical coatings are commonly used in a wide range of applications, including lenses, mirrors, and other optical components. They can be used to improve the performance of optical systems, as well as to create new and innovative optical devices.

1.3.1.5 Electrical coatings

Electrical coatings are designed to modify the electrical properties of materials, such as their conductivity or resistivity. These coatings are typically composed of materials such as metals, oxides, and nitrides, which are deposited onto the surface of the material using techniques such as PVD or CVD. Electrical coatings are commonly used in a wide range of applications, including electronic components, sensors, and other electrical devices. They can be used to improve the performance of these devices, as well as to create new and innovative electrical systems.

1.3.1.6 Polymer-based coatings

Polymer-based coatings are organic materials made up of long chains of repeating units, which provide a barrier that prevents corrosive agents from reaching the metal surface. Polymer-based coatings can be applied as thin films and are generally easy to apply using various techniques, including spray coating, dip coating, and electrodeposition. These coatings can also be tailored to exhibit specific properties, such as self-healing, anti-fouling, and anti-corrosion properties.

There are two types of polymer-based coatings: thermoplastic and thermosetting coatings. Thermoplastic coatings can be heated and reformed multiple times without losing their mechanical properties. These coatings typically have low chemical resistance and may not be able to withstand high temperatures. Thermosetting coatings are cross-linked and cured at high temperatures, resulting in a more durable coating with excellent chemical resistance and thermal stability.

One example of a polymer-based coating is epoxy coating. Epoxy coatings are thermosetting coatings that are widely used in the industry due to their excellent corrosion resistance and adhesion properties. They are typically used in applications where the metal substrate is exposed to harsh environments, such as marine and industrial environments.

1.3.1.7 Organic–inorganic hybrid coatings

Organic–inorganic hybrid coatings combine the advantages of organic and inorganic materials to create a unique coating with enhanced properties. These coatings are composed of both organic and inorganic components, such as a sol–gel matrix with an organic functional group attached to it. Organic–inorganic hybrid coatings can be synthesized using various methods, such as sol–gel processing, CVD, and electrodeposition. These coatings exhibit excellent adhesion to the metal substrate and provide superior corrosion protection due to their unique structure and properties.

One example of an organic-inorganic hybrid coating is silane-based coating. Silane-based coatings are synthesized through a sol–gel process, where a silane precursor is hydrolyzed and polymerized to form a three-dimensional network structure. The organic functional group attached to the silane molecule provides adhesion to the metal substrate, while the inorganic network provides corrosion protection.

1.3.1.8 Self-assembled monolayers

Self-assembled monolayers (SAMs) are composed of a single layer of molecules that are chemically adsorbed onto a metal substrate. SAMs are formed through a process called self-assembly, where the molecules arrange themselves into an ordered pattern on the metal surface.

SAMs can be tailored to exhibit specific properties, such as hydrophobicity, hydrophilicity, and anti-corrosion properties. They are typically synthesized using a solution-phase deposition method, where the metal substrate is immersed in a solution containing the molecules of interest. One example of SAM coating is the thiol-based coating. Thiol-based SAMs are formed by chemically adsorbing thiol molecules onto a metal substrate. The thiol functional group provides excellent adhesion to the metal substrate, while the organic molecule attached to the thiol provides the desired properties such as anti-corrosion properties.

1.4 Development of safer alternatives

Chromate-free thin films were developed as a safer option to improve the corrosion resistance of metallic components in the late 1990s. To improve film performance and ensure more uniform cross-linking during fixing, considerable attention has been paid into the statistical design and functionality of polymers when creating these coatings. Sol–gel was one of the novel methods for producing thin films that was found to adhere well to both organic top coats and metallic substrates. A key consideration when selecting sol–gel coatings for corrosion protection is economically viable manufacture [20]. In the industry, organic coatings are frequently employed to shield metals from corrosion. The fact that ceramic coatings have strong thermal and electrical properties and are more resistant to oxidation, corrosion, erosion, and wear than metals in high temperature conditions makes them particularly desirable as coating materials. Coatings made of TiO_2 or Al_2O_3 are widely employed [21]. Metal surfaces have been protected against corrosion using PANI (polyaniline) thin film coatings [22]. PANI coatings can offer corrosion protection through a variety of ways. The barrier-protection mechanism that separates the metal surface from its surroundings is the simplest mechanism for an organic polymer. PANI may be used as a barrier layer component or as a primer.

Micro-arc oxidation (MAO) is a unique technique for growing a ceramic oxide film on a magnesium alloy; the resulting coating has a rough microstructure and is firmed to the magnesium substrate [23–25]. The structure and attributes of the steel coating layer produced by micro-arc oxidation were studied in order to extend the service life of steel in complicated environments. The resistance of the substrate can be improved by the ceramic film, which can also successfully reduce the impact stress overload [26]. The chemical etching technique is a good competitor for creating super hydrophobic coatings because it does not require expensive machinery, unique circumstances, or difficult steps. Through the redox reaction between the metal oxidant and the metal itself, chemical etching creates the rough structure on the Mg surface. Then, low surface energy substances can be modified to create the super hydrophobic coating [27]. However, the physical barrier coatings suffer deterioration over time in some complex service situations [28]. Currently, the majority of the damaged coatings require manual repair or replacement, which is expensive. Researchers discovered that one potential solution to this issue is to create a self-healing covering that alters its performance in response to external stimuli [29, 30]. A self-healing coating is one that has undergone scientific advancement on the basis of regular physical barrier coating to acquire bionic self-healing function. When the coating is damaged, the affected portion of the coating can be restored with little to no assistance from outside sources. Self-healing properties can increase protection capacity of coatings and increase the useful life of magnesium alloys [31]. For the corrosion shield of Mg substrate, it is crucial to design a cutting-edge protective coating with intelligent self-healing capabilities. The most popular types of material used to address the numerous issues caused by corrosion are coatings and film-based barriers. These polymer-clay nanocomposite (PCN) thin films are produced by a well-known layer-by-layer deposition method [32, 33], [34] that results in highly organized composite thin films with oriented layers of exfoliated clay platelets. Numerous physical qualities can be increased by mixing inorganic clay platelets with a polymer to create a polymer clay nanocomposite (PCN) covering without considerably raising the cost [35].

Since being discovered by Novoselov and collaborators in 2004 through mechanical exfoliation, graphene and composites based on it have attracted great attention in recent years [36]. Due to its hexagonally structured sp^2 linked carbon atom network and one-atom thick planar sheet-like shape, graphene has superior electrical, thermal, and mechanical capabilities as well as chemical stability. Graphene oxide (GO) is produced when graphite is subjected to sonication and oxidation. The basal planes of GO contain functional groups including hydroxyl, epoxy, and carboxyl that aid in the dispersion of GO in epoxy matrix and offer covalent interaction with epoxy networks [37, 38]. GO nano-sheets are excellent for functionalization with both organic and inorganic materials due to their oxygenated groups. Graphene and hexagonal boron nitride (h-BN) have recently been identified as two-dimensional, layered materials with a number of applications. The purpose of this study is to demonstrate how h-BN can be used to prevent corrosion in marine coatings. Using electrochemical techniques in seawater media, the performance of h-BN/polymer coatings on stainless steel was evaluated. Due to the hydrophobic, inert, and dielectric properties of boron nitride, h-BN/polymer coating exhibits effective corrosion protection with low corrosion current density of $5.14 \ 10^{-8}$ A/cm^2 and corrosion rate of 1.19 10–3 mm/year. The study revealed that coated stainless steel had better corrosion resistance [39]. To investigate the resistance to electrochemical corrosion, a simple sol–gel technique was proposed to successfully deposit Ni-doped ZnO thin films on low carbon steel. Scanning Electron Microscopy (SEM), energy dispersive X-ray spectroscopy (EDX), atomic force microscopy (AFM) were used to determine the morphology, composition, and crystalline phase of the surface coating. According to the electrochemical data, thick coatings of Ni-doped ZnO film act as insulator and showed greater resistances and lower capacitances than other samples [40]. For ZrO_2, TiO_2SiO_2, and $Al_2O_3SiO_2$ coatings, the chemical resistance of 316-L stainless steel has been confirmed. Weight and loss, electrochemical methods, X-ray Diffraction (XRD) and Fourier Transform Infrared Spectroscopy (FTIR), and SEM spectroscopic techniques have all been used to investigate the anticorrosion behavior of the films in both acidic and basic solutions. Data analysis shows that the films serve as geometric blocking layers to protect the substrate from corrosive substances and significantly extend substrate lifespan [41]. New protective coating techniques using nanocomposites and carbon-based materials are being developed to lessen corrosion. Due to their numerous industrial uses and practicality, conducting polymers have received a lot of attention recently. As polymers have long-chain carbon bonds, they may adsorb and effectively block a significant portion on metal surfaces. The thin layers that have been adsorbed on the metal substrate act as a barrier between the metal and its surroundings [42]. The feasibility of atomic layer deposition to create coatings that prevent corrosion was examined using a number of techniques. On stainless steel substrates, thin films of the oxides Al_2O_3, TiO_2, Ta_2O_5, and Al_2O_3-TiO_2 were deposited at temperatures between 150 and 400 °C. In 3.5 wt% NaCl and 0.1 and 1 mol/L HCl solutions, the corrosion behavior of the samples was investigated using electrochemical impedance spectroscopy. In NaCl, Al_2O_3 was discovered to protect the material from corrosion for a limited duration. Al_2O_3 and TiO_2 together in a multilayer structure provide superior protection than either material acting alone [43].

1.5 Conclusion

Corrosion of metals in an acidic medium is inevitable and affects the world's economy in a very serious way. Technically and economically, solutions to the problem of metal corrosion are becoming very important. In conclusion, functionalized thin film coatings provide an effective solution in preventing corrosion in metal substrates. These coatings can be tailored to exhibit specific properties, such as self-healing, anti-fouling, and anti-corrosion properties, depending on the application. The various types of functionalized thin film coatings discussed in this article include metallic coatings, ceramic coatings, composite coatings, polymer-based coatings, organic–inorganic hybrid coatings, and self-assembled monolayers. Each type of coating offers unique advantages and disadvantages, depending on the application requirements. Overall, functionalized thin film coatings have proven to be a valuable tool in the fight against corrosion and have a wide range of applications in various industries.

References

[1] Callister WD, Rethwisch DG. Materials science and Engineering: An Introduction. New York: Wiley; 2018.

[2] ISO D. Corrosion of metals and alloys–basic terms and definitions.

[3] Fontana MG, Greene ND. Corrosion Engineering.

[4] Tait WS. Electrochemical corrosion basics. In: Handbook of Environmental Degradation of Materials. William Andrew Publishing, pp. 97–115; 2018

[5] Hossain N, Asaduzzaman Chowdhury M, Kchaou M. An overview of green corrosion inhibitors for sustainable and environment friendly industrial development. J Adhes Sci Technol 2021;35 (7):673–90.

[6] Fayomi OS, Akande IG, Odigie S. Economic impact of corrosion in oil sectors and prevention: An overview. In: In Journal of Physics: Conference Series, Vol. 1378, No. 2, IOP Publishing, p. 022037; 2019.

[7] Koch G. Cost of corrosion. Trends in oil and gas corrosion research and technologies. pp. 3–0, 2017.

[8] Haque J, Srivastava V, Verma C, Quraishi MA. Experimental and quantum chemical analysis of 2-amino-3-((4-((S)-2-amino-2-carboxyethyl)-1H-imidazol-2-yl) thio) propionic acid as new and green corrosion inhibitor for mild steel in 1 M hydrochloric acid solution. J Mol Liq 2017;225:848–55.

[9] Rajagopalan K. In: 10th International Congress on Metallic Corrosion, Vol. 2, Chennai: Oxford & IBH, New Delhi, p. 1765; 1987.

[10] Khanna A. News Letter. India: NACE, p. 3; 1997.

[11] Boopathy R, Daniels L. Effect of pH on anaerobic mild steel corrosion by methanogenic bacteria. Appl Environ Microbiol 1991;57(7):2104–08.

[12] Prawoto Y, Ibrahim K, Wan Nik WB. Effect of pH and chloride concentration on the corrosion of duplex stainless steel. Arabian J Sci Eng 2009;34(2):115.

[13] Khodair ZT, Khadom AA, Jasim HA. Corrosion protection of mild steel in different aqueous media via epoxy/nanomaterial coating: preparation, characterization and mathematical views. J Mater Res Technol 2019;8(1):424–35.

[14] Parapurath S, Jacob L, Gunister E, Vahdati N. Effect of microstructure on electrochemical properties of the EN S275 mild steel under chlorine-rich and chlorine-free media at different pHs. Metals 2022;12(8):1386.

[15] Nowak WB. Thin metallic films for corrosion control. Surf Coat Technol 1991;49(1–3):71–77.

[16] Hofer AM, Mori G, Fian A, Winkler J, Mitterer C. Improvement of oxidation and corrosion resistance of Mo thin films by alloying with Ta. Thin Solid Films 2016;599:1–6.

[17]   Jantaping N, Schuh CA, Boonyongmaneerat Y. Influences of crystallographic texture and
       nanostructural features on corrosion properties of electrogalvanized and chromate conversion
       coatings. Surf Coat Technol 2017;329:120–30.
[18]   Zhang X, Sloof WG, Hovestad A, Van Westing EP, Terryn H, De Wit JH. Characterization of chromate
       conversion coatings on zinc using XPS and SKPFM. Surf Coat Technol 2005;197(2–3):168–76.
[19]   Alanazi NM, Leyland A, Yerokhin AL, Matthews A. Substitution of hexavalent chromate conversion
       treatment with a plasma electrolytic oxidation process to improve the corrosion properties of ion
       vapour deposited AlMg coatings. Surf Coat Technol 2010;205(6):1750–56.
[20]   Kumar A, Singh R, Bahuguna G. Thin film coating through sol-gel technique. Res J Chem Sci
       2016;6:65.
[21]   Shen GX, Chen YC, Lin CJ. Corrosion protection of 316 L stainless steel by a TiO2 nanoparticle coating
       prepared by sol–gel method. Thin Solid Films 2005;489(1–2):130–36.
[22]   Abu-Thabit NY. Electrically conducting polyaniline smart coatings and thin films for industrial
       applications. In: In Advances in Smart Coatings and Thin Films for Future Industrial and Biomedical
       Engineering Applications, Elsevier, pp. 585–617; 2020.
[23]   Zhang ZQ, Wang L, Zeng MQ, Zeng RC, Kannan MB, Lin CG, Zheng YF. Biodegradation behavior of
       micro-arc oxidation coating on magnesium alloy-from a protein perspective. Bioactive Mater 2020;5
       (2):398–409.
[24]   Lin Z, Wang T, Yu X, Sun X, Yang H. Functionalization treatment of micro-arc oxidation coatings on
       magnesium alloys: A review. J Alloys Compd 2021;879:160453.
[25]   Kaseem M, Ramachandraiah K, Hossain S, Dikici B. A review on LDH-smart functionalization of
       anodic films of Mg alloys. Nanomaterials 2021;11(2):536.
[26]   Wang W, Feng S, Li Z, Chen Z, Zhao T. Microstructure and properties of micro-arc oxidation ceramic
       films on AerMet100 steel. J Mater Res Technol 2020;9(3):6014–27.
[27]   Wang Q, Zhang B, Qu M, Zhang J, He D. Fabrication of super hydrophobic surfaces on engineering
       material surfaces with stearic acid. Appl Surf Sci 2008;254(7):2009–12.
[28]   Li B, Zhang Z, Liu T, Qiu Z, Su Y, Zhang J, Lin C, Wang L. Recent Progress in Functionalized Coatings
       for Corrosion Protection of Magnesium Alloys – A Review. Materials 2022;11:3912.
[29]   Liu S, Li Z, Yu Q, Qi Y, Peng Z, Liang J. Dual self-healing composite coating on magnesium alloys for
       corrosion protection. Chem Eng J 2021;424:130551.
[30]   García SJ, Fischer HR, Van Der Zwaag S. A critical appraisal of the potential of self healing polymeric
       coatings. Prog Org Coat 2011;72(3):211–21.
[31]   Anjum MJ, Zhao J, Ali H, Tabish M, Murtaza H, Yasin G, Malik MU, Khan WQ. A review on self-healing
       coatings applied to Mg alloys and their electrochemical evaluation techniques. Int J Electrochem Sci
       2020;15:3040–53.
[32]   Cussler EL, Hughes SE, Ward WJ, III, Aris R. Barrier membranes. J Membr Sci 1988;38(2):161–74.
[33]   Yoo D, Wu A, Lee J, Rubner MF. New electro-active self-assembled multilayer thin films based on
       alternately adsorbed layers of polyelectrolytes and functional dye molecules. Synth Met 1997;85
       (1–3):1425–26.
[34]   Decher G, Eckle M, Schmitt J, Struth B. Layer-by-layer assembled multicomposite films. Curr Opin
       Colloid Interface Sci 1998;3(1):32–39.
[35]   Percival SJ, Melia MA, Alexander CL, Nelson DW, Schindelholz EJ, Spoerke ED. Nanoscale thin film
       corrosion barriers enabled by multilayer polymer clay nanocomposites. Surf Coat Technol
       2020;383:125228.
[36]   Deyab MA, De Riccardis A, Mele G. Novel epoxy/metal phthalocyanines nanocomposite coatings for
       corrosion protection of carbon steel. J Mol Liq 2016;220:513–17.
[37]   Zhang XF, Chen RJ, Hu JM. Super hydrophobic surface constructed on electrodeposited silica films by
       two-step method for corrosion protection of mild steel. Corros Sci 2016;104:336–43.

[38] Pourhashem S, Vaezi MR, Rashidi A. Investigating the effect of SiO2-graphene oxide hybrid as inorganic nanofiller on corrosion protection properties of epoxy coatings. Surf Coat Technol 2017;311:282–94.

[39] Husain E, Narayanan TN, Taha-Tijerina JJ, Vinod S, Vajtai R, Ajayan PM. Marine corrosion protective coatings of hexagonal boron nitride thin films on stainless steel. ACS Appl Mater Interfaces 2013;5 (10):4129–35.

[40] Zhang D, Wang MM, Jiang N, Liu Y, Yu XN, Zhang HB. Electrochemical corrosion behavior of Ni-doped ZnO thin film coated on low carbon steel substrate in 3.5% NaCl solution. Int J Electrochem Sci 2020;15:4117–26.

[41] Atik M, De lima neto P, Avaca LA, Aegerter MA. Sol-gel thin films for corrosion protection. Ceram Int 1995;21(6):403–06.

[42] Ates M. A review on conducting polymer coatings for corrosion protection. J Adhes Sci Technol 2016;30(14):1510–36.

[43] Matero R, Ritala M, Leskelä M, Salo T, Aromaa J, Forsén O. Atomic layer deposited thin films for corrosion protection. Le J de Phys IV 1999;9(PR8):Pr8–493.

Saumya Jaithlia, Manash Protim Mudoi

2 Functionalized thin film coatings: fabrication strategies, synthesis, and characterization

Abstract: A thin film is a layer of substance with a thickness of several micrometers to fractions of a nanometer. The controlled synthesis of thin films (a process referred to as thin film deposition) is a fundamental step in many applications. Applications for thin film coatings are numerous and serve various functions, and they can be used to make lenses with a certain amount of reflection, shield displays from damage, or add metallization layers to semiconductor wafers. This chapter will further discuss different types of coatings and their properties, applications of thin film coating, different technologies and methods that can be used to apply thin film coatings, and an array of tools and equipment that can be used to streamline or enhance the thin film deposition process; we will also discuss different properties of a thin film such as surface roughness, chemical composition, adhesion, and stress.

Keywords: Corrosion, Thin Film, Coating, Fabrication, Functionalization, Synthesis

2.1 Introduction

A metallic material's surface may corrode if certain environmental factors are present. Wet corrosion involves the presence of water in contact with the bare metallic surface and often takes place at low temperatures. Hot corrosion, which occurs at high temperatures, involves the presence of oxygen or other oxidizing agents in contact with the metal. Therefore, keeping water or oxygen away from sensitive parts is one of the potential tactics for preventing or reducing corrosion to an acceptable pace. This is where thin films play a role in the prevention of corrosion. Although there are numerous other technologies such as painting, the use of inhibitors, active protection, etc., that can be used for the same objective, the use of thin films for corrosion protection is one of the more adaptable, practical, and efficient options in the market right now.

To achieve this, the thin layer must be able to withstand the penetration of the gaseous and/or liquid chemicals, which can lead to corrosion and eventually come into contact with the exposed part. To offer the greatest level of protection while maintaining the visual appeal of the product, the protective thin layer must be able to expand and contract with the coated metal surface without cracking. A variety of thin film coatings and deposition processes have been developed to satisfy these criteria. In this chapter, we discuss some of the advanced techniques and principles of thin film depositions, their characterization, etc.

https://doi.org/10.1515/9783111016160-002

2.1.1 Deposition techniques of different types of coatings

There are many distinct kinds of thin film coatings such as optical, electrical or electronic, magnetic, chemical, mechanical, thermal, and thin films made of other techniques. They have numerous valuable applications in our daily lives; for materials to last longer, hold their value, and protect lives and property, corrosion must be avoided and thus thin film technology and a wide range of thin film deposition methods provide the basis for numerous types of protective coatings. We shall explore the wide range of thin film deposition processes in more detail in this chapter.

2.1.1.1 Physical vapor deposition

Physical vapor deposition (PVD) techniques is one of the most used approaches, which has seen numerous modifications and advancements in recent years. Researchers are working hard to boost coating qualities and testimony rates [1]. Evaporation and sputtering are the primary methods by which the particle can be extracted from the material and are the two main methods used to categorize PVD [2].

2.1.1.1.1 Evaporation techniques
Since the recent discovery of brand-new high-critical temperature (Tc) bulk superconductors, numerous evaporation techniques, including thermal, electron beam, and others, have been used to create superconducting thin films on a variety of substrates [3].

– Thermal evaporation
A common method for putting thin films onto a substrate is thermal evaporation (TE). In order to create a thin film on a substrate, it includes heating a source material in a vacuum chamber until it vaporizes and condenses. This method is frequently employed in the creation of decorative coatings on glass and metal surfaces, as well as coatings for electronic equipment like flat-panel displays and solar cells. The preparation of the source material, usually a solid or liquid that can vaporize in vacuum, is the first step in the procedure. An electron beam, resistive heating, or another heat source is used to heat the material in a crucible or boat to a high temperature. The material vaporizes and condenses onto a substrate that is close to the source material when it is heated. Usually, the substrate is kept at a lower temperature than the source material, allowing the vaporized substance to condense and create a thin coating on its surface. By altering the source material's temperature, its distance from the substrate, and the deposition period, it is possible to regulate the film's thickness and homogeneity. The fact that thermal evaporation is a relatively easy and affordable method that can be used to deposit a variety of materials, including metals, semiconductors, and organic compounds, is one of its benefits. The inability to deposit highly

reactive or volatile compounds and the risk of contamination by leftover gases or contaminants in the vacuum chamber are some of its drawbacks.

– Electron beam evaporation

Similar to thermal evaporation, electron beam evaporation is a method for forming thin coatings onto a substrate. The source material is vaporized using an electron beam, which has a number of benefits over other deposition methods. In electron beam evaporation, the source material, which is often a solid or powder is targeted by a high-energy electron beam. The material is heated by the beam to a high temperature; it vaporizes and then condenses onto a substrate in the vacuum chamber. In order to allow the vaporized material to condense and form a thin film, the substrate is normally maintained at a lower temperature than the source material.

The ability to deposit a variety of materials, including metals, semiconductors, and ceramics, with outstanding film quality and uniformity is one benefit of electron beam evaporation. Additionally, it may deposit materials that cannot be deposited using conventional methods due to their high reactivity, such as titanium and aluminum. The ability to precisely and thickly deposit films is another benefit of electron beam evaporation, which makes it helpful for producing electrical products like thin film transistors and solar cells. The method also enables deposition of multilayer films with regulated composition and thickness, enabling the construction of intricate device architectures. However, there are significant drawbacks to electron beam evaporation, including the need for a high vacuum environment and the possibility that the electron beam could harm delicate substrates. Compared to other deposition processes, the equipment and maintenance expenses may also be higher. Table 2.1 shows the advantages and disadvantages of electron beam sintering (EBS).

Table 2.1: Advantages and disadvantages of the EBS technique.

Advantages	Disadvantages
Controlling the deposition rate precisely between nm/min and m/min (usually 0.11 m/min).	The possibility of harming the target by ionizing or decomposing its surface.
Uniform film thickness.	Due to the extensive use of X-rays and electrical energy, this method is expensive.
Deposition of many layers by altering the source material.	
For this technique, a comparatively low temperature is used.	
High-material utilization efficiency.	

– Ion plating

In the PVD process known as ion plating or ion vapor deposition, the substrate material is bombarded with atom-sized energized particles, resulting in the deposition of atoms on the substrates in the form of films. By depositing a variety of materials, ion plating has established a strong reputation in the industry. Ion plating is used on X-ray tube bearings, strip steel, aviation engine parts, etc. [1]

2.1.1.2 Cathodic arc physical vapor deposition

In the Cathodic Arc Physical Vapor Deposition (CAPVD) process, low-voltage and high-density electric current is passed between two separate electrodes. The cathodic substance is simultaneously vaporized and ionized during this procedure, which is carried out in vacuum, and this causes the creation of plasma. Unlike other PVD techniques, here, high kinetic energy results in coatings that are intermixed layers with improved adherence [5].

2.1.1.3 Sputtering

One of the earliest and most widely used methods of thin film deposition for industrial purposes is sputtering. Thin films of metal alloys, refractory materials (like oxides and nitrides), metals, and non-metals like polytetrafluoroethylene (PTFE) can all be deposited by sputtering. Sputtering is a method for depositing thin films that have benefits such as consistent thickness across a vast area, simple thickness control, etc. As opposed to the original target composition, the alloy composition is maintained in the case of alloy sputtering.

Sputtering is a glow-discharge process, which means that glow discharge produces the energetic particles used to attack the target. Similar to etching, sputtering relies on the momentum transfer of bombarded ions to surface atoms to cause the ejection of surface atoms from the electrode surface. Microparticles of a solid material are ejected from its surface during sputtering after the solid substance has been blasted by intense plasma or gas particles. An exchange of momentum takes place when energetic ions (typically inert gas species like Ar) collide with target material atoms. These "incident ions" cause collision cascades in the target, hence the name.

There are several different types of sputtering:

2.1.1.3.1 Diode sputtering

Metals and metal alloys can be deposited as thin layers using diode sputtering. Two electrodes are enclosed in a vacuum chamber with adequate pumping and gas flow in the diode DC Sputtering system (typically argon). The material to be sputtered, which is a plate electrode at a negative potential, is represented as the target or cathode. The substrate represents the anode, which can be grounded, floating, or biassed. When using grounded substrates, the deposits build up throughout the chamber as the sputtering process occurs. The target is bombarded by positive argon ions (Ar1), which expel primarily neutral target atoms that subsequently condense onto the substrate. Although a straightforward technology, diode sputtering addresses a number of drawbacks, such as low deposition rates caused by low plasma densities, high gas densities, and high discharge voltages, using only a small amount of application on conducting targets. Due to the target being poisoned by the compound dielectric layers on the target surface, reactive sputtering cannot be employed for non-conductive dielectric materials.

2.1.1.3.2 RF diode sputtering

Although DC diode plasmas are simple to build and run, they are infrequently used in real-world applications, particularly in the semiconductor manufacturing industry. This is because insulating or dielectric targets cannot be successfully sputtered using DC sputtering. This is because the target surface develops a positive charge that repels incident positive ions. As a result, the voltage needed to maintain the glow discharge cannot be reached, and as a result, the plasma will shut off after 1 s. Similar circumstances arise when reactive sputtering is combined with DC. Metal oxides will accumulate on the cathode surface and reduce the amount of current flowing through it; as a result, the rate of accretion will be too slow to be useful.

2.1.1.3.3 Triode sputtering

A thermionic electron emitter filament is used in triode sputtering to increase the quantity of electrons released from the cathode (target). Higher deposition rates (up to several thousand/min) at lower target voltages (%500 V) and lower pressures are possible because of the enhanced electron density, which improves control over the ion energy and flux (0.51 mT). The non-uniform coating across vast surface areas, complexity and difficulty of scaling, lower filament lifetime, and filament interaction with reactive gas species are all drawbacks of triode sputtering.

2.1.1.3.4 Magnetron sputtering

Chapin received a patent in August 1979 for the planar magnetron concept. The cathode is given a magnetic field in the fundamental design, which gives the electrons an E3B drift route to follow. A cross-product of E and B is E3B. This causes electrons to migrate perpendicular to the magnetic and electric fields, known as E 3 B drift (of the gyration or "guiding center"). This raises the possibility of collisions between electrons traveling in a helical path as opposed to a straight line and gas atoms. The drift is azimuthally closed when the planar magnetron shape is taken into account, allowing the trapping and "recycling" of the electron to take place. As a result, a magnetic field that aids in confining electrons close to the target raises the proportion of electrons that results in ionizing collisions. In magnetron sputtering, alternating-pole permanent magnets are positioned behind the cathode such that the magnetic and electric fields intersect at a small height (1,020 mm, for example) above the cathode surface. The target-generated secondary electrons in this modification to the original sputtering technique do not bombard substrates because they are trapped in cycloidal trajectories close to the target and do not contribute to the increased substrate temperature or radiation damage. The benefit that follows is the ease with which temperature-sensitive substrates, like polymers and materials used in metal-oxide-semiconductor devices, can be coated. In comparison to conventional deposition techniques, magnetron sputtering is performed at lower deposition pressure, and the sources yield higher deposition rates. As a result, there is increased economic value and scalability for numerous large-scale industrial applications. Planar, cylindrical, and conical magnetron sources are three alternative configurations with advantages and disadvantages. Figure 2.1 shows the sputtering method with the magnetron technique for applying a coating layer on the silicon substrate.

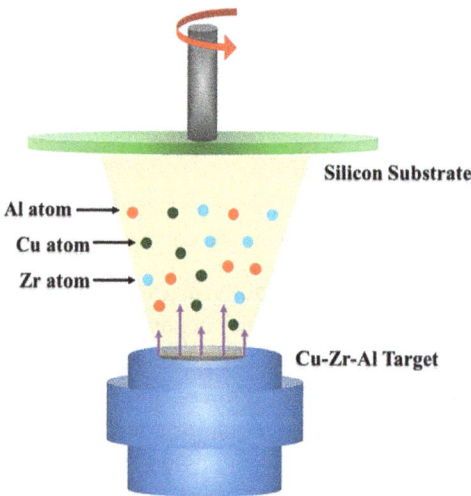

Figure 2.1: Technique for spin-coating: (a) coating application on the substrate; (b) spreading of coating material with rapid substrate rotation; and (c) final deposition of thin film [6]. Source attained from Open Access (CCBY).

2.1.1.3.5 Reactive sputtering

Reactive sputtering creates a wide range of useful compound thin film coatings by adding a reactive gas (like oxygen) to the sputtered material. Typically, the inert or background gas is combined with the reactive gas (invariably, Ar). The sputtering apparatus can be run in the magnetron, magneto-diode, or RF modes. Reactive sputtering has been used for coating glass, cutting tools, microelectronic devices, and micro-electro-mechanical systems (MEMS) since the 1950s, following the deposition of TaN for hybrid circuits. Fluorine is preferentially sputtered and lost in reactive sputtering with fluorides, either from the fluoride target or from the forming film, exactly like in ion-assisted processes. However, target poisoning is a significant issue in reactive sputtering because the reacting gas frequently reacts with both the target and the growing film. By using the plasma ions to scour the cathode's whole surface, target poisoning might be removed.

2.1.1.3.6 Ion-beam sputtering

Neutral ion beams from a broad-beam source are used in ion-beam sputtering (IBS) to sputter the target material, which solves several issues with the prior sputtering methods. As a result, target-charging issues are avoided, making it possible to sputter solely on dielectric materials. Two ion sources are used in a dual-ion beam sputtering machine: the first is used for sputtering and the second is used to bombard the substrate before coating or during deposition (in this instance, the process is known as ion-assisted deposition or IAD). The plasma from which the ions are produced is in a separate chamber in this setup, preventing the sputtered material from moving from the target to the substrate. In what is known as dual-ion beam sputtering, a second ion source is typically used directly to bombard the developing layer. Further adjusting and control of the film properties, in particular intrinsic stress, is possible thanks to this ion treatment. IBS creates films of the highest caliber. Controlling ion energy, flow, species, and angle of incidence is one of the main benefits of IBS. Another aspect is the containment and separation of the ion source's gas discharge (plasma) from the rest of the deposition system. The useful coated area is smaller, and the IBS process is somewhat slower than other sputtering methods. In comparison to other sputtering techniques, the key drawback of the IBS process is its challenging scaling (e.g., magnetron process). Due to geometrical restrictions, the beam-sputter arrangement frequently needs to be compact in order to offer acceptable deposition rates.

2.1.1.4 Chemical vapor deposition

In chemical vapor deposition (CVD), a thin solid coating is created by chemically reacting gases, as the name suggests. It involves methods for depositing amorphous, polycrystalline, and epitaxial films in the broadest sense. [7–9]. The way these strategies are physically implemented can also vary greatly. In order to deposit a thin film with a certain structure and composition, a variety of techniques such as thermally driven CVD, photo-assisted CVD, and plasma-enhanced CVD (PECVD) are used [10]. Figure 2.2 shows a CVD technique used for nanocomposite coating.

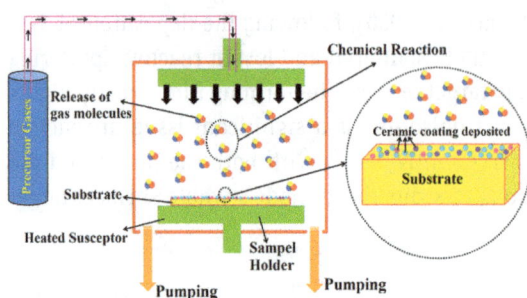

Figure 2.2: Chemical vapor deposition technique for nanocomposite coating [11]. Source attained from Open Access (CCBY).

In CVD processes, precursors interact chemically to produce a final thin film product [4]. The precursors are supplied into the reactor by using a stream of carrier gas. There are a large range of chemical processes that are used in CVD, including pyrolysis (thermal decomposition), hydrolysis, disproportionation, reduction, carburization, oxidation, and nitridation [4]. These reactions can either be used singly or in a combination. These can be activated by various methods, such as:

- **Thermal activation**: This typically occurs at high temperatures (900 °C), although the temperature can also be considerably lowered if metalorganic precursors are used (MOCVD).
- **Plasma activation**: This method occurs at much lower temperatures, 300 °C–500 °C.
- **Photon activation**: This method, usually with shortwave ultraviolet radiation, can occur by directly activating a reactant or by activating an intermediate.

The utilization of the horizontal and vertical CVD reactor layouts depends on the intended uses of the CVD process. The reactor could be hot-walled or cold-walled. In cold-walled reactors, the steep thermal gradients over the heated substrate can form complex buoyancy-driven-recirculation flows. In a vertical configuration, reactants are fed from the top. The horizontal reactor, with its simple design and ability to accommodate multiple wafers, is the mainstay for the production of dielectrics, polycrystalline silicon, and passivation films [10].

There are several different types of CVD techniques, each with a unique way of introducing the reactants into the reactor and a unique set of circumstances for the reaction to occur. The following are some of the most typical CVDs:

- Atmospheric pressure CVD (APCVD): This method involves introducing the reactant gases into the reactor at atmospheric pressure, and the reaction occurs at relatively low temperatures (between 600 and 1,000 °C).
- Low-pressure CVD (LPCVD): In this method, the reaction occurs at high temperatures (about 500–800 °C) and low pressures (about 0.01–10 torr).

- PECVD: In this method, the reactant gases are excited by a plasma to speed up the process. Several techniques, such as radio frequency (RF) or microwave discharge, can be used to produce plasma.
- Hot-wire CVD (HWCVD): In this method, the reactant gases are broken down by a hot filament often made of tantalum or tungsten heated to temperatures above 1,800 °C and.
- (MOCVD: This method uses metalorganic compounds as the precursor molecules, and the reaction occurs at high temperatures (about 600–1,200 °C).
- Laser-assisted CVD (LACVD): In this method, the reactant gases are broken down, and the reaction is accelerated. The gas phase or the substrate can both be the focus of the laser.

Each CVD type has advantages and disadvantages, depending on the specific application and the properties of the material being deposited.

2.1.1.5 Sol–gel technique

One of the most famous wet-chemical processes is the sol–gel method, which is primarily employed to synthesize the oxide materials. The oxide networks are produced in the sol–gel method by a series of condensation events involving molecular precursors in a liquid medium. There are two basic techniques to make sol–gel coatings: organic and inorganic approaches. The organic technique uses organic monomers with functionally formed double bonds, whereas the inorganic route uses inorganic precursors and the networks through a colloidal suspension (often oxides) and gelation of the sol to form a network in a continuous liquid phase. Metalloid alkoxide precursors or monomeric metals are first dissolved in alcohol or another low-molecular-weight organic solvent. The sol–gel process consists of four main steps: (a) hydrolysis; (b) particle growth; (c) agglomeration of the polymer structures; and (d) condensation and polymerization of monomers to form chains and particles. These four steps are followed by the formation of networks that extend throughout the liquid causing the liquid medium to thicken and eventually form a gel [12]. After the formation of the sol–gels, thin film coatings are made, usually by two techniques, which are spray coating and dip coating [13, 14]. Figure 2.3 shows the schematic of the spray- and dip-coating methods.

The spin-coating method enables the formation of ultra-thin coatings with homogeneous layers [14]. The thickness of the layer can be controlled by the viscosity of the sol and the rotational speed of the spin-coater [15, 16]. However, only flat surfaces with limited dimensions are convenient for spin-coating. The size and shape limitation of spin coating can be solved by applying dip coating, which allows the coating of multidimensional and complies with substrates [14]. The withdrawal speed of the substrates from the coating solutions influences the morphology and thickness of the coating based on the viscosity of the coating solution.

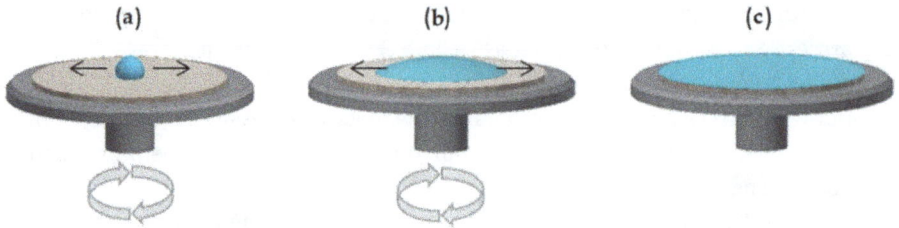

Figure 2.3: Spin-coating technique: (a) applying coating material on the substrate; (b) rapid rotation of the substrate for spreading the coating material; and (c) final deposition of thin film [13]. Source attained from Open Access (CCBY).

2.1.1.6 Co-precipitation method

This technique creates thin film coatings with numerous biomedicine, optics, and electronics uses. The following steps are involved in the co-precipitation process for producing thin film coatings:

- Precursor solution preparation: The precursor solutions are made by combining the necessary quantity of metal salts with a suitable solvent. The mixture of the liquids creates a homogenous solution.
- Surfactants are added to the precursor solution to stabilize the nanoparticles and stop them from clumping together while being coated.
- The precursor solution is deposited onto a substrate using a number of methods, including spin coating, dip coating, and spray coating. The solvent is subsequently evaporated from the substrate, starting the co-precipitation process.
- Drying and calcination: To create a solid material with the necessary qualities, the thin film is next dried and calcined at high temperatures.

The co-precipitation technique for producing thin film coatings has a number of benefits, including homogeneity, control over film thickness, and the capacity to create films on a range of substrates. However, the technique can call for different procedures to perfect the coating procedure, like post-annealing or surface alterations, to improve the film's adherence to the substrate.

2.1.1.7 Anion-exchange method

Anion-exchange resins, which have the capacity to exchange anions in solution with anions on their surface, are used in this process. A thin coating of the required material can be applied on a substrate using this procedure. The anion-exchange method for thin film coatings typically involves the following steps:

- The substrate is cleaned and prepared to make sure it is free of impurities and has an appropriate surface for deposition.
- The anion-exchange resin is typically prepared in a solution, and after that, it is washed to get rid of unwanted impurities.
- The thin film is then deposited on the substrate by first covering it with an anion-exchange resin solution, which causes anions in the solution to exchange with anions on the resin's surface.
- Post-treatment: The thin film coating may undergo post-treatment to enhance certain aspects, including its surface roughness, thickness, and adherence to the substrate.

The anion-exchange approach is superior to other thin film coating techniques in a number of ways, including ease of use, adaptability, and the capacity to deposit homogeneous coatings across huge surfaces. On a number of substrates, including glass, silicon, and metals, it can be used to deposit a wide range of materials, including metals, oxides, and polymers. The anion-exchange approach is a viable technology for large-scale manufacturing applications because it is also compatible with current industrial processes.

2.2 Characterization of different types of coatings

Several experiments have been designed to prove the ability of metallic structures to withstand years of exposure in harsh environments. Some of those experiments are:

2.2.1 Rutherford backscattering spectroscopy

Rutherford Backscattering Spectrometry (RBS) is an ion scattering technique used for thin film analysis without using reference standards. RBS analysis involves passing high-energy ions over the sample and energy distribution while measuring the back-scattered yield at a specific angle. In order to investigate the thickness of the corrosion product film, its protective or non-protective nature, or its inhibitory nature, RBS is a very potent approach for corrosion study. RBS could serve as an alternative to ellipsometry, an optical-based technique for measuring film thickness [17].

2.2.2 Auger electron spectroscopy

The atoms on a sample's surface are excited using a highly concentrated, finely focused electron beam. A core-level electron emerges and creates a singly ionized excited atom when the beam collides with a solid atom. An outer level electron fills the resulting core level vacancy. Auger electron spectroscopy (AES) offers knowledge on the elemental and chemical states of a surface, usually up to 5 nm. In most cases, ion milling or sputtering are applied simultaneously for depth profile investigations. The kinetic energy of Auger electrons and Auger intensity peaks are measured in order to identify the chemical. Investigating the surface's chemical condition is aided by peak shape and position [17].

2.2.3 Scanning electron microscopy

Scanning electron microscopy (SEM) can provide information on surface topography, crystalline structure, chemical composition, and electrical behavior of the top 1 urn or so specimen. To enable behavior under various situations to be evaluated, several specialized stages (such as hot, cold, or intended to permit in situ mechanical testing) can be installed. For instance, cathodoluminescence (light emission) is much brighter at temperatures close to absolute zero than it is at normal temperature, resulting in images made from the light generated by a cold specimen having substantially lower noise levels [18]. The basis of SEM is the detection of high-energy electrons that are emitted from a sample's surface after exposure to a tightly concentrated electron beam from an electron gun. Using the SEM objective lens, this electron beam is concentrated to a narrow area on the sample surface. Figure 2.4 shows the SEM images of silica coatings.

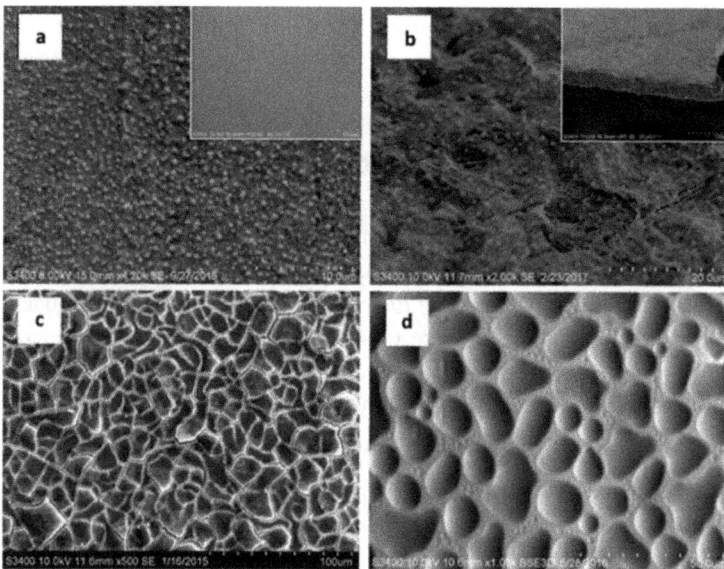

Figure 2.4: SEM micrographs: (a) stainless steel surface with SiO_2 coating; (b) titanium alloy substrate with SiO_2 coating; (c) cracked surface of SiO_2 coating on steel; and (d) bubbles on steel and coated with SiO_2 [13]. Source attained from Open Access (CCBY).

2.2.4 Energy spectroscopy (EDS)

The most popular chemical analysis instrument used in failure analysis is energy dispersive spectroscopy, often known as energy dispersive X-ray analysis (EDX) or EDS. It offers a number of enormous benefits. It is used as an attachment for the SEM, which is available in almost all the failure analysis labs. The analysis is completed in a short period of time. The spectra are simple to understand. Spatial resolution is good. It also has several limitations as a tool for analysis. Sensitivity is only effective at concentrations in the sampled volume that are of the order of 0.1%. A second drawback is that the sampled volume is very large when compared to the thickness of semiconductor thin films and deep submicron particles. One other drawback is that it only delivers atomic information, not molecular.

In this process of the formation of a vacancy caused by the ejection of an inner shell electron, the distinctive X-ray generating process is initiated. An upper shell electron enters the inner shell vacancy from this excited state. An X-ray is generated with an energy equal to the difference between the energies of the electron shell. The majority of the electron beam's interaction volume with the sample leads to the generation of X-rays [19].

2.2.5 X-ray diffraction

The crystallographic structure of a material can be determined using the X-ray diffraction (XRD) analysis method. XRD measures the X-ray intensities and scattering angles that exit a material after being exposed to incident X-rays. XRD analysis is primarily used to identify materials based on their diffraction pattern. XRD provides information on phase identification as well as how internal stresses and flaws affect the actual structure and how it differs from the ideal one. XRD is used to determine the following:
– Identify crystalline phases and orientation
– Determine structural properties like lattice parameters, epitaxy, strain, grain size, phase composition, preferred orientation
– Measure the thickness of thin films and multilayers
– Determine atomic arrangement

X-rays can be thought of as electromagnetic radiation waves, whereas crystals are regular arrangements of atoms. Crystal atoms typically scatter incident X-rays via interactions with their electrons. The electron is the scatterer in this event, and this phenomenon is referred to as elastic scattering. A regular array of spherical waves results from a regular array of scatterers. According to Bragg's law, these waves add together constructively in a small number of specified directions but cancel each other out in most directions through destructive interference. The specific directions appear as spots on the diffraction pattern called reflections. Consequently, XRD patterns result from electromagnetic waves impinging on a regular array of scatterers.

2.2.6 Fourier transform infrared spectroscopy

In Fourier transform infrared (FTIR) spectroscopy, an interference wave interacts with the sample in contrast to a dispersive instrument where the interacting energy assumes a well-defined wavelength range. The interference wave is produced in an interferometer, the most common of which is the Michelson interferometer. A computer is used to control the interferometer, to collect and store data, and to perform the Fourier transformation [20].

In this technique, a sample is exposed to infrared light from the FTIR instrument that ranges in wavelength from 10,000 to 100 cm^{-1}, some of which is absorbed and some of which passes through. The sample molecules transform the absorbed radiation into rotational and/or vibrational energy. The signal that is produced at the detector appears as a spectrum, typically ranging from 4,000 cm^{-1} to 400 cm^{-1}, and it serves as the sample's molecular signature. As each molecule or chemical structure will provide a distinct spectral fingerprint, FTIR analysis is a fantastic technique for identifying specific chemicals.

FTIR analysis is used to:
- identify and characterize unknown materials (e.g., films, solids, powders, or liquids)
- identify contamination on or in a material (e.g., particles, fibers, powders, or liquids)
- identify additives after extraction from a polymer matrix
- identify oxidation, decomposition, or uncured monomers in failure analysis investigations

2.2.7 X-ray photoelectron spectroscopy

An approach for examining the surface chemistry of a material is called X-ray photoelectron spectroscopy (XPS), sometimes referred to as electron spectroscopy for chemical analysis (ESCA). A material's atomic state, chemical composition, and electronic state can all be determined using XPS. By exposing a solid surface to an X-ray beam and measuring the kinetic energy of the electrons that are released from the top 1–0 nm of the material, XPS spectra are obtained. By counting ejected electrons over a variety of kinetic energies, a photoelectron spectrum is recorded. All surface elements (except hydrogen) can be identified and measured thanks to the energies and intensities of the photoelectron peaks.

Through these methods, thin films are specifically engineered for optimum corrosion protection, as well as to maximize coating adhesion, generate excellent impact resistance, and improve abrasion, flexibility, and slip resistance in the most challenging conditions [21].

(XPS is a technique used to analyze the chemical composition of a material's surface. It involves irradiating a material with X-rays, which causes electrons to be emitted from the material's surface. These electrons are then analyzed to determine the elements present in the material, as well as their chemical states.

Lux et al. [20] used XPS measurements to determine the absolute ratio of sp^2/sp^3 by the method of C 1s-peak deconvolution [20]. Figure 2.5 represents the XPS surface scan along with the high resolution of deconvoluted compounds of C 1s peaks.

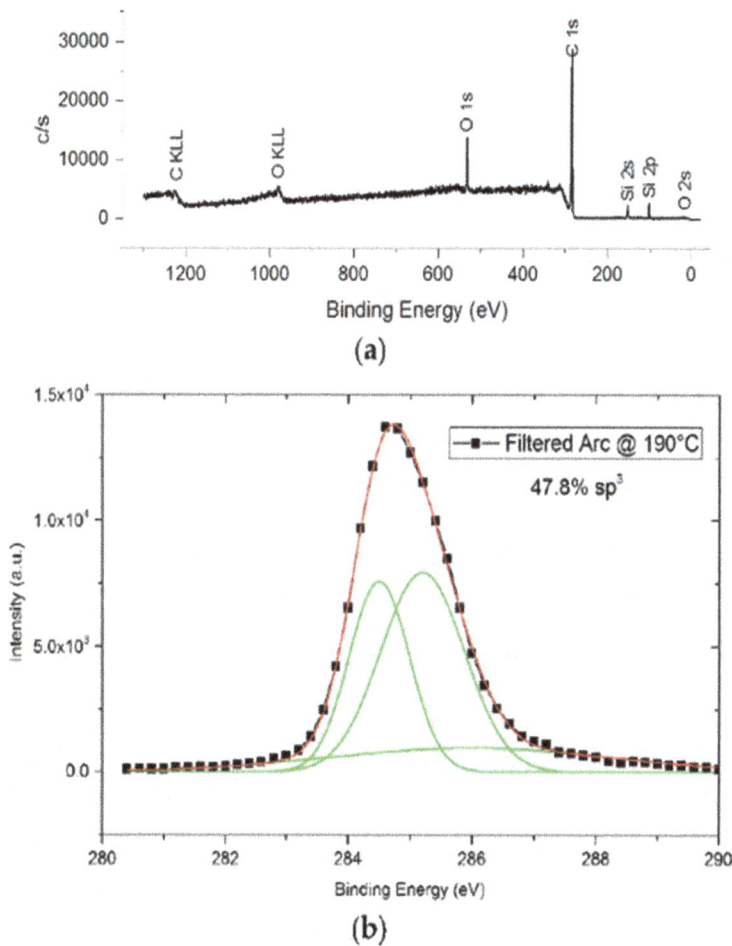

Figure 2.5: XPS scan: (a) DLC and (b) C 1s peak from DLC with deconvolution of sp^3-carbon, sp^2-carbon, and C–O contamination compounds [20]. Source attained from Open Access (CCBY).

The authors observed that higher substrate temperature caused the lowering of the sp³-content in the coating. They found that magnetic-filtered arc coating resulted in higher sp³-content than the cage-coating method. XPS can provide information about the chemical composition and electronic structure of a material's surface, as well as the depth distribution of chemical elements. It is commonly used in materials science, surface chemistry, and nanotechnology research to study the surface chemistry of materials, such as semiconductors, metals, polymers, and ceramics. XPS works by measuring the kinetic energy and number of electrons emitted from the material's surface as a result of X-ray irradiation. The electrons are separated by a magnetic field and detected by a detector, which records the number of electrons with a particular kinetic energy. By analyzing the energies of the electrons, the XPS instrument can determine the elements present in the material and their chemical states.

In XPS, the kinetic energy of the emitted electrons is related to the binding energy of the electrons in the material. The binding energy is determined by the chemical environment of the atom from which the electron is emitted. Thus, by measuring the binding energy of the emitted electrons, XPS can provide information about the chemical environment of the atoms at the surface of the material.

XPS is a non-destructive technique, meaning that the sample is not damaged during the analysis. The X-ray source used in XPS is typically a monochromatic X-ray beam, which means that the X-rays have a well-defined energy. The energy of the X-rays used in the analysis can be adjusted to optimize the sensitivity and resolution of the analysis.

2.3 Conclusions

This chapter discusses various fabrication techniques of thin film coatings and their characterization methods. Some methods, like sol–gel technology, have reached the mature stage with industrial implementation; some can be applied with modifications, whereas some (like chitosan) are still at a very early stage of pilot-scale validation [22]. The different systems show different control ability and corrosion protection capability. Parameters like adhesion to the substrate and then to the polymeric layers can significantly influence the implementation of these technologies on an industrial scale.

Fundamental research is still required to protect the multitude of composite and metal materials used in automobile vehicles from corrosion. The integrated and coupled mechanism of the corrosion process in these materials will demand development of newer material components of coating as well as newer methods of coating fabrication. Still, many new solutions need to be validated within the industrial environment. However, the urgent need will be the development of sustainable, Cr (VI)-free solutions that are more effective for corrosion protection of aluminum alloys. The selection of materials that will be environment-friendly, cheap, and readily available will continue to be the major challenge.

References

[1]   Gupta G, Tyagi RK, Rajput SK, Saxena P, Vashisth A, Mehndiratta S PVD based thin film deposition
      methods and characterization / property of different compositional coatings – A critical analysis.
      In: Materials Today: Proceedings. Elsevier Ltd; 2020. pp. 259–64.

[2]   Baptista A, Silva FJG, Porteiro J, Míguez JL, Pinto G, Fernandes L. On the physical vapour deposition
      (PVD): Evolution of magnetron sputtering processes for industrial applications. In: Procedia
      Manufacturing. Elsevier BV; 2018, pp. 746–57.

[3]   Singh RK, Holland OW, Narayan J. Theoretical model for deposition of superconducting thin films
      using pulsed laser evaporation technique. J Appl Phys 1990;68(1):233–47.

[4]   Abu-Thabit NY, Makhlouf ASH. Fundamental of smart coatings and thin films: Synthesis, deposition
      methods, and industrial applications. In: Advances in Smart Coatings and Thin Films for Future
      Industrial and Biomedical Engineering Applications 2019. Elsevier, 3–35.

[5]   Ikeda T, Satoh H. Phase formation and characterization of hard coatings in the TiAlN system
      prepared by the cathodic arc ion plating method. Thin Solid Films 1991;195(1–2).

[6]   Wei X, Ying C, Wu J, Jiang H, Yan B, Shen J. Fabrication, corrosion, and mechanical properties of
      magnetron sputtered Cu-Zr-Al metallic glass thin film. Materials 2019;12(24).

[7]   Hoyos-Palacio LM, Cuesta Castro DP, Ortiz-Trujillo IC, Botero Palacio LE, Galeano Upegui BJ, Escobar
      Mora NJ, et al. Compounds of carbon nanotubes decorated with silver nanoparticles via in-situ by
      chemical vapor deposition (CVD). J Mater Res Technol 2019;8(6).

[8]   Leem M, Lee H, Park T, Ahn W, Kim H, Lee E, et al. Intriguing morphological evolution during
      chemical vapor deposition of HfS2 using HfCl4 and S on sapphire substrate. Appl Surf Sci 2020;509.

[9]   Meng WJ, Zhang XD, Shi B, Jiang JC, Rehn LE, Baldo PM, et al. Structure and mechanical properties of
      Ti-Si-N ceramic nanocomposite coatings. Surf Coat Technol 2003;163–64.

[10] Briones BA. Wiley Encyclopedia of Electrical and Electronics Engineering. Charleston Advisor 2019;21(2).

[11] Sabzi M, Mousavi Anijdan SH, Shamsodin M, Farzam M, Hojjati-Najafabadi A, Feng P, et al. A Review
      on Sustainable Manufacturing of Ceramic-Based Thin Films by Chemical Vapor Deposition (CVD):
      Reactions Kinetics and the Deposition Mechanisms. Coatings 2023;13(1):188.

[12] Wang D, Bierwagen GP. Sol-gel coatings on metals for corrosion protection. Prog Org Coat
      2009;64:327–38.

[13] Gasiorek J, Szczurek A, Babiarczuk B, Kaleta J, Jones W, Krzak J. Functionalizable sol-gel silica
      coatings for corrosion mitigation. Materials 2018;11.

[14] Brinker CJ, Scherer GW. Sol-Gel Science: The Physics and Chemistry of Sol-Gel Processing. 2013.

[15] Zhang FJ, Di CA, Berdunov N, Hu Y, Hu Y, Gao X, et al. Ultrathin film organic transistors: Precise
      control of semiconductor thickness via spin-coating. Adv Mater 2013;25(10).

[16] Uzum A, Fukatsu K, Kanda H, Kimura Y, Tanimoto K, Yoshinaga S, et al. Silica-sol-based spin-coating
      barrier layer against phosphorous diffusion for crystalline silicon solar cells. Nanoscale Res Lett
      2014;9(1).

[17] Dwivedi D, Lepkova K, Becker T Emerging surface characterization techniques for carbon steel
      corrosion: A critical brief review. In: Proceedings of the Royal Society A: Mathematical, Physical and
      Engineering Sciences. Vol. 473, Royal Society; 2017.

[18] Parry VK. Scanning Electron Microscopy : An introduction. III-Vs Review 2000;13(4).

[19] Ngo PD. Energy Dispersive Spectroscopy. In: Wagner, LC eds Failure Analysis of Integrated Circuits.
      The Springer International Series in Engineering and Computer Science. Vol. 494, Springer, Boston,
      MA, 1999.

[20] Lux H, Edling M, Lucci M, Kitzmann J, Villringer C, Siemroth P, et al. The role of substrate temperature and magnetic filtering for DLC by cathodic arc evaporation. Coatings 2019;9(5).
[21] Cao Y, Zheng D, Zhang F, Pan J, Lin C. Layered double hydroxide (LDH) for multi-functionalized corrosion protection of metals: A review. J Mater Sci Technol Chin Soc Met 2022;102:232–63.
[22] Galvão TLP, Bouali A, Serdechnova M, Yasakau KA, Zheludkevich ML, Tedim J. Anticorrosion thin film smart coatings for aluminum alloys. In: Advances in Smart Coatings and Thin Films for Future Industrial and Biomedical Engineering Applications. 2019.

Sheerin Masroor

3 Functionalized thin film coatings: basics, properties, and applications

Abstract: Ancient people considered the discovery of thin films to be of great significance because of their potential for anti-corrosion, and their decorative and optical applications. The variety of applications of thin films has grown to the point where every industry now uses them to meet the unique chemical and physical needs of surface materials in bulk. Their development supports the advancement of Hoover technology and electric power plants. The most cutting-edge applications of today, such as microelectronics and biomedicine, are made possible by the ability to vary the film characteristics by altering the microstructure via the deposition parameters used in a particular deposition process. The relationship between all stages of the creation of thin films – specifically the deposition parameters, morphology and features – is not totally precise despite this great improvement. The development of complex models for an accurate prediction of film properties has been hindered, among other things, by the lack of characterization equipment suitable for probing films with a thickness of less than a single atomic layer and a lack of physics expertise. Advanced structures like quantum wells and wires, however, continue to have some challenges with mass production and are still rather expensive to deposit. After overcoming these obstacles, thin film technology will be more competitive for cutting-edge technological applications.

Keywords: Functionalization, Thin Films, Basics, Coatings, Metals

3.1 Introduction

A few decades ago, it was discovered that the most frequently used materials for issues linked to corrosion were coatings and film–based obstacles [1, 2]. It has been shown that these coatings primarily consist of polymers or occasionally epoxy composite films [3–5]. The primary purpose of these enhancements is to strengthen and weigh down the exposed surfaces. The coatings must deal with numerous challenges that make it challenging to apply to various shapes in order to achieve all of these facts. However, the production of coatings can produce a lot of toxic waste. Additionally, it costs a lot of money to manufacture thin films using alternative methods. Other considerably more expensive methods are therefore available and ready for use. With the use of diverse materials, such as nanocomposites, we have recently noticed that there is a much greater tendency toward the development of thin coatings. The ideal coatings that have anti-corrosion qualities are those with thin layers. A single atom to many micrometers can be used to measure thin films. When coating a

https://doi.org/10.1515/9783111016160-003

substrate or specimen, a single layer of a certain material can be used, enhancing the physical and chemical properties that are not present in bulk materials with the same chemical composition [6]. Chemical vapor deposition (CVD) and physical vapor deposition (PVD) are two methods we can use to make thin films [7].

In CVD, a chemical reaction is started by the deposition method when a thin/slim layer of the intended specimen or substance is applied to the substrate surface, while in PVD, the target material is vaporized and then condensed onto the substrate surface to produce a film. The procedure entails vacuum or thermal evaporation, sputtering, and ion plating [8]. For these, two instruments – a challenging vacuum chamber and deposition system – are needed.

3.1.1 Morphological aspects of thin films

The microstructure of thin film formation is influenced by deposition features, growth mechanisms, and most modern manufacturing-related technologies. The following are the three categories of microstructures used in the production of thin films:
–   Those which form the perfect crystalline lattices and are called, amorphous.
–   Those which form micro or nanocrystallites of varying size ranges.
–   Those which form the solid crystalline film structure with latticework.

The application of thin film mainly depends upon different parameters such as deposition techniques, morphology, film thickness and growth structure, and atom adsorption film growths.

3.1.2 Basic properties of thin film coatings

In the use of contemporary thin films and their corresponding coatings, a number of features are crucial. Specific safety measures and environmental conditions are required for each of the processes used to make these films so that they can be used in subsequent applications. The sort of content a given material or substrate possesses to improve its long-term qualities greatly influences the deposition of thin film coatings on those materials or substrates.

But it is also important to note that produced thin film coatings have characteristics that are significantly different from those of their bulk elements. Thin film coatings can be used in a variety of industries and technologies due to their fundamental applicability. During the coating deposition process, the manufacturers employ a number of techniques to improve the chemical and mechanical properties of the target materials. Typical examples include anti-reflective, anti-ultraviolet or anti-infrared, anti-scratch, and lens polarization. The film coatings are used in a variety of ways, including:

- Reduction of rust and deterioration in pipes, engine components, bearings, etc. using anti-corrosion coatings.
- Improvement of cutting tool longevity by lowering friction by using hard coatings.
- Improvement of energy efficiency and cutting power expenses in workplaces and skyscrapers with architectural glazing.

3.2 Making of thin films

Chemical and physical deposition, which are further classified into the following kinds, are the two main processes that have been used to produce thin film coatings. It can be distilled as depicted in Figure 3.1.

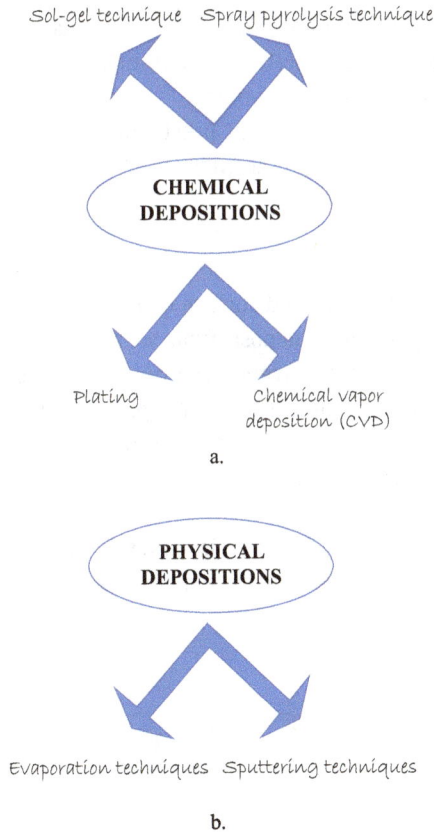

Figure 3.1: Broad classification of thin films deposition technique: (a) chemical deposition and (b) physical deposition.

3.2.1 Chemical deposition methods

Chemical deposition techniques are widely employed because it is essential to generate high-quality thin films at a fair price. These methods are inexpensive and produce films of high caliber. Most of them do not need pricey equipment. The chemistry of the solution, pH level, viscosity, and other components all play a significant role in chemical deposition. The sol–gel method, chemical bath deposition, electrodeposition, CVD, and spray pyrolysis techniques have all been used to create the most typical chemical depositions. Only the sol–gel and chemical bath deposition techniques are covered in this section since they produce high-quality films with minimal equipment requirements.

Oxide compounds are frequently created using the sol–gel method [9]. One of the well-known wet chemical processes is the sol–gel method. For multicomponent materials, it produces improved homogeneity while operating at lower processing temperatures. The processes of creating a colloidal suspension and converting it into viscous gels or solids, respectively, are referred to as "sol" and "gel". Transition metal oxides (TMOs) can be produced using one of the following two techniques:
- using aqueous solutions of inorganic salts to create inorganic precursors
- Making metal alkoxide precursors by dissolving metal alkoxides in non-aqueous solvents

The dip-coating method is almost universally used to produce clear oxide coatings on transparent substrates with excellent levels of planarity and surface quality [10]. Another method is to use extra substrates. It is possible to deposit films with a precise thickness of up to 1 m. You can combine many additive layers. According to Scriven [11], the five steps of the dip-coating procedure are immersion, start-up, deposition, drainage, and evaporation. Therefore, as seen in Figure 3.1, evaporation usually comes after the start-up, deposition, and draining processes.

After the precursor solution has been created, another process called spin coating or spinning can be applied. An equal distribution of the solution is achieved by dripping it onto a rotating substrate. The spinning method is least expensive, yet best suited for coating tiny discs or lenses.

3.2.2 Physical deposition techniques

3.2.2.1 Evaporation procedures

Using evaporation processes, materials are typically deposited as thin-layer films. By altering the phase, the fundamental workings of these techniques are revealed. On the specific substrate, the material is converted from solid to vapor and back to solid form. Either a controlled environment or a hoover is used.

3.2.2.1.1 Vacuum thermal evaporation technique

The easiest method for creating amorphous thin films, particularly chalcogenide films like Ge-Te-Ga [12], MnS [13], and CdSSe [14], is vacuum evaporation. Chalcogenide materials can generally be employed for solar applicatory-switching phase-change materials [17, 18], and memory-switching applications. The application of a potential difference to the substrate under medium- or higher-vacuum levels ranging from 105–109 mbar and the use of thermally evaporated material are two key factors in the thermal evaporation process [19].

3.2.2.1.2 Electron beam evaporation

Another physical deposition method is this type of evaporation, which works by producing an intense electron beam from a filament and guiding it through magnetic and electric fields before impacting the target and vaporizing it. The thin films created by electron beam evaporation are of high quality and clarity [20]. The electron beam evaporation method can be used to generate a wide variety of materials, such as metals [22], oxides [23], amorphous and crystalline semiconductors [21], and molecular compounds [24].

3.2.2.1.3 Laser beam evaporation (pulsed-laser deposition)

For the thin film coating system, pulsed-laser deposition (PLD) is another physical deposition method. To vaporize the item, various types of laser sources are used. The Nd-YAG (248 nm), KrF (308 nm), and XeCl lasers are the sources that are employed the most frequently. A plume that could deposit on different substrates is created when the laser beam collides with the target substance. In addition to ionized species, the generated plume may also contain neural- and ground-state atoms. Metal oxides created with the help of oxygen make up metal oxide thin films. The PLD's thin film quality is influenced by the laser's power, pulse duration, ambient gas pressure, and target-to-substrate distance [25]. Using laser-induced fluorescence [26], laser ablation molecular isotopic spectroscopy [27], and optical emission spectroscopy [28], one can regulate and monitor the ablation process during the deposition. The temperature of the substrate also affects the form of the thin films that are deposited.

The Frank-van der Merwe, Stranski–Krastanov, and Volmer–Weber techniques are used in PLD to coat thin films [29, 30]. Due to its quick deposition time and compatibility with oxygen and other inert gases, PLD offers a few advantages over other physical deposition processes.

3.2.3 Sputtering technique

Sputtering is often used to create metal and oxide layers by modifying surface roughness and crystalline structure [31]. In order to produce a glow discharge in the chamber's residual gas, the simplest sputtering technique includes enclosing a metallic anode and cathode. Additionally, many KeV of applied voltage and a pressure greater than 0.01 mbar are needed for the deposition of films. The cathode must be blasted by the discharge ions in order for the cathode molecules to be liberated with increased kinetic energy during the sputtering phase. Sputtering has a number of benefits. Creating materials with a high melting point is straightforward with sputtering. The composition of the original material is also present in the deposited films. Applications requiring extremely high temperatures can employ the sputtering method. The sputtering sources are compatible with oxygen and other reactive gases. On the other hand, it is difficult to deposit evenly on complex structures and it is impossible to build thick coats. Direct current (DC) and radio frequency (RF) are the two most popular kinds of sputtering procedures. On targets composed of materials that conduct electricity, the first one commonly uses DC power. With affordable options, control is simple. Most dielectric materials used in RF sputtering use RF power. Aluminum nitride films are a popular type of sputtered film. Both the structural and optical qualities of these films, which were created using DC- and RF-sputtering techniques, were evaluated [32, 33].

3.2.4 Nucleation

The crucial growing stage of nucleation has an impact on the ultimate structure of a thin film. Nucleation control (atomic layer deposition) is a crucial element of numerous growth techniques, including atomic-layer epitaxy. By describing the surface processes of adsorption, desorption, and surface diffusion, nucleation can be predicted [34].

3.2.4.1 Adsorption and desorption

Adsorption is the process of an atom or molecule from a vapor coming into contact with a substrate surface. The interaction is defined by the sticking ratio and the proportion of incoming species that has reached thermal equilibrium with the surface. Desorption, which is the opposite of adsorption, is the process when a previously adsorbed molecule overcomes the binding energy and leaves the substrate surface. The degree of the atomic contacts determines whether physisorption or chemisorption occurs. A stretched or bent molecule's Van der Waals bond with the surface denoted by the adsorption energy, *Ep*, is referred to as physisorption. Evaporating molecules form connections with atoms on the surface, which lowers their free energy and causes them to lose kinetic energy quickly. The adsorption energy, Ec, describes chemisorption as a strong electron transfer between substrate atoms and a molecule (through an ionic or covalent interaction).

Consider the potential energy as a function of distance to picture how physisorption and chemisorption work. Chemisorption and physisorption have a greater equilibrium separation from the surface. The effective energy barrier, Ea, controls when a condition changes from being physiosorbed to being chemisorbed [34].

On crystal surfaces with higher Ea values, vapor molecules would preferentially occupy specific bonding sites in order to lower the overall free energy. On step edges, voids, and screw dislocations, these stable regions are frequently observed. The importance of the adatom–adatom (vapor molecule) interaction increases when the locations with the highest degree of stability are filled [35].

3.2.4.2 Nucleation models

Modeling nucleation kinetics only needs to take into account adsorption and desorption. Imagine a situation in which interactions with step edges, clustering, or other adatom-related phenomena are absent. The given expression represents the rate at which the surface density of an adatom changes. This rate depends on several factors, including the net flux represented by J, the mean surface lifespan of the adatom before desorption and its sticking coefficient as follows:

$$\frac{dn}{dt} = J\sigma - \frac{n}{\tau_a}$$

$$n = J\sigma\tau_a \left[1 - \exp\left(\frac{-t}{\tau_a}\right) \right] \quad n = J\sigma\tau_a \left[\exp\left(\frac{-t}{\tau_a}\right) \right]$$

Different isotherms, including the Langmuir and Brunauer-Emmett-Teller (BET) models, can also be used to simulate adsorption. The interaction of a vapor adatom and a vacancy on the substrate surface results in an equilibrium constant, b, for the Langmuir model. In order to permit atom deposition on adatoms that have previously been adsorbed without interfering with nearby atom piles, the BET model can also be expanded. The surface coverage that arises is influenced by the equilibrium vapor pressure as well as the applied pressure.

Langmuir model where P_A is the vapor pressure of adsorbed adatoms:

$$\theta = \frac{bP_A}{(1 + bP_A)}$$

BET model where p_e is the equilibrium vapor pressure of adsorbed adatoms and p is the applied vapor pressure of adsorbed adatoms:

$$\theta = \frac{Xp}{(p_e - p)\left[1 + (X - 1)\frac{p}{p_e}\right]}$$

In lowering total free electronic and bond energies brought on by surface-level broken bonds, surface crystallography differs from bulk crystallography in an important way. This may lead to the preservation of the parallel bulk lattice symmetry at a new equilibrium location called "selvedge." Theoretical nucleation calculations may become inaccurate as a result of this phenomenon.

3.2.4.3 Surface diffusion

Surface diffusion refers to the lateral migration of atoms that have adhered to the surface of a substrate between energy minima. Diffusion is more easily accomplished in the regions with the fewest barriers separating them. Glancing-angle ion scattering allows for the computation of surface diffusion. The following best sums up the typical interval between events:

$$\tau_d = (1/\nu_1)\exp(E_d/kT_s)$$

Adatom clusters can migrate, merge, or decline in addition to migrating. In order to lower the system's total surface energy, processes including sintering and Ostwald ripening take place, which lead to cluster coalescence. Ostwald repining is the term used to describe the growth of larger adatom islands at the expense of smaller ones. When the islands come into contact and unite, sintering is the mechanism for coalescence [34].

3.3 Applications of thin films

Thin film technology has historically been employed in a variety of applications, initially beginning with aesthetic ones and subsequently progressing to optical ones. This is a result of the advent of enhanced deposition procedures made possible by the swift developments in hoover technology and electrical power. Thin films are frequently used to improve the usability of bulk materials by depositing a layer with the necessary physical and chemical properties. In the section that follows, the main technologically significant sectors of use for thin films are briefly discussed.

3.3.1 Thin films derived from advanced electronics: optoelectronic devices

Many different kinds of thin films are used in contemporary electronics and optoelectronic devices. The technological advancement in semiconductor thin films is mostly being driven by metal-oxide semiconductor field-effect transistor (MOSFET) and complementary metal-oxide semiconductor (CMOS). To separate the conducting channel from the gate during MOSFET production, silicon dioxide (SiO_2) thin films are necessary [36, 37]. This thin film has been employed because it is simple to make, has high impedance from a wide band gap, is chemically resistant, and operates at high temperatures. Additionally, metallic coatings are necessary for the production of a number of microelectronic, optoelectronic, and optical devices [38, 39].

In order to replace traditional flash memory with non-volatile memory, thin films have also been crucial in data storage devices [40]. This is due to their strong magnetic characteristics. For this use, thin films such as $BiFeO_3$, lead-zirconium titanate films, amorphous silicon, and organic compounds are being investigated as potential foundation materials [41–44].

Thin film technology is seen in the photovoltaic (PV) industry as a potential means of lowering the cost per watt of electricity generation. Due to the accelerated achievement of better efficiencies and the development of thin film structures with enhanced stability, this industry has been seeing rapid market penetration. In fact, record efficiencies of about 22.1% and 23.3% were attained using copper indium gallium selenide (CIGS) and CdTe thin films as the base materials, respectively [45].

Many sectors, including optics, have used thin films and coatings to increase the functioning and longevity of bulk materials by improving their mechanical and chemical properties. Optical thin films are widely used in eyeglasses to enhance eyesight by employing an optical component made of a polymer that is applied to the eyewear. Additionally, by using coatings materials capable of absorbing wavelengths lower than 400 nm and by using anti-reflective coatings, which are usually formed of dielectric materials, unwanted transmission of UV light and unwanted reflection are limited [46].

TiN and TiC ceramic coatings can be applied in many layers to provide coatings that improve cutting tool wear resistance and reduce friction [47]. Many different industries use coatings for corrosion resistance, including the coating of pipes with SiC, the coating of engine parts with high temperature corrosion protection materials like $MoSi_2$, the coating of stainless steel parts with oxides like SiO_2 and Al_2O_3, and more [48].

Organic thin films have attracted a lot of attention because of their unique properties, particularly their flexibility and affordable material processing, which are required to increase the number of applications for different technologies. For instance, albeit still modest, the efficiency of organic polymer-based modules in photovoltaics has grown dramatically from 0.04% to roughly 8.3% [49]. The low cost of material processing, such as printing, spraying, and the capacity to produce flexible modules, is still a major driving force behind this advancement.

The right biological and mechanical characteristics for implants to be employed in neural applications have also been achieved using composite thin films. For example, amorphous silica nanoparticles loaded with aluminum, silicon dioxide, or silver have been used to coat silicone implants to act as a microbiological barrier. On flexible substrates, inorganic thin films having piezoelectric capabilities are being studied for use in creating nano-generators and nano-sensors for biomedical purposes [50]. By converting the mechanical energy generated by internal organ movement into electrical energy, these piezoelectric devices can power devices like pacemakers or nano-sensors. These tools can be used to track the nanoscale deformation of cells since cells are so sensitive to mechanical movement. Perovskite materials such as $BaTiO_3$, PZNT, and PMN-PT have been used to produce piezoelectric devices with improved performance [51].

Applications for thin films are numerous, which necessitate the use of unique physical, mechanical, and chemical characteristics. These characteristics are related to the final morphology and structure and, as a result, are dependent on the deposition methods and parameters used. There are still a lot of issues that need to be resolved in order to fully understand how the different stages of thin film manufacture relate to one another. A still-expensive end product, lack of reproducibility, inappropriate bonding of the film to the substrate, high deposition temperatures that forbid the use of less expensive substrates, and limited control over the final attributes are issues with various CVD and PVD deposition techniques. In order to manage the microstructure and lower the cost, deposition technology must therefore advance with higher precision and a faster deposition rate suitable for large-area deposition. Even though some of the nanostructured thin films have the potential for novel applications in the fields of microelectronics, optics, photovoltaics, and biomedicine, some of them need to be transferred to specific substrates in order to function properly. Thin film technology must therefore be used to its full potential to improve the current subpar transferring approach. For instance, characterizing X-ray or Raman diffraction could be hampered by the substrate's inherent contribution to the resulting spectra. Complex models for data analysis are therefore needed to extract the relevant information from the data. Understanding protein adsorption in substrates is important for the right material selection in biomedical applications, but the characterization techniques for this purpose are still in their infancy and are based on intricate models for data processing. Therefore, more complex and narrowly focused in vitro models may facilitate the quicker identification of suitable thin films. Despite significant improvements in deposition and characterization techniques, it is still very difficult to predict film properties as a function of microstructure. This is due to complex transport properties, which include, but are not limited to, interface scattering, many defects, grain boundaries, material phases, and quantum confinement effects in extremely thin films. Therefore, to establish the appropriate link between microstructure and film properties and advance thin film technology, comprehensive models that account for all of these structural changes are needed.

3.4 Conclusion

In conclusion, functionalized thin film coatings have emerged as a promising technology with broad applications in diverse fields, ranging from electronics to biomedical engineering. These coatings can be tailored to exhibit specific chemical and physical properties, such as hydrophobicity, super hydrophobicity, and anti-fouling properties, making them highly versatile and adaptable to a wide range of applications. The development of functionalized thin film coatings is largely driven by advances in materials science and surface chemistry, which have led to the synthesis of a variety of functional materials, including polymers, nanoparticles, and biomolecules. These materials can be precisely engineered to exhibit specific properties and functional groups that can be used to modify the surface of substrates. The properties of functionalized thin film coatings are largely determined by the composition, structure, and thickness of the coating. In addition, the method of deposition also plays a critical role in determining the properties of the coating. Several methods, such as CVD, sol–gel deposition, and plasma deposition, have been developed for depositing thin film coatings with specific properties.

One of the most important applications of functionalized thin film coatings is in the field of electronics, where these coatings can be used to enhance the performance of electronic devices. For example, functionalized coatings can be used to modify the surface of electrodes to enhance their conductivity, and to reduce the buildup of biofilms on medical implants. Moreover, the use of functionalized coatings in biomedical engineering has also shown great promise, with applications in drug delivery, tissue engineering, and biosensors. Functionalized coatings also have great potential in the field of energy, where they can be used to improve the efficiency of solar cells, fuel cells, and batteries. For example, functionalized coatings can be used to modify the surface of electrodes in batteries to enhance their performance and to prevent the formation of dendrites that can lead to short circuits.

In conclusion, functionalized thin film coatings have shown great promise in a variety of applications, ranging from electronics to biomedical engineering, energy, and environmental applications. The development of these coatings is driven by advances in materials science and surface chemistry, and can be tailored to exhibit specific properties and functionalities. With further research and development, functionalized thin film coatings are expected to play an increasingly important role in a wide range of applications in the future.

References

[1] Macdonald DD. Parameters for electrochemical systems that exhibit pseudoinductance. J Electrochem Soc 1978;125:2062–64. https://doi.org/10.1149/1.2131363.

[2] Epelboin I, Keddam M, Takenouti H. Use of impedance measurements for the determination of the instant rate of metal corrosion. J Appl Electrochem 1972;2:71–79. https://doi.org/10.1007/BF00615194.

[3] Amand S, Musiani M, Orazem ME, Pébère N, Tribollet B, Vivier V. Constant-phase-element behavior caused by inhomogeneous water uptake in anti–corrosion coatings. Electrochimica Acta 2013;87:693–700. https://doi.org/10.1016/j.electacta.2012.09.061.

[4] Musiani M, Orazem ME, Pébère N, Tribollet B, Vivier V. Determination of resistivity profiles in anticorrosion coatings from constant-phase–element parameters. Prog Org Coat 2014;77:2076–83. https://doi.org/10.1016/j.porgcoat.2013.12.013.

[5] Musiani M, Orazem ME, Pébère N, Tribollet B, Vivier V. Constant-phase-element behavior caused by coupled resistivity and permittivity distributions in films. J Electrochem Soc 2011;158:C424–C28. https://doi.org/10.1149/2.039112jes.

[6] Vyas S. A short review on properties and applications of zinc oxide based thin films and devices: ZnO as a promising material for applications in electronics, optoelectronics, biomedical and sensors. Johnson Matthey Technol Rev 2020;64(2):202–18. https://doi.org/10.1595/205651320×15694993568524.

[7] Oluwatosin Abegunde O, Titilayo Akinlabi E, Philip Oladijo O, Akinlabi S, Uchenna Ude A. Overview of thin film deposition techniques. AIMS Mater Sci 2019;6(2):174–99. https://doi.org/10.3934/matersci.2019.2.174.

[8] Oluwatosin Abegunde O, Titilayo Akinlabi E, Philip Oladijo O, Akinlabi S, Uchenna Ude A. Overview of thin film deposition techniques. AIMS Mater Sci 2019;6(2):174–99. https://doi.org/10.3934/matersci.2019.2.174.

[9] Livage J, Sanchez C, Henry M, Doeuff S. The chemistry of sol-gel process. Solid State Ion 1989;32–33(2):633–38. doi: 10.1016/0167-2738(89)90338-X.

[10] Klein LC. Sol-Gel Technology for Thin Films, Fiber, Preform, Electronics and Specialty Shapes, Park Ridge, NJ, USA: Noyes Publications; 1987. doi: 10.1002/pi.4980210420.

[11] Scriven LE. Physics and applications of dip coating and spin coating. In: Brinker, CJ, Clark, DE, Ulrich, DR (eds). Better Ceramics Through Chemistry, 3rd ed., Pittsburgh, PA: Materials Research Society; 1988, pp. 712–29.

[12] Wang G, Nie Q, Shen X, Chen F, Li J, Zhang W, Xu T, Dai S. Phase change and optical band gap behaviour of Ge-Te-Ga thin films prepared by thermal evaporation. Vacuum 2012;86(10):1572–75. doi: 10.1016/j.vacuum.2012.03.036.

[13] Hannachi A, Segura A, Meherzi HM. Growth of manganese sulfide (α-MnS) thin films by thermal vacuum evaporation: Structural, morphological, optical properties. Mater Chem Phys 2016;181:326–32. doi: 10.1016/j.matchemphys.2016.06.066.

[14] Hassanien AS, Akl AA. Effect of Se addition on optical and electrical properties of chalcogenide CdSSe thin films. Superlattices Microstruct 2016;89:153–69. doi: 10.1016/j.spmi.2015.10.044.

[15] Salomé PMP, Alvarez HR, Sadewasser S. Incorporation of alkali metals in chalcogenide solar cells. Sol Ener Mater Sol Cells 2015;143:9–20. doi: 10.1016/j.solmat.2015.06.01.

[16] Orava J, Kohoutek T, Wagner T. Deposition techniques for chalcogenide thin films. In: Adam, JL, Zhang, X (eds). Chalcogenide Glasses Preparations, Properties and Applications. Oxford: Woodhead Publishing Series; 2014. doi: 10.1533/9780857093561.1.265.

[17] Rafea MA, Farid H. Phase change and optical band gap behaviour of Se0.8S0.2 chalcogenide glass films. Mater Chem Phys 2009;113(2–3):868–72. doi: 10.1016/j.matchemphys.2008.08.045.

[18] Sangeetha BG, Joseph CM, Suresh K. Preparation and characterization of Ge1Sb2Te4 thin films for phase change memory applications. Microelect Eng 2014;127:77–80. doi: 10.1016/j.mee.2014.04.032.

[19] Orava J, Kohoutek T, Wagner T. Deposition techniques for chalcogenide thin films. In: Adam, JL, Zhang, X (eds). Chalcogenide Glasses Preparations, Properties and Applications, Oxford: Woodhead Publishing Series, 2014. doi: 10.1533/9780857093561.1.265.

[20] Lokhande AC, Chalapathy RBV, He M, Jo E, Gang M, Pawar SA, Lokhande CD, Kim JH. Development of Cu_2SnS_3 (CTS) thin film solar cells by physical techniques: A status review. Sol Ener Mater Sol Cells 2016;153:84–107. doi: 10.1016/j.solmat.2016.04.003.

[21] Barranco A, Borras A, Elipe ARG, Palmero A. Perspectives on oblique angle deposition of thin films: From fundamentals to devices. Progress Mater Sci 2016;76:59–153. doi: 10.1016/j.pmatsci.2015.06.00.

[22] Mukherjee S, Gall D. Structure zone model for extreme shadowing conditions. Thin Solid Films 2013;527:158–63. doi: 10.1016/j.tsf.2012.11.007.

[23] Schulz U, Terry SG, Levi CG. Microstructure and texture of EB-PVD TBCs grown under different rotation modes. Mater Sci Eng A 2003;360(1–2):319–29. doi: 10.1016/S0921-5093(03)00470-2.

[24] Yang B, Duan H, Zhou C, Gao Y, Yang J. Ordered nanocolumn- array organic semiconductor thin films with controllable molecular orientation. Appl Surf Sci 2013;286:104–08. doi: 10.1016/j.apsusc.2013.09.028.

[25] Ashfold MNR, Claeyssens F, Fuge GM, Henley SJ. Pulsed laser ablation and deposition of thin films. Chem Soc Rev 2004;33(1):23–31. doi: 10.1039/B207644F.

[26] Lynds L, Weinberger BR, Potrepka DM, Peterson GG, Lindsay MP. High temperature superconducting thin films: The physics of pulsed laser ablation. Physica C 1989;159(1–2):61–69. doi: 10.1016/0921-4534(89)90104-4.

[27] Russo RE, Mao X, Gonzalez JJ, Zorba V, Yoo J. Laser ablation in analytical chemistry. Analy Chem 2013;85(13):6162–77. doi: 10.1021/ac4005327.

[28] Geyer TJ, Weimer WA. Parametric effects on plasma emission produced during excimer laser ablation of YBa2Cu3O7-x. Appl Spectros 1990;44(10):1659–64. doi: 10.1366/0003702904417454.

[29] Karl H, Stritzker B. Reflection high-energy electron diffraction oscillations modulated by laser pulse deposition YBa2Cu3O7-x. Phys Rev Lett 1992;69(20):2939–42. doi: 10.1103/PhysRevLett.69.2939.

[30] Lippmaa M, Nakagawa N, Kawasaki M, Ohashi S, Inaguma, Itoh M, Koinuma H. Stepflow growth of SrTiO3 thin films with a dielectric constant exceeding 104. Appl Phys Lett 1999;74(23):3543–45. doi: 10.1063/1.124155.

[31] Angusmacleod H. Recent developments in deposition techniques for optical thin films and coatings. In: Piegari A, Flory F editors. Optical Thin Films and Coatings from Materials to Applications, Oxford: Woodhead Publishing Series; 2013, pp. 3–25. 10.1533/9780857097316.1.3.

[32] Morosanu C, Dumitru V, Cimpoiasu E, Nenu C. Comparison between DC and RF magnetron sputtered aluminum nitride films. In: Prelas, MA, Benedictus, A, Lin, LTS, Popovici, G, Gielisse, P (eds). Diamond Based Composites and Related Materials. 1st ed., Petersburg, Russia: Springer Science+Business Media Dardrecht; 1997, pp. 127–32. doi: 10.1007/978-94-011-5592-2_9.

[33] Dumitru V, Morosanu C, Sandu V, Stoica A. Optical and structural differences between RF and DC AlxNy magnetron sputtered films. Thin Solid Films 2000;359:17–20. doi: 10.1016/S0040-6090(99)00726-9.

[34] Ohring M. Materials science of thin films: deposition and structure, 2nd ed., San Diego, CA: Academic Press; 2002, ISBN. 9780125249751.

[35] Venables JA. Introduction to Surface and Thin Film Processes, 1st ed., Cambridge University Press; 2000. doi: 10.1017/cbo9780511755651. ISBN, 978-0-521-78500-6.

[36] Chiu F-C, Pan T-M, Kumar Kundu T, Shih C. Thin Film Applications in Advanced Electron Devices. Adv Mater Sci Eng 2014;2014:927358.

[37] Rao MC, Shekhawat MS. A brief survey on basic properties of thin films for device application. Int J Mod Phys: Conf Ser 2013;22:576–82.

[38] Usui T, Nasu H, Takahashi S, Shimizu N, Nishikawa T, Yoshimaru M, Shibata H, Wada M, Koike J. Highly reliable copper dual-damascene interconnects with self-formed MnSi/sub x/O/sub y/ barrier Layer. IEEE Trans Electron Devices 2006;53:2492–99.

[39] Choy KL. Chemical vapour deposition of coatings. Prog Mater Sci 2003;48:57–170.

[40] Setter N, Damjanovic D, Eng L, Fox G, Gevorgian S, Hong S, Kingon A, Kohlstedt H, Park NY, Stephenson GB, Stolitchnov I, Taganstev AK, Taylor DV, Yamada T, Streiffer S. Ferroelectric thin films: Review of materials, properties, and applications. J Appl Phys 2006;100:051606.

[41] Catalan G, Scott JF. Physics and applications of bismuth ferrite. Adv Mater 2009;21:2463–85.

[42] Juan P-C, Hu Y-P, Chiu F-C, Lee JY-M. The charge trapping effect of metal-ferroelectric (PbZr0.53Ti0.47O3)-insulator (HfO2)-silicon capacitors. J Appl Phys 2005;98:044103.

[43] Chiu F-C, Shih W-C, Feng J-J. Conduction mechanism of resistive switching films in MgO memory devices. J Appl Phys 2012;111.

[44] Chiu F-C, Li P-W, Chang W. Reliability characteristics and conduction mechanisms in resistive switching memory devices using ZnO thin films. Nanoscale Res Lett 2012;7:178.

[45] Sundaram S, Shanks K, Upadhyaya H. In: A Comprehensive Guide to Solar Energy Systems. Elsevier, pp. 361–70; 2018.

[46] Larruquert J. In Optical Thin Films and Coatings. In: Piegari, A, François, F (eds). Woodhead Publishing, 2013, pp. 290–356.

[47] Choy KL. Chemical vapour deposition of coatings. Prog Mater Sci 2003;48:57–170.

[48] Choy KL. Chemical vapour deposition of coatings. Prog Mater Sci 2003;48:57–170.

[49] Krebs FC. Fabrication and processing of polymer solar cells: A review of printing and coating techniques. Solar Energy Mater Solar Cells 2009;93:394–412.

[50] Hwang G-T, Byun M, Kyu Jeong C, Jae Lee K. Flexible piezoelectric thin-film energy harvesters and nanosensors for biomedical applications. Adv Healthc Mater 2015;4:646–58.

[51] Gong J, Xu T. In Revolution of Perovskite: Synthesis, Properties and Applications. In: Arul, NS, Devaraj, NV (eds) Singapore: Singapore: Springer; 2020, pp. 95–116.

Sanjukta Zamindar, Surya Sarkar, Manilal Murmu, Priyabrata Banerjee
4 Functionalized superhydrophobic surface and thin film coatings for corrosion protection

Abstract: Functionalized superhydrophobic surfaces and thin-film coatings have emerged as ecological, environment-friendly, and efficient corrosion-inhibiting surface coatings in the domain of metal protection from adverse environments. Surfaces with high water contact angle values that are facile to facilitate the sliding away of water droplets or similar solvents are referred to as superhydrophobic surfaces, and these surfaces are non-wettable. Fabrication of superhydrophobic surfaces and application of functionalized thin films act as water-repelling as well as a barrier layer, which prevents corrosive electrolytes from penetrating toward the exposed metal surface, thereby protecting it from further degradation. Accordingly, there is an urgent need to understand the specific surface properties as well as corrosion inhibition effectiveness of the protective thin film by imparting superhydrophobicity. In this chapter, the most recent developments of highly efficient functionalized superhydrophobic surface and thin film coating to protect metallic surfaces from various adverse corrosive conditions have been outlined and discussed insightfully with explanation of the associated mechanism.

Keywords: Superhydrophobic Surface, Thin Film, Coating, Corrosion Inhibition, Barrier Property

4.1 Introduction

Superhydrophobic surfaces are typically described as the surfaces having contact angle values greater than 150° [1, 2]. The idea of superhydrophobicity was inspired by nature, e.g., lotus leaves, rose petals, etc. [3, 4]. The primary cause of the superhydrophobic quality of lotus leaves is hierarchical surface roughness. Using the idea of superhydrophobicity, the surface characteristics of lotus leaves can be imitated on di-

Acknowledgments: PB and NCM are very thankful to Department of Higher Education, Science and Technology and Biotechnology, Govt. of West Bengal, India for providing financial assistance to carry out this research work [vide sanction order no. 78(Sanc.)/ST/P/S&T/6 G-1/2018 dated 31.01.2019 and project no. GAP-225612]. SZ acknowledges Department of Science and Technology, Ministry of Science and Technology, Govt. of India for her DST-INSPIRE fellowship (vide IF 200407). MM would like to acknowledge Ministry of Tribal Affairs, New Delhi, India for his National Fellowship for Higher Education of Scheduled Tribes candidates, NFST (vide award letter no. F1-17.1/2014-15/RGNF-2014-15-ST-JHA-71,559).

https://doi.org/10.1515/9783111016160-004

verse metallic substrates to prevent corrosion. Metals are not obtained in their purest form in nature. They are generally obtained as oxides, sulfides or ore forms, which are energetically more stable. That is why pure metals are prone to be oxidized to gain its more stable form; consequently, corrosion of metal takes place [5, 6]. Corrosion is eventually caused by several environmental phenomena like temperature, presence of moisture in the air, reaction of metal with the surrounding atmosphere, *etc.* Several unfavorable conditions like acid rain, pollution, etc., also cause corrosive degradation of metallic structures as well as equipment. In industry, acid cleaning, acid descaling, and several other processes are practiced, which also leads to corrosion of the nascent metallic materials coming in contact with harsh conditions [7–9]. The devastating effects of corrosion have recently raised severe concerns all around the world. Recent analysis has estimated that ~ 4% of the global gross domestic product is squandered as a result of devastating corrosive attacks [10, 11]. Coating and inhibitors are frequently used to address the threat of corrosion because of their ease of application. The use of coating technology can significantly enhance the service lives of the metal or alloys. Nevertheless, the penetration of several corrosive species like water through the coating leads to delamination, deterioration, as well as reduced adhesion of the coating. Consequently, the cost of maintenance is increased [12]. This problem can be solved by introducing superhydrophobic nature to the as-formulated and applied coating. The water contact angle value is mainly responsible for the wetting property of the surface. When the contact angle value is less than 90°, it is called hydrophilic surface, i.e., the surface can be wetted easily. Similarly, when the contact angle value is greater than 90° but less than 150°, it indicates hydrophobic surface, i.e., the surface can repel the water or water like solvents. Surprisingly, the surface is called superhydrophobic surface when the water contact angle value is greater than 150° and that feature enables the repelling property toward water or water-like solvents very strongly [13–15]. In this chapter, the unique properties, several preparative methods and recent application of superhydrophobic coating to retard the rate of corrosion have been vividly overviewed and described.

4.2 Concept of contact angle, their properties, and associated theories

Superhydrophobic surface and thin film coatings have prevalent usages due to their superior and unique properties. The properties of such surfaces, contact angles, their characteristic properties, and associated theories are discussed in the subsequent sections.

4.2.1 Wettability

Wettability is one of the major properties of a liquid or fluid. Wettability is the term used to describe the capacity of water as a liquid to spread out on a surface [16]. This phenomenon is mainly caused because of the interactions between water molecules and substrate surfaces. Wetting phenomena happen when the free energy of the system goes down. When the solid is wet, the molecules of the droplet migrate over the solid surface to reduce the surface tension. The surface of the solid may be hydrophilic, hydrophobic, or superhydrophobic in nature. When a liquid is exposed to superhydrophobic surface, the interaction among the molecules of the liquid is greater than that of the interaction with the solid surface. The degree of wetting is determined by the balance between the adhesive and cohesive forces.

4.2.2 Contact angle value

Contact angle is defined as the angle formed between the solid surface of the substrates on which the droplet rests and the interior edge of the droplet. Actually, it is the angle between the tangent to the liquid or fluid and the surface of the solid as shown by angle θ in Figure 4.1.

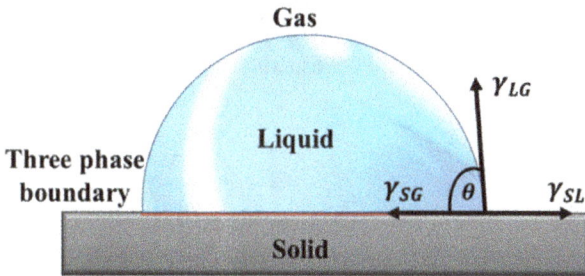

Figure 4.1: Diagram showing the forces acting at a boundary-phase contact line of a liquid onto a solid surface.

The formation as well as variation of the contact angle is attributed to the surface energy and surface tension of a liquid. The circumference of a liquid droplet is dictated by the limit of three phases, e.g., solid, liquid, and gas, defining three phase boundaries, and having the components γ_{SG}, γ_{LS}, and γ_{LG} defined as solid–gas interface, liquid–solid interface, and liquid–gas interface as shown in Figure 4.1.

In 1805, Thomas Young provided the following explanation regarding the equilibrium conditions on an ideally smooth surface as shown in eq. (4.1) known as Young's equation:

$$Cos\theta_Y = \frac{\gamma_{SG} - \gamma_{LS}}{\gamma_{LG}} \qquad (4.1)$$

where θ_Y, γ_{SG}, γ_{LS}, and γ_{LG} define the Young's water contact angle surface tension at solid–gas interface, liquid–solid interface, and liquid–gas interface, respectively. According to Young's equation, there are three aspects to classify the wetting behaviour of the surface. When, the contact angle value lies as $0° < \theta \leq 90°$ for a surface in which the droplet has been dispensed, the surface is called hydrophilic surface. But, if the contact angle value lies as $90° < \theta \leq 150°$, it indicates hydrophobic nature of the surface. When, θ lies as $150° < \theta \leq 180°$ for a surface, it is called superhydrophobic surface. A droplet of liquid is dispensed onto the solid surface with the help of a micro-syringe for the experimentation purposes. Then, a tensiometer is used to measure the contact angle of a liquid surface.

4.2.3 Surface roughness

The unevenness of the surface of the substrates is known as roughness. The wettability of liquid or water also depends upon the surface of the substrates in addition to its surface tension, viscosity, and other parameters. Superhydrophobicity of a material can be improved by changing the surface roughness and topology. The presence of surface defects and deformities on a sizable portion of a surface causes surface roughness to rise. Superhydrophobicity can be explained by using two main theories of surface roughness. They are Wenzel model and Cassie–Baxter model as explained thoroughly in the subsequent sections.

4.2.3.1 Wenzel model

According to the Wenzel model, the increase in surface contact area is caused by the penetration of droplets inside the microscopic grooves within the surface [17]. It implies that when the surface roughness of the substrates is increased, the surface contact angle increases simultaneously. It is interesting to note that the micro and nano structures can change the wetting behaviour of the solid surface. Therefore, superhydrophobicity is controlled geometrically. The Wenzel equation can be expressed as follows, *vide*, eq. (4.2).

$$Cos\theta_W = r\, Cos\theta_Y \qquad (4.2)$$

where, r is called roughness factor and r 1. θ_W is the Wenzel's contact angle, and θ_Y is the Young's contact angle. The roughness of the surface is mainly dictated by r, as shown in eq. (4.3):

$$r = \frac{Actual\ surface\ area}{Projected\ surface\ area} \qquad (4.3)$$

The actual surface area refers to the area of the local surface, whereas, the projected surface area refers to the area formed on the line of apparent solid surface, considering the rough surface of the solid. For the rough surface, the value of $r>1$ is considered, while, for an ideal smooth surface, the value of $r=1$. With increasing surface roughness, contact angle value also increases, which incorporates hydrophobicity to the surface. But the disadvantage of this model is that this model is applicable for only homogenous interface. For understanding the heterogeneous surface, another model was proposed by Cassie and Baxter.

4.2.3.2 Cassie–Baxter model

According to Cassie and Baxter, water droplets form spherical shapes and they stay on the micro-structured and nano-structured fibrous surfaces [18]. They introduced the concept of the area of fraction (f) to modify the aforementioned Wenzel model, which is known as Cassie–Baxter (CB) model. The equation used by this model is called Cassie–Baxter equation, as formulated in eq. (4.4):

$$Cos\theta_{BC} = f_1 Cos\theta_1 + f_2 Cos\theta_2 \qquad (4.4)$$

where θ_{BC} is the Cassie–Baxter contact angle, θ_1 and θ_2 are the intrinsic contact angles for area of fraction f_1 and f_2, respectively.

4.3 Different functionalization strategies to get superhydrophobic surface and thin film anti-corrosive coatings

Chemical compositions and surface roughness are the two significant factors for formulating superhydrophobic materials [19]. There are primarily four procedures to synthesize functionalized superhydrophobic coating material, viz., sol–gel, dip-coating, etching, chemical and electrical deposition processes. These four main processes along with their advantages and disadvantages are outlined in a flow chart as shown in Figure 4.2, while, the explanations are discussed in the subsequent sections.

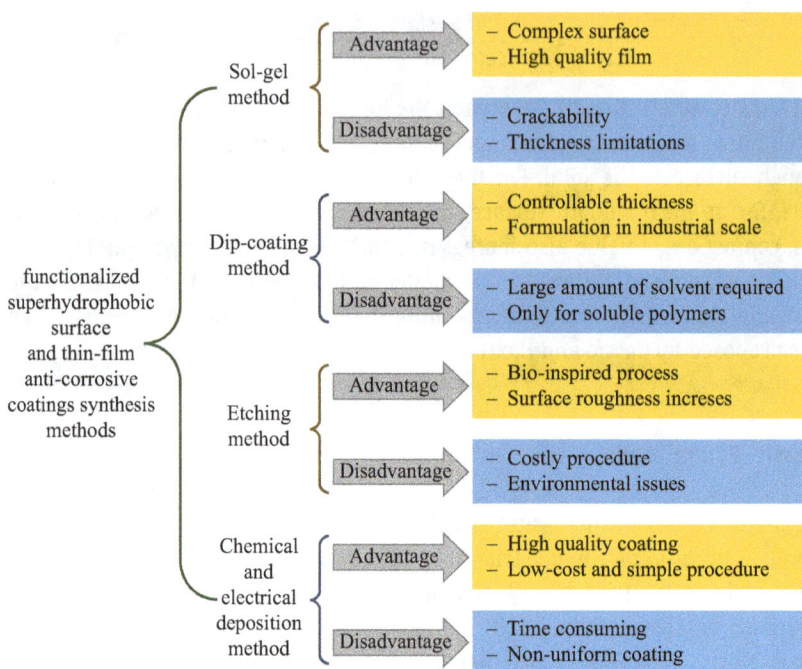

Figure 4.2: Flow chart of general synthetic procedures for the formulation of functionalized superhydrophobic surface and thin film anti-corrosive coatings, their advantages, and disadvantages.

4.3.1 Sol–gel process

Sol–gel process is a well-known wet chemical process for formulating polymeric superhydrophobic coating materials. The main advantage of this process is that the superhydrophobic coating materials obtained using this process possesses excellent thermal resistance [20–23]. In this process, "soft chemistry" is applied, which deals with the reactions performed at ambient condition in open reaction vessels mimicking the reactions occurring in the biological systems. Heat treatment followed by hydrolysis or polycondensation reactions are involved in this process to cover a surface of a material such as metal, glasses, woods, etc. [24]. But, there are also a few limitations of this process. The main disadvantage of this process is that the coating formulated by this process sometimes shows crackability and there are also thickness limitations of the as-formulated coating.

4.3.2 Dip-coating process

Dip-coating process is a renowned two-step process for the formulation of superhy-drophobic coating materials [25, 26]. In the first step, the targeted substrate is dipped within a solution containing nanoparticles followed by pulling the substrate upward with a steady and regulated speed. In the second step, after being removed from the solution, the substrates are subsequently coated with a hydrophobic substrate. The pull-up rates affect the thickness of the underlying nano-coating. The rate of solvent evaporation is greater than the rate of plate shrinkage for low pull-up rate while the reverse phenomenon is observed for high pull-up rate. The advantage of this method is that it provides a quick surface that can be used for several types of coated surfaces. The main disadvantages of this process are that a large amount of solvent is required and that it is applicable only for soluble polymers.

4.3.3 Etching process

Chemical and physical etching is a prominent process for formulating superhydro-phobic coating materials [27, 28]. The main advantage of this bio-inspired process is to enhance the surface roughness. But, it is a costly procedure and that is the main dis-advantage of this process. Some environmental issues also occur due to the usages of harsh conditions like strong acids. Furthermore, sometimes non-uniformity in the as-formulated coating surface is obtained.

4.3.4 Chemical and electrical deposition process

Chemical and electrical deposition is a widespread and easily affordable technique for the formulation of superhydrophobic coating materials in order to change the sur-face behaviour of as-desired structure [29, 30]. A thin layer of metal or metallic alloys can be obtained with a variety of morphologies comprising rod, needle, ribbon, tube, and so on. The principle of this process is mainly based on the aim of diminishing the amount of cations within the electrodeposited metals using an electrical current. The main advantages of this process are the incredible stability of the coating over time and application of this coating at room temperature. In this process, the thickness of the coating with reduced surface roughness can be easily customized.

4.3.5 Other processes

Apparently, several other techniques like spray coating, chemical vapor deposition technique, spin coating, electrodeposition, co-electrodeposition, hydrolytic condensation method, air-assisted spray method, one-step condensation reaction, etc. are also used for producing superhydrophobic coating [31]. There are several pros and cons in each technique. For instance, spray-coating technique is appropriate for treated surfaces as well as external applications though it provides contact angle of about 160°. Contrarily, superhydrophobic coating produced by spin-coating method provides minimal rugosity as well as precise thickness, and it can be applied on small surfaces suitably.

4.4 Recent promising applications of functionalized superhydrophobic surface and thin film anti-corrosive coatings

For a "high-energy society," it is crucial to improve the lifetime of a metal body. The two major issues in the maritime sectors are corrosion and bio-fouling of under-sea metallic parts. The most crucial step in retarding the cathodic and anodic reaction rates on metal surfaces is mitigation of corrosion. Coating technology, especially, functionalized superhydrophobic surface and thin film anti-corrosive coatings have grabbed much attention due to their excellent protective nature against corrosion. Some of the important works reporting functionalized superhydrophobic surface and thin film anti-corrosive coatings have been tabulated in Table 4.1.

4.4.1 Anti-corrosion property

In 2018, Ye et al. fabricated a novel aniline trimer-modified siliceous superhydrophobic anti-corrosive hybrid coating and applied on Q235 steel via one-step electrodeposition of a mixed solution of triethoxysilylpropyl isocyanate functionalized aniline trimer, designated as M-AT [32]. It was observed that aniline trimer containing silica coating exhibits high anti-corrosive performance along with better superhydrophobicity than that of pure silica coating. It has been reported that the superhydrophobic nature of the coating creates an efficient barrier against corrosive medium. Additionally, aniline trimer facilitates quick formation of the passive oxide-films on metal substrate, which helps in protecting the underlying metal surface. Cyclic voltammetry (CV) study revealed the redox catalytic nature of the aniline trimer. The two peaks obtained in CV specified the conversion between emeraldine state (oxidized state) and leucoemeraldine state (reduced state). Electrochemical tests like open circuit potential (OCP), Tafel, and electro-

Table 4.1: Recent promising applications of functionalized superhydrophobic surface and thin film anti-corrosive coatings.

Sl. no.	Coating composite	Coating process	Substrate	Medium	Studies performed[#]	Remarks	Ref.
1	Tetraethoxy silane and triethoxysilylpropyl isocyanate modified aniline trimer (M-AT)	Electrodeposition	Q235 steel	3.5 wt% NaCl	FTIR, ^{1}H-NMR, AFM, SEM, EDX, XPS, XRD, CV, OCP, EIS, Tafel, SVET, Adhesion test, CA analysis	Single step electrodeposition coating process, superhydrophobic nature with CA at 160.2°; prohibits the penetration of corrosive species and helps in formation of passive metal oxide films on metal substrate.	[32]
2	Tetraethyle orthosilicate and N,N′-bis(4′-(3-triethoxysilylpropylureido)-phenyl)-1,4-quinonenediimine@epoxy	Co-electrodeposition	Q235 steel	3.5 wt% wt. NaCl	UV-Vis, FE-SEM, AFM, CA analysis, EDX, XRD, CV, Tafel, EIS, SVET	Single step electrodeposition coating process, superhydrophobic nature with CA at 160.7°.	[33]
3	Perfluorodecyltrichlorosilane (FDTS) modified poly (dimethylsiloxane) (PDMS) -ZnO coating	Sol-gel process	Q235 steel	3.5 wt% NaCl	FESEM, AFM, EDS, TGA, EIS, Salt spray analysis	FDTS-modified PDMS/ZnO (1:1) coating shows better corrosion inhibition effectiveness and superhydrophobicity with CA at 150°.	[34]
4.	ZIF-8/POTS/EP coating [ZIF-8 = zeolitic imidazolate frameworks – 8, POTS = 1 H,1 H,2 H,2 H-perfluorooctyltriethoxysilane, EP = epoxy]	Hydrolytic condensation method	Q235 steel	3.5 wt% NaCl	CA, XRD, TEM, FT-IR, FE-SEM, EDS, EIS measurements	The effect of superhydrophobic composite coating shows excellent performance in self-cleaning, corrosion resistance, and anti-icing.	[35]

(continued)

Table 4.1 (continued)

Sl. no.	Coating composite	Coating process	Substrate	Medium	Studies performed[#]	Remarks	Ref.
5.	Ni-TiO$_2$/TMPSi composite	Electron deposition	Copper	NaCl	FESEM, AFM, EIS, FTIR, WCA, SA	TiO$_2$ nanoparticles with Ni after addition of trimethoxy(propyl)silane (TMPSi); the corrosion resistance significantly increased.	[36]
6.	PDMS/GO-ZnO [PDMS = polydimethyl siloxane, GO = graphene oxide, ZnO = zinc oxide]	Air assisted spray method	Carbon steel	3.5 wt% NaCl	FESEM, AFM, EIS, FTIR, WCA, EDS	The corrosion current in virgin PDMS is higher than that of ternary PDMS/GO-ZnO, which means corrosion reduces.	[37]
7.	Polyhedral oligomeric silsesquioxane-graphene oxide coating	One step condensation reaction	Q235 steel	3.5 wt% NaCl	FTIR, UV-vis, Raman and XPS	The coating shows superhydrophobic and anti-corrosion activity in saline solution over Q235 steel surface.	[38]
8.	ZnO nanoparticles-based epoxy coating	Facile one step approach	Glass slide	Water	CA, SA, XRD, TEM, FT-IR, FE-SEM, EDS,	The mechanical toughness and self-cleaning property of ZnO nanoparticles (NP) superhydrophobic coating on glass surface were studied.	[39]
9.	EP/FEP/m-ZnO [EP = epoxy resin, FEP = fluorinated ethylene propylene, m-ZnO = modified zinc oxide]	Spraying process	Aluminum and steel plate	3.5 wt% NaCl	SEM, FTIR, EDX, EIS	The coating material can clear all the contaminants from the metal surface and prevent it from the corrosion damage.	[40]

No.	Coating	Method	Substrate	Medium	Studies performed	Description	Ref.
10.	PDMS-β–MnO$_2$ nanorod composite	Electron deposition	Metal surface	Seawater	XRD, TEM, FT-IR, FE-SEM, XPS	Superhydrophobic coating of Polydimethylsiloxane (PDMS)β–MnO$_2$ nanorod composite coating exhibits high thermal stability, is corrosion resistive, and has ultraviolet resistance capabilities.	[41]
11.	Polyethersulfone (PES)-based coating	Scale-up spraying process	Aluminum	3.5 wt% saline solution	WCA, SA, FTIR, UV-Vis, EIS, SEM	Superhydrophobic polyethersulfone(PES) composite coating shows better anti-corrosion activity and stable in all weathers.	[42]
12.	Ni-SiO$_2$ coating	Electron deposition	Aluminum substrate	Acidic and alkaline environments	SEM, FTIR, EDX, EIS, CA, SA	The coating materials were examined in different temperature cycles between −35 °C and 100 °C	[43]

#Studies performed: AFM = atomic force microscopy; CA = contact angle measurement; CV = cyclic voltammetry; EDS = energy dispersive spectroscopy; EDX = energy dispersive X-ray; EIS = electrochemical impedance spectroscopy; ESR = electron spin resonance; FE-SEM = field-emission scanning electron microscopy; FTIR = Fourier-transform infrared spectroscopy; OCP = open circuit potential; SA = Sliding angle; SEM = scanning electron microscopy; SVET = scanning vibrating electrode technology; TEM = transmission electron microscopy; TGA = thermogravimetric analysis; UV-Vis study = UV-visible spectroscopy study; WCA = water contact angle measurement; XPS = X-ray photoelectron spectroscopy; XRD = X-ray diffraction.

chemical analysis for inhibition of corrosion efficiency test were done with four different concentrations, viz., 0, 2, 5, and 10 wt% of M-AT with tetraethoxysilane. The 5 wt % M-AT-containing silica coating showed greater semicircle loop (suggesting higher charge transfer resistance) in Nyquist plot along with 160.2° contact angle value, indicating better corrosion inhibition behaviour and hydrophobicity nature. This research team further prepared five types of epoxy coating with and without oligoaniline-modified silica coating (M-SiO$_2$), electrodeposited silica coating (E-SiO$_2$), and electrodeposited oligoaniline modified silica coating (E-M-SiO$_2$), respectively [33]. The applications of these newly synthesized coatings were made on Q235 steel surface via electrodeposition method using electrochemical workstation (CHI-660E model). Electrochemical studies like OCP, EIS, Tafel, and SVET were performed in 3.5 wt% saline medium. It was found that E-M-SiO$_2$/epoxy coating showed more positive OCP value, higher impedance value, and less amount of corrosion current value compared to other formulated coatings. E-M-SiO$_2$/epoxy coating exhibited 160.7° contact angle value indicating its superhydrophobic nature, which is sufficient to hinder the diffusion of corrosive species through the coating. The highest corrosion inhibition efficiency was 99.94%. In the same year, Arukalam et al. formulated perfluorodecyltrichlorosilane (FDTS)-modified poly(dimethylsiloxane) (PDMS)–ZnO composite coating [34]. Composite coating with three different ratios of PDMS and ZnO were synthesized. It was found that 1:1 ratio of PDMS:ZnO showed better corrosion inhibition effectiveness having impedance value of 10^9 $\Omega.cm^2$. This as-formulated coating showed excellent water repellence, i.e., superhydrophobicity with CA value of 150°. Chen et al. describes the performance of coating with superhydrophobic property for anti-corrosion, self-cleaning, and anti-freezing application in Q235 steel in saline environment [35]. The coating of ZIF-8/POTS/EP (say, type 2) showed better hydrophobic property than that of ZIF-8/POTS coating (say, type 1), where POTS and ZIF-8 are abbreviated for 1 H,1 H,2 H,2 H-perfluorooctyltriethoxysilane and zeolitic imidazolate frameworks-8, respectively. They performed the FESEM, EDS, and EIS study to show the active anti-corrosive performance of the composite material. Here, the superhydrophobic paint was formed according to the schematic diagram shown in Figure 4.3, in which initially the POTS were dissolved in ethanol. Subsequently, ZIF-8 was added into it and sonicated to obtain type-1 paint formulation, which was then applied on the substrates through spray coating. On the other way, EP and polyamide (PA) were dissolved in acetone to obtain a homogeneous solution, which was applied on substrate as primer and over this coated surface, the ZIF-8/POTS spray coating was applied to get type-2 coating. This as-formulated coating showed better corrosion resistance, anti-icing, and water-repelling property.

Measurements of the static water contact angle (CA) were used to inspect the wettability of the bare Q235 steel surface and the coated specimen surfaces, as presented in Figure 4.4. From Figure 4.4 (a), it is clear that the bare polished Q235 steel sample surface is hydrophilic as confirmed from the measured CA, which is 74.5°, while the ZIF-8 coated steel sample surface is more hydrophilic as its CA is 26.2° (vide Figure 4.4 (b)).

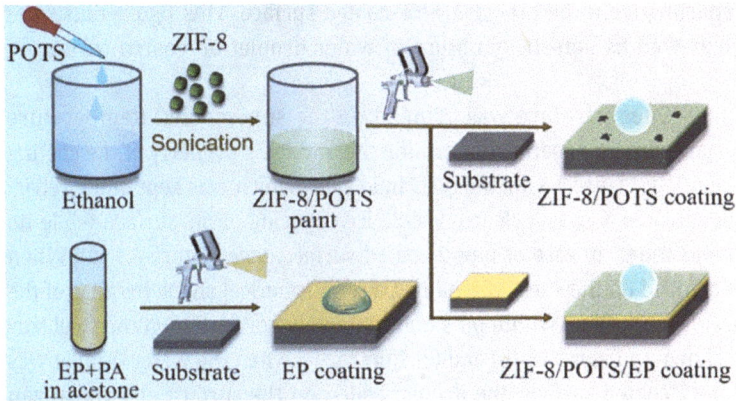

Figure 4.3: The synthesis process of the superhydrophobic coatings, namely, ZIF-8/POTS and ZIF-8/POTS/EP (reused with permission from [35], © 2021 Elsevier Ltd.).

Again, Figures 4.4 (c) and (d) show that the POTS-coated steel surface and the EP-coated steel surface have 82.5° and 63.7° CA, respectively, which signify that both of these surfaces are hydrophilic. But, it was surprisingly noted that the CA for the ZIF-8/POTS-coated steel surface was 164.4° suggesting that the surface is superhydrophobic in nature (vide Figure 4.4 (e)). The superhydrophobicity of the surface was increased after adding ZIF-8/POTS with EP as the CA of type-2 coating was observed as 162.8° as shown in Figure 4.4 (f). Figure 4.4 (g) presents the slide angle (SA) of this (type-2) coating is ~2°, which indicates that water droplets are easily able to roll on the surface. Furthermore, Figure 4.4 (h) depicts the steps of a water droplet in approaching, contacting, squeezing,

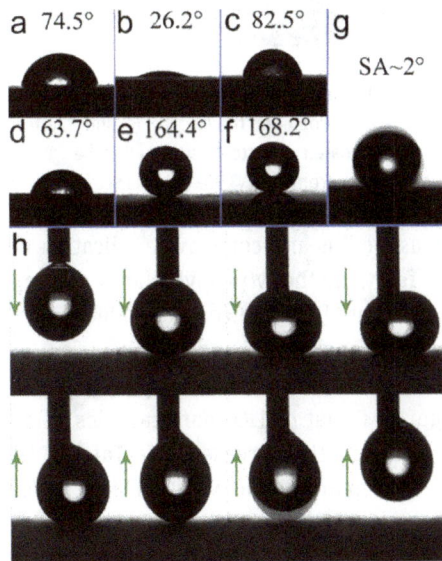

Figure 4.4: CA measurements of (a) bare Q235 steel, (b) ZIF-8-coated steel sample, (c) POTS-coated steel sample, (d) EP-coated steel sample, (e) ZIF-8/POTS-coated steel sample, and (f) ZIF-8/POTS/EP-coated steel sample; (g) the SA measurement of ZIF-8/POTS/EP-coated steel sample, and (h) the steps of a water droplet in approaching, contacting, squeezing, elevating, and departing from ZIF-8/POTS/EP-coated surface using CA meter (adapted with permission from [35], © 2021 Elsevier Ltd.).

elevating, and departing from the ZIF-8/POTS/EP-coated surface. This figure illustrates the non-adhesive as well as non-sticky nature of water droplet on coated surface of type-2.

Water having 0 °C temperature was dripped on the supercooled (temperature around −20 °C) surface of type-2 coating to examine the anti-icing property of the coating as shown in Figure 4.5. In Figure 4.5 (a), the Q235 mild steel, which was kept in the refrigerator for two hours at −20 °C, has a clearly visible ice covering on its surface while no such observation was found in case of type-2 coated surface, *vide*, Figure 4.5 (d). When the water droplet dripped on the surface, it gained a semi-spherical shape because of the hydrophilic nature of the surface within 0.1 s and ultimately got frozen on the steel surface within 2 s as shown in Figure 4.5 (b) and (c). Contrarily, when the water droplet was dripped on the type-2 coated surface, the droplet rolled on the surface and got frozen outside the coated surface, which in turn indicates the water-repelling as well as anti-icing properties of the type-2 coating as pictorially represented in Figure 4.5 (e) and (f). It has been also found that this coating possesses good stability as it sustained its CA value at 156.9° after the anti-freezing tests of 100 dripping droplets as shown in Figure 4.5 (g).

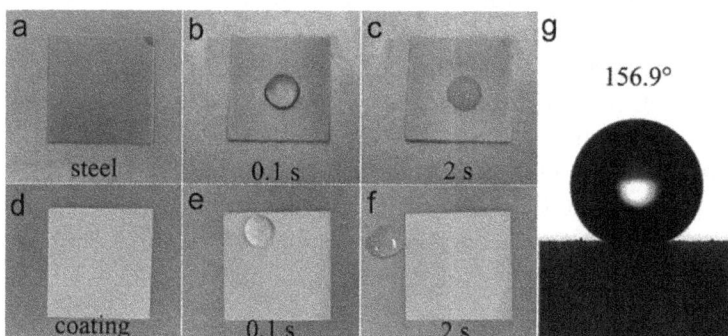

Figure 4.5: The images taken afterward placing (a) the Q235 mild steel and (d) type-2-coated mild steel at −20 °C after two hours; (b) and (e) after 0.1 s of dripping a water droplet on the surface; (c) and (f) after 2 s of dripping a droplet on the surface, and (g) CA of the type-2-coated mild steel surface after the anti-freezing tests of 100 dripping droplets (reused with permission from [35], © 2021 Elsevier Ltd.).

Salehi et al. described the superhydrophobic as well as anti-corrosive application of Ni-TiO$_2$/TMPSi (where TMPSi is abbreviated for trimethoxy(propyl)silane) coating with nanocomposite material on the copper substrate [36]. The coating material possesses high-water repelling behavior having contact angle 151.6° and sliding angle 6.2°, which proves the effectivity of the material. The corrosion resistance property was studied in NaCl solution. It is noted from Figure 4.6 that in TiO$_2$ nanoparticles, after addition of TMPSi, the corrosion resistance significantly increased. The Tafel of Ni coating, Ni-TiO$_2$ nanocomposite coating, and superhydrophobic Ni-TiO$_2$/TMPSi nanocomposite coating presented in Figure 4.6 (a) show that there is a huge shifting in the corrosion potential toward more positive direction, suggesting more shielding property of Ni-TiO$_2$/TMPSi nanocomposite coating against corrosion. The Bode plot (vide Figure 4.6 (b)) and Nyquist plots (vide Figure 4.6 (c)) of all the three as-fabricated coatings revealed that the incorporation of TiO$_2$ enhances the hydrophobicity as well as the corrosion retardance property of the coating. On the other hand, the incorporation of TMPSi within the nanocomposite coating increases the superhydrophobicity and corrosion resistivity of the coating materials. When the Ni-TiO$_2$ nanocomposite- coated materials are dipped into ethanolic solution of deionized water, the activated -OH functional groups are formed on the surface. Further, these samples with active -OH groups are immersed into TMPSi-hexane solution. The oligomeric hydrogens tend to bond formation with the active -OH groups on the coated surface as shown in Figure 4.6 (d). Ultimately, dehydration condensation process leads to form covalent bonds between the coated surface and the TMPSi molecules via loss of water. The copper sample was coated via electrodeposition process (vide Figure 4.6 (e)). It has been reported that the charged TiO$_2$ nanoparticles drift toward the copper surface, which is the working electrode, while the Ni^{2+} ions get co-deposited on the copper surface via discharging process. A re-entrant morphology based on TiO$_2$ and Ni-nanoparticles is generated on the copper surface, which efficiently upsurges the water-repellent property of the surface.

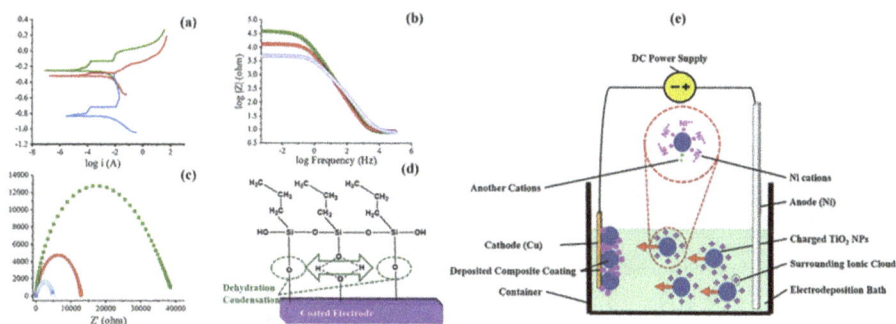

Figure 4.6: (a) Potentiodynamic polarization curves, (b) Bode plots, and (c) Nyquist plots of Ni coating (in blue color), Ni-TiO$_2$ nanocomposite coating (in red color) and superhydrophobic Ni-TiO$_2$/TMPSi nanocomposite coating (in green color), respectively, (d) TMPSi treated with coated copper, and (e) electrodeposition of Ni and TiO$_2$ nanoparticles on Cu substrate (adapted with permission from [36], © 2019 Elsevier Ltd.).

Selim et al. examined the superhydrophobic and corrosion barrier activity of polydimethylsiloxane (PDMS)/ graphene oxide-ZnO nanocomposite over the carbon steel in 3.5 wt% NaCl solution [37]. This is seen from Figure 4.7; the nanocomposites of PDMS/ GO-ZnO were synthesized by ex situ method where the percentage of nanofillers varied accordingly as 0.5, 1, 2.5, and 5 wt%. The water contact angle was 108°, which reveals that the coating material acted as a superhydrophobic surface. GO-ZnO nanocomposite was synthesized in one-step chemical deposition method through the reaction of GO nanosheets and ZnO nanoparticles as shown in Figure 4.7 (a). Then, GO-ZnO nanocomposite was dissolved in PDMS solution via solution casting method. Air-assisted spray method was used to coat the sample surface. Figure 4.7 (b) presents how the groove, created on the surface of the nanocomposite can trap air which provides self-cleaning and water-repellent property. The change in water CA and surface free energy (SFE) with the change in the percentage of nanofillers have been represented in Figure 4.8.

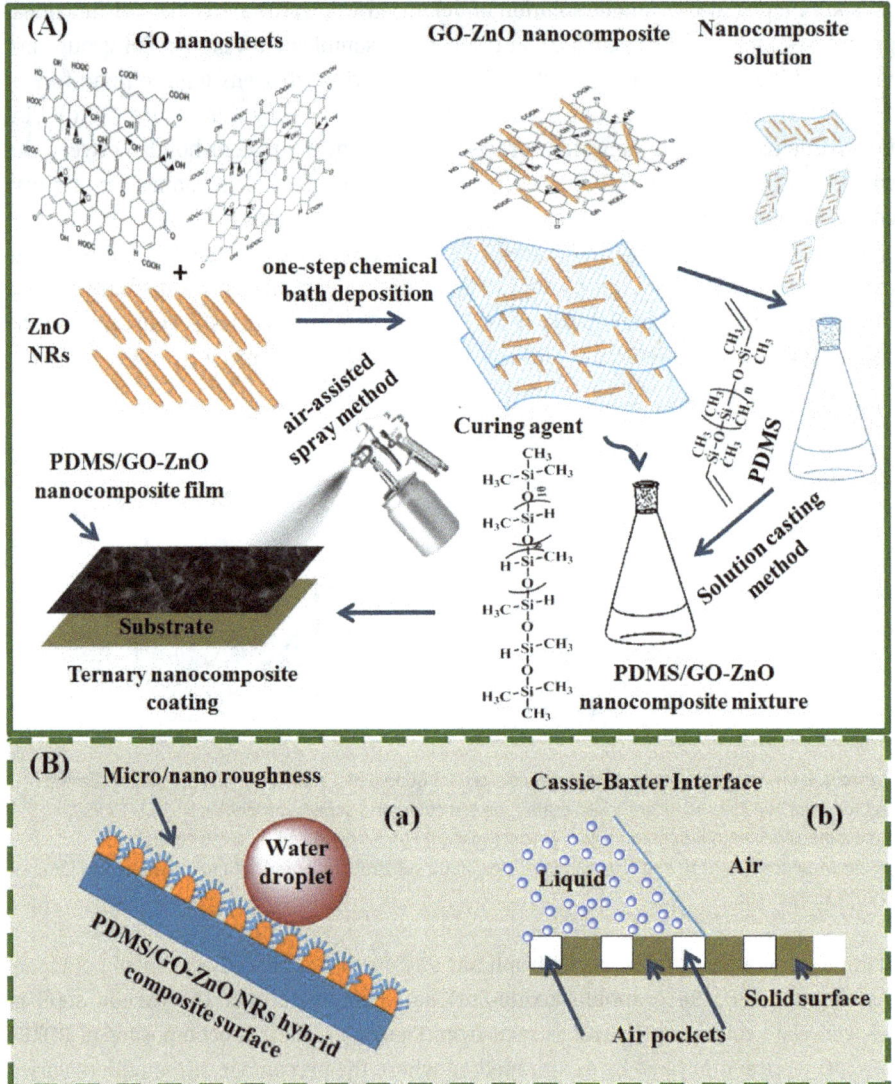

Figure 4.7: Schematic representation of (A) Procedure of PDMS/GO-ZnO nanocomposite coating, and (B) hydrophobically modified rough structure for the well-dispersed PDMS/GO-ZnO nanocomposite showing micro/nano roughness and Cassie–Baxter interface (reused with permission from [37], © 2021 Elsevier Ltd.).

The SFE of the unfilled PDMS film was 20.13 mN/m. Additionally, it was found that the SFE was reduced to 12.65 mN/m followed by the addition of GO-ZnO hybrid nanofiller up to 1 wt%. It has also been observed that the CA values increased primarily with the increase of nanofillers (up to 1 wt%) due to enhanced surface area and volume ratio. However, after addition of more than 1 wt% of nanofillers, the CA value tends to decrease while the SFE tends to increase because of agglomeration.

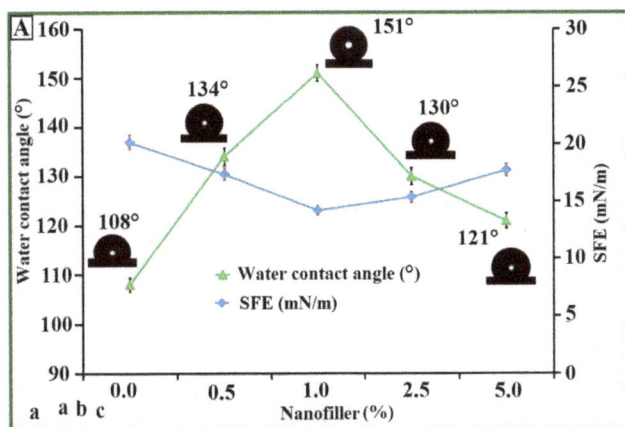

Figure 4.8: Change in water CA along with SFE with the change of nanofillers wt% (reused with permission from [37], © 2021 Elsevier Ltd.).

According to Ye et al. the polyhedral oligomeric silsesquioxane-graphene oxide coating shows superhydrophobic and anti-corrosion activity in saline solution on Q235 steel surface [38]. Initially, the functionalized graphene oxide (GP) was synthesized through the reaction of graphene oxide (GO) and polyhedral oligomeric silsesquioxane (POSS) – NH_2 in a single mouth flask (vide Figure 4.9).

Ethanol was used as solvent wherein a certain quantity of condensation agent (DCC) was used. Various processes including ultrasonic (for 15 min), circulating reflux (for 48 h at 75 °C), rotary evaporation (for 15 min at 75 °C), heat treatment (for 10 h at 120 °C), filtration, and desiccation (for 3 h at 80 °C) in respective sequence were followed to get the as-desired GP. Due to the vacuolization of the coating, some particles were dispersed around the edges of all wear grooves. Figure 4.10 shows the excellent anti-wear ability at 0.5 wt% GP/EP because the area of cross section and depth of the rust portion over the specimen is the least here. For EP specimen, the extreme wear track depth and sectional area were about 7.82 m and 3,176 m^2, respectively (vide Figure 4.10 (a)).

Under the same experimental conditions, wear track depth and sectional area were found to be reduced in 0.5 wt% GP/EP specimen than that of GO/EP specimen. Hence, an admirable anti-wear feature of 0.5 wt% GP/EP specimen was observed as shown in Figure 4.10 (b) and (c). Contrarily, the wear track depth and sectional area

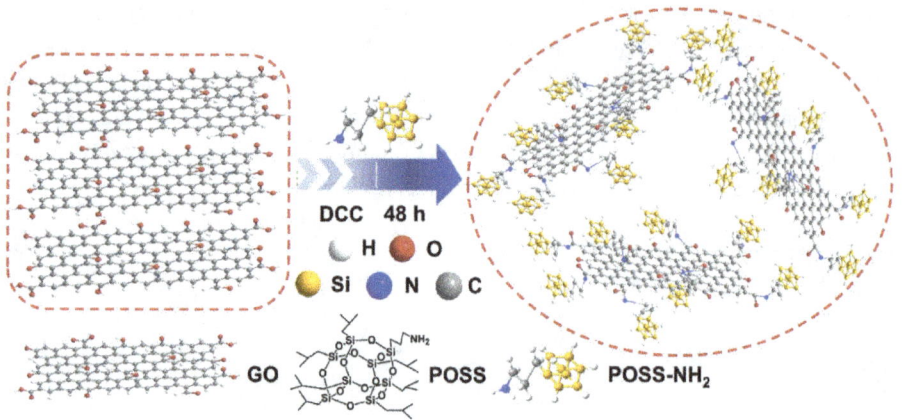

Figure 4.9: Schematic representation of the synthetic procedure of functionalized graphene oxide (reused with permission from [38], © 2019 Elsevier Ltd.).

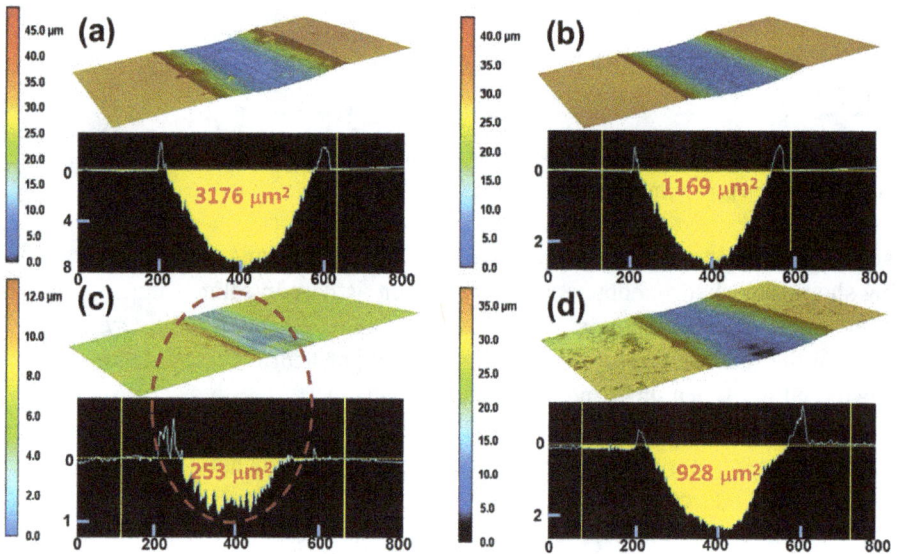

Figure 4.10: Image of area of cross section and 3D topographic of rust depth in every specimen (a) EP; (b) 0.5 wt% GO/EP; (c) 0.5 wt% GP/EP; and (d) 1 wt% GP/EP (reused with permission from [38], © 2019 Elsevier Ltd.).

were found to be increased for 1 wt% GP/EP specimen, which was consistent with the findings of wear rate analysis tests.

The lubrication mechanism test of 0.5 wt% GP/EP specimen was performed and the wear model was displayed as presented in Figure 4.11. The 0.5 wt% GP/EP specimen was allowed to slide against a 316 L ball, while some chippings were transported to the edge and left a small pit, which acted as a corrosion channel to allow corrosive media to penetrate. Some chippings were mechanically compacted and adsorbed on the surface; subsequently a pit was left partially as shown in Figure 4.11 (a). Alternatively, the addition of GP within the epoxy coating enhanced the hydrophobicity, that is, it produced more hindrance against the penetration of corrosive medium as depicted in Figure 4.11 (b). From a lubrication perspective, the huge specific surface area of GP aids in the creation of a lubricating transfer film as shown in Figure 4.11 (c). It is also to be noted that the incorporation of POSS nanospheres have also enhanced the lubricating property significantly (vide Figure 4.11 (d)).

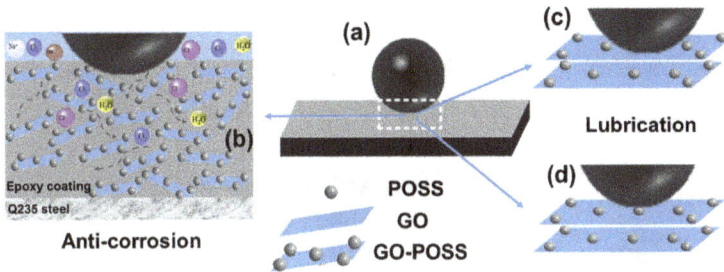

Figure 4.11: (a, c, d) Lubrication property and (b) anti-corrosive mechanism of 0.5 wt% GP/EP sample (reused with permission from [38], © 2019 Elsevier Ltd.).

4.4.2 Self-cleaning property

Self-cleaning property is another property of the applied coating on the metal substrates. Yap et al. studied the action of self-cleaning and mechanochemical robustness of ZnO nanoparticles (NP) coating with superhydrophobic property on glass surface [39]. They performed FTIR, FESEM, contact angle, sliding angle, and XRD studies to validate and investigate the effectiveness of the as-formulated coating material. The reported procedure of making superhydrophobic coating is depicted in Figure 4.12, where steric acid serves as a constituent having a polar head and non-polar hydrophobic tail. The dispersion of ZnO nanoparticles in ethanol solution reduces the surface energy of the coating material. Epoxy plays a vital role to make a molecular structure with the nanoparticles as it imparts a high mechanical stability.

In self -cleaning test, the distribution of carbon powder on tilted glass slide is noticed. It is seen that the contaminants were removed by the precipitation of water drop in ZnO/EP surface (Figure 4.13(a, b, c)). The powder of carbon was placed on the coated sample and this was tilted at 10° angle and 10 ml water was used to clear the powder. But, the effect of self-cleaning is more prominent in Figure 4.13 (c). The pollutants were totally rolled off along with the water droplets rolling on the superhydrophobic surface of ZnO/EP. The adhesion between contaminants and droplets is

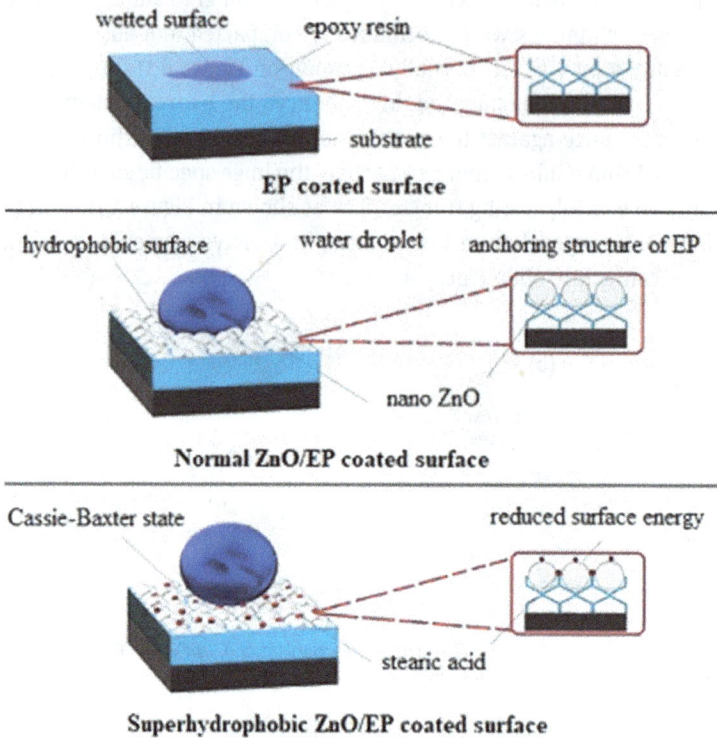

Figure 4.12: Description of the mechanism of long-lasting superhydrophobic coating of EP/SA (reused with permission from [39], © 2021 Elsevier Ltd.).

greater than that between contaminants and superhydrophobic surfaces. Thus, the superhydrophobic surfaces facilitates self-cleaning by the dropped water droplets on the coated substrates.

The adhesion test of water particles on the superhydrophobic surface is depicted in Figure 4.14 (a). The water pocket presented on the surface due to lower surface energy clears the water droplet from the surface. In pure epoxy-coated surface and ZnO/EP coating, the water droplets are sustained as shown in Figure 4.14 (b, c) but on the superhydrophobic surface (vide Figure 4.14 (d)) the droplets are removed fully without the surface wetting.

Wang et al. successfully constructed a multifunctional EP/FEP/modified ZnO coating for corrosion resistance and superhydrophobic property [40]. Herein, the fabrication of ZnO and modified ZnO with steric acid for constructing the superhydrophobic surface has been shown in Figure 4.15 (I–II). The aluminum surface of 40 × 40 ×1 mm was taken as the substrate and polished with abrasive paper to get a rough surface Figure 4.15(a–b). After the pre-treatment, the EP/FEP/modified ZnO coating formulation was spray-coated on the metal substrate as shown in Figure 4.15(c–d). Then fi-

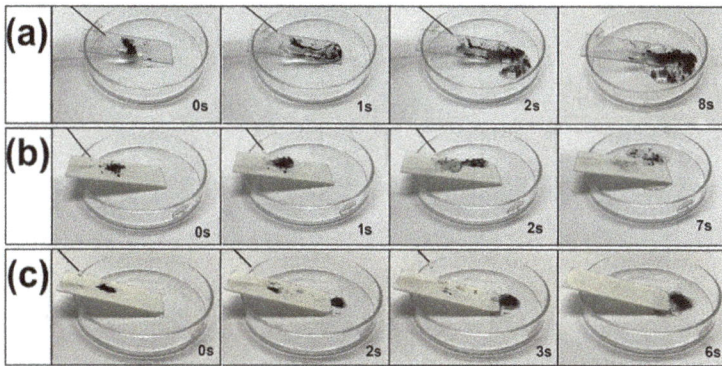

Figure 4.13: Self-cleaning test on the (a) EP-coated surface, (b) coated with ZnO/EP, and (c) coated with superhydrophobic ZnO/EP (reused with permission from [39], © 2021 Elsevier Ltd.).

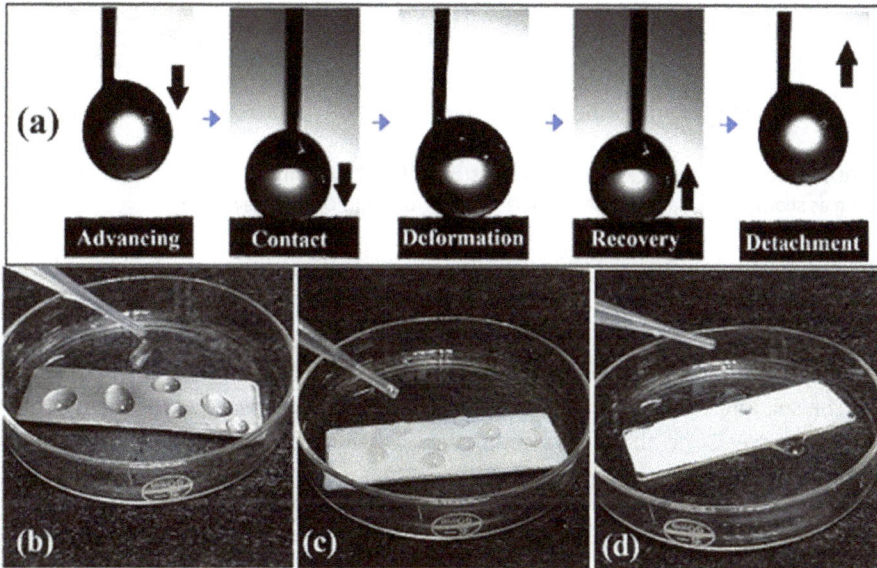

Figure 4.14: Water adhesion test applying different coatings over the sample surface (a–d) (reused with permission from [39], © 2021 Elsevier Ltd.).

nally it was heated and kept at room temperature to construct multifunctional (EP)/FEP)/modified ZnO coating as shown in Figure 4.15 (e).

In Figure 4.16, the variation of WCA and SA with modified ZnO and NH_4HCO_3 containing the superhydrophobic coating have been shown, which revealed a vital role for increasing hydrophobicity and the corrosion inhibition efficiency of the coating material. The WCA and SA are 140° and 155°, respectively, in Figure 4.16 (a). But, when the coating material was modified by ZnO and NH_4HCO_3 the WCA and SA become

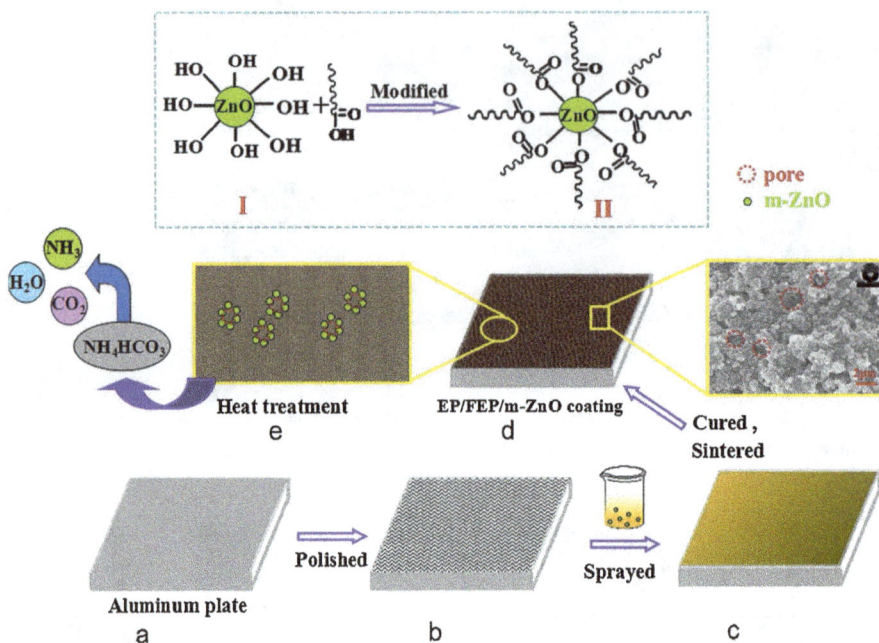

Figure 4.15: Constructing process of on different stages of multifunctional (EP)/(FEP)/modified ZnO coating as shown in (I–II, a–e) (reused with permission from [40], © 2019 Elsevier Ltd.).

160° and 2°, respectively, in Figure 4.16 (b), which confirmed the superhydrophobicity of the material.

Furthermore, it can be noted from Figure 4.17 that the coating material can clear all the contaminants from the metal surface and prevent it from the corrosion damage. The soil and fly ash are the impurities on the coating surface of the plate of aluminum and steel. The ability of cleaning the surface by EP/FEP/modified ZnO coating improves the durability of the material.

4.4.3 Thermal stability

According to Selim et al., the superhydrophobic coating of polydimethylsiloxane (PDMS)β–MnO_2 nanorod composite exhibits high thermal stability, corrosion resistance, as well as ultraviolet resistance [41]. The metal substrates coated with various concentration levels of coating materials are examined and it is seen that with higher concentration of the additives the corrosion inhibition effectiveness increases. In Figure 4.18, it is seen that the virgin film makes the surface smooth. But, the roughness increases significantly when β–MnO_2 is added into it. This also helps increase the WCA and self-cleaning property. The heterogeneous structure removes the water droplets and increases the self-cleaning property.

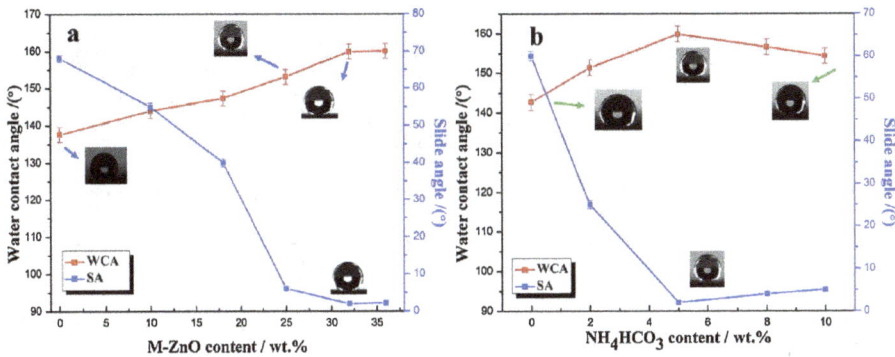

Figure 4.16: Variation of WCA and SA in M-ZnO and NH$_4$HCO$_3$ (a, b) (reused with permission from [40], © 2019 Elsevier Ltd.).

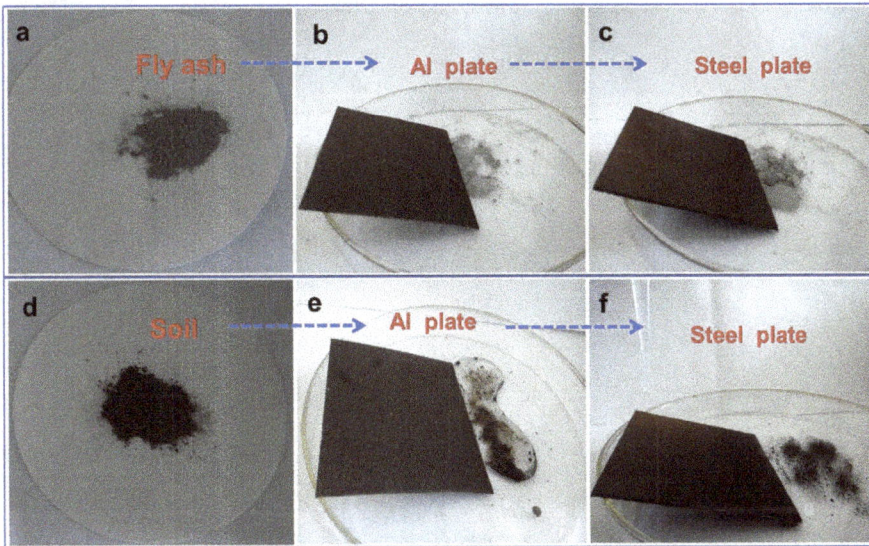

Figure 4.17: Removal of contaminants of fly ash (a) and soil (d) on the coated sample of Al plate (b, e) and steel plate (c, f) (reused with permission from [40], © 2019 Elsevier Ltd.).

Zhao et al. successfully created the superhydrophobic polyethersulfone (PES) composite made from polyimide (PI) particle, octadecylamine (ODA)-functionalized carbon nanofiber (CNF) and room temperature vulcanized silicone rubber (RTVSR) abbreviated as PES/RTVSR/PI/ODA-CNF for thermal and all-weather stable anti-corrosion coating [42]. It has been reported that the results of heat and weathering test of the coating material indicate the stability of the coating in extreme temperature ranges from −30–375 °C and snowfall time (vide Figure 4.19). The excellent thermal stability is indicated by the

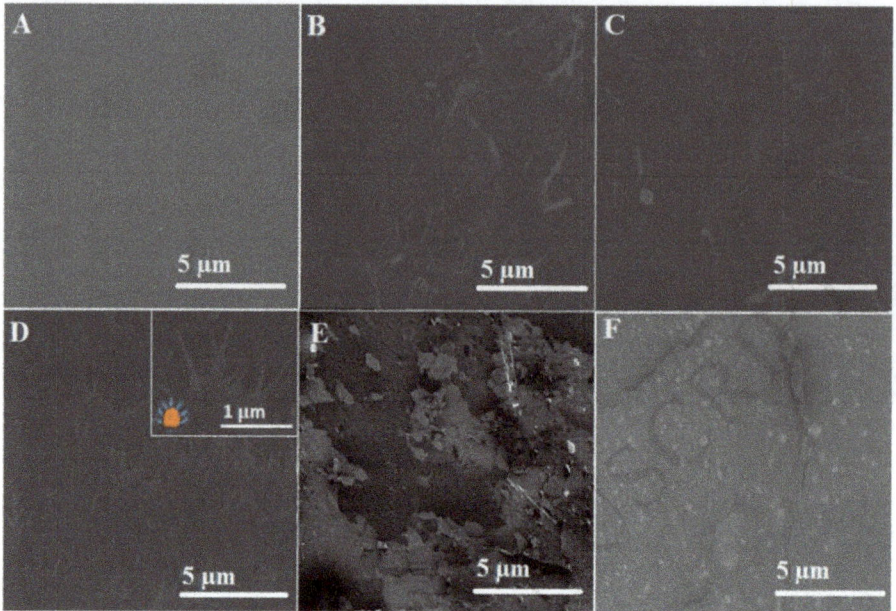

Figure 4.18: The FESEM images of different virgin-coated surfaces (a–f) (reused with permission from [41], © 2019 Elsevier Ltd.).

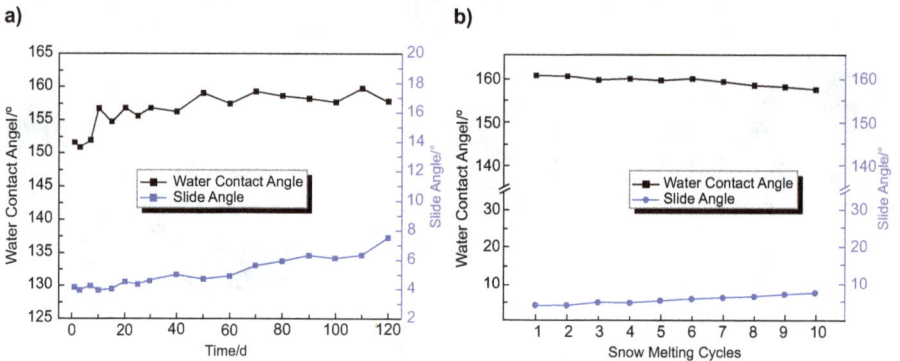

Figure 4.19: (a) The variation of WCA and SA with diverse weather conditions outdoors at −20 °C, (b) changes of SA and WCA for different snowfall times (reused with permission from [42], © 2019 Elsevier Ltd.).

values of WCA and SA, which also suggest that these coatings protect the material against oxidization at high temperatures.

Furthermore, from Figure 4.20, it can be seen that the heat transfer is reduced by the coating material. An interesting property of the material is that here the WCA and SA do not change up to the temperature of 375° C. But, after the values of these two parameters remain fixed even after increasing the temperature. Consequently, the thermal stability of the as-formulated coating material is demonstrated well.

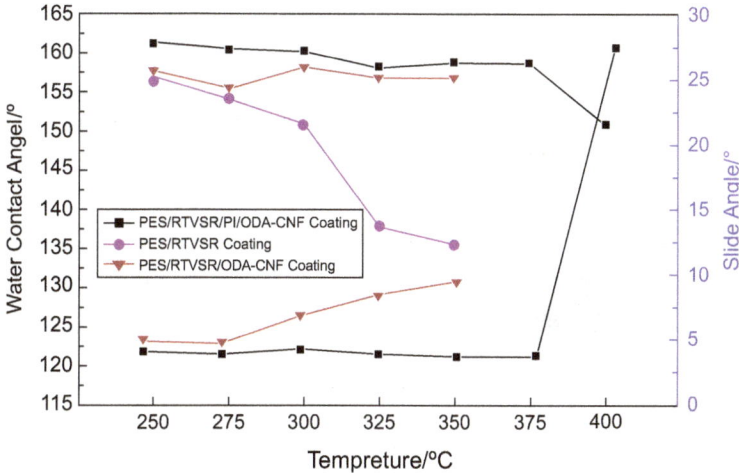

Figure 4.20: The wettability changes of PES/RTVSR coating, PES/RTVSR/ODA-CNF coating and PES/RTVSR/PI/ODA-CNF coating under different temperatures (reused with permission from [42], © 2019 Elsevier Ltd.).

The corrosion resistance performance was studied by potentiodynamic polarization method through which we can easily calculate the corrosion potential and corrosion current. Before applying the coating material, the corrosion potential and current density were −731 mV and 4.86×10^{-6} A/cm^2; after using the coating substrate the values change significantly as −542 mV and 7.93×10^{-10} A/cm^2 as evidenced from the Tafel's plot in Figure 4.21(a). The high anti-corrosion activity of PES/RTVSR/PI/ODA-CNF coating aluminum substrate is attributed to the presence of trapped air cushion in the composite and the crack-reducing nanofillers, which are able to form an air-shielding layer, inhibiting the diffusion of corrosive Cl$^-$ and O$_2$ penetrating along with water as depicted in Figure 4.21(b). Furthermore, the variation in the WCA and the SA, that is the wettability with varying pH and influence of different immersion times on the wettability with pH ranging from 1~13 depicted in Figure 4.21(c, d) reveals its high hydrophobicity. Thus, it proved to be efficient as corrosion inhibition coating.

Superhydrophobic Ni-SiO$_2$ coating was successfully fabricated by Li et al. and applied on aluminum surface [43]. The coating material has the superior properties of adhesion strength, corrosion resistance, flexibility, and hardness. In the Figure 4.22(a, b), it is seen that the coating materials were examined in different temperature cycles between −35 °C and 100 °C. The contact and sliding angles were measured in every cycle and the values are $163.5 \pm 3.5°$ and $5.5 \pm 1.5°$, which implies better performance of coating

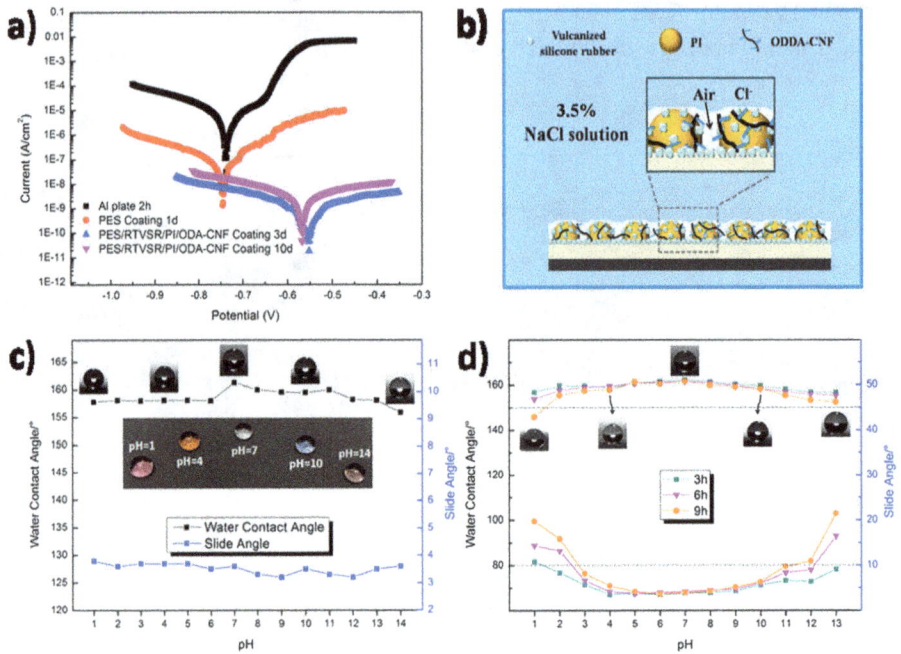

Figure 4.21: (a) Plots of linear polarization curves of the coatings and uncoated aluminum, (b) the schematic presentation of the anti-corrosion property of the as-prepared superhydrophobic coating surface, (c) influence of different pH values on the wettability of the prepared superhydrophobic coating, and (d) influence of different immersion times on the wettability under pH ranging from 1 to ~13 (reused with permission from [42], © 2019 Elsevier Ltd.).

material in all temperatures. It exhibited high corrosion inhibition both in diluted H_3PO_4 (i.e., acidic) as well as NaOH (i.e., alkaline) medium. The applied superhydrophobic $Ni-SiO_2$ coating was highly stable on the aluminum alloy as evidenced from the adhesion strength, wear resistance, robustness, and flexibility analysis. This coating also exhibited superhydrophobic coating as evidenced from Figure 4.22(a, b), which is expected to find great potential in the industries.

4.4.4 Water-repelling property

Krishnan et al. critically reviewed the various properties of superhydrophobic surface in their review article [44]. They highlighted the progress and recent applications of durable superhydrophobic surface in various segments. They show the water-repelling property of this surface by CA measurements. A diagrammatic representation of theories describing superhydrophobicity of surfaces has been shown in Figure 4.23, where, the WCA $\theta \cong 0\,^{\circ}C$ refers to the super-hydrophilic surface, $0\,^{\circ}C < \theta < 90\,^{\circ}C$ refers to hy-

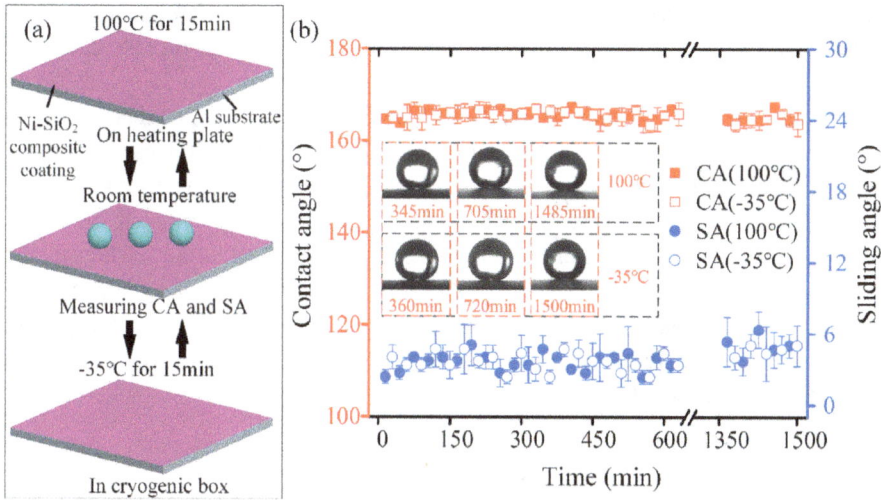

Figure 4.22: (a) Calculation of CA and SA of composite coating of Ni-SiO$_2$ at T = 100 °C and −35 °C and (b) curves of CA and SA of the Ni–SiO$_2$ composite coating at same temperature (reused with permission from [43], © 2020 Elsevier Ltd.).

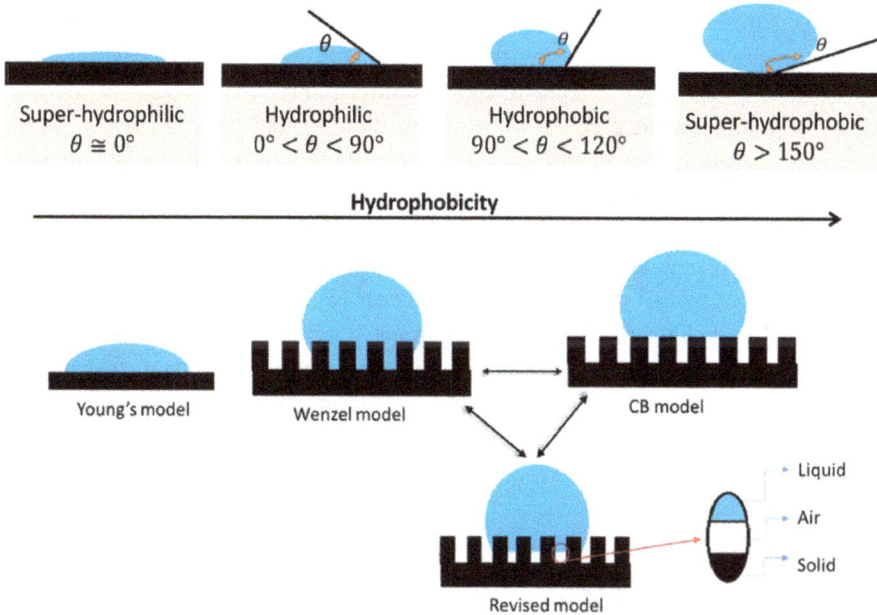

Figure 4.23: Diagrammatic representation of theories describing superhydrophobicity of surfaces (reused with permission from [44], © 2021 Elsevier Ltd.).

drophilic surface, refers to hydrophobic surface and $\theta > 150\ ^oC$ refers to superhydrophobic surface [44]. Figure 4.23 also describes the diagrammatic representation of Young's model, Wenzel model, CB model, and the modified or revised CB model. In the Young's model the droplet may be assumed to be dropped on the smooth surface getting diffused over it, whereas in the Wenzel model the droplet is supposed to be diffused through the surface roughness or inhomogeneity, displacing the air between the droplet and the surface in contact. In CB model the droplets are assumed to be placed on the top of the rough surface or the tiny peaks with trapped air pockets, whereas, the revised CB model suggests that the droplets dropped on the top of the rough surface may diffuse to the peaks to some extent and do not allow the trapped air to escape and remained between the locked liquid–solid interface. It is also noted that rolling angle of water droplet always tends to less than 10°.

4.5 Conclusion and future perspectives

In summary it is noted that superhydrophobic surface has excellent properties, and for this reason, it becomes a very interesting area for the current researchers. From the discussions in this chapter we get a clear idea about the various properties of this surface. One of the magical properties is that in this surface the water droplets make a contact angle>150° and sliding angle<10°, which show the better repelling nature of water droplets over the substrate surface. Another interesting property is that the superhydrophobic surface acts as a corrosion barrier and provides a corrosion-resistive layer over the metal surface. It demonstrates the anti-icing and self-cleaning properties as well. The graphene-based superhydrophobic surface has excellent antibacterial property; hence it is used as virus-protecting mask too. Here we discussed different techniques in constructing the superhydrophobic surface such as chemical vapor deposition, sol–gel method, electrodeposition, electrospinning, etc. The suitable coating method modifies the surface of the targeted substrate by using the superhydrophobic particles or powder. We believe that this chapter can give a clear insight for improvement and wider application of superhydrophobic coating on metal surfaces in the future.

References

[1]   Zhang D, Wang L, Qian H, Li X. Superhydrophobic surfaces for corrosion protection: a review of recent progresses and future directions. J Coat Technol Res [Internet] 2016;13(1):11–29, Available from. http://link.springer.com/10.1007/s11998-015-9744-6.

[2]   Sharma V, Sharma V, Goyat MS, Hooda A, Pandey JK, Kumar A, et al. Recent progress in nano-oxides and CNTs based corrosion resistant superhydrophobic coatings: A critical review. Prog Org Coatings [Internet] 2020;140:105512. Available from. https://doi.org/10.1016/j.porgcoat.2019.105512.

[3]   Barthlott W, Neinhuis C. Purity of the sacred lotus, or escape from contamination in biological surfaces. Planta [Internet] 1997;202(1):1–8, Available from. http://link.springer.com/10.1007/s004250050096.

[4]   Feng L, Zhang Y, Xi J, Zhu Y, Wang N, Xia F, et al. Petal effect: A superhydrophobic state with high adhesive force. Langmuir [Internet] 2008;24(8):4114–9. Available from, https://pubs.acs.org/doi/10.1021/la703821h.

[5]   Zarras P, Stenger-Smith JD. Smart inorganic and organic pretreatment coatings for the inhibition of corrosion on metals/alloys. Intell Coat Corros Control [Internet]. Elsevier2015;59–91. Available from. https://linkinghub.elsevier.com/retrieve/pii/B9780124114678000039.

[6]   Tripathy DB, Murmu M, Banerjee P, Quraishi MA. Palmitic acid based environmentally benign corrosion inhibiting formulation useful during acid cleansing process in MSF desalination plants. Desalination [Internet] 2019;472:114128. Available from. https://linkinghub.elsevier.com/retrieve/pii/S0011916419311543.

[7]   Mahato P, Mishra SK, Murmu M, Murmu NC, Hirani H, Banerjee P. A prolonged exposure of Ti-Si-B-C nanocomposite coating in 3.5 wt% NaCl solution: Electrochemical and morphological analysis. Surf Coat Technol [Internet] 2019;375:477–88. Available from. https://linkinghub.elsevier.com/retrieve/pii/S0257897219307728.

[8]   Cao M, Liu L, Fan L, Yu Z, Li Y, Oguzie E, et al. Influence of Temperature on Corrosion Behavior of 2A02 Al Alloy in Marine Atmospheric Environments. Materials (Basel) [Internet] 2018;11(2):235. Available from, http://www.mdpi.com/1996-1944/11/2/235.

[9]   Murmu M, Saha SK, Bhaumick P, Murmu NC, Hirani H, Banerjee P. Corrosion inhibition property of azomethine functionalized triazole derivatives in 1 mol L−1 HCl medium for mild steel: Experimental and theoretical exploration. J Mol Liq [Internet] 2020;313:113508. Available from. https://linkinghub.elsevier.com/retrieve/pii/S0167732219366516.

[10]   Mandal S, Zamindar S, Sarkar S, Murmu M, Guo L, Kaya S, et al. Quantum chemical and molecular dynamics simulation approach to investigate adsorption behaviour of organic azo dyes on TiO 2 and ZnO surfaces. J Adhes Sci Technol [Internet] 2022;1–17. Available from. https://www.tandfonline.com/doi/full/10.1080/01694243.2022.2086199.

[11]   Nine MJ, Cole MA, Tran DNH, Losic D. Graphene: a multipurpose material for protective coatings. J Mater Chem A [Internet] 2015;3(24):12580–602. Available from. http://xlink.rsc.org/?DOI=C5TA01010A.

[12]   Fihri A, Bovero E, Al-Shahrani A, Al-Ghamdi A, Alabedi G. Recent progress in superhydrophobic coatings used for steel protection: A review. Colloids Surf A Physicochem Eng Asp [Internet] 2017;520:378–90. Available fromhttps://linkinghub.elsevier.com/retrieve/pii/S0927775716311001.

[13]   Zhang P, Lv FY. A review of the recent advances in superhydrophobic surfaces and the emerging energy-related applications. Energy [Internet] 2015;82:1068–87. Available from. https://linkinghub.elsevier.com/retrieve/pii/S0360544215000857.

[14]   X-m L, Reinhoudt D, Crego-Calama M. What do we need for a superhydrophobic surface? A review on the recent progress in the preparation of superhydrophobic surfaces. Chem Soc Rev [Internet] 2007;36(8):1350. Available from. http://xlink.rsc.org/?DOI=b602486f.

[15]   Nosonovsky M, Bhushan B. Superhydrophobic surfaces and emerging applications: Non-adhesion, energy, green engineering. Curr Opin Colloid Interface Sci [Internet] 2009;14(4):270–80, Available from. https://linkinghub.elsevier.com/retrieve/pii/S1359029409000399.

[16]   Lu Z, Wang P, Zhang D. Super-hydrophobic film fabricated on aluminium surface as a barrier to atmospheric corrosion in a marine environment. Corros Sci [Internet] 2015;91:287–96. Available from. https://linkinghub.elsevier.com/retrieve/pii/S0010938X14005502.

[17]   Wenzel RN. Resistance of solid surfaces to wetting by water. Ind Eng Chem [Internet] 1936;28
        (8):988–94. Available from. https://pubs.acs.org/doi/abs/10.1021/ie50320a024.

[18]   Cassie ABD, Baxter S. Wettability of porous surfaces. Trans Faraday Soc [Internet] 1944;40:546.
        Available from. http://xlink.rsc.org/?DOI=tf9444000546.

[19]   Su B, Tian Y, Jiang L. Bioinspired Interfaces with Superwettability: From Materials to Chemistry. J Am
        Chem Soc [Internet] 2016;138(6):1727–48, Available from. https://pubs.acs.org/doi/10.1021/jacs.
        5b12728.

[20]   Nguyen-Tri P, Tran HN, Plamondon CO, Tuduri L, D-vn V, Nanda S, et al. Recent progress in the
        preparation, properties and applications of superhydrophobic nano-based coatings and surfaces: A
        review. Prog Org Coatings [Internet] 2019;132:235–56. Available from. https://linkinghub.elsevier.
        com/retrieve/pii/S030094401930092X.

[21]   Al-Daraghmeh MY, Hayajneh MT, Almomani MA. Corrosion Resistance of TiO2-ZrO2 Nanocomposite
        Thin Films Spin Coated on AISI 304 Stainless Steel in 3.5 wt. % NaCl Solution. Mater Res [Internet]
        2019;22(5):Available from. http://www.scielo.br/scielo.php?script=sci_arttext&pid=S1516-
        14392019000500231&tlng=en.

[22]   Casula MF, Corrias A, Falqui A, Serin V, Gatteschi D, Sangregorio C, et al. Characterization of FeCo
        −SiO 2 Nanocomposite Films Prepared by Sol−Gel Dip Coating. Chem Mater [Internet] 2003;15
        (11):2201–7. Available from, https://pubs.acs.org/doi/10.1021/cm0217755.

[23]   Nagarajan S, Mohana M, Sudhagar P, Raman V, Nishimura T, Kim S, et al. Nanocomposite Coatings
        on Biomedical Grade Stainless Steel for Improved Corrosion Resistance and Biocompatibility. ACS
        Appl Mater Interfaces [Internet] 2012;4(10):5134–41. Available from, https://pubs.acs.org/doi/
        10.1021/am301559r.

[24]   Sheen Y-C, Huang Y-C, Liao C-S, Chou H-Y, Chang F-C. New approach to fabricate an extremely
        super-amphiphobic surface based on fluorinated silica nanoparticles. J Polym Sci Part B: Polym Phys
        [Internet] 2008;46(18):1984–90, Available from. https://onlinelibrary.wiley.com/doi/10.1002/polb.
        21535.

[25]   Zhang Z, Wang H, Liang Y, Li X, Ren L, Cui Z, et al. One-step fabrication of robust superhydrophobic
        and superoleophilic surfaces with self-cleaning and oil/water separation function. Sci Rep [Internet]
        2018;8(1):3869. Available from. http://www.nature.com/articles/s41598-018-22241-9.

[26]   Nguyen-Tri P, Nguyen TA, Carriere P, Ngo Xuan C. Nanocomposite Coatings: Preparation,
        Characterization, Properties, and Applications. Int J Corros [Internet] 2018;2018:1–19. Available from.
        https://www.hindawi.com/journals/ijc/2018/4749501/.

[27]   Khodaei M, Shadmani S. Superhydrophobicity on aluminum through reactive-etching and TEOS/
        GPTMS/nano-Al2O3 silane-based nanocomposite coating. Surf Coat Technol [Internet]
        2019;374:1078–90. Available from. https://linkinghub.elsevier.com/retrieve/pii/S0257897219306978.

[28]   Kumar A, Gogoi B. Development of durable self-cleaning superhydrophobic coatings for aluminium
        surfaces via chemical etching method. Tribol Int [Internet] 2018;122:114–8. Available from.
        https://linkinghub.elsevier.com/retrieve/pii/S0301679X18301191.

[29]   Walsh FC, Wang S, Zhou N. The electrodeposition of composite coatings: Diversity, applications and
        challenges. Curr Opin Electrochem[Internet] 2020;20:8–19. Available from. https://linkinghub.elsev
        ier.com/retrieve/pii/S2451910320300181.

[30]   Emarati SM, Mozammel M. Efficient one-step fabrication of superhydrophobic nano-TiO2/TMPSi
        ceramic composite coating with enhanced corrosion resistance on 316L. Ceram Int [Internet]
        2020;46(2):1652–61, Available from. https://linkinghub.elsevier.com/retrieve/pii/
        S0272884219326689.

[31]   Celia E, Darmanin T, Taffin de Givenchy E, Amigoni S, Guittard F. Recent advances in designing
        superhydrophobic surfaces. J Colloid Interface Sci [Internet] 2013;402:1–18. Available from.
        https://linkinghub.elsevier.com/retrieve/pii/S0021979713002865.

[32] Ye Y, Zhao H, Wang C, Zhang D, Chen H, Liu W. Design of novel superhydrophobic aniline trimer modified siliceous material and its application for steel protection. Appl Surf Sci [Internet] 2018;457:752–63. Available from. https://linkinghub.elsevier.com/retrieve/pii/S0169433218317021.

[33] Ye Y, Liu Z, Liu W, Zhang D, Zhao H, Wang L, et al. Superhydrophobic oligoaniline-containing electroactive silica coating as pre-process coating for corrosion protection of carbon steel. Chem Eng J [Internet] 2018;348:940–51Available from. https://linkinghub.elsevier.com/retrieve/pii/S138589471830247X.

[34] Arukalam IO, Oguzie EE, Li Y. Nanostructured superhydrophobic polysiloxane coating for high barrier and anticorrosion applications in marine environment. J Colloid Interface Sci [Internet] 2018;512:674–85. Available from. https://linkinghub.elsevier.com/retrieve/pii/S0021979717312584.

[35] Chen H, Wang F, Fan H, Hong R, Li W. Construction of MOF-based superhydrophobic composite coating with excellent abrasion resistance and durability for self-cleaning, corrosion resistance, anti-icing, and loading-increasing research. Chem Eng J [Internet] 2021;408:127343. Available from. https://linkinghub.elsevier.com/retrieve/pii/S1385894720334677.

[36] Salehi M, Mozammel M, Emarati SM. Superhydrophobic and corrosion resistant properties of electrodeposited Ni-TiO2/TMPSi nanocomposite coating. Colloids Surf A Physicochem Eng Asp [Internet] 2019;573:196–204. Available from https://linkinghub.elsevier.com/retrieve/pii/S0927775719303279.

[37] Selim MS, El-Safty SA, Abbas MA, Shenashen MA. Facile design of graphene oxide-ZnO nanorod-based ternary nanocomposite as a superhydrophobic and corrosion-barrier coating. Colloids Surf A Physicochem Eng Asp [Internet] 2021;611:125793. Available from https://linkinghub.elsevier.com/retrieve/pii/S0927775720313868.

[38] Ye Y, Zhang D, Li J, Liu T, Pu J, Zhao H, et al. One-step synthesis of superhydrophobic polyhedral oligomeric silsesquioxane-graphene oxide and its application in anti-corrosion and anti-wear fields. Corros Sci [Internet] 2019;147:9–21. Available from. https://linkinghub.elsevier.com/retrieve/pii/S0010938X18312903.

[39] Yap SW, Johari N, Mazlan SA, Hassan NA. Mechanochemical durability and self-cleaning performance of zinc oxide-epoxy superhydrophobic coating prepared via a facile one-step approach. Ceram Int [Internet] 2021;47(11):15825–33, Available from. https://linkinghub.elsevier.com/retrieve/pii/S0272884221005319.

[40] Wang H, Di D, Zhao Y, Yuan R, Zhu Y. A multifunctional polymer composite coating assisted with pore-forming agent: Preparation, superhydrophobicity and corrosion resistance. Prog Org Coat [Internet] 2019;132:370–8. Available from. https://linkinghub.elsevier.com/retrieve/pii/S03009044018313316.

[41] Selim MS, Yang H, El-Safty SA, Fatthallah NA, Shenashen MA, Wang FQ, et al. Superhydrophobic coating of silicone/β–MnO2 nanorod composite for marine antifouling. Colloids Surf A Physicochem Eng Asp. [Internet] 2019;570:518–30. Available from. https://linkinghub.elsevier.com/retrieve/pii/S0927775719302201.

[42] Zhao Z, Wang H, Liu Z, Zhang X, Zhang W, Chen X, et al. Durable fluorine-free superhydrophobic polyethersulfone (PES) composite coating with uniquely weathering stability, anti-corrosion and wear-resistance. Prog Org Coat [Internet] 2019;127:16–26Available from. https://linkinghub.elsevier.com/retrieve/pii/S0300944018308105.

[43] Li X, Yin S, Luo H. Fabrication of robust superhydrophobic Ni–SiO2 composite coatings on aluminum alloy surfaces. Vacuum [Internet] 2020;181:109674. Available from. https://linkinghub.elsevier.com/retrieve/pii/S0042207X20305364.

[44] Krishnan A, Krishnan A, Ajith A, Shibli SMA. Influence of materials and fabrication strategies in tailoring the anticorrosive property of superhydrophobic coatings. Surf Interfaces [Internet] 2021;25:101238. Available from. https://linkinghub.elsevier.com/retrieve/pii/S2468023021003151.

Sukdeb Mandal, Manilal Murmu, Naresh Chandra Murmu,
Priyabrata Banerjee

5 Porous materials-based functionalized thin film coatings as corrosion inhibitors

Abstract: Porous materials have received significant interest in several fields of emerging research and development. Porous materials such as metal-organic frameworks (MOFs) are a relatively new and rapidly expanding class of porous materials. MOFs are widely recognized for having a remarkably large surface area and highly flexible capacity due to the variety of ligands and metal nodes that are accessible and that may be tuned and used alternately to confer new desired features. MOFs are explored and employed for various applications such as sensing, catalysis, gas storage, drug delivery, water treatment, corrosion resistance, and so on. Nevertheless, there is scant attention on their performances in anti-corrosive coating applications. Anti-corrosive coating is one of the facile strategies in combating the corrosive dissolution of metallic materials from adverse environments and assists in reducing the economic losses owing to the metallic corrosion and the maintenance of the materials. In this chapter, special attention has been given to viewing the exploitation of MOFs for the research and development of thin film anti-corrosive coating applications on metal substrates and attempt has been made to delve deeper into the mechanism of film formation, the exhibition of barrier properties, and corrosion inhibition activities.

Keywords: Porous Materials, Metal-Organic frameworks, Thin Film, Coatings, Corrosion Inhibitors

5.1 Introduction

Equipment made of metals and its alloys are used by people during their daily lives. But, most of these metallic materials inevitably corrode in their working environments. The result of corrosion is irreversible damage to the metallic materials or structures such as bridges, pipelines, ships, buildings, etc. These irreversible damages owing to corrosion not only threaten human safety but also harm the environment,

Acknowledgments: PB is very thankful to Department of Higher Education, Science & Technology and Biotechnology, Govt. of West Bengal, India for providing financial assistance to carry out this research work [vide sanction order no. 78(Sanc.)/ST/P/S&T/6G-1/2018 dated 31.01.2015 and project no. GAP-225612]. SM acknowledges the University Grants Commission, Government of India, New Delhi, India for his fellowship [212/CSIR-UGC NET DEC.2017]. MM would like to acknowledge Ministry of Tribal Affairs, New Delhi, India for his National Fellowship for Higher Education of Scheduled Tribes candidates, NFST, (vide award letter no. F1-17.1/2014-15/RGNF-2014-15-ST-JHA-71,555).

https://doi.org/10.1515/9783111016160-005

causing pollution or contamination and catastrophic accidents leading to huge economic losses across the globe [1–6]. Although corrosion cannot be completely avoided, it may be minimized by using corrosion protection techniques such as corrosion inhibitors, barrier coatings, anodic and cathodic protection methods, plating, anti-corrosion coatings, etc. [7–13]. One of the most popular ways to improve the corrosion resistance properties of metallic materials is to apply a protective anti-corrosion coating that efficiently isolates the materials from destructive substances or environments. Over past centuries, protective anti-corrosion coatings comprised of chromate and phosphate were extensively used as anti-corrosion agents. However, the release of toxic substances during the prolonged service time has caused serious environmental threats. As a result, it has become necessary to develop environment-friendly coating substances to prevent the release of toxic substances during prolonged service. Metals and alloys can also be effectively protected from corrosion in aggressive environments by using high-efficiency corrosion inhibitors. The use of long-term corrosion protection will prevent 20%–30% of corrosion losses. The hazardous nature of typical corrosion protection compounds makes it difficult to produce an environmentally acceptable property with long-term protective capabilities. For enhancing the anti-corrosion performance of metallic materials, there is growing interest in using nanomaterials with corrosion-inhibitory and/or barrier properties [13].

Metal-organic frameworks (MOFs; the term first coined by Prof. Omar M. Yaghi in 1555) are hybrid materials having well-organized pores formed by metal ions or clusters (as nodes or centers) and electron-donating organic ligands (as linkers) [14, 15]. These MOFs have sparked significant attention due to their unique features, such as flexibility, softness, and spatial and electrical structural variety [15, 16]. A variety of applications for MOFs has been investigated, including gas storage, sensors, drug delivery, catalysis, and other areas due to their huge specific surface area, abundance of active sites, and adaptable structures [17–22]. As a result of this adaptability, there are many opportunities in diverse fields. Additionally, several MOFs exhibit hydrophobic characteristics with strong water stability as well as corrosion inhibition characteristics. MOFs have received comparatively less attention for their capacity to prevent corrosion compared to their other uses [23]. MOFs provide corrosion protection through four main strategies. These are: (i) MOFs as corrosion inhibitors: the corrosion prevention feature of various MOF species is widely thought to be due to the presence of numerous heteroatoms and π-electron moieties in MOF structures; (ii) MOFs-based protective coatings: in this class, the MOFs serve as nanofillers that are dispersed in organic or inorganic protective coating materials; (iii) MOFs membrane: in situ nucleation and/or self-adsorption can be used to produce continuous and intergrown MOF membranes for use on metal surfaces; and (iv) Composite coating: using MOFs as nanocontainers for corrosion inhibitors, robust materials can be incorporated into this process. Many different methods, including the hydrothermal/solvothermal approaches, have been used to synthesize MOF-based anti-corrosion materials, each of which has advantages and limitations. In situ growth and seeded synthesis are two important techniques for generating MOF membranes as anti-corrosive coatings. The substrates are submerged in the precursor solution in the in situ approach, allowing nucleation sites to develop on the substrate surface and produce MOF crystals. However, there are challenges in producing continuous MOF films because of inadequate heterogeneous nucleation, which also affects the coverage of MOF coatings. By strategically developing MOF-based coatings on the substrate surface, the detrimental effects of substrate surface chemistry may be effectively reduced. For the development of MOFs via seeded synthesis, seed crystals must be adhered to the surface of the substrate and then submerged in the precursor solution. The various manufacturing processes have a substantial influence on the characteristics of the synthesized MOFs in that situation.

In this chapter, the recent advancements in corrosion protection by the application of MOFs-based thin film coating have been overviewed and an attempt has been made to delve deeper into the mechanism of film formation, exhibition of barrier properties, and corrosion inhibition activities for different substrates in different corrosive and adverse media. Additionally, based on the literature overviewed, the future perspectives and the scope of innovation of MOFs-based corrosion-mitigating coating in the realm of corrosion inhibition have been outlined.

5.2 Fabrication methods of MOF-based thin film

MOF-based thin films (MOF-TFs) manufacturing, characterization, and application receive a lot of attention in the scientific community [16, 24, 25]. MOF-TFs placed on diverse substrates enable a variety of applications. There are three main different approaches of MOF-based thin film fabrication: (i) direct growing or depositing inside the solution, (ii) successive layer fabrication on substrates, and (iii) integration of these performed carefully on chosen nanocrystals of optimum size and shape. The phases of the precursors in the synthesis reaction are used to categorize the manufacturing techniques, which include solid, liquid, gel, and vapor. The common steps of fabrication for MOF-based thin film synthesis procedures include liquid–liquid, liquid–solid, or additional synthetic procedures.

5.2.1 Liquid–liquid approach

Mostly, the MOF-based thin film syntheses are performed employing liquid-phase reactions, wherein, the reactions are initiated by dissolving both the organic and metal ingredients within the media. During direct solvothermal synthesis such as layer-by-layer deposition, the dissolved precursors are either sequentially applied in contact with the substrates or thoroughly combined to form a mother solution before adding the substrates. The large concentration of ionized reactants (deprotonated organic linkers and metal cations) within solution phases facilitates homogenous reactions. At the substrate surface, however, heterogeneous responses also appear, and these two-phase processes compete with one another. The heterogeneous reaction leads to form MOF-TFs on the surface of the substrates, whereas, the homogeneous reaction yields MOF crystals in the solvent [26]. These two processes are in direct competition with one another; therefore, increasing reaction through heterogeneous pathways while inhibiting homogeneous ones becomes the viable strategy for attaining the production of high-quality MOF-TFs in an inexpensive as well as ecologically responsible manner.

5.2.1.1 Direct synthesis

One of the most used techniques for synthesizing MOFs on large scales is the one-pot hydro/solvothermal synthesis process. Both hydrothermal and solvothermal synthesis procedures are performed in a pressure vessel and a sealed reactor like an autoclave. A conventional hydro/solvothermal method involves placing the substrate in a combination of MOF in the precursor solutions and subjecting it to a reaction at a high temperature, and after a certain time the crystals of MOFs are generated. A schematic illustration of MOF synthesis via hydrothermal or solvothermal synthesis route has been depicted in Figure 5.1. It depicts that initially, the organic linkers and metal salts are taken in desired solvents and mixed properly; then this reaction mixture is transferred to a Teflon liner in which the PTFE inner chamber protects it from corroding, and it is placed inside the autoclave reactor. The reaction conditions are maintained at 220 °C with a maximum pressure of 3 MPa following the heating and the cooling rate of 5 °C/min. Upon completion of the reaction, the products are washed and dried to collect MOFs crystals (vide Figure 5.1).

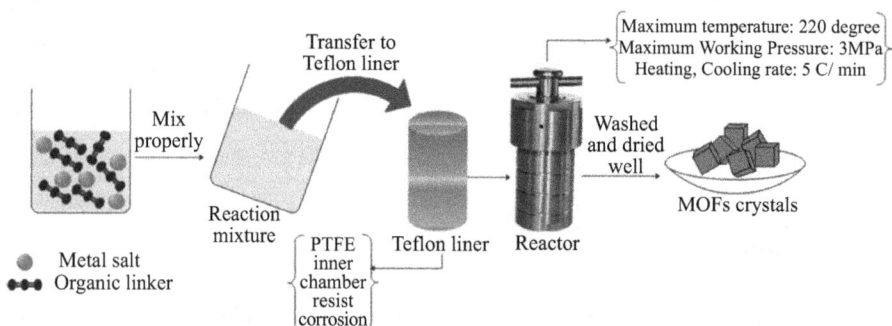

Figure 5.1: A schematic representation of MOF synthesis via hydrothermal or solvothermal synthesis route (reprinted with permission from ref. [27]).

5.2.1.2 Secondary approach

It is difficult to construct incessant MOF-TFs on the pristine substrate, despite the effectiveness of traditional hydro/solvothermal synthetic procedures for synthesizing MOF-TFs. This is because heterogeneous nucleation of MOF structures makes this mechanism typically ineffective at initiating MOF structures [28]. Furthermore, solution-grown MOF structures find it difficult to attach to target surfaces because they do not contain sites for adherence. The most effective way for promoting heterogeneous nucleation of MOF-TFs is to modify substrate surfaces [29]. Such a secondary growth approach for synthesizing MOF-TFs frequently involves functionalizing the substrate surface with the addition of a functional layer to the surface prior to synthesizing thin films [30]. According to this method, functional group-ended surfaces were thoroughly examined to facilitate heterogeneous nucleation of MOF-TFs, more effectively bonding the metal nodes and organic linkers to the substrate surface. The surface of a substrate may be changed using a variety of functional layers, including inorganic functional layers like MOFs and metal oxide NPs as well as organic functional layers like polymers and self-assembled monolayers (SAMs) [31]. The deposition of the microporous amino-functionalized [Al_4 $(OH)_2(OCH_3)_4(H_2N\text{-}bdc)_3$].$xH_2O$ (as CAU-1, CAU: Christian-Albrechts-University) on SAM-functionalized gold substrates was explored by Hinterholzinger et al. [32].

5.2.1.3 Layer-by-layer approach

The conventional hydrothermal or solvothermal synthesis of MOF-TFs has been well-established for many years, but there are still many challenges to be overcome before it can be widely applied. There are several challenges associated with controlling the fabrication method, such as the complexity of controlling the film thickness and discontinuity of film formation [33]. Additionally, the large number of reactants used results in a high cost. There are several synthetic approaches already in existence to deal with these issues, and the layer-by-layer deposition process (vide Figure 5.2) has great control over layer thickness and surface roughness [34, 35]. During the manufacturing process, the substrate is immersed in each solution containing the precursor sequentially, resulting in the formation of thin films through the deposition of alternating layers of species with opposing charges, such as metal cations and deprotonated organic ligands. Layer-by-layer deposition method is advantageous in synthesizing orientated and distinct MOF-TFs, particularly for surface-mounted MOFs (abbreviated as SURMOFs) [36]. In order to modify the surfaces of silicon substrates with a thin layer of amine groups, three MOF-building components, i.e., **P1** = [5,15-di(4-pyridylacetyl)-10,20-diphenyl], **L1** = [1,2,4,5-tetrakis(4-carboxyphenyl)benzene], zinc(II) acetate, and 3-aminopropyl-trimethoxysilane(3-APTMS) were used. The hydroxyl-terminated surfaces of the silicon samples were initially derivatized by refluxing in a 1:100 (v:v) solution of 3-APTMS in octanol for 20 min. It was then dried in an oven at 70 °C for 15 min after being rinsed with hexanes and water. It was observed that the introduction of the amine layer was a vital step, without which the development of Diphenylporphinato-Metal Organic Framework (DA-MOF) was ineffective. By capturing Zn^{2+}, the 3-APTMS promotes the formation of succeeding MOF structures (step-I in Figure 5.2). Following the capture of Zn^{2+}, 3-APTMS-modified silicon substrate is sequentially soaked in solutions containing **L1** and, lastly, **P1** (step-II and step-III in Figure 5.2). The substrate was cleaned in ethanol before each soaking to remove any precursor ions or molecules that had not yet started to react, ensuring uniform growth of the film [37].

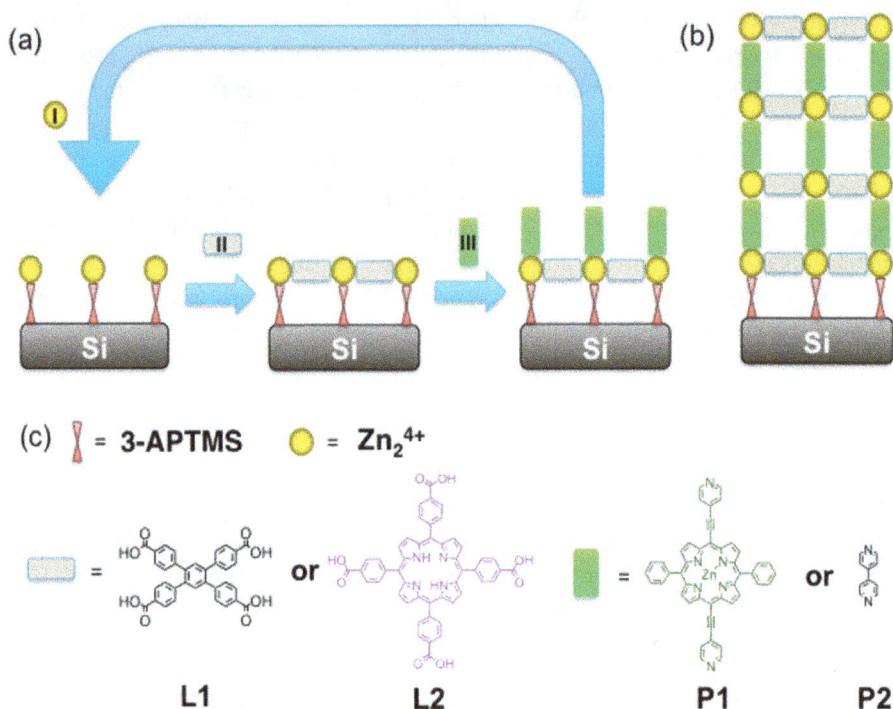

Figure 5.2: The schematic representation of developing MOFs thin film on a substrate via layer-by-layer approach (reprinted with permission from ref. [37]).

5.2.1.4 Dip-coating approach

Dip coating is a straightforward, affordable, as well as repeatable technique for creating TFs and is widely applied in various industries [38]. The organic precursor and dissolved metal are properly put together to get a colloidal suspension, which is then combined with a substrate to form a thin film. It is necessary to refill the precursor solution blend after a set amount of time when MOF crystals are grown continuously in suspension to maintain sufficient reactant concentration for film development [39]. Depending on reaction kinetics it takes variable reaction times for different MOF substances. The thickness of the thin film may be modified for layer-by-layer deposition by adjusting the immersion duration and cycles along with withdrawal speed [40].

5.2.1.5 Spin coating approach

Spin coating is a typical process for developing a homogeneous thin film onto the planar solid surface that may also be used to fabricate MOF-based thin film coating. In this technique of manufacturing, various MOF precursor solutions are sprayed onto the center of a spinning horizontal plane operated by a spinning device at a predetermined spin rate in addition to spin duration in order to produce MOF-based thin films. Evenly distributed precursor on the desired surface is achieved by carefully controlling the amount of solutions. This fabrication approach may produce a dense and homogeneous MOFs thin film (thickness from micron to nanoscale) in a short time. However, this technique may lead to structural defects in the thin film [40, 41]. Figure 5.3 illustrates a schematic manufacturing procedure for MOF thin films using the spin-coating method and the liquid-phase epitaxy methodology. In this procedure, four small syringes on a fully automated spin coating apparatus are used to apply microdroplets of a metal cation precursor solution to the substrate over a predefined

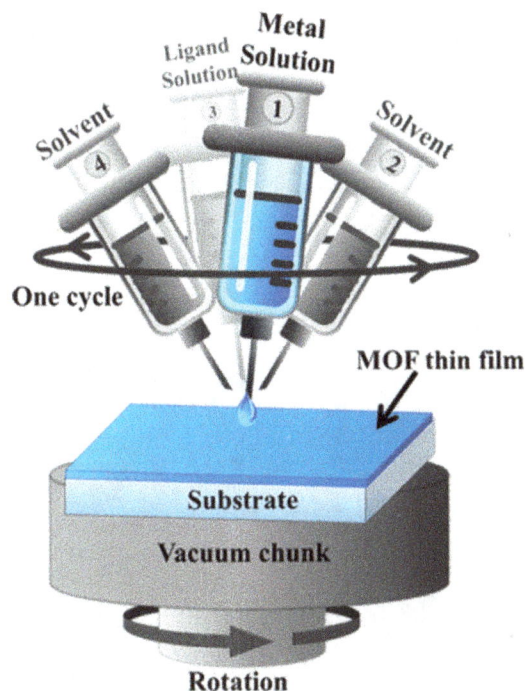

Figure 5.3: Fabrication of MOF thin films using the spin-coating technique and a liquid-phase epitaxy methodology (reprinted with permission from ref. [41]).

period of time while the apparatus is spinning constantly at a speed of 500 rpm. Owing to the centrifugal force the entire fluid evenly spreads across the substrate. The substrate is then cleaned by injecting tiny droplets of solvent into it. The four-step process (steps **1–4 in** Figure 5.3) that is described is thought of as one cycle and is repeated to produce thicker thin films.

5.2.1.6 Interfacial synthesis

By using these techniques, MOF thin films can be formed at the interfaces of immiscible phases, such as water and air, etc. (vide Figure 5.4). Interfacial synthesis of MOF thin film might be achieved at liquid–liquid interface [26]. In order to prevent water-insoluble monomers or even nanoparticles from freely rotating and moving in lateral directions, the air–water interface (Figure 5.4a) is used. If the particles are not packed firmly enough to form a monolayer, they are free to move in all directions. In most cases, a measured quantity of monomer "A" or ligand (represented by crosses in Figure 5.4a) is added to aqueous solution of monomer B or salts (shown by dots in Figure 5.4a) in an organic solvent. At the air–water interface, polymerization can occur as organic solvent evaporates, leaving monomer "A" on aqueous layer. Special techniques are used in this method: (i) functionalization with hydrophilic moieties guarantees their adequate dispersion on the surface of the water, where the hydro-

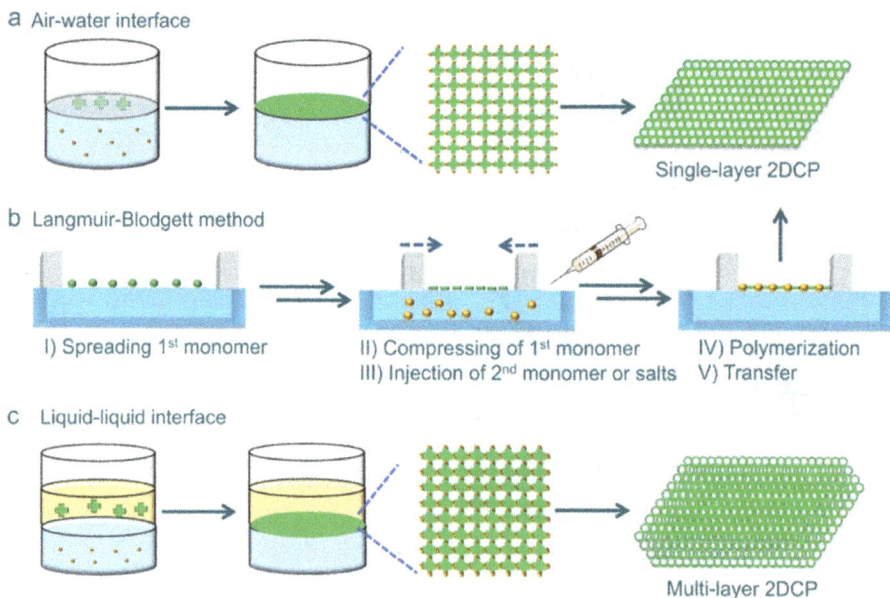

a Air-water interface

Single-layer 2DCP

b Langmuir-Blodgett method

I) Spreading 1st monomer   II) Compressing of 1st monomer   IV) Polymerization
                            III) Injection of 2nd monomer or salts   V) Transfer

c Liquid-liquid interface

Multi-layer 2DCP

Figure 5.4: A pictorial illustration of interfacial synthesis approach of MOF thin film (a) at air–water interface; (b) the LB technique; and (c) at liquid–liquid interface (reprinted with permission from ref. [42]).

philic parts are bonded to the H_2O molecules and the comparatively hydrophobic components face the air; (ii) to enable additional 2D polymerization, the reactive parts of the monomers/precursors must layer tightly along the water surface.

With control over the thickness of the monomer stacking, tightly packed monolayers may be more readily synthesized using the Langmuir–Blodgett (LB) technique, which also establishes the air–water interface. Consequently, changing the composition and structural characteristics of 2D materials by making use of covalent and non-covalent interactions at the molecular level is conceivable. In the first phase of the LB technique, the water-insoluble monomer spreads across the water surface (vide Figure 5.4b). When the monomers were compacted into a thick film with a highly ordered internal structure, a solution of the second monomer or metal salts was added into the aqueous phase in the subsequent stage. The movement of a second water-soluble monomer from the bulk of the solution to the interface causes a significant area of targeted single layers to form during the final step of 2D polymerization. Layer-stacked 2D materials might be synthesized through repeating transfer procedure. Additionally, the MOF-based thin film preparation has also been investigated in the liquid–liquid interface system. It is possible to establish multilayer structures at the liquid–liquid interface by confining monomers and their polymerization in a two-dimensional space. The choice of solvent depends upon the monomer's ability to be properly dissolved in either the upper or lower phase.

5.2.1.7 Gel-layer approach

Without using hydro/solvothermal precursor solutions, gel-layer synthesis preserves a high proportion of reactants during the formation of heterogeneous films. The thicknesses of the MOF-TFs are modulated by gel molecular weight along with concentrations of metal cations inside it. This technique should be generally applicable with the correct gel matrices; however, based on MOFs and gel matrices used, it may take a long time [43]

5.2.1.8 Contra-diffusion approach

There were numerous techniques developed to accelerate the heterogeneous nucleation of MOFs on surfaces, but these techniques frequently enhance the difficulty of the process [44, 45]. Then, the contra-diffusion synthesis method was discovered, which is a viable process to fabricate MOF-TFs. A contra-diffusion synthesis is carried out in the same manner as interfacial synthesis of MOF-TFs. In contrast to interfacial synthesis, a contra-diffusion synthesis must have porous substrates that differentiate these two precursor solutions [46–50]. When precursors are encountered on the substrate, they diffuse in opposite directions, and thereby, the MOFs are formed. Contrary to the traditional diffusion approach, the contra-diffusion method may embed MOF-TFs onto porous substrates, creating a strong bond between the porous substrate and the MOF thin film [48, 51]. Herein, the pictorial representation showing the fabrication of a polycarbonate track-etched membrane (PCTM)-supported 1D hollow superstructure by polydopamine(PDA)-mediated contra-diffusion approach yielding the desired PCTM@PDA@ZIF-8 followed by its subsequent efficient separation membrane capable of separating uranyl ions from the solutions containing other ions in the solution is shown in Figure 5.5.

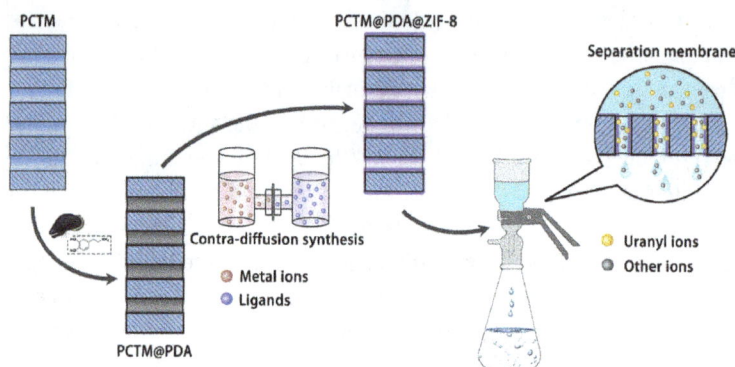

Figure 5.5: Preparation of PCTM-supported 1D hollow superstructures by PDA-mediated contra-diffusion approach (reprinted with permission from ref. [52]).

5.2.1.9 Evaporation approach

Through an evaporation-induced crystallization process, the evaporation approach works well for the controlled synthesis of MOF-TFs [53, 54]. In this process the solvent is slowly evaporated away, and crystallization occurs, which causes MOF crystals to develop locally on the substrate with high accuracy [55, 56].

5.2.2 Liquid–solid approach

Despite the fact that liquid–liquid synthesis techniques find wide applicability for fabricating MOF-TFs, the challenges in scaling production hinder large-scale manufacturing. It may greatly enhance heterogeneous MOF-TFs nucleation and growth on surfaces when used in conjunction with the liquid–solid method, one of several synthesis techniques designed to reduce costs and increase environmental sustainability.

5.2.2.1 Electrochemical deposition method

One new synthesis approach used in the fabrication of MOF-based thin film is the electrochemical deposition technique. An electrochemical cell (vide Figure 5.6) is used to deposition MOF-TF in a traditional process of this deposition approach, which consists of metals, organic links, and electrolytes [57, 58]. Deprotonated organic ligands are coordinated with metal ions at the electrode surface via anodic dissolution, cathodic reduction, or charge driving, resulting in the formation of an MOF-TF [59–63]. Using this technique, it is possible to create MOF-TFs with adjustable thicknesses by continuously measuring the amount of charge passed. Additionally, the electrochemical characterization of this method allows for the in-place correction of flaws like fractures and pinholes. However, this technique can only be used to create nonconductive MOF-TFs on conductive surfaces, which limits the range of applications. Additionally, organic linkers may be oxidized and inert metal ions on the cathode may separate.

Figure 5.6 illustrates the deposition of both anodic and cathodic films of UiO-66 and Zr(IV)-based MOFs under anodic and cathodic conditions. A synthesis modulation technique using acetic acid is used to regulate the development of zirconium oxide films on the anode, the speed of cathodic deposition, and the shape of the UiO-66 film. This set of deposition conditions enables the deposition of a wide range of Zr-based MOFs on a variety of conductive substrates. Before applying a voltage to the electrodes, the synthesis solution was heated to a temperature of 383 K, which is the temperature at which the film deposition was carried out. In order to ensure a constant current density, a potential difference ranging from 6 to 5 V was applied.

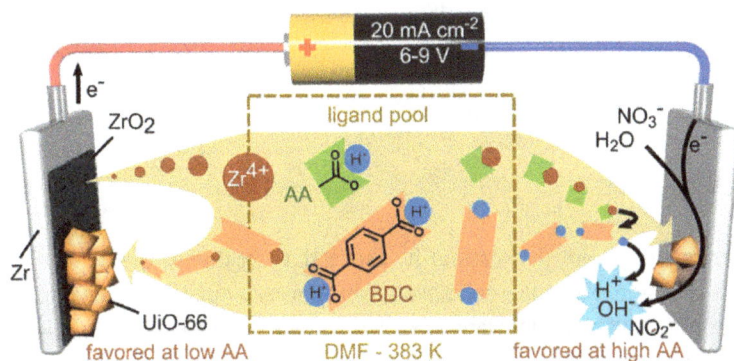

Figure 5.6: Electrochemical deposition technique of MOF thin film (reprinted with permission from ref. [64] © 2016 ACS).

5.2.2.2 Self-sacrificing template synthesis

The MOF thin film may be produced using this approach, which uses metal, metal-oxide, or metal-hydroxides. In self-sacrificing template synthesis, the reaction solution solely provides organic linkers. The template used, includes the essential MOF's metal species, functions as the substrate and delivers metal cations creating MOF-TFs. In such a synthesis method, MOF-TFs may be quickly built on a template under suitable reaction conditions, preventing the growth of free MOF crystals in liquid solutions. In addition, this reaction is restricted by metal ion dissolution and can self-terminate when thick MOF-TFs are produced between the metal template and organic precursor solution [65, 66].

5.2.3 Other synthetic approaches

Recently, several synthesis approaches for different MOFs have been developed. Among these approaches, the solid–solid approach, vapor–solid approach, gel–vapor approach, and post-assembly methods have been outlined briefly.

5.2.3.1 Solid–solid approach

Fabrication of MOF-TFs without using solvents has emerged as the newly developing strategy that can achieve green synthesis and avoid the limitations of solution-processing procedures. By substituting metal oxides or hydroxides for the commonly used metal salts reacting with organic linkers, it is possible to fully avoid the production of acids, because this synthesis approach only necessitates straightforward acid-base neutralization with water as exclusive side product [67].

5.2.3.2 Vapor–solid approach

To apply MOF-TFs to equipment that cannot be processed wet due to the danger of corrosion and contamination, a vapor–solid synthesis approach appears promising. A variety of techniques, including atomic layer deposition (ALD), chemical vapor deposition (CVD), and physical vapor deposition (PVD), can be used to make MOF-TFs by employing vapor–solid synthesis approach.

5.2.3.3 Gel–vapor approach

Using the benefits of both techniques, this approach combines vapor deposition and sol–gel coating to produce MOF-TFs with a customizable thickness without the need for solvents or additional processing time [68].

5.2.3.4 Post-assembly methods

A phase where the framework is constructed and another phase where chemical reactions are occurring are often involved in the majority of MOF-TF synthesis procedures, which make use of diverse metal and organic linker precursors. An alternative approach that requires the framework assembly phase to produce MOF-TFs uses pre-made MOF particles on substrates [69]. A post-assembly procedure may be used to create MOF-TFs using a number of techniques. For the creation of thick MOF-TFs, the layer-by-layer deposition approach is typically used to construct prefabricated MOF nanocrystals. The LB deposition process is a well-established method for generating ordered monolayers on liquid–substrate interfaces. On the basis of prefabricated MOF nanocrystals, Langmuir–Schäfer (LS) deposition process is equally used for producing thin MOF films. In order to assemble MOF crystals on substrates, drop-casting and spin-coating procedures have also been employed; however, resulted film morphologies are inferior to that achieved by LS or LB approach [70–74].

5.3 Application of MOF thin film in corrosion inhibition

The use of protective anti-corrosive coatings, which may effectively prevent direct interaction with corrosive environment, is one of the most crucial methods to increase metal and alloy corrosion resistance. To create a thin film anti-corrosion coating based on MOF by adding a Cu-based MOF, CuBTTri, to the polyamide layer, Wen et al. were able to fabricate an anti-biofouling thin film nanocomposites membrane. CuBTTri was created by reacting 1,3,5-tris(1 H-1,2,3-triazol-5-yl) benzene(H_3BTTri) with $CuCl_2.H_2O$. The MOF has exceptional water stability. Their research demonstrated that MOFs greatly boosted membrane water permeability without reducing selectivity when included in the TFN active layer. The MOF-based thin film also exhibited antibacterial characteristics attributed to the addition of CuBTTri [75].

Dehghani et al. have developed metal organic complex film for the protection of mild steel surface based on samarium nitrate and [bis(phosphomethyl)amino] methyl phosphonic acid (ATMP). They investigated how well corrosion was inhibited in simulated seawater and found that after 120 h of exposure to a combination of 200:600

ppm ATMP:Sm, 58% of the corrosion was inhibited. The corresponding EIS and Tafel analysis are shown in Figures 5.7 and 5.8, respectively [76]. From the Nyquist plot it has been observed that for pure 800 ppm ATMP the R_{ct} value did not follow a straightforward pattern with increasing metal exposure period, although the diameter of the semicircle did not significantly increase under the best inhibition condition. This demonstrates the inadequate inhibitory performance of the ATMP in aggressive saline medium. A small decrease in the R_s values of 800 ppm ATMP and/or Sm in comparison to the blank sample clearly demonstrated the poor electrical conductivity of the solution. This decrease is most likely due to ATMP and/or Sm molecule desorption in the NaCl medium. Additionally, the potent inhibitory action of ATMP:Sm in saline medium can be explained by a notable rise in R_{ct} values (particularly for 200:600 ppm ATMP:Sm). The result that the R_f values for all ratios increased with immersion time provided additional support for the strong adsorption and exceptionally dense barrier that formed on the steel surface. As can be observed from the ATMP:Sm(200:600 ppm)sample, the investigational findings were fitted using two different electrical equivalent circuit models. In order to fit the data, two-time constants were employed up to 48 h after the start of the MS exposure; however, for 120 h, two-time constants were used in a series, which shows that a porous barrier coating forms on the metal substrate after a 120h exposure.

According to the Bode plots, the protective film is thought to have a relaxation time that originated in high frequencies, and a second time constant that originated in low frequencies (for 200:600 sample ratios) indicating the existence of a double layer at the metal–solution interface. It is crystal clear that the excellent hindrance behavior of the ATMP:Sm protective layer in NaCl medium accounts for increased separation between the impedance results between ATMP:Sm (200:600 ppm ratio) and other samples, as the metal immersion duration increases (particularly after 120 h exposure). and the significant chelation between the ATMP molecules and Sm ions are primarily responsible for the high phase angle values, which validate these findings. Again, all Bode diagrams with a single time constant demonstrated that corrosion was dependent on the charge transfer method, with the exception of the ATMP:Sm (200:600 ppm ratio)

A MOF (ZIF-8)-based super hydrophobic coating with outstanding erosion resistance and durability has been developed by Chen et al. for studies on self-cleaning, resistance to corrosion, anti-icing, and loading increase (vide Figure 5.9) [77]. In order to develop a water-repellent coating, the synthesized ZIF-8 particles were treated with 1 H,1 H,2 H,2 H-perfluorooctyltriethoxysilane (POTS). This procedure is depicted in Figure 5.9. Resulted ZIF-8/POTS was spray-coated on the substrate giving a super hydrophobic coating attributed to uneven textured ZIF-8 nanomaterials as well as POTS with minimal surface energy. Furthermore, the subsequent addition of an epoxy resin (EP) layer enabled the ZIF-8/POTScoating to create a composite coating that is super hydrophobic by overcoming feeble abrasion, bonding, and robustness.

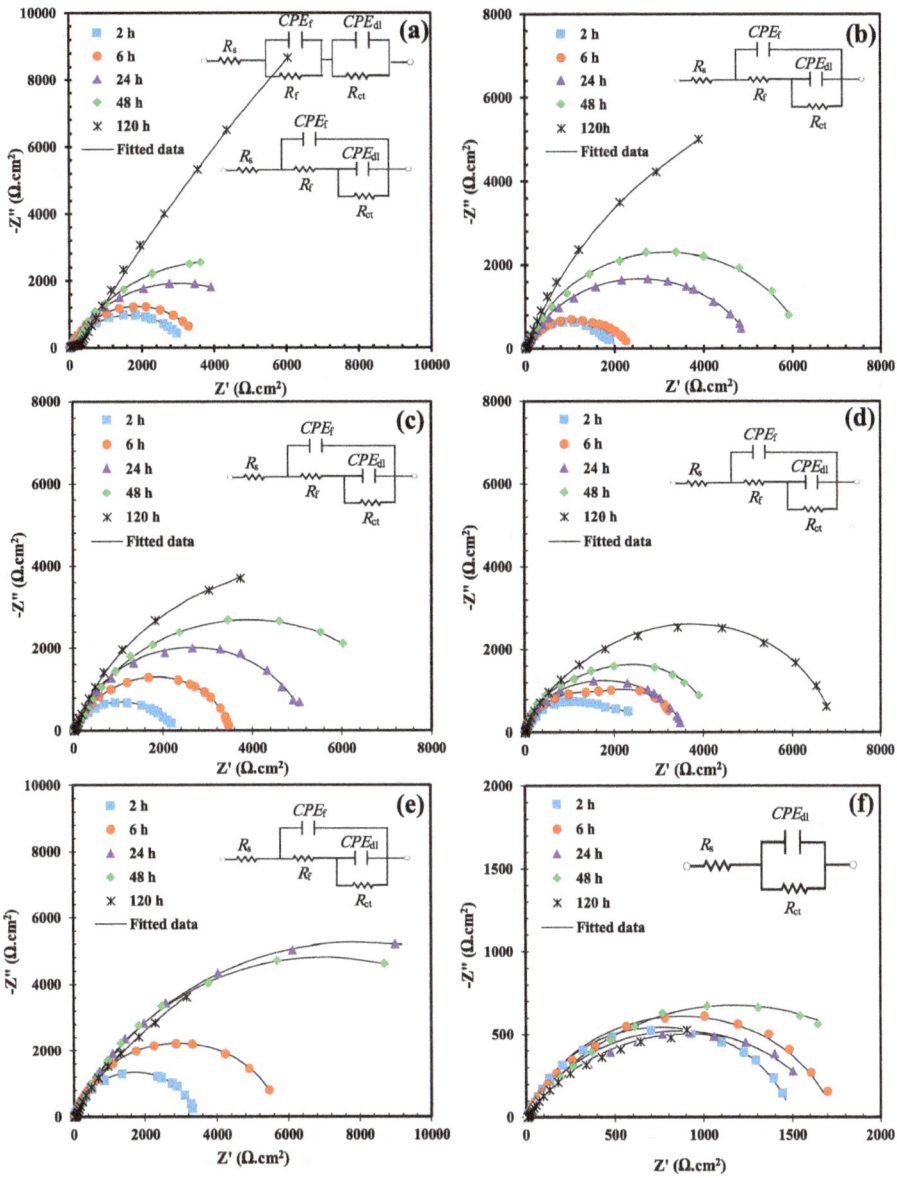

Figure 5.7: Nyquist plots in 3.5 wt% NaCl media for (a) ATMP:Sm 200:600 ppm ratio, (b) ATMP:Sm 400:400 ppm ratio, (c) ATMP:Sm(600:200 ppm ratio), (d) ATMP (800 ppm), (e) Sm (800 ppm), and (f) without inhibitor (reprinted with permission from ref. [76]).

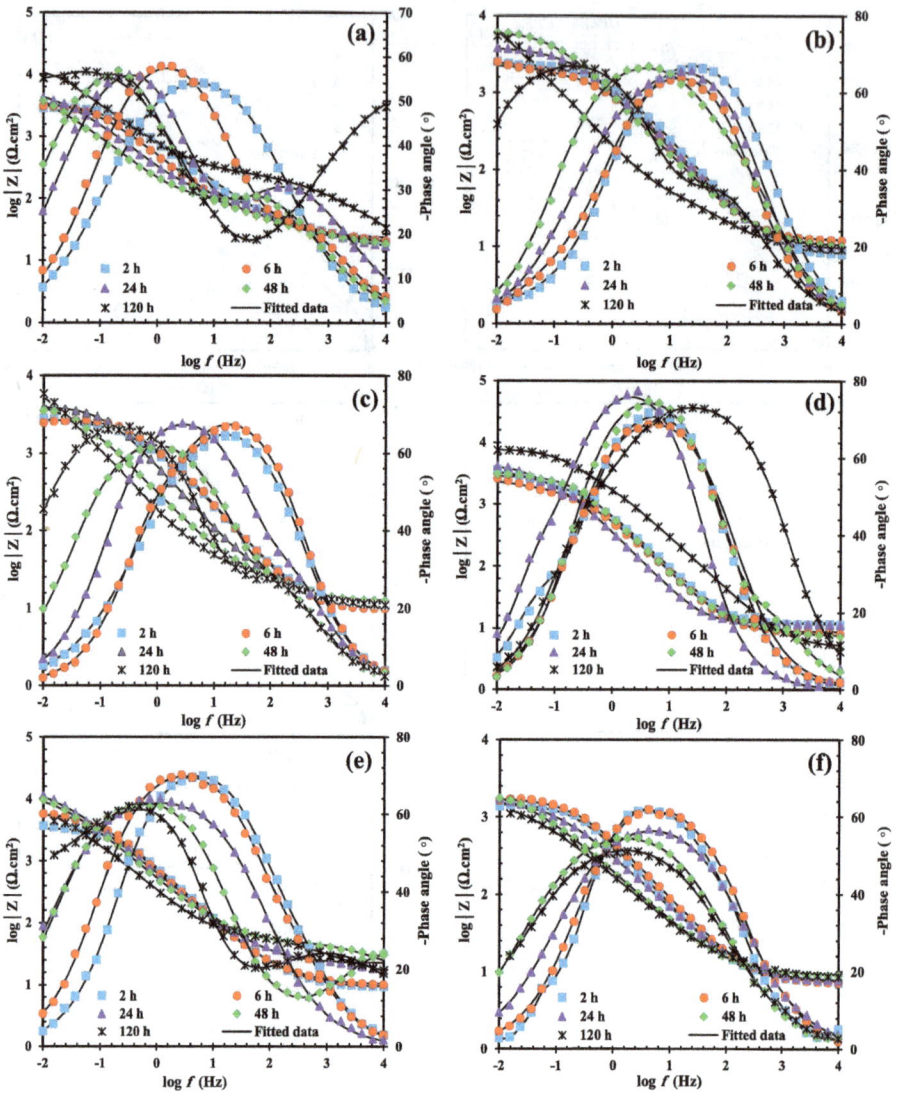

Figure 5.8: Bode plot in 3.5 wt% NaCl media for (a) ATMP:Sm 200:600 ppm ratio, (b) ATMP:Sm(400:400 ppm ratio), (c) ATMP:Sm(600:200 ppm ratio) (d) ATMP (800 ppm), (e) Sm(800 ppm), and (f) without inhibitor (reprinted with permission from ref. [76] © 2020 Elsevier).

The super hydrophobic composite coating may be applied directly to a variety of substrates, including steel, copper, glass, ceramic, etc. Through the use of dripping water droplets, the water repellency of the coated substrate on various substrates was examined (vide Figure 5.10). Figure 5.10 shows that whereas water droplets on the composite coatings are typically spherical, they have compressed hemisphere morphologies

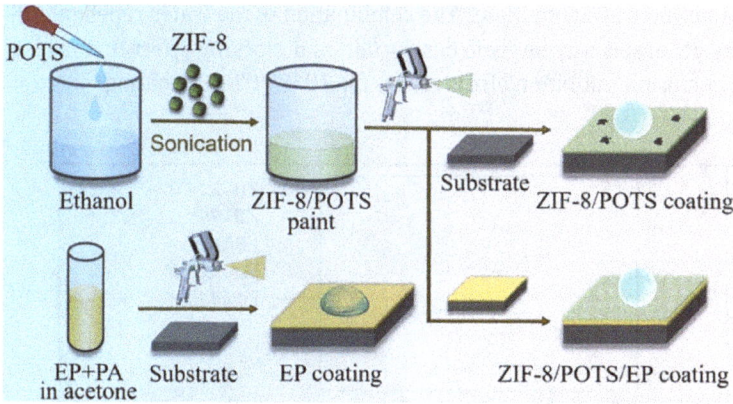

Figure 5.9: Schematic illustrations for stepwise ZIF-8/POTS/EP coating (reprinted with permission from ref. [77]).

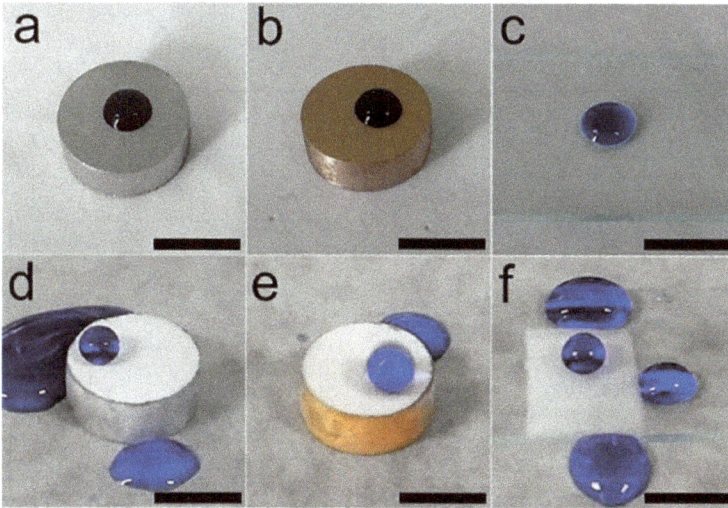

Figure 5.10: Water repellency test on (a) Q235 steel surface, (b) copper surface (c) glass surface, (d) ZIF-8/POTS/EP-coated Q235 steel substrate(e) ZIF-8/POTS/EP-coated copper surface, and (f) ZIF-8/POTS/EP-coated glass surface (reprinted with permission from ref. [77]).

on all bare surfaces. This finding suggests that when applied to a variety of hard substrates, the composite coating can maintain its super hydrophobicity.

It has been investigated through a study of the EIS analysis of the coating substrate to determine its corrosion resistance behavior (vide Figure 5.11). It is important to note that all capacitive loops are wider compared to steel substrate; in addition, these loops are gradually increasing for the coated substrate as shown in Figure 5.11. As a result of the larger capacitive loop, there is likely to be better corrosion protection under the same corrosive conditions. The combination of the water-repellent ZIF-8/POTS and epoxy generates a synergistic obstruction and blocking effect that considerably increases corrosion inhibition effectiveness for ZIF-8/POTS/EP coating.

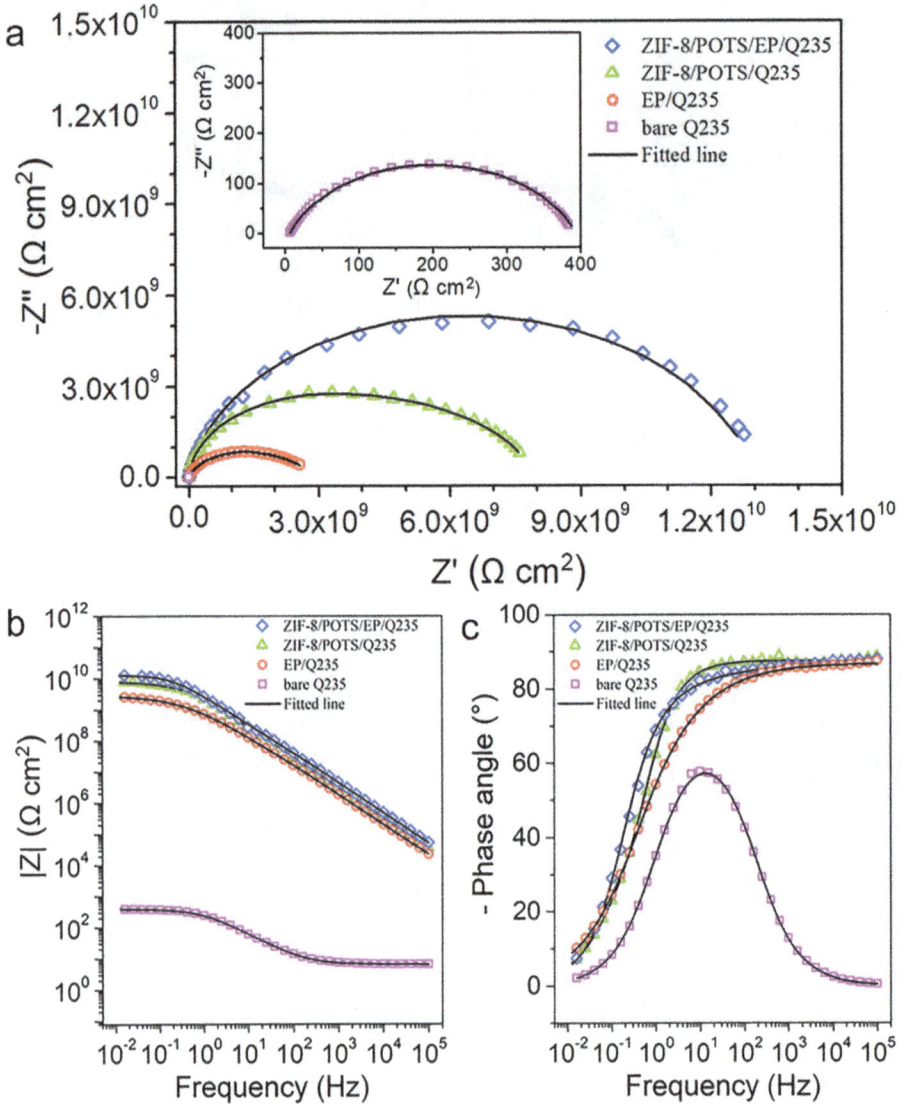

Figure 5.11: EIS analysis of different samples in 3.5 wt% NaCl solution: (a) Nyquist plot, (b, c) Bode plot (reprinted with permission from ref. [77]).

Chen et al. demonstrated a novel synthetic strategy for producing uniform ZIF-7 film oriented in the (110) plane by combining an in-plane epitaxial growth process with a microwave heating approach [78]. It was found that the epitaxial growth technique applied in single-mode microwave heating had a significant impact on the development of ZIF-7 films with improved microstructure. Outstanding corrosion resistance for aluminum plates is also shown by the resultant film. The rod-shaped ZIF-7 seed was manufactured initially under carefully controlled solvothermal conditions (reagents: zinc chloride and b-Im; solvent: DMF; reaction conditions: 50 °C for 24 h). Then, using the dynamic air–liquid interface-assisted self-assembly (ALIAS) method (dispersant: isopropanol; additives: linoleic acid (LA) and polyvinylpyrrolidone (PVP); injection rate: 2 μL/min), the rod-shaped ZIF-7 crystals were quickly assembled into a closely packed (110) oriented single seed layer. After that, single-mode microwave heating and in-plane epitaxial growth were used to produce a highly orientated (110) uniform ZIF-7 film. With the use of a realistic microwave synthesis approach, this study proposes experimental techniques for the quick fabrication of high-quality MOF films and offers suggestions for the upcoming development of novel MOF–based corrosion protective materials. They have carried out DC polarization experiments for the investigation of corrosion inhibition that is effective. According to the experimental findings, the corrosion current density (I_{corr}), which is determined by finding the point where the tangent lines of the cathode and anode polarization curves connect, was 10^{-4} A/cm^2 for the bare Al plate. I_{corr} was reduced by two orders of magnitude when the Al plate surface was coated with a c-oriented ZIF-7 coating made from either ZnCl$_2$ or ZnBr$_2$ sources. I_{corr} was further lowered for strongly (110)-oriented ZIF-7 film-modified Al plates. The excellent anti-corrosion efficacy of the firmly (110) oriented ZIF-7 layer is considered to be due to the parallel orientation of straight pore channels on the Al plate surface, which effectively prevents direct contact between the corrosive solution and the Al plate surface [78]. Tehrani et al. described the deposition of an effective, long-lasting MOF thin layer on a mild steel surface using divalent zinc cations extracted from *Malva sylvestris* (M.S) [79]. Electrochemical analysis was used to determine how well the produced layer prevented corrosion in saline media. EIS analysis of bare steel data at various intervals (1, 6, 48, 56, 152, and 240 h) indicated no corrosion resistance more than 1,810 cm^2. The potentiodynamic polarization plot, which has been shown in Figure 5.12, demonstrates that the Zn:M.S considerably reduced both the cathodic and anodic branches, indicating a decrease in the rate of corrosion. The shifted curve indicates that Zn:M.S has an impact, particularly on the anodic branches. It is evident from the results that the Zn concentration in the 1,000 ppm range is the only one to exceed 85 mV. In accordance with the literature, anodic protection is indicated by a difference larger than 85 mV (depending on the kind of switch), whereas synergistic samples exhibit a mixed control inhibitory mechanism.

Using *Malva sylvestris* and zinc metal ions, a long-lasting anti-corrosive layer with a corrosion inhibition of 170 kΩ.cm^2 was developed in this research. Figure 5.13 shows the surface morphology study of steel samples immersed for 152 h in chloride

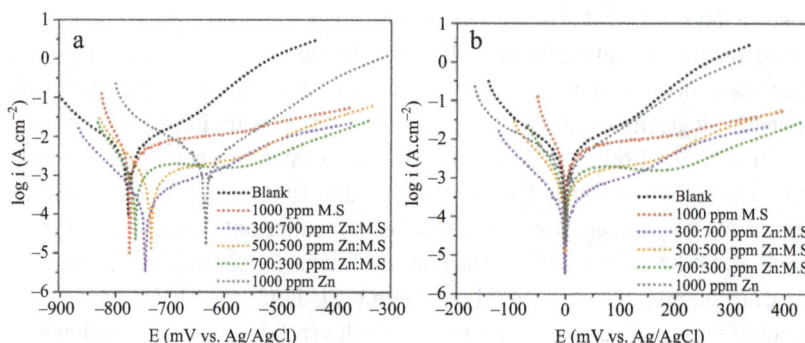

Figure 5.12: (a) Potentiodynamic polarization plots for steel substrates submerged in chloride medium with differing quantities of Zn:M.S and (b) E_{corr} transition to Equilibrium (reprinted with permission from ref. [75]).

medium containing Zn and M.S inhibitors with different ratios. In the absence of any inhibitor, the surface morphology of the blank sample following immersion in the saline solution reflects severe corrosion. On the other hand, when zinc ions are added to the solution, the surface is covered with small particles that resemble flowers and are identified as the zinc hydroxide components. Although the produced components cover the steel surface, the porous structure of the developed layer makes it unable to effectively stop the corrosion process. Additionally, the introduction of the M.S causes the formation of various morphological structures. The entire surface of the metal is covered in the worm-like structure components. The coordinate linkage between the metal surface and the organic M.S. materials is involved with this layer, which is produced via the donor-acceptor action. According to Zheng et al., the corrosion resistance of magnesium alloy may be increased by applying a composite coating comprised of polycaprolactone (PCL) and a copper-based MOF modified with folic acid (FA) [80]. The modified Cu-MOF (HKUST-1) was homogeneously distributed throughout the PCL matrix by hydrogen bonding, enhancing the compactness of the coating (vide Figure 5.14). AZ31 Mg alloy was coated with a mixed matrix coating (PCL-MOF) made of folic acid-modified MOFs (MOF-FA), which was then prepared by first dissolving the powder in PCL solution. The schematic representation is shown in Figure 5.14. Here, MOFs function as fillers and improve the PCL matrix's compactness, which reduces the deterioration of the magnesium alloy. Along with coating deterioration, Cu ions were gradually released, promoting osteoblast differentiation and proliferation. Overall, it was determined that the mixed matrix coating offers a potential alternative for use on alloys based on magnesium since it provides these alloys superior biocompatibility and corrosion inhibition. According to electrochemical analysis, corrosion resistance significantly increased as a result of a reduction in corrosion current density from $7.18 \pm 3.243 \times 10^{-7}$ to $1.10 \pm 0.537 \times 10^{-10}$ A/cm^2. Additionally, osteoblastic cell growth

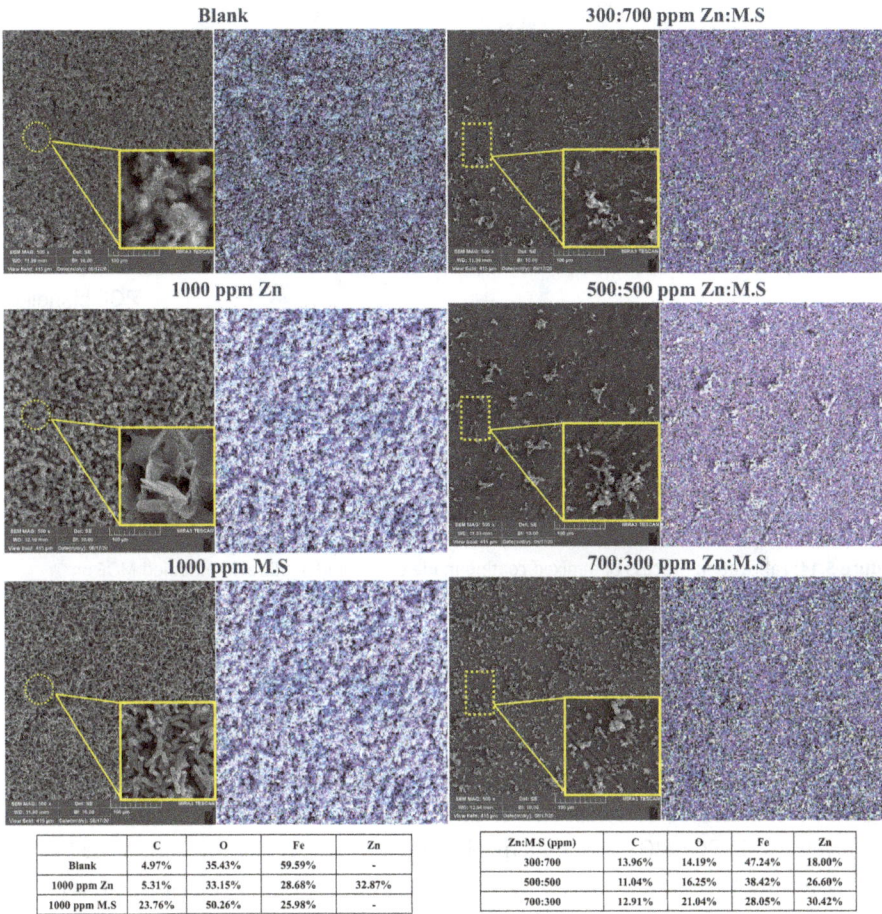

	C	O	Fe	Zn
Blank	4.97%	35.43%	59.59%	-
1000 ppm Zn	5.31%	33.15%	28.68%	32.87%
1000 ppm M.S	23.76%	50.26%	25.98%	-

Zn:M.S (ppm)	C	O	Fe	Zn
300:700	13.96%	14.19%	47.24%	18.00%
500:500	11.04%	16.25%	38.42%	26.60%
700:300	12.91%	21.04%	28.05%	30.42%

Figure 5.13: Surface morphology evaluations of steel samples immersed in chloride environment containing various concentrations of Zn and M.S. inhibitors for 152 h (reprinted with permission from ref. [75]).

and development were promoted by copper ions that were gradually released from the composite coating.

After the as-produced MOF powders were blended with a PCL-containing DCM (Dichloromethane) solution, the amounts of MOF and PCL were adjusted to 0.5 wt% and 6 wt%, respectively. The AZ31 alloys were then equally coated with 50 μL of the combined solutions, which were then dried at 37 °C. The magnesium surface was exposed to PCL DCM solution (6 wt%) in the same manner. In phosphate buffer saline (PBS) (composition of the existing salt is as follows: KH_2PO_4: 0.48 g/L, NaCl: 8 g/L, NaH_2PO_4.-$12H_2O$: 3.81 g/L, and KCl: 0.24 g/L) electrochemical studies were carried out. According to the Nyquist plot depicted in Figure 5.15a and 5.15b, the capacitive arc observed at high frequencies was generated by the charge transfer mechanism in the double-layer

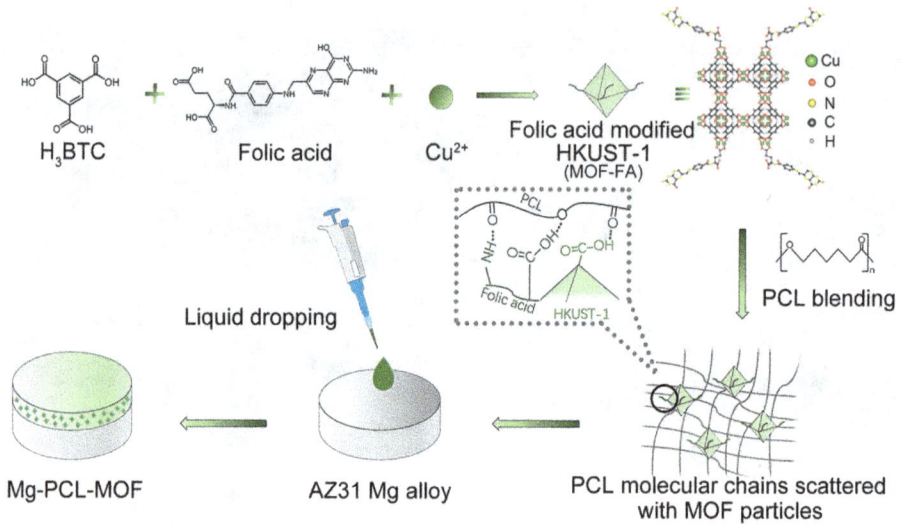

Figure 5.14: Fabrication process of a mixed coating made of PCL and a folic acid-modified MOF on AZ31 Mg alloy (reprinted with permission from ref. [80]).

architecture, whereas the capacitive arc observed at medium frequencies was driven on by coating resistance. A greater semicircular size often indicates a higher coating resistance, which indicates a superior anti-corrosion capacity. All polymer-coated samples had a much bigger capacitive loop than the naked AZ31 Mg alloy, demonstrating the outstanding barrier capabilities of the coatings and the largest impedance was displayed by Mg-PCL-MOF. An inductive-like loop, which is often linked to the formation of pitting corrosion, could be observed on the untreated Mg alloy in the low-frequency region (Figure 5.15b). However, in samples with improved surfaces, the inductive arc was practically undetectable, highlighting the coating's superior anti-corrosion properties.

The Mg-PCL-MOF (1.48×10^8 Ω cm^2) had a substantially greater $|Z|$ value than the pure PCL coated (3.75×10^7 Ω cm^2) samples, according to the Bode plot (Figure 5.15c). Solution resistance is low for only blank Mg alloy; strong H_2 evolution and increase in local pH take place during electrochemical testing for bare magnesium alloy, generating screening effect that influences solution resistance. On the other hand, Mg-PCL as well as Mg-PCL-MOF had larger solution resistance as H_2 evolution and pH rise were greatly reduced. Two distinct time constants can be noticed in the curves of all samples on the Bode phase angle plot (Figure 5.15d).

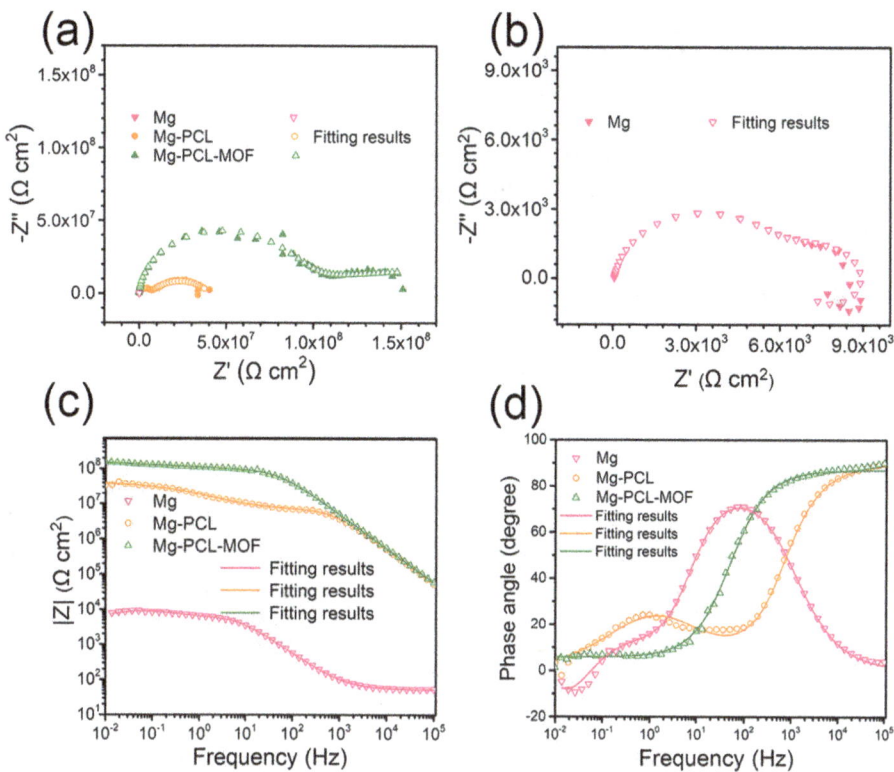

Figure 5.15: Electrochemical analysis results: (a, b) Nyquist plot; (c, d) Bode plot (reprinted with permission from ref. [80]).

5.4 Conclusion

For corrosion protection applications, MOF materials have received a lot of attention due to their rapid development and considerable advantages. The most recent developments in MOFs used for corrosion protection have been presented in this chapter, with a focus on manufacturing methods and corrosion protection applications for MOF-based thin film coating. It includes a concise overview of the significant research progress over the last ten years, as well as the corrosion protection processes of each MOF-based anti-corrosion material. The major objective is to provide direction for future research on MOFs for corrosion prevention. Corrosion resistance being one of the greatest challenges of MOF-based materials, their application is still limited. This article outlines the direction MOF development will take in the future to meet the growing need for corrosion prevention. The development of novel corrosion protection compounds first and foremost requires in-depth knowledge of the mechanism of its formation and structural features of MOFs. Design, development, as well as use of MOF-based anti-corrosive materials can pave a new path in forthcoming research using new machine learning techniques, which is an effective approach for material evaluation. Lastly, rather than being restricted to lab-scale research, many investigations of materials based on MOFs aim toward true commercial use; nonetheless, MOFs are still impracticable for industrial applications due to their high production cost and small-scale synthesis procedure. As a result, extensive research should concentrate on several critical aspects, such as large-scale synthesis using simple technology, cheap cost, and environmentally friendly nature. In order to satisfy these demands, novel MOF materials could be developed and reactions designed to minimize waste in an atom-efficient way to expand the use of commercial anti-corrosion products.

References

[1] Jiang L, Dong Y, Yuan Y, Zhou X, Liu Y, Meng X. Recent advances of metal–organic frameworks in corrosion protection: From synthesis to applications. Chem Eng J [Internet]. 2022 Feb [cited 2023 Mar 15];430:132823. Available from: https://linkinghub.elsevier.com/retrieve/pii/S1385894721044004

[2] Murmu M, Saha SKr, Murmu NC, Banerjee P. Corrosion Inhibitors for Acidic Environments. In: Hussain CM, Verma C, editors. ACS Symposium Series [Internet]. Washington, DC: American Chemical Society; 2021 [cited 2023 Mar 15]. p. 111–62. Available from: https://pubs.acs.org/doi/abs/10.1021/bk-2021-1403.ch007

[3] Saha SKr, Dutta A, Ghosh P, Sukul D, Banerjee P. Adsorption and corrosion inhibition effect of Schiff base molecules on the mild steel surface in 1 M HCl medium: a combined experimental and theoretical approach. Phys Chem Chem Phys [Internet]. 2015 [cited 2023 Jan 30];17(8):5675–50. Available from: http://xlink.rsc.org/?DOI=C4CP05614K

[4] Murmu M, Saha SKr, Murmu NC, Banerjee P. Effect of stereochemical conformation into the corrosion inhibitive behaviour of double azomethine based Schiff bases on mild steel surface in 1 mol L−1 HCl medium: An experimental, density functional theory and molecular dynamics simulation study. Corros Sci [Internet]. 2015 Jan [cited 2023 Jan 30];146:134–51. Available from: https://linkinghub.elsevier.com/retrieve/pii/S0010538X17314464

[5] Saha SKr, Dutta A, Ghosh P, Sukul D, Banerjee P. Novel Schiff-base molecules as efficient corrosion inhibitors for mild steel surface in 1 M HCl medium: experimental and theoretical approach. Phys Chem Chem Phys [Internet]. 2016 [cited 2023 Jan 30];18(27):17858–511. Available from: http://xlink.rsc.org/?DOI=C6CP01553E

[6] Sengupta S, Murmu M, Mandal S, Hirani H, Banerjee P. Competitive corrosion inhibition performance of alkyl/acyl substituted 2-(2-hydroxybenzylideneamino)phenol protecting mild steel used in adverse acidic medium: A dual approach analysis using FMOs/molecular dynamics simulation corroborated experimental findings. Colloids Surf Physicochem Eng Asp [Internet]. 2021 May [cited 2023 Jan 30];617:126314. Available from: https://linkinghub.elsevier.com/retrieve/pii/S0527775721001837

[7] Saha SKr, Murmu M, Murmu NC, Banerjee P. Benzothiazolylhydrazine azomethine derivatives for efficient corrosion inhibition of mild steel in acidic environment: Integrated experimental and density functional theory cum molecular dynamics simulation approach. J Mol Liq [Internet]. 2022 Oct [cited 2023 Mar 15];364:120033. Available from: https://linkinghub.elsevier.com/retrieve/pii/S0167732222015715

[8]   Mandal S, Bej S, Banerjee P. Insights into the uses of two azine decorated d10-MOFs for corrosion inhibition application on mild steel surface in saline medium: Experimental as well as theoretical investigation. J Mol Liq [Internet]. 2023 July [cited 2023 Sep 15];381:121789. Available from: https://www.sciencedirect.com/science/article/abs/pii/S0167732223005925

[9]   Aslam R, Mobin M, Zehra S, Aslam J. A comprehensive review of corrosion inhibitors employed to mitigate stainless steel corrosion in different environments. J Mol Liq [Internet]. 2022 Oct [cited 2023 Mar 15];364:115552. Available from: https://linkinghub.elsevier.com/retrieve/pii/S0167732222015306

[10]   Murmu M, Saha SKr, Murmu NC, Banerjee P. Amine cured double Schiff base epoxy as efficient anticorrosive coating materials for protection of mild steel in 3.5% NaCl medium. J Mol Liq [Internet]. 2015 Mar [cited 2023 Jan 30];278:521–35. Available from: https://linkinghub.elsevier.com/retrieve/pii/S016773221831555X

[11]   Sukul D, Pal A, Mukhopadhyay S, Saha SKr, Banerjee P. Electrochemical behaviour of uncoated and phosphatidylcholine coated copper in hydrochloric acid medium. J Mol Liq [Internet]. 2018 Jan [cited 2023 Mar 15];245:530–40. Available from: https://linkinghub.elsevier.com/retrieve/pii/S0167732217328623

[12]   Mahato P, Mishra SK, Murmu M, Murmu NC, Hirani H, Banerjee P. A prolonged exposure of Ti-Si-B-C nanocomposite coating in 3.5 wt% NaCl solution: Electrochemical and morphological analysis. Surf Coat Technol [Internet]. 2015 Oct [cited 2023 Mar 15];375:477–88. Available from: https://linkinghub.elsevier.com/retrieve/pii/S0257857215307728

[13]   Sastri VS. Green Corrosion Inhibitors: Theory and Practice [Internet]. Hoboken, NJ, USA: John Wiley & Sons, Inc.; 2011 [cited 2023 Mar 15]. Available from: http://doi.wiley.com/10.1002/5781118015438

[14]   Yaghi OM, Li H. Hydrothermal Synthesis of a Metal-Organic Framework Containing Large Rectangular Channels. J Am Chem Soc [Internet]. 1555 Oct [cited 2023 Mar 15];117(41):10401–02. Available from: https://pubs.acs.org/doi/abs/10.1021/ja00146a033

[15]   Zhou HC, Long JR, Yaghi OM. Introduction to Metal–Organic Frameworks. Chem Rev [Internet]. 2012 Feb 8 [cited 2023 Jan 30];112(2):673–74. Available from: https://pubs.acs.org/doi/10.1021/cr300014x

[16]   Li WJ, Tu M, Cao R, Fischer RA. Metal–organic framework thin films: electrochemical fabrication techniques and corresponding applications & perspectives. J Mater Chem A [Internet]. 2016 [cited 2023 Mar 15];4(32):12356–65. Available from: http://xlink.rsc.org/?DOI=C6TA02118B

[17]   Mondal U, Bej S, Hazra A, Mandal S, Pal TK, Banerjee P. Amine-substituent induced highly selective and rapid "turn-on" detection of carcinogenic 1,4-dioxane from purely aqueous and vapour phase with novel post-synthetically modified d^{10}-MOFs. Dalton Trans [Internet]. 2022 [cited 2023 Jan 30];51(5):2083–53. Available from: http://xlink.rsc.org/?DOI=D1DT03576H

[18]   Bej S, Mandal S, Mondal A, Pal TK, Banerjee P. Solvothermal Synthesis of High-Performance d^{10}-MOFs with Hydrogel Membranes @ "Turn-On" Monitoring of Formaldehyde in Solution and Vapor Phase. ACS Appl Mater Interfaces [Internet]. 2021 Jun 2 [cited 2023 Mar 15];13(21):25153–63. Available from: https://pubs.acs.org/doi/10.1021/acsami.1c05558

[19]   Alezi D, Belmabkhout Y, Suyetin M, Bhatt PM, Weseliński ŁJ, Solovyeva V, et al. MOF Crystal Chemistry Paving the Way to Gas Storage Needs: Aluminum-Based **soc**-MOF for CH_4, O_2, and CO_2 Storage. J Am Chem Soc [Internet]. 2015 Oct 21 [cited 2023 Mar 15];137(41):13308–18. Available from: https://pubs.acs.org/doi/10.1021/jacs.5b07053

[20]   Hazra A, Mondal U, Mandal S, Banerjee P. Advancement in functionalized luminescent frameworks and their prospective applications as inkjet-printed sensors and anti-counterfeit materials. Dalton Trans [Internet]. 2021 [cited 2023 Jan 30];50(25):8657–70. Available from: http://xlink.rsc.org/?DOI=D1DT00705J

[21] Hasan MdN, Bera A, Maji TK, Pal SK. Sensitization of nontoxic MOF for their potential drug delivery application against microbial infection. Inorganica Chim Acta [Internet]. 2021 Aug [cited 2023 Mar 15];523:120381. Available from: https://linkinghub.elsevier.com/retrieve/pii/S0020165321001377

[22] Pal TK, De D, Neogi S, Pachfule P, Senthilkumar S, Xu Q, et al. Significant Gas Adsorption and Catalytic Performance by a Robust Cu^{II} –MOF Derived through Single-Crystal to Single-Crystal Transmetalation of a Thermally Less-Stable Zn^{II} –MOF. Chem – Eur J [Internet]. 2015 Dec 21 [cited 2023 Mar 15];21(52):15064–70. Available from: https://onlinelibrary.wiley.com/doi/10.1002/chem.201503163

[23] Shahini MH, Mohammadloo HE, Ramezanzadeh M, Ramezanzadeh B. Recent innovations in synthesis/characterization of advanced nano-porous metal-organic frameworks (MOFs); current/future trends with a focus on the smart anti-corrosion features. Mater Chem Phys [Internet]. 2022 Jan [cited 2023 Mar 15];276:125420. Available from: https://linkinghub.elsevier.com/retrieve/pii/S0254058421012037

[24] Zhang Y, Chang CH. Metal–Organic Framework Thin Films: Fabrication, Modification, and Patterning. Processes [Internet]. 2020 Mar 24 [cited 2023 Mar 15];8(3):377. Available from: https://www.mdpi.com/2227-5717/8/3/377

[25] Zacher D, Shekhah O, Wöll C, Fischer RA. Thin films of metal–organic frameworks. Chem Soc Rev [Internet]. 2005 [cited 2023 Mar 15];38(5):1418. Available from: http://xlink.rsc.org/?DOI=b805038b

[26] Ameloot R, Vermoortele F, Vanhove W, Roeffaers MBJ, Sels BF, De Vos DE. Interfacial synthesis of hollow metal–organic framework capsules demonstrating selective permeability. Nat Chem [Internet]. 2011 May [cited 2023 Mar 15];3(5):382–87. Available from: http://www.nature.com/articles/nchem.1026

[27] Sud D, Kaur G. A comprehensive review on synthetic approaches for metal-organic frameworks: From traditional solvothermal to greener protocols. Polyhedron [Internet]. 2021 Jan [cited 2023 Mar 15];153:114857. Available from: https://linkinghub.elsevier.com/retrieve/pii/S0277538720305544

[28] Van Vleet MJ, Weng T, Li X, Schmidt JR. In Situ, Time-Resolved, and Mechanistic Studies of Metal–Organic Framework Nucleation and Growth. Chem Rev [Internet]. 2018 Apr 11 [cited 2023 Mar 15];118(7):3681–721. Available from: https://pubs.acs.org/doi/10.1021/acs.chemrev.7b00582

[29] Liu J, Wöll C. Surface-supported metal–organic framework thin films: fabrication methods, applications, and challenges. Chem Soc Rev [Internet]. 2017 [cited 2023 Mar 15];46(15):5730–70. Available from: http://xlink.rsc.org/?DOI=C7CS00315C

[30] Brower LJ, Gentry LK, Napier AL, Anderson ME. Tailoring the nanoscale morphology of HKUST-1 thin films via codeposition and seeded growth. Beilstein J Nanotechnol [Internet]. 2017 Nov 3 [cited 2023 Mar 15];8:2307–14. Available from: https://www.beilstein-journals.org/bjnano/articles/8/230

[31] Bradshaw D, Garai A, Huo J. Metal-organic framework growth at functional interfaces: thin films and composites for diverse applications. Chem Soc Rev [Internet]. 2012 [cited 2023 Mar 15]; 41(6):2344–81. Available from: http://xlink.rsc.org/?DOI=C1CS15276A

[32] Hinterholzinger F, Scherb C, Ahnfeldt T, Stock N, Bein T. Oriented growth of the functionalized metal–organic framework CAU-1 on –OH- and –COOH-terminated self-assembled monolayers. Phys Chem Chem Phys [Internet]. 2010 [cited 2023 Mar 15];12(17):4515. Available from: http://xlink.rsc.org/?DOI=b524657f

[33] Li J, Wu Q, Wu J. Synthesis of Nanoparticles via Solvothermal and Hydrothermal Methods. In: Aliofkhazraei M, editor. Handbook of Nanoparticles [Internet]. Cham: Springer International Publishing; 2016 [cited 2023 Mar 15]. p. 255–328. Available from: https://link.springer.com/10.1007/578-3-315-15338-4_17

[34] Zhao J, Gong B, Nunn WT, Lemaire PC, Stevens EC, Sidi FI, et al. Conformal and highly adsorptive metal–organic framework thin films via layer-by-layer growth on ALD-coated fiber mats. J Mater Chem A [Internet]. 2015 [cited 2023 Mar 15];3(4):1458–64. Available from: http://xlink.rsc.org/?DOI=C4TA05501B

[35] Xiao FX, Pagliaro M, Xu YJ, Liu B. Layer-by-layer assembly of versatile nanoarchitectures with diverse dimensionality: a new perspective for rational construction of multilayer assemblies. Chem Soc Rev [Internet]. 2016 [cited 2023 Mar 15];45(11):3088–121. Available from: http://xlink.rsc.org/?DOI=C5CS00781J

[36] Wang Z, Wöll C. Fabrication of Metal–Organic Framework Thin Films Using Programmed Layer-by-Layer Assembly Techniques. Adv Mater Technol [Internet]. 2015 May [cited 2023 Mar 15];4(5):1800413. Available from: https://onlinelibrary.wiley.com/doi/10.1002/admt.201800413

[37] So MC, Jin S, Son HJ, Wiederrecht GP, Farha OK, Hupp JT. Layer-by-Layer Fabrication of Oriented Porous Thin Films Based on Porphyrin-Containing Metal–Organic Frameworks. J Am Chem Soc [Internet]. 2013 Oct 23 [cited 2023 Mar 15];135(42):15658–701. Available from: https://pubs.acs.org/doi/10.1021/ja4078705

[38] Jiang D, Burrows AD, Xiong Y, Edler KJ. Facile synthesis of crack-free metal–organic framework films on alumina by a dip-coating route in the presence of polyethylenimine. J Mater Chem A [Internet]. 2013 [cited 2023 Mar 15];1(18):5457. Available from: http://xlink.rsc.org/?DOI=c3ta10766c

[39] Horcajada P, Serre C, Grosso D, Boissière C, Perruchas S, Sanchez C, et al. Colloidal Route for Preparing Optical Thin Films of Nanoporous Metal-Organic Frameworks. Adv Mater [Internet]. 2005 May 18 [cited 2023 Mar 15];21(15):1531–35. Available from: https://onlinelibrary.wiley.com/doi/10.1002/adma.200801851

[40] Huang Y, Tao C an, Chen R, Sheng L, Wang J. Comparison of Fabrication Methods of Metal-Organic Framework Optical Thin Films. Nanomaterials [Internet]. 2018 Aug 30 [cited 2023 Mar 15];8(5):676. Available from: http://www.mdpi.com/2075-4551/8/5/676

[41] Chernikova V, Shekhah O, Eddaoudi M. Advanced Fabrication Method for the Preparation of MOF Thin Films: Liquid-Phase Epitaxy Approach Meets Spin Coating Method. ACS Appl Mater Interfaces [Internet]. 2016 Aug 10 [cited 2023 Mar 15];8(31):20455–64. Available from: https://pubs.acs.org/doi/10.1021/acsami.6b04701

[42] Wang L, Sahabudeen H, Zhang T, Dong R. Liquid-interface-assisted synthesis of covalent-organic and metal-organic two-dimensional crystalline polymers. Npj 2D Mater Appl [Internet]. 2018 Sep 3 [cited 2023 Mar 15];2(1):26. Available from: https://www.nature.com/articles/s41655-018-0071-5

[43] Schoedel A, Scherb C, Bein T. Oriented Nanoscale Films of Metal-Organic Frameworks By Room-Temperature Gel-Layer Synthesis. Angew Chem Int Ed [Internet]. 2010 Aug 20 [cited 2023 Mar 15];45(40):7225–28. Available from: https://onlinelibrary.wiley.com/doi/10.1002/anie.201001684

[44] Scherb C, Williams JJ, Hinterholzinger F, Bauer S, Stock N, Bein T. Implementing chemical functionality into oriented films of metal–organic frameworks on self-assembled monolayers. J Mater Chem [Internet]. 2011 [cited 2023 Mar 15];21(38):14845. Available from: http://xlink.rsc.org/?DOI=c0jm04526h

[45] Szelagowska-Kunstman K, Cyganik P, Goryl M, Zacher D, Puterova Z, Fischer RA, et al. Surface Structure of Metal–Organic Framework Grown on Self-Assembled Monolayers Revealed by High-Resolution Atomic Force Microscopy. J Am Chem Soc [Internet]. 2008 Nov 5 [cited 2023 Mar 15];130(44):14446–47. Available from: https://pubs.acs.org/doi/10.1021/ja8065743

[46] Kwon HT, Jeong HK. *In Situ* Synthesis of Thin Zeolitic–Imidazolate Framework ZIF-8 Membranes Exhibiting Exceptionally High Propylene/Propane Separation. J Am Chem Soc [Internet]. 2013 Jul 24 [cited 2023 Mar 15];135(25):10763–68. Available from: https://pubs.acs.org/doi/10.1021/ja403845c

[47] Kwon HT, Jeong HK. Improving propylene/propane separation performance of Zeolitic-Imidazolate framework ZIF-8 Membranes. Chem Eng Sci [Internet]. 2015 Mar [cited 2023 Mar 15];124:20–26. Available from: https://linkinghub.elsevier.com/retrieve/pii/S000525051400308X

[48] Yao J, Dong D, Li D, He L, Xu G, Wang H. Contra-diffusion synthesis of ZIF-8 films on a polymer substrate. Chem Commun [Internet]. 2011 [cited 2023 Mar 15];47(5):2555. Available from: http://xlink.rsc.org/?DOI=c0cc04734a

[49] Shamsaei E, Lin X, Low ZX, Abbasi Z, Hu Y, Liu JZ, et al. Aqueous Phase Synthesis of ZIF-8 Membrane with Controllable Location on an Asymmetrically Porous Polymer Substrate. ACS Appl Mater Interfaces [Internet]. 2016 Mar 5 [cited 2023 Mar 15];8(5):6236–44. Available from: https://pubs.acs.org/doi/10.1021/acsami.5b12684

[50] Barankova E, Tan X, Villalobos LF, Litwiller E, Peinemann KV. A Metal Chelating Porous Polymeric Support: The Missing Link for a Defect-Free Metal-Organic Framework Composite Membrane. Angew Chem Int Ed [Internet]. 2017 Mar 6 [cited 2023 Mar 15];56(11):2565–68. Available from: https://onlinelibrary.wiley.com/doi/10.1002/anie.201611527

[51] Hoseini SJ, Bahrami M, Nabavizadeh SM. ZIF-8 nanoparticles thin film at an oil–water interface as an electrocatalyst for the methanol oxidation reaction without the application of noble metals. New J Chem [Internet]. 2015 [cited 2023 Mar 15];43(35):15811–22. Available from: http://xlink.rsc.org/?DOI=C5NJ02855B

[52] Yu B, Ye G, Chen J, Ma S. Membrane-supported 1D MOF hollow superstructure array prepared by polydopamine-regulated contra-diffusion synthesis for uranium entrapment. Environ Pollut [Internet]. 2015 Oct [cited 2023 Mar 15];253:35–48. Available from: https://linkinghub.elsevier.com/retrieve/pii/S0265745115323322

[53] Ameloot R, Gobechiya E, Uji-i H, Martens JA, Hofkens J, Alaerts L, et al. Direct Patterning of Oriented Metal-Organic Framework Crystals via Control over Crystallization Kinetics in Clear Precursor Solutions. Adv Mater [Internet]. 2010 Mar 8 [cited 2023 Mar 15];22(24):2685–88. Available from: https://onlinelibrary.wiley.com/doi/10.1002/adma.200503867

[54] Zhuang JL, Ceglarek D, Pethuraj S, Terfort A. Rapid Room-Temperature Synthesis of Metal-Organic Framework HKUST-1 Crystals in Bulk and as Oriented and Patterned Thin Films. Adv Funct Mater [Internet]. 2011 Apr 22 [cited 2023 Mar 15];21(8):1442–47. Available from: https://onlinelibrary.wiley.com/doi/10.1002/adfm.201002525

[55] Zhuang JL, Ar D, Yu XJ, Liu JX, Terfort A. Patterned Deposition of Metal-Organic Frameworks onto Plastic, Paper, and Textile Substrates by Inkjet Printing of a Precursor Solution. Adv Mater [Internet]. 2013 Sep 6 [cited 2023 Mar 15];25(33):4631–35. Available from: https://onlinelibrary.wiley.com/doi/10.1002/adma.201301626

[56] Al-Kutubi H, Gascon J, Sudhölter EJR, Rassaei L. Electrosynthesis of Metal-Organic Frameworks: Challenges and Opportunities. ChemElectroChem [Internet]. 2015 Apr 15 [cited 2023 Mar 15];2(4):462–74. Available from: https://onlinelibrary.wiley.com/doi/10.1002/celc.201402425

[57] Stassen I, Burtch N, Talin A, Falcaro P, Allendorf M, Ameloot R. An updated roadmap for the integration of metal–organic frameworks with electronic devices and chemical sensors. Chem Soc Rev [Internet]. 2017 [cited 2023 Mar 15];46(11):3185–241. Available from: http://xlink.rsc.org/?DOI=C7CS00122C

[58] Campagnol N, Van Assche T, Boudewijns T, Denayer J, Binnemans K, De Vos D, et al. High pressure, high temperature electrochemical synthesis of metal–organic frameworks: films of MIL-100 (Fe) and HKUST-1 in different morphologies. J Mater Chem A [Internet]. 2013 [cited 2023 Mar 15];1(15):5827. Available from: http://xlink.rsc.org/?DOI=c3ta10415b

[59] Li WJ, Feng JF, Lin ZJ, Yang YL, Yang Y, Wang XS, et al. Patterned growth of luminescent metal–organic framework films: a versatile electrochemically-assisted microwave deposition method. Chem Commun [Internet]. 2016 [cited 2023 Mar 15];52(20):3551–54. Available from: http://xlink.rsc.org/?DOI=C6CC00515E

[60] Ameloot R, Stappers L, Fransaer J, Alaerts L, Sels BF, De Vos DE. Patterned Growth of Metal-Organic Framework Coatings by Electrochemical Synthesis. Chem, Mater [Internet]. 2005 Jul 14 [cited 2023 Mar 15];21(13):2580–82. Available from: https://pubs.acs.org/doi/10.1021/cm500065f

[61] Li M, Dincă M. Reductive Electrosynthesis of Crystalline Metal–Organic Frameworks. J Am Chem Soc [Internet]. 2011 Aug 24 [cited 2023 Mar 15];133(33):12526–5. Available from: https://pubs.acs.org/doi/10.1021/ja2041546

[62] Hod I, Bury W, Karlin DM, Deria P, Kung CW, Katz MJ, et al. Directed Growth of Electroactive Metal-Organic Framework Thin Films Using Electrophoretic Deposition. Adv Mater [Internet]. 2014 Sep [cited 2023 Mar 15];26(36):6255–300. Available from: https://onlinelibrary.wiley.com/doi/10.1002/adma.201401540

[63] Zhu H, Liu H, Zhitomirsky I, Zhu S. Preparation of metal–organic framework films by electrophoretic deposition method. Mater Lett [Internet]. 2015 Mar [cited 2023 Mar 15];142:15–22. Available from: https://linkinghub.elsevier.com/retrieve/pii/S0167577X1402103X

[64] Stassen I, Styles M, Van Assche T, Campagnol N, Fransaer J, Denayer J, et al. Electrochemical Film Deposition of the Zirconium Metal-Organic Framework UiO-66 and Application in a Miniaturized Sorbent Trap. Chem Mater [Internet]. 2015 Mar 10 [cited 2023 Mar 15];27(5):1801–07. Available from: https://pubs.acs.org/doi/10.1021/cm504806p

[65] Zou X, Zhu G, Hewitt IJ, Sun F, Qiu S. Synthesis of a metal–organic framework film by direct conversion technique for VOCs sensing. Dalton Trans [Internet]. 2005 [cited 2023 Mar 15];(16):3005. Available from: http://xlink.rsc.org/?DOI=b822248g

[66] Li J, Cao W, Mao Y, Ying Y, Sun L, Peng X. Zinc hydroxide nanostrands: unique precursors for synthesis of ZIF-8 thin membranes exhibiting high size-sieving ability for gas separation. CrystEngComm [Internet]. 2014 [cited 2023 Mar 15];16(42):5788–51. Available from: http://xlink.rsc.org/?DOI=C4CE01503G

[67] Stassen I, Campagnol N, Fransaer J, Vereecken P, De Vos D, Ameloot R. Solvent-free synthesis of supported ZIF-8 films and patterns through transformation of deposited zinc oxide precursors. CrystEngComm [Internet]. 2013 [cited 2023 Mar 15];15(45):5308. Available from: http://xlink.rsc.org/?DOI=c3ce41025k

[68] Li W, Su P, Li Z, Xu Z, Wang F, Ou H, et al. Ultrathin metal–organic framework membrane production by gel–vapour deposition. Nat Commun [Internet]. 2017 Sep 1 [cited 2023 Mar 15]; 8(1):406. Available from: https://www.nature.com/articles/s41467-017-00544-1

[69] Makiura R, Motoyama S, Umemura Y, Yamanaka H, Sakata O, Kitagawa H. Surface nano-architecture of a metal–organic framework. Nat Mater [Internet]. 2010 Jul [cited 2023 Mar 15];5(7):565–71. Available from: https://www.nature.com/articles/nmat2765

[70] Makiura R, Kitagawa H. Porous Porphyrin Nanoarchitectures on Surfaces. Eur J Inorg Chem [Internet]. 2010 Aug [cited 2023 Mar 15];2010(24):3715–24. Available from: https://onlinelibrary.wiley.com/doi/10.1002/ejic.201000730

[71] Motoyama S, Makiura R, Sakata O, Kitagawa H. Highly Crystalline Nanofilm by Layering of Porphyrin Metal–Organic Framework Sheets. J Am Chem Soc [Internet]. 2011 Apr 20 [cited 2023 Mar 15]; 133(15):5640–43. Available from: https://pubs.acs.org/doi/10.1021/ja110720f

[72] Makiura R, Konovalov O. Bottom-up assembly of ultrathin sub-micron size metal–organic framework sheets. Dalton Trans [Internet]. 2013 [cited 2023 Mar 15];42(45):15531. Available from: http://xlink.rsc.org/?DOI=c3dt51703a

[73] Rubio-Giménez V, Tatay S, Volatron F, Martínez-Casado FJ, Martí-Gastaldo C, Coronado E. High-Quality Metal–Organic Framework Ultrathin Films for Electronically Active Interfaces. J Am Chem Soc [Internet]. 2016 Mar 2 [cited 2023 Mar 15];138(8):2576–84. Available from: https://pubs.acs.org/doi/10.1021/jacs.5b05784

[74] Benito J, Sorribas S, Lucas I, Coronas J, Gascon I. Langmuir–Blodgett Films of the Metal–Organic Framework MIL-101(Cr): Preparation, Characterization, and CO_2 Adsorption Study Using a QCM-Based Setup. ACS Appl Mater Interfaces [Internet]. 2016 Jun 25 [cited 2023 Mar 15];8(25):16486–52. Available from: https://pubs.acs.org/doi/10.1021/acsami.6b04272

[75] Wen Y, Chen Y, Wu Z, Liu M, Wang Z. Thin-film nanocomposite membranes incorporated with water stable metal-organic framework CuBTTri for mitigating biofouling. J Membr Sci [Internet]. 2015 Jul [cited 2023 Mar 15];582:285–57. Available from: https://linkinghub.elsevier.com/retrieve/pii/S0376738815307008

[76]   Dehghani A, Poshtiban F, Bahlakeh G, Ramezanzadeh B. Fabrication of metal-organic based complex film based on three-valent samarium ions-[bis (phosphonomethyl) amino] methylphosphonic acid (ATMP) for effective corrosion inhibition of mild steel in simulated seawater. Constr Build Mater [Internet]. 2020 Apr [cited 2023 Mar 15];235:117812. Available from: https://linkinghub.elsevier.com/retrieve/pii/S0550061815332655

[77]   Chen H, Wang F, Fan H, Hong R, Li W. Construction of MOF-based superhydrophobic composite coating with excellent abrasion resistance and durability for self-cleaning, corrosion resistance, anti-icing, and loading-increasing research. Chem Eng J [Internet]. 2021 Mar [cited 2023 Mar 15];408:127343. Available from: https://linkinghub.elsevier.com/retrieve/pii/S1385854720334677

[78]   Chen S, Sun Y, Chen S, Gao Y, Wang F, Li H, et al. Facile fabrication of a highly (110)-oriented ZIF-7 film with rod-shaped seeds. Chem Commun [Internet]. 2021 [cited 2023 Mar 15];57(17):2128–31. Available from: http://xlink.rsc.org/?DOI=D0CC07810G

[79]   Tehrani MEHN, Ramezanzadeh M, Ramezanzadeh B. A highly-effective/durable metal-organic anti-corrosion film deposition on mild steel utilizing Malva sylvestris (M.S) phytoextract-divalent zinc cations. J Ind Eng Chem [Internet]. 2021 Mar [cited 2023 Mar 15];55:252–304. Available from: https://linkinghub.elsevier.com/retrieve/pii/S1226086X21000022

[80]   Zheng Q, Li J, Yuan W, Liu X, Tan L, Zheng Y, et al. Metal–Organic Frameworks Incorporated Polycaprolactone Film for Enhanced Corrosion Resistance and Biocompatibility of Mg Alloy. ACS Sustain Chem Eng [Internet]. 2015 Nov 4 [cited 2023 Mar 15];7(21):18114–24. Available from: https://pubs.acs.org/doi/10.1021/acssuschemeng.5b05156

Pragnesh N. Dave, Pradip M. Macwan

6 Functionalized conductive polymer-based thin film coating as corrosion inhibitors

Abstract: Corrosion is the degradation of a material caused by the environment either by itself or in combination with mechanical forces. Corrosion is an electrochemical phenomenon and often involves oxidation of metals. Metal corrosion is one of the significant concerns facing the manufacturing industry. Many corrosion regulator techniques involve conversion layers and coatings of conductive polymers, which include poisonous and ecologically dangerous chemicals, particularly chromium compounds. New protective coating techniques have been created as a result of these goals. Conducting polymers have attracted interest recently, owing to their numerous industrial uses and feasibility from an economic viewpoint. Polymers have long-chain carbon bonds, which allow them to block a significant portion of the decaying metal surfaces, upon adsorption. A hurdle effect is created between the metal and its surroundings by the thin layers that have been adsorbed on the metal substrate. The majority of the examined polymeric materials are poor corrosion inhibitors. To increase the inhibitory capacity of polymers, a number of initiatives have been attempted, including copolymerization, introduction of chemicals that have synergistic effects, cross-linking, blending, and, of late, the introduction of inorganic nanoparticles. The implementation of conducting polymers, polymer composites, and nanocomposites for the prevention of corrosion of various industrial metal substrates is examined and explained in this chapter, based on the findings of experiments. There have also been several shortcomings noted and future directions in this field are outlined.

Keywords: Corrosion, Conducting Polymer, Coatings, Nanocomposites

6.1 Introduction

Conducting polymers have been used as linear actuators because they are known to physically react to electrochemical stimuli. Ionic entry and egress have been recognized to be the best operational method of actuation as yet, but molecular interactions and methods based on conformational changes have also been suggested. It is important to develop, synthesize, and analyze novel materials, spanning scientific fields from synthetic chemistry to materials and mechanical engineering, in the search for new conducting polymer actuators. Intrinsically conducting polymers were discovered around three decades ago, and this discovery attracted researchers' interest in the scientific community due to the variety of possibilities of use for these polymers.

https://doi.org/10.1515/9783111016160-006

As a result of the strong electrical conductivity of these materials, which is comparable to metals, they are also called as synthetic metals. As shown in Figure 6.1, polyacetylene, polyfuran, polypyrrole, and polythiophene are examples of different conducting polymers (CPs), all of which are insulators in their neutral form. By the doping of various salts by chemical and electrochemical redox processes, the polymers' insulating property can be changed into conducting behavior [1].

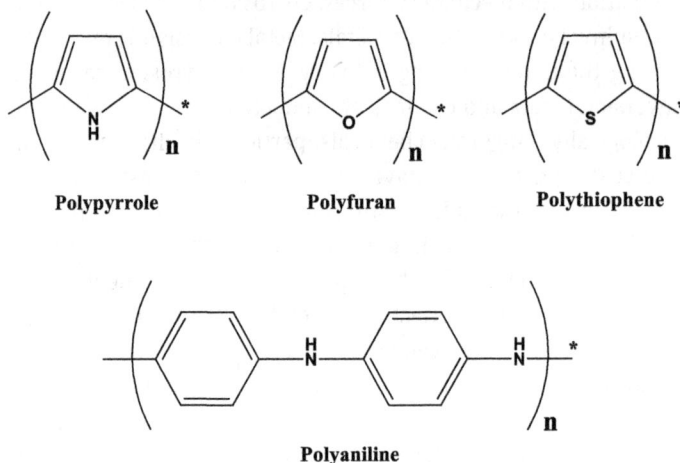

Polypyrrole   Polyfuran   Polythiophene

Polyaniline

Figure 6.1: Schematic representation of intrinsically conducting polymers.

MacDiarmid was the first to propose conductive polymers for corrosion prevention (1985). The classes of polyanilines, polyheterocycles, and poly(phenylene vinylenes) encompass nearly all of the CPs utilized in corrosion prevention. CPs can be created chemically or electrochemically. According to studies, the majority of CPs may be electrochemically formed by anodic oxidation, allowing one to make a conducting film, straight on an exterior. A number of doping techniques can be used to convert CPs from an insulating state to a conducting state, including charge transfer chemical doping, electrochemical doping, acid–base chemistry doping (only polyaniline is susceptible to this type of doping), photodoping, and charge injection at the metal–semiconducting polymer interface [2].

In 1977, the Chemical Communication journal recognized and published their research on doping polyacetylene with halogen compounds. The three researchers were awarded the 2000 Nobel Prize in Chemistry for developing conducting polymers (CPs). After learning about conducting polyacetylene, researchers were motivated to produce additional conducting polymers such as polythiophene, polyaniline, polypyrrole, and polyfuran.

Conjugated π-electrons, delocalized along the polymer backbone or aromatic rings, are used to create intrinsically conducting polymers or ICPs. They are special materials because of the physio-electrochemical features they combine. For the creation of CP nanotubes and nanowires, several synthetic approaches, i.e., several lithography techniques, including electrospinning, hard physical template-guided synthesis, soft chemical template synthesis, and others, have been suggested. Intrinsically conducting polymers, doped conducting polymers, extrinsically conducting polymers, and coordination conducting polymers are the different types of CPs. The conductance of intrinsically CPs is supported by a strong backbone comprised of a heavily conjugated structure. Doping and the existence of conjugated double bonds are the two requirements for polymers to become inherently conducting. In most cases, doping refers to the introduction or removal of electrons from the materials. Redox doping (oxidative doping), charge injection doping, and non-redox doping are all methods for doping polymers. P-doping involves oxidation and a positive charge is formed on the polymer chain, increasing its conductivity [3].

What first attracted attention to CPs was the probability of developing CP-based "smart" coatings that may inhibit metallic corrosion, even in damaged areas where the basic metal exterior is exposed to the corrosive atmosphere. CPs may readily switch between their two possible states (oxidation-conductive state and the reduction-non-conductive state) under the right circumstances. CPs go through redox processes, which make it possible for counter ions (dopants) to be bound and released in response to changes in the metal's external potential brought on by native electrochemical reactions, which are brought on by corrosion. Depending on the local corrosive circumstances, the CP may insert or expel dopants. Upon release, these dopants are frequently inhibitors that stop the local corrosion process. This is one of the methods recommended for using the benefits offered when a CP is employed as a significant component of a corrosion-resistant covering [4].

A family of organic materials, known as conductive polymers (CPs), has unique electrical and optical features that are analogous to those of inorganic semiconductors and metals. Simple, adjustable, and economical strategies may be customary to create CPs. Simple electro-polymerization techniques can be used to easily assemble them into supramolecular structures with multiple functionalities. In order to integrate and interface the CPs with biomedical applications like biomaterials and biosensors, a broad variety of approaches have been created to change and modify the CPs. A variety of biomedical devices' characteristics such as their electrical sensitivity, speed, and stability as well as their interactions with biological tissues have been improved through the use of CPs. Because various CP types are recognized for their interaction with biological samples while preserving their biocompatibility, one may assume that CPs might be certified as excellent candidates for use in a range of biological and medical applications. There are many different kinds of CPs and they are categorized according to the electric charge they carry, such as ions, conductive nanomaterials, and delocalized π-electrons [5].

Applying one or more organic coatings to the metal is now the most popular corrosion control method. An oxidant must be present at the metal surface of an active corrosion cell, and the corrosion cell must also include a mechanism for ion transport over the surface between the anodic and cathodic sites (to preserve charge equilibrium). Such an ion motion at the contact typically takes place inside a thin electrolyte layer that develops on the metal surface. By slowing down the pace at which these vital components (dioxygen, water, and ions like H^+) access the surface, coatings slow down corrosion. Additionally, the coating increases the ion movement resistance at the contact, resulting in a slower rate of corrosion. Eventually, the coating is penetrated by ions, water, and dioxygen from the environment, eventually reaching the metal contact. This process is accelerated by coating flaws. As a result, a coating system technique is often utilized, in which a primer coating is applied to the metal, followed by a topcoat with the desired barrier and, maybe, aesthetic attributes. The primer is a coating, selected for having strong adherence to the metal, and frequently containing dynamic chemicals to further lower the rate of corrosion once the barrier has been broken. Electroactive conducting polymers (ECPs) may be crucial for the success of this latter function [6].

In a variety of industrial uses, including the food sector, corrosion of stainless steel is a severe problem. For the construction of food equipment, higher grades of stainless steel with greater anti-corrosion properties (e.g., 304 and 316 types) are frequently employed. However, prolonged contact with corrosive substances like chloride in cleaning or processing products, chlorine and oxidizing sanitizers, food additives (such meat or blood), etc., speeds up the deterioration of steel. Organic coating is the method most frequently used to prevent corrosion in stainless steel. Different organic coatings, such as acrylic-based paints and epoxy, are applied to the surface. Additionally, electroactive conducting polymers, including polypyrrole, polyaniline, and others, are a family of substances that are extensively employed in a broad range of application areas, including energy storage, catalysts, biosensors, and other similar fields in a number of diverse morphologies. Due to their electrical conductivity and their simplicity in chemical and electrochemical production, they are also the ideal prospects for anti-corrosive materials. Because conducting polymer coatings are difficult to produce, electrochemical deposition is the most effective method for creating them [7]. The usage of conductive polymer-based coatings to stop the corrosion of metals, mentioned in recent advancements, is explored in this chapter.

6.2 Corrosion basics

Corrosion primarily occurs due to an electrochemical heterogeneous process that involves the metal and its surrounding environment. [4]. Metals corrode when they come into contact with an electrolyte because a corrosion cell is formed when regions with a greater free energy or greater potential act as anodes and regions with a minor free energy or minor potential act as cathodes. At the anode, metal ions are created, and they disintegrate into the electrolyte. The environment is affected by the electrons as they travel through the metal to the nearby cathode locations. The transfer of the associated charges from the cathode to the anode via the electrolyte and the passage of electrons from the anode to the cathode constitute the corrosion current. As a result, the corrosion current is linked to the corrosion rate. The following reactions can be used to illustrate the corrosion process.

6.2.1 Anodic reaction

It is possible to express the electrochemical process at the anode or metal suspension as follows:

$$M = M^{n+} + ne$$

As the liberated electrons go through the metal towards the cathode, corrosion current is created.

6.2.2 Cathodic reaction

The environment has an impact on the type of reaction that occurs at the cathode, where the electrons generated during anodic suspension are spent. The following are the cathodic reactions that occur most frequently in corrosion:

(a) Oxygen reduction (Acid solution)

$$O_2 + 4H^+ + 4e = 2H_2O$$

(b) Oxygen reduction (Neutral or basic solution)

$$O_2 + H_2O + 2e = 2(OH)^-$$

(c) Hydrogen evolution

$$2H^+ + 2e = H_2$$

(d) Metal ion reduction: Metal ions present in the solution may be reduced:

$$M^{n+} + 2e = M^{(n-1)+}$$

This is possible only in the presence of a significant amount of M^{n+} ions . By receiving an electron, the metal ion lowers its valence state in this process.

(e) Metal deposition

A metal's ionic state can be converted to a neutral metallic state:

$$M^{n+} + e = M$$

Due to the frequent occurrence of acidic media, hydrogen evolution is widespread. Oxygen reduction is also frequent since it can be triggered by any aqueous solution that comes into contact with air. Less frequent reactions include metal deposition and metal ion reduction. These partial reactions, such as the corrosion of simple carbon steel in water, may be utilized to recognize the bulk of the corrosion processes. Corrosion takes place in two phases. In contrast to other regions that are open to air, areas enclosed by a water droplet are first starved of oxygen and turn anodic. Plain carbon steel becomes anodic when it comes into contact with any electrolyte that has a low concentration of dissolved oxygen, as opposed to plain carbon steel that is in contact with an electrolyte that is high in dissolved oxygen. One of the main causes is that the following reaction will occur in an environment with water that is high in oxygen:

$$O_2 + 2H_2O + 4e^- = 4(OH)^-$$

Because it absorbs electrons, this reaction is cathodic. Near the water droplet's edge, where there will be more oxygen available, hydroxyl ions develop, making this region cathodic. These regions serve as anodes because they are located in another portion of the metal where there is less oxygen. The reaction produces electrons in the anodic region below the water droplet's centre:

$$Fe = Fe^{2+} + 2e^-$$

The $(OH)^-$ anions will typically go in the direction of the anode whereas the Fe^{2+} cations will move through the electrolyte in the direction of the cathode. The reduction product $(OH)^-$ from the cathode and the oxidative product (Fe^{2+}) from the anode travel separately until they associate to produce ferrous hydroxide.

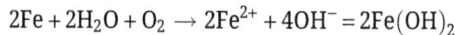

$$2Fe + 2H_2O + O_2 \rightarrow 2Fe^{2+} + 4OH^- = 2Fe(OH)_2$$

Due to its insolubility, ferrous hydroxide $(Fe(OH)_2)$ precipitates from the oxygenated aqueous solution and collects as precipitates, close to the cathode region. In another step, access to dissolved oxygen at the $Fe(OH)_2$ layer's outer surface causes the ferrous hydroxide to change into the hydrous and insoluble ferric hydroxide as follows:

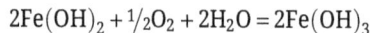

$$2Fe(OH)_2 + 1/2O_2 + 2H_2O = 2Fe(OH)_3$$

Typically referred to as red rust, the reddish brown ferric hydroxide, $Fe(OH)_3$ that develops during the early stages of corrosion progressively transforms into ferric oxyhydroxide, FeOOH. Red rust is another name for many ferric oxy-hydroxides. $Fe(OH)_3$ dehydrates to ferric oxide when exposed to air:

$$2\text{Fe(OH)}_3 = \text{Fe}_2\text{O}_3 + 3\text{H}_2\text{O}$$

When there is insufficient oxygen available, this process produces black magnetite (Fe_3O_4). The area beneath the permeable deposit loses oxygen as soon as a rust deposit forms on the steel surface, turning it anodic, in contrast to the bare metal. Air and water may travel through the rust layers, making them neither protective nor impermeable. Therefore, below the coating of rust, the corrosion of steel might proceed unnoticed.

6.3 Corrosion prevention mechanisms

In comparison to conventional organic coatings, the corrosion protection process is more intricate [3]. According to the inhibitory mechanism, corrosion inhibitors are categorized as follows.

6.3.1 Anodic protection mechanism

Using anodes that are incorporated into the repair patches, embedded reinforcement is protected against current and potential corrosion without the need for wiring or external devices.

6.3.2 Controlled inhibitor release mechanism (CIR)

The CIR model suggests that when applied to a base metal substrate, the oxidized and doped type of particular ICPs, such PANI, provide the anion dopant, following reduction, due to the coupling to the base metal through coating defects.

6.3.3 An energetic electronic block at the metallic surface

An electric field is anticipated to arise any time a metal comes in contact with a doped semiconductor or an electrically conducting polymer, preventing or decreasing corrosion by restricting the passage of electrons from the metal to an oxidizing species.

6.3.4 Barrier protection mechanism

On a base metal, the coatings of CPs provide a solid, adhering, low porosity barrier that preserves a neutral environment, restricts oxidant access, and prevents oxidation of the metal surface. Instead of acting as a basic barrier, CPs' coatings provide active protection, and the effectiveness of the barrier is improved by increased adhesion, decreased porosity, and increased compactness.

6.4 Conducting polymers

Organic polymers, known as conducting polymers, as their name suggests, conduct electricity. They are composed of single carbon-carbon bonds that connect aromatic rings like phenylene, naphthalene, anthracene, pyrrole, and thiophene. They are also called as intrinsically conducting polymers (ICPs). Polyacetylene, polypyrrole, polyaniline, etc. are few examples.

6.4.1 Properties of conducting polymers

Doped conducting polymers and non-doped conducting polymers are the two types of conducting polymers. The existence of a conductivity band that resembles a metal is what gives non-doped conjugated polymers their conductivity. Three of the four valence electrons in a conjugated polymer undergo significant localized sp^2 hybridization to generate strong sigma bonds. Each carbon atom's remaining unpaired electron is still in a P_z orbital. It creates a π-connection by crossing over to a nearby P_z orbital – conjugated P_z orbitals electrons form an extended P_z orbital system that permits free movement of electrons (delocalization of electrons). There is some conductivity in non-doped polymers, although it is negligible. In the case of doped conjugated polymers, one electron is either supplied to the conducting band by reduction or is removed from the valence band by oxidation (p-doping) (n-doping). P-doping makes the polymer extremely conductive and enhances the mobility of electrons in these delocalized orbitals.

6.4.2 Conducting polymers for corrosion protection

The usage of conducting polymers (CPs) for the prevention of metal corrosion is currently receiving more attention. CPs are special because they combine characteristics of traditional metals, semiconductors, and polymers [8]. Due to their insolubility in aqueous media, which is brought on by the stiffness of their backbone structure, the use of CPs has hitherto been constrained. Chemically modified CPs now have better solubility due to modifications made to their molecular structures. This kind of polymer has been made into thin coatings that have been applied to metal substrates, used for things like antistatic shields [9], light emitting diodes, electrochromatic devices, etc. DeBerry originally mentioned the potential for using CPs in the field of metal corrosion prevention in 1985. CPs, in particular polyaniline, polypyrrole, and their substituted compounds, have been used for decades to protect metals against corrosion in a variety of hostile situations. The conducting polymers studied for corrosion prevention are concisely defined in Table 6.1.

Table 6.1: An overview of several conductive polymers tested for corrosion resistance.

Sr. no.	Label	Contraction	Possessions
1.	Polyaniline	PANI	Leucoemeraldine, emeraldine (green for the emeraldine salt, blue for the emeraldine base), and (per)nigraniline are the three distinct colors of PANI, a semi-flexible rod polymer. Owing to its great steadiness at room temperature, PANI in the emeraldine state is the most often used form. The emeraldine salt that has been doped has a high electrical conductivity. Pernigraniline and leucoemeraldine are weak conductors.
2.	Polypyrrole	PPy	Amorphous, yellowish PPy darkens in air as a result of oxidation. The oxidized version has excellent electrical and conductive qualities. The doped films can be either blue or black, depending on the amount of polymerization and film width. PPy can swell, yet is unsolvable in diluents.
3.	Polyindole	PIN	PIN is a polymer made from indole, an aromatic bicyclic molecule with a six-membered benzene ring bonded to a five-membered pyrrole ring, which contains nitrogen. Possesses excellent blue photoluminescent properties, shows extremely steady redox activity, and has quick switchable electrochromic abilities. The doped version is stable in air and has strong electrical conductivity.
4.	Polythiophene	PTh	Thiophenes, a sulfur heterocycle molecule, are polymerized to form PTh. When the polymer is oxidized, it displays electrical and conductive conductivity. The conjugated backbone of PTh confers intriguing optical characteristics. With fascinating color variations in reaction to variations in solvent, temperature, applied voltage, and addition of further molecules, the optical characteristics react to environmental stimuli.
5.	Poly (phenylenediamine)	PPD	PPD is an aromatic semiconductive polymer with a backbone made up of pyrazine and phenazine rings. PPDs are created utilizing o-, m-, and p-phenylenediamines.

6.4.3 The usage of conducting polymers as corrosion preventing films

A substance known as a metal corrosion inhibitor has the ability to significantly reduce corrosion, when added in small quantities to caustic media [8]. Organic and inorganic inhibitors make up the two main categories of inhibitors. In contrast to their organic counterparts that obstruct by employing their heteroatoms and/or unsaturated bonds as the adsorption centre, corrosion inhibitors for inorganic metals primarily act by oxidizing a metal exterior to produce an impermeable oxide layer. As a result of their many adsorption sites, polymers are known to interact with metal surfaces quite well. In reality, this is not always the case. The methylene group (CH$_2$) that separates the repeating unit (monomer) in maximum polymer molecules gives them their hydrophobicity. The main obstacle to the usage of unchanged polymers as metal corrosion inhibitors is their insolubility in aqueous media. Umoren and Solomon [10] are correct when they point out that solubility is a crucial prerequisite for corrosion inhibition. Research on CPs as a metal corrosion inhibitor is quite limited, according to a study of the corrosion literature. The insolvability of unchanged CPs' non-aqueous solutions would not be a far-fetched explanation for this discovery. In a 0.5 M H$_2$SO$_4$ solution at room temperature, Jeyaprabla et al. [11] found that 10 ppm of polyaniline could only provide 53% security to the iron exterior. Jeyprabla et al. used 0.001 M ceric ions to improve the presentation of polyaniline due to the little inhibition of iron corrosion that it provided, and 88% corrosion inhibition effectiveness was attained. It has been established that poly (methoxy aniline) may be 89% more effective in preventing iron corrosion in 1 M HCl solution. The methoxy suspended group has an impact on the polyaniline's solubility.

6.4.4 Classification of conducting polymers

Conducting polymers are those that have the ability to conduct electricity. Polymers typically made using standard techniques are almost insulators. Some polymers, nevertheless, could function as conductors. These are further classified as shown below in Figure 6.2.

6.4.4.1 Fundamentally conducting polymers

These kinds of polymers feature a strong, extended conjugated structure that serves as their backbone and this gives them their conductivity. They come in two varieties:

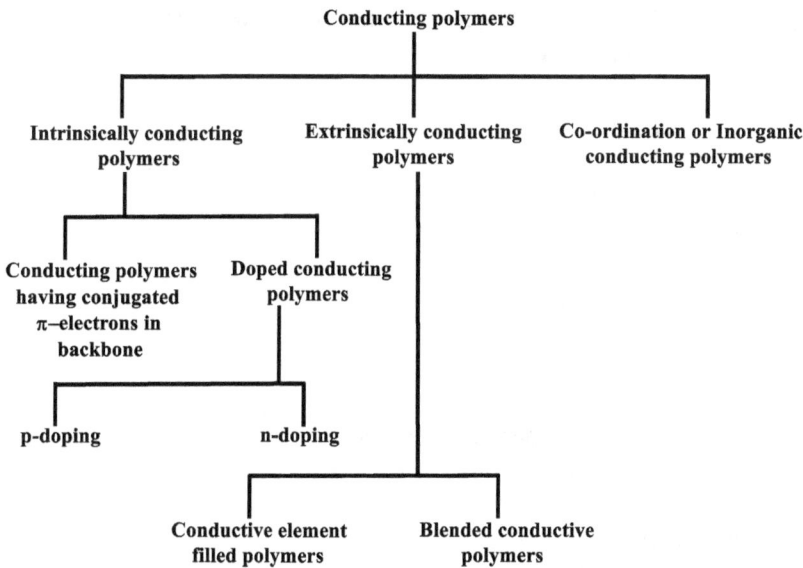

Figure 6.2: Classification of conducting polymers.

6.4.4.1.1 Conducting polymers having conjugated π-electrons in the backbone

The conjugated π-electron backbones of these polymers often generate an electrical charge within them. When an electrical field is present, conjugated π-electrons in the polymer get energized and are then able to move across the solid polymer. Moreover, the orbitals of conjugated π-electrons overlap throughout the whole backbone, resulting in the establishment of valence and conduction bands that encompass the entire polymer molecule. Conjugated electrons improve a material's conductivity (e.g. Polypyrrole).

6.4.4.2 Doped conducting polymers

Doped conducting polymers are those that are made when a charged transfer agent is introduced to the polymer, either in the gas phase or in the solution. Doping is the process of making the polymer's backbone negatively or positively charged by oxidation or reduction in order to boost conductivity.

Doping includes two types:

– **p-Doping:** It is accomplished through oxidation. A Lewis acid is used in this procedure to treat the conducting polymer.

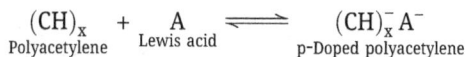

$$\underset{\text{Polyacetylene}}{(CH)_x} + \underset{\text{Lewis acid}}{A} \rightleftharpoons \underset{\text{p-Doped polyacetylene}}{(CH)_x^- A^-}$$

– n-Doping: Reduction is used to accomplish this doping. A Lewis base is used in this procedure to treat the conducting polymer.

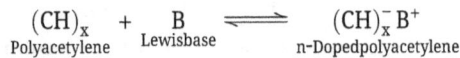

$$(CH)_x \ + \ B \ \rightleftharpoons \ (CH)_x^- B^+$$
$$\text{Polyacetylene} \quad \text{Lewisbase} \quad\quad \text{n-Dopedpolyacetylene}$$

Benefits of intrinsically conducting polymers
(i)   Their conductivity.
(ii) Their capability to stock a charge.
(iii) Their capacity to undertake ion altercation.
(iv) They can absorb visible light to give colored yields.
(v)   They are translucent to X-rays.

Restrictions of intrinsically conducting polymers
(i)   Their conductivities are inferior to metals.
(ii) Their improcessability.
(iii) Their deprived mechanical strength.
(iv) They are less steady at high temperatures.
(v)   On storing, their conductivity is affected.

6.4.4.3 Extrinsically conducting polymers

Extrinsically conducting polymers are those conducting polymers that are conductive due to the presence of components that have been added from outside the polymer. There are two varieties of such polymers:

6.4.4.3.1 Conductive-element-filled polymer
In this kind, a polymer is used as a binder to bond the conducting components to a solid structure. The percolation threshold is the lowest concentration of conductive filler that the polymer requires to start conducting. The following vital characteristics are present in these polymers:
(a) They have strong majority conductivity.
(b) They cost less.
(c) They are lightweight in nature.
(d) They are strong and mechanically resilient.
(e) They can be processed readily in a variety of forms, profiles, and dimensions.

6.4.4.3.2 Blended conducting polymers
A normal polymer and a conducting polymer are combined physically or chemically to make these polymers. Better physical, chemical, and mechanical characteristics are present in such polymers, which are also simple to produce.

6.4.4.3.3 Coordination or inorganic conducting polymers

These polymers, which have charge transfer complexes, are made by fusing metal atoms with polydentate ligands.

Applications of conducting polymers

Conducting polymers are broadly used in:

1. rechargeable batteries
2. creating analytical sensors for pH, O_2, SO_2, NH_3, glucose, etc.
3. the groundwork of ion exchangers
4. the controlled release of drugs
5. optical filters
6. photovoltaic devices
7. telecommunication systems
8. microelectronic devices
9. biomedical applications

Applications of Conducting Polymers

1. They are employed in the production of corrosion inhibitors, electromagnetic shielding, antistatic coatings, and chemical sensors.
2. They are also utilized in lasers, light-emitting diodes (LEDs), and polymer-based transitions, which are small electrical devices.
3. They are employed to cover stealth aircraft with microwave- and radar-absorbent materials, in particular.
4. They are employed in the production of printed circuit boards because they shield copper from oxidation and keep them from soldering.

6.5 Mechanism of action of polymers in a corrosion inhibitor

Numerous researchers have proposed a variety of theories to explain the action of these green inhibitors [12]. It was suggested that organic molecules adsorb on the cathodic sites of metal surfaces and impede the cathodic process because they form onium ions in acidic environments. The corrosion-inhibiting ability of natural compounds has been attributed to a number of different modes of action. In-depth research has been carried out on the utilization of organic chemicals that include oxygen, sulphur, and, particularly, nitrogen to lessen the assault of corrosion on steel. Based on the data that is currently available, it is learnt that the majority of organic inhibitors adsorb on the metal surfaces by transferring water molecules and creating a dense fence. It is simpler to transfer electrons from the inhibitor to the metal because inhibitor molecules include non-bonded and p electrons. Transferring electrons from the inhibitor to the metal exterior allows for the formation of a coordinate covalent bond. The strength of the chemisorption bond is dependent on the electron density on the donor atom and the polarizability of the functional group. A H-atom bonded to the C in the ring can be substituted with a substituent group (NH_2, NO_2, CHO, or COOH), which slows down the cathodic or anodic process by altering the metal's electron density at the attachment point. At the cathode, electrons are used up, and at the anode, they are supplied. Corrosion is therefore delayed. We have looked at straight chain amines with three to fourteen carbons. The capacity to stop corrosion improves with the number of carbons in the chain, up to around 10 carbons, but there is no change with more members. This is explained by the fact that as the distance of the hydrocarbon chain rises, its solubility in water decreases.

But if the molecule included a hydrophilic functional group, the inhibitors would be more soluble. The electron density at the donor atom, the nature of the p orbital, the electronic structure of the molecule, and the chemical structure and physicochemical characteristics of the compound-like functional groups all have an impact on an organic inhibitor's efficacy. The following factors may contribute to the inhibition: (1) on cathodic or anodic sites, the adsorption of the molecules or their ions; (2) an upsurge in cathodic or anodic over voltage; and (3) the creation of an obstruction layer for protection. Because certain polymers form complexes with metal ions on the exteriors of metals, thanks to their functional groups, some studies have suggested that they can be employed as corrosion inhibitors. These complexes cover a substantial amount of the surface, shielding metals from the corrosive elements in the solution. The ability of these polymers to hinder adhesion is structurally related to the cyclic rings – heteroatom (O and N) – that serve as the majority of active adsorption sites. It has also been discovered that some natural polymers are effective corrosion inhibitors. Agricultural polymers made of extracellular polysaccharides released by *Leuconostoc mesenteroides* have been demonstrated by Finkenstadt et al. to reduce corrosion on corrosion-sensitive metals. The overall structure of the essentially pure exopolysaccharide consists of an alpha (1-6)–linked D–glucose backbone with 3% to 4% branching of alpha (1-3) links.

6.6 Synthetic methods of conducting polymers

CPs can be created by any one of the following methods [13]:

6.6.1 Chemical method

CPs have been created chemically by polymerizing the matching monomers after they have undergone oxidation or reduction. The potential for affordable mass manufacturing is one of its benefits. Numerous readings have been utilized to upsurge the manufacturing process's yield and product quality. Chemical route principles do not require electrochemical procedures to be used. For instance, the well-known and widely researched CP poly(3-hexylthiophene) is virtually always created chemically. Chemical methods can be used to manufacture polypyrrole (PPy) and polyaniline (PANI); however, electrochemical methods typically result in variations with higher conductivity and mechanical qualities. After conjugation, stability is the primary need while getting ready for chemical polymerization. Oligomers and small molecular weight polymers must be sufficiently sensitive and solvable in order for high molecular weight polymerization to be successful. Although this is getting less and less common as the quantity of monomer and the reactive polymer decline, polymerization should still proceed using a heterogeneous method if an oligomer precipitates from the solution. The reaction vessel walls would have a mechanically unstable coating if a chemical polymerization failed because it would cease before the required molecular weight is attained. Nevertheless, chemical polymerization ensures the precise selection of an oxidant to selectively yield cation radicals at the proper spot on the monomer in a sufficiently soluble environment.

6.6.2 Electrochemical method

Electrochemical synthesis of CPs is one of the many synthesis approaches that have been discussed. It is important because it is simple, inexpensive, reproducible, and produces films with the required thickness and uniformity. The electrochemical process that is used the most commonly to make ECPs is anodic oxidation of appropriate electroactive functional monomers; cathodic reduction is used far fewer normally. The oxidation-induced doping of counter ions and the development of a polymer layer happen instantaneously in the previous scenario. Moreover, the capacity for monomer oxidation leading to polymerization is better than the capability for charging oligomeric intermediary polymers. An electroactive monomer, such as pyrrole or thiophene, was polymerized utilizing an expedited electropolymerization procedure that alternated chemical and electrode reaction phases. For example, during the potentio-dynamic electro-polymerization of thiophene, a radical cation is likely to be generated in the initial phase of the thiophene electro-oxidation at a significantly positive potential. Subsequently, it is expected to undergo anodic peak and form a protonated dimer of the radical cation in the following chemical reaction stage. In the subsequent electrode reaction step, the radical cation's protonated dimer is electro-oxidized to form the di-cation.

6.6.3 Photochemical method

The primary techniques for locating polymers in industry and academic research facilities have been chemical and ECP approaches. However, throughout the past two decades, photochemical groundwork, being extensively studied, has been claimed to have minimal benefits due to its speed, low cost, and environmental friendliness. The technique can be used to fabricate certain CPs. As an illustration, pyrrole has been successfully polymerized to PPy by exposure to visible light, while acting as both a suitable electron acceptor and a photosensitizer. Recently, hydrogen peroxide has been employed to initiate oxidative-free radical coupling processes that would polymerize aniline, when present. Aniline polymerization is possible under more hospitable environmental circumstances as compared to chemical and electrochemical processes.

6.6.4 Methathesis method

The interchange of one component from each substance to create a new one is known as metathesis, and it occurs when two chemicals interact chemically. The three kinds of metathesis polymerization are ring opening cyclo-olefin metathesis, acyclic or cyclic alkynes, and di-olefin metathesis.

6.6.5 Concentrated emulsion method

The water section, the latex unit section, and the monomer drop division may all be classed as segments of the emulsion polymerization technique, which is a heterophase polymerization process. Radical polymerization is its primary mechanism. The monomer serving as the initiator and the solvent are both present in the same segment in the bulk and solution polymerization procedures. The synthesized polymer stays soluble in the solvent or the monomer till significant modification. In this procedure, a combination of a water-soluble initiator, a water-insoluble monomer, and a micelle-forming surfactant is commonly used. When compared to suspension polymerization, latex particles and monomer-swollen micelles are the main loci of polymerization. Thus, the term "emulsion polymerization" is misleading since the final output is a dispersion of latex particles, but the initial point is an emulsion of monomer droplets in water. The monomer droplets in microemulsion polymerization are typically very tiny and serve as the site of polymerization. Inverse emulsion polymerizations can also be carried out when the constant segment is an organic grouping with a water-soluble monomer in tiny water droplets. It is only used to create masterpieces made of modacrylic in the acrylic fibre business.

6.6.6 Inclusion method

Inclusion polymerization is frequently used at the atomic or molecular level manufacture of composite materials. Hence, this kind of polymerization can lead to the development of exceptional, low-dimensional combined materials with massive promise. For instance, a molecular wire might be made using an electroconductive polymer.

Based on inclusion, these polymers have been combined with organic hosts to form composites. Miyata et al. claim that this polymerization may also be viewed as a conventional space-dependent polymerization and should not be just viewed in terms of stereoregular polymerization.

6.6.7 Solid-state method

In the absence of oxygen and water, solid-state polymerization lengthens the polymer chain lengths by removing the reaction by-products with heat, hoover, or elimination with an inert gas. The reaction is influenced by the following factors: temperature, pressure, and the diffusion of waste materials from the basic to the shell of the tablet. It is a crucial method that is routinely used to boost the mechanical and rheological properties of polymers before inoculation blow moulding after melt polymerization. The commercial production of PET films, high-tech industrial fibres, and bottle-compatible fibres all benefit greatly from this method.

6.6.8 Plasma polymerization

Plasma polymerization is an innovative approach to produce thin films from a wide range of organic and organometallic starting materials. Strongly cross-linked, pin-hole-free plasma-polymerized films are abrasion-resistant, thermally stable, chemically inert, and insoluble. These films also adhere incredibly well to a wide range of substrates, such as those constructed of common polymers, glass, and metal surfaces. They have been widely used in recent years for a number of claims, including perm selective membranes, protective shells, biological materials, electrical, optical, and adhesive supports, due to their outstanding properties.

6.7 Synthesis of conducting polymers (CPs)

6.7.1 Synthesis of polyaniline (PANI)

PANI has received the greatest attention among the CPs for corrosion prevention [2]. PANI is superior to most current CPs in a number of ways. Its benefits include: (a) simple chemical and electrochemical polymerization; (b) straightforward doping and de-doping by treatment with regular aqueous acids and bases; and (c) great resilience to environmental deterioration. PANI is created chemically by polymerizing aniline (monomer) with $(NH_4)_2S_2O_8$ in a hydrogen chloride solution. Another method for making PANI is electro-polymerization (Figure 6.3).

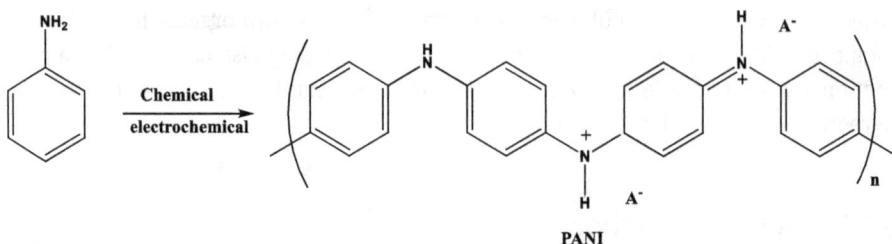

Figure 6.3: Synthesis of PANI.

PANI has always been a challenging substance to process. The polymer backbone is altered by the addition of several functional groups, such as alkoxy, amino, alkyl, aryl, and sulfonyl groups, to create a more processable version of PANI. An alternate method to get a soluble version of PANI is to utilize new acid additions such camphor sulfonic acid. PANI is a peculiar CP since it may exist in four distinct oxidation states and is generated chemically and electrochemically. Leucoemeraldine, pernigraniline, emeraldine base (EB), an intermediary form, and emeraldine salt, an electrically conductive form, are among these states (Figure 6.4).

By using chemical and electrochemical electrochemical polymerization techniques, PANI is frequently converted into the emeraldine salt (ES). By treating ES with a base, the EB form may be created. The procedures utilized to upsurge and reduce the electrical conductivity of PANI are protonation (doping) and de-protonation (de-doping), respectively (Figure 6.5).

6.7.2 Synthesis of poly(heterocyclics)

Another category of monomers that can be polymerized to create CPs is the heterocyclic monomers (Figure 6.6). Pyrrole (X = NH), the most prevalent of them, has been researched for corrosion prevention. Due to their high oxidation potential (>1.7 V vs. Ag/Ag$^+$), furans (X = O) have not been researched as much as thiophenes (X = S) for their electrical, electro-optical, and corrosion characteristics. The irreversible oxidation of the polymer, brought on by the high oxidation potential of the furans, results in extremely inferior films. Both chemical and electrochemical approaches have been used to polymerize polypyrrole (PPy) and polythiophene (PTh), which may then be doped to produce comparatively high conductivities. Due to the abundance of electrons in polyheterocycles, these materials quickly oxidize and become stable in their oxidized state. Due to its extraordinary stability and ease of structural change for increased processability, polypyrrole and polythiophene have both been studied for corrosion protection. Angeli (1916) initially described polypyrrole, which was synthesized by chemically oxidizing pyrrole with hydrogen peroxide. At that time, PPy was referred to as "pyrrole black." The most straightforward technique of making PPy is still chemical polymeriza-

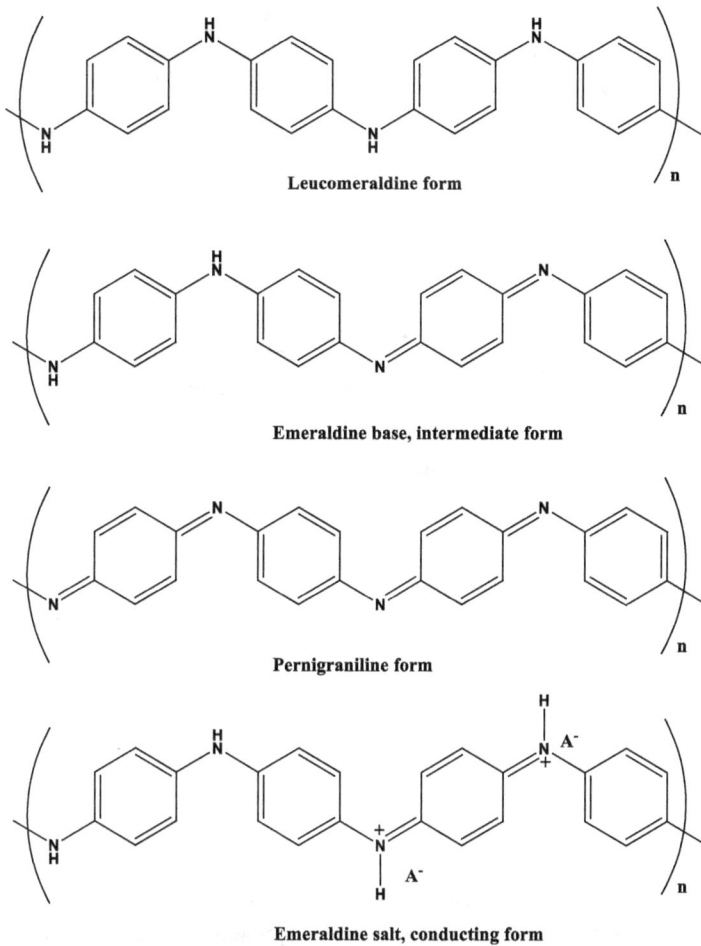

Leucomeraldine form

Emeraldine base, intermediate form

Pernigraniline form

Emeraldine salt, conducting form

Figure 6.4: Different oxidation states of PANI.

tion; however, electrochemical polymerization is the most significant and useful process, creating processable films with a rather high conductivity (~10^2 S/cm).

6.7.3 Synthesis of poly(phenylene vinylene)

McDonald and Campbell created the first synthetic version of poly(phenylene vinylene) (PPV). The materials from these early PPVs could not be processed. The most popular precursor approach, which permits better solubility and processability (Figure 6.7), was created by R. A. Wessling in 1985 [14]. This process begins with the creation of the 1,4-bis(chloromethyl)benzene bis-sulfonium salt, which is then eliminated with sodium hy-

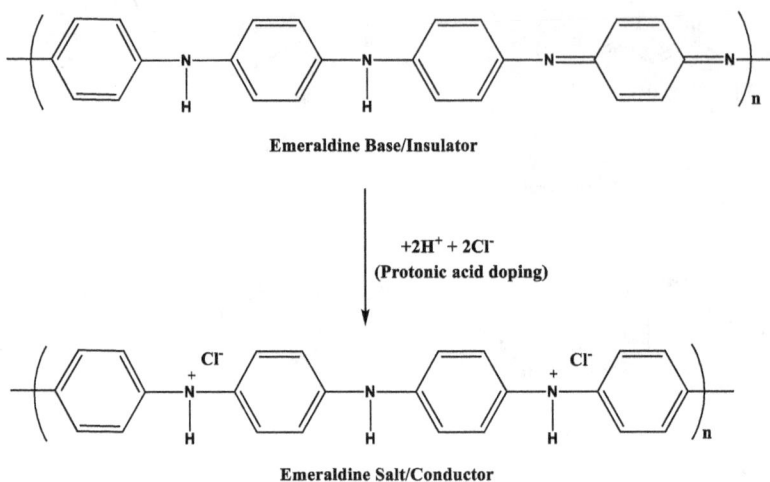

Emeraldine Base/Insulator

+2H⁺ + 2Cl⁻
(Protonic acid doping)

$+2H^+ + 2Cl^-$
(Protonic acid doping)

Emeraldine Salt/Conductor

Figure 6.5: Protonic acid doping.

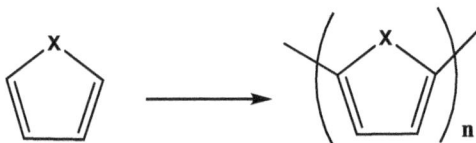

where X = O, NH, S

Figure 6.6: Heterocyclic structure.

droxide, and polymerized at low temperatures to create an aqueous solution of a fore-runner polymer. In addition to being utilized to make films, foams, or fibres, this soluble precursor polymer may also be thermally removed to yield PPV.

6.8 Properties of conducting polymers

The features of CPs include conduction, electrical, magnetic, wetting, mechanical, optical, and microwave absorption [13]. The following list includes some of the CP's attributes:

6.8.1 Electrical conducting properties

The doping level, chain arrangement, conjugation length, and sample purity all affect the conductivity of polymers. ECPs lack long-range organization and are molecular in

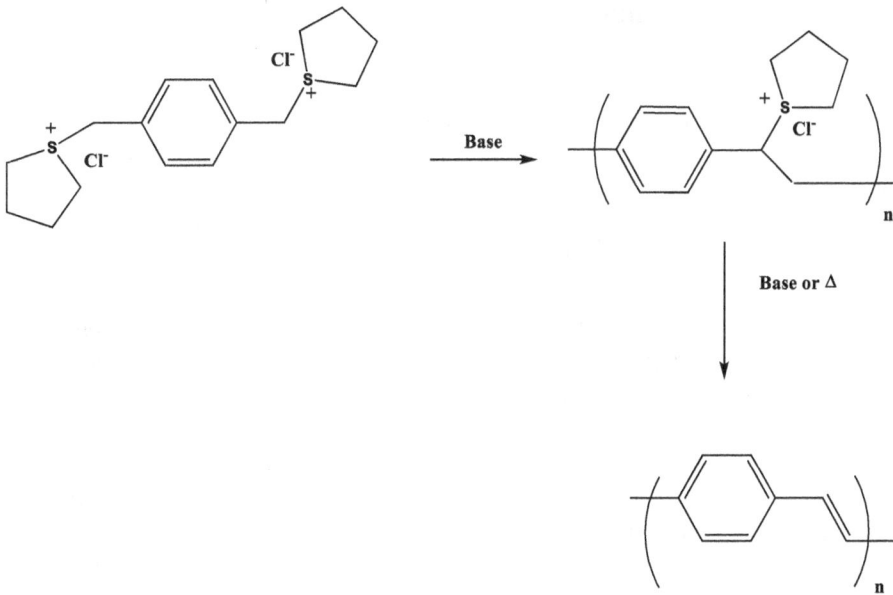

Figure 6.7: Synthesis of poly(phenylene vinylene) (PPV).

origin. Electronic motion occurs around the individual macromolecules because polymers are molecular. For polymers and inorganic semiconductors, several processes are used to achieve high conductivity. The emergence of self-localized excitons such solitons, polarons, and bipolarons is associated with greater conductivities, which are reliant on doping in the polymers. These particles are the end product of an influential contact between the charges on the chain that doping allowed. Charge carriers in CPs with degenerate ground states, like trans-polyacetylene, are charged solitons, whereas polarons are commonly produced during doping in CPs with non-degenerate ground states, such cis-PA, PPy, PTh, or PPV. These polarons then unite to form spinless bipolarons, which function as charge carriers. These are highly sought-after materials for electrically conductive applications because of the low cost of the polymers and their ability to be molecularly designed with the right properties.

Features that affect the electrical conductivity are
- Density of the charge carriers
- Their mobility
- The direction
- Presence of doping materials
- Temperature

6.8.2 Magnetic properties

Due to their exceptional magnetic characteristics and technological implications, CP's magnetism is of great interest. In order for nanomaterials with magnetic and structural capabilities to be incorporated into polymer medium, transition metal oxide nanoparticles are also essential. Nandapure et al. [15] examined the magnetic and transport properties of conducting PANI/nickel oxide. It was also demonstrated that when the wt% of nickel oxide in PANI grew, the conductivity of the PANI/nickel oxide nanocomposites declined and their magnetic enhanced. Anodic oxidation was used by Yan [16] to produce the CP/ferromagnet film. By using p-dodecyl benzenesulfonic acid sodium salt as a surfactant and dopant, a chemical process was used to create a composite of PPy with ferromagnetic behavior. The final composites' magnetic characteristics demonstrated ferromagnetic behavior such as high saturation magnetization. The best synthetic conditions allowed for the greatest conductivity of 10 S/cm to be achieved. By in situ polymerization, particles of $NiFe_{1.95}Gd_{0.05}O_4$ composites with various ferrite/PANI proportions have been effectively produced. In order to demonstrate magnetic characteristics with rising quantities of PANI, Aphesteguy [17] chose various ferrite/aniline ratios.

6.8.3 Mechanical properties

Their magnetic properties have been well explored because CPs offer valuable insights on the charge-carrying group and unpaired spins. Sulong et al. [18] looked at the mechanical properties of a nanocomposite made of carbon nanotubes, graphite, and polypropylene that was used in conducting polymer composites (CPC). The same author examined the impact of chemical functionalization and filler concentrations on the mechanical properties of the resultant CPCs. The functionalized CPC outperforms the unmodified CPC in terms of strength and elongation on the tensile and flexural tests. It has a maximum tensile strength of 35 MPa and a maximum flexural strength of 80 MPa. Moreover, the functionalized CPC possesses tougher traits than the original CPC. The instantaneous thermal isolation and uniaxial stretching of the poly sulphonium salt source has been shown to produce a thin film with uniaxial orientation that exhibits mechanical characteristics in both machine and diagonal directions. The degree of molecule orientation determines its anisotropic properties. The Young's modulus varies with draw ratio between 2.3 and 37 GPa in the machine direction and 2.3 and 0.5 GPa in the diagonal direction.

6.8.4 Microwave radiation absorbing properties

CPs have been investigated as intriguing new microwave materials because of their small density, simplicity of fabrication, and low cost. In order to examine the microwave radiation absorption properties, the team employed the free space technique to calculate the complex permittivity, complex permeability, and reflection damage in the microwave frequency region. The same author also showed how PANI accumulation may be used to achieve considerable absorption across a broad frequency range. The microwave radiation absorption properties of PANI nanocomposites incorporating TiO_2 nanoparticles were studied by Wai et al. [19,20] Using surface-initiated polymerization, PANI nanocomposites strengthened with tungsten oxide nanoparticles and nanorods may be made quickly. Magnesium alloy with microwave-absorbing shells was analyzed electrochemically using an impedance spectrometer by Guo[21] in a solution of 3.5 wt% NaCl.

6.8.5 Anticorrosion properties

Conducting polymers, along with their composites, are now employed as corrosion inhibitors on metal surfaces[1]. The ability of these composites to prevent corrosion-causing chemicals from moving over metal surfaces is what gives them their corrosion-protecting properties. According to studies, the polyaniline coating on steel surfaces creates a passive layer that stops corrosion. Steel coatings with a polyaniline epoxy mix have been the subject of studies to prevent corrosion. Composites made of polyaniline/polypyrrole and polyaniline-polypyrrole phosphotungstate were used as corrosion inhibitors on the surface of mild steel. Comparable to bare polyaniline and polypyrrole, composite films offer higher corrosion resistance. When utilized as corrosion inhibitors, polyanilines doped with TiO_2 nanoparticles (PTC) were more effective than those that were not. It has been seen that oxalate and tungstate doped-PPy are employed to prevent corrosion on aluminum. In comparison to pure PANI coating, it has been found that PANI $-$ MoO$_4^{2-}$ coating works as a superior corrosion inhibitor. Corrosion-inhibiting capabilities are present in polyaniline and its composite films. Zinc phosphate-doped PPy has been shown to provide greater corrosion resistance than undoped PPy in studies. Mild steel is protected against corrosion by composites made of poly-6-amino-m-cresol and nanoscale copper. When compared to pure polymers, these composites perform better. We observed the corrosion behaviour of samples made of 7,075 aluminum, copper-modified aluminum, polypyrrole-modified aluminum, and copper-/polypyrrole-modified aluminum.

6.9 Polymer-based nanocomposites coatings (PNCs)

Polymer-based nanocomposites are the materials of the twenty-first century, each with its unique properties, structures, and uses [22]. These abilities set them apart from customary composites. Even though they were first proposed in 1992, nothing is known about their properties and how they work. It is generally recognized that the qualities of nanofillers are significantly dependent on their dimensions, with "their critical size" providing the most benefit. In light of this, the surface area-to-volume ratio of nanocomposites is a noteworthy and vital aspect that adds to our understanding of the link between their structural traits and qualities. Because of their versatility, superior chemical compatibility, adherence (to metallic surfaces), and other qualities, polymer-based varnishes have been broadly used to guard metals against corrosion and aging throughout the years. Polymer coverings principally act as a physical barrier for the underlying metallic substrates in order to reduce corrosion, by blocking destructive (corrosive) species from retrieving the substrate exterior. Normal polymers are typically challenging to use as anti-corrosive coatings because they (1) lack the necessary porosity to prevent corrosive ions from passing through, and (2) are vulnerable to attire, surface abrasion, and scrapes. In order to survive stress, high temperatures, various chemical conditions, and hostile atmospheres without cracking or deforming, anti-corrosive polymer coatings must be created to persist and have strong adherence to the metal substrate underneath them as well as great toughness and flexibility. Engineers and scientists have experimented with a range of approaches to attain these excellent traits, changing the microstructure of polymer coatings through the design of copolymers, structural variation of polymers, hybridization, and the development of nanocomposite materials. In terms of adhesion, flexibility, and mechanical properties, schemes including physical alteration and hybrid polymers have revealed some enhancement.

6.10 Polymer nanocomposite coating (PNC) methods

The inherent qualities of the polymer nanocomposite and workable ways of applying the material to various surfaces are both necessary for coating success [22]. In this context, "workable" refers to a set of conditions that must be satisfied to ensure qualities of high coating quality, including surface consistency, intersection adhesion, width regulator, and material non-toxicity. For large-scale manufacturing, fabrication ability is an additional critical feature to reflect. In research and construction, an extensive series of coating methods have been used. Nonetheless, not all of them can be used with nanofluid polymers. By using vaporizing techniques that are commonly carried out at extremely high temperatures or energy, polymer chains can be willingly disrupted. For polymer nanocomposites, a low deposition temperature is often necessary. There are mainly four coating methods to select from for polymer nanocomposites: roll-to-roll (R2R) processing deposition, chemical and electrochemical deposition, physical vapor deposition (PVD), and chemical vapor deposition (CVD).

6.10.1 Chemical vapor deposition (CVD) method

Chemical reactions happen in the vapor phase during chemical vapor deposition, which is based upon a heating process that turns a solid objective material into a vapor (CVD). As a result, there is either a homogeneous gas phase or a deposition with particles or filaments due to the heterogeneous chemical reactions that take place on or near the heated surface. Often, CVD is performed at temperatures as high as 1,000 °C to encourage chemical reactions in the vapor phase. The solid target's chemical makeup is altered by these enormously large temperatures, changing the properties of their original material. As a consequence, a low-temperature CVD approach, operating between 350 and 700 °C was developed, utilizing inorganic and metallo-organic precursors. Activation temperatures between 200 and 400 °C are now practical when using a plasma. Hot-wall CVD (HWCVD) and hot-filament CVD (HFCVD) reactors are the most often used CVD devices for transporting a high-temperature stage because they are less costly than other coating techniques. Nevertheless, because HWCVD does not require the high temperatures needed for HFCVD, it is more suitable for the deposition of polymer nanocomposites. With the exception of very particular conditions like the deposition of nanodiamonds, these coating processes are seldom used on polymer nanocomposites.

6.10.2 Chemical and electrochemical deposition

The use of chemical and electrochemical depositions for solution analysis and electroplating has significant theoretical and practical implications. The coating goals should not be impacted by the low ionization temperature. Precursors are transformed into active species in chemical and electrochemical depositions utilizing reducing agents or input voltage applied to the polyelectrolyte medium. The four conductive polymer matrices that are most frequently used in nanocomposites are polyaniline (PANI), polythiophene (PTh), polypyrrole (PPy), and poly(3,4-ethylenedioxythiophene) (PEDOT). The cathodic reaction is divided into anodic and cathodic sites when using the electrochemical method, which requires the injection of a peripheral current. In chemical deposition, the anodic and cathodic processes are induced and occur concurrently on the workpiece. Furthermore, because redox reactions may only occur on surfaces that are catalytically active, freshly coated metallic exteriors must be sufficiently catalytically active to promote redox reactions. Electrochemical deposition using conductive polymer nanocomposites has frequently been utilized to immobilize nanofillers like dense metal colloids or specific enzymes inside the conducting polymer matrix.

Moreover, by employing bioactive thin films, biotransistors that convert analogue bio-signals into electrical signals may be made.

6.10.3 Roll-to-roll (R2R) processing deposition

Smooth electronics, including organic light-emitting diodes (OLEDs), photovoltaics (PV), and electrophoresis presentations, have been produced using roll-to-roll process-ing deposition (R2R), a low-cost industrial coating technique (EPDs). Coyle et al. [23] claim that the fundamental principle of a roll coating is the flow of fluid amongst two revolving rollers that control the wideness and uniformity of the coating. Several terms have been used in the industry for specific applications including reverse-roll coating and gravure coating.

6.10.4 Physical vapor deposition (PVD) method

The physical vapor deposition (PVD) process is often utilized to create inorganic mediums and inorganic nanoparticles for inorganic/inorganic polymer nanocomposite varnishes. A number of PVD techniques, including laser ablation, thermal evaporation, ion beam deposition, ion implantation, laser-assisted deposition, and atom beam co-sputtering, were used to generate these coatings. PVD is the name given to thin-film installation pro-cesses that need condensation of solids that have been vaporized on the surface of solids in a partially vacuum environment. Atoms or molecules are physically cleared, con-densed, and nucleated onto a substrate using the atomistic deposition technique, known as PVD, in a vacuum, low-pressure gaseous, or plasma atmosphere. Typically, plasma or ions make up the vapor phase. Reactive deposition is the addition of a reactive gas while a vapour is being deposited. In a hoover with low-pressure gaseous environment or plasma environment, the atoms or molecules are carried to the substrate's exterior in the form of a vapor. Thin films with widths ranging from a rare nanometer to a thou-sandth of a nanometer are routinely deposited using PVD methods.

6.11 Other coating methods

6.11.1 Sol–gel method

The sol–gel technique is an addition to physical deposition techniques for producing superior coatings of up to micron thickness. There are a few restrictions on using sol–gel to cover metallic subtracts, though. Numerous issues with this method exist, including cracking and thickness restrictions. The second phase of the sol–gel may be used for inorganic nanofillers such as inorganic/inorganic coatings. Inorganic sol–gel forerunners such as metal alkoxides of silicon (Si), titanium (Ti), aluminium (Al), and zirconium (Zr) are utilized to make nanocomposite coatings. Low molar mass organic molecules and a variety of oligomers have both been seen in organic phase preparations. A covering comprising silica nanoparticles or nanophases can occasionally be produced when silanes and organic molecules are combined. An acrylated phenyl phosphine oxide oligomer, urethane acrylate resin, methacryloxypropyltrimethoxysilane (MAPTMS), and tetraethyl orthosilicate (TEOS) have all been reported to be used in the production of silica nanocomposites (APPO). Inorganic nanofillers can be incorporated into either the polymeric organic medium or the inorganic medium using the electrodeposition technique, in combination with the sol–gel approach.

6.11.2 In-situ polymerization method

PNCs were created utilizing this technique, employing organic matrices such as conducting polymer or other monomers with initiators. Metals or metal oxides were used as fillers at the nanoscale. Electrodeposition, oxidising agents, or photopolymerization can all be used to produce polymers. Similar procedures include the emulsion polymerization of latex or organic matrices. Guo et al. [24] utilized in situ polymerization to produce graphene and functionalized graphen oxide (GO)-epoxy nanocomposites (GN). The packing was initially discrete in acetone using ultrasound prior to beginning the production. The dispersal was next supplemented to the epoxy medium, which had already been heated in a hoover to 50 °C. The m-phenylenediamine (MDP) was supplementary, and the solvent was vigorously stirred, after 80% of the solvent had evaporated. The liquid was placed in a stainless steel mould to create the composites, dehydrated at 60 °C for 5 h to eliminate any leftover solvent, precured in an oven for 2 h at 80 °C, and then postcured for 2 h at 120 °C. Two more GO composites are polypyrrole (PPy)/GO and PMMA/GO, which have also been described. To create PPy/GO composites, liquid–liquid interfacial polymerization was utilized. Because it is slower and more regulated than the more conventional in situ polymerization procedure, the authors chose this approach. Large volumes can also be prepared using this method.

6.11.3 Solution dispersion

PNCs supplemented with nanofillers such metal oxides, nanoclays, and CNTs are commonly made via solution dispersion. This strategy merged customary magnetic/mechanical moving methods with ultrasound-assisted churning to improve the dispersion of nanofillers in polymer mediums. Bian et al. [25] utilized solution dispersion to generate the nanocomposites poly(propylene carbonate) and (PPC)/modified GO (MGO). MGO was circulated for 30 min in 25 mL of dimethylformamide (DMF) and mechanically swirled for 10 min. The dispersion was then given PPC, and it was disturbed for 24 h at 40 °C. The solvent was let to evaportate at room temperature in a Petri plate while it was under vacuum. As hydrophobic PPC and hydrophilic GO could not cohabit, the latter had to be changed. The GO surface was modified with hydroxyl groups to improve interfacial adhesion and encourage the creation of nanocomposite materials.

6.11.4 Sprig coating and twist coating methods

The procedures of sprig coating and spin coating are typically used to confront PNCs. Using an atomizer for spray coating allowed for the enhancement of the characteristics of nanocomposite coatings. Atomized spray plasma deposition, a kind of thermal spray, may be employed with the atomizer. Spin coatings offer flat substrates uniform thin layers of material. Centrifugal force is used to disseminate the coating ingredients on the substrate while it rotates quickly. Coatings made of thin-film nanocomposite materials can be created using this method.

6.11.5 Dip Coating

In a commonly used procedure called dip coating, a substrate is submerged in a polymer nanocomposite solution and pulled up at a steady and controlled pace. Nanocomposite is applied to the substrates as they are detached from the solution. The forced pulled up rate also regulates how much polymer nanocomposite is present on the substrate surface. The benefit of this method is that it can provide a flat surface for any sort of coated substrate. This method is very beneficial for industrial applications since the solution may be recycled until the solute is exhausted or evaporates.

6.12 Different types of polymer nanocomposite coating (PNC) systems

PNCs of various forms, including (i) coatings made of carbon-based polymers, (ii) metal oxide nanocomposite varnishes, (iii) polymer-inorganic nanocomposite varnishes, and (iv) amalgam polymer nanocomposite varnishes, are made by utilizing nanofillers.

6.12.1 Polymer-inorganic nanocomposite coating system

The surfaces of nanoparticles can be changed to make them more tolerant of the polymer matrix by polymerizing polymers. A modest esterification process was used by Feng et al. [26] to graft poly(methylmethacrylate) (PMMA) onto the exterior of SiO_2. They recommended grafting nanoparticles with different polymers using this simple esterification approach. The dispersity and similarity of nanofillers in polymers have been improved by in situ polymerization of nanoparticles. By producing inorganic nanoparticles directly inside the polymer matrix's pores, the in situ particle creation technique increases the miscibility of nanofillers. Many inorganic oxide nanoparticles may be produced in polymer nets using the sol–gel process, which is a popular in situ particle-making approach. The capacity of the polymer matrix to preserve the evenly dispersed state of the nanoparticles efficiently inhibits agglomeration. Silica, titania, or alumina nanoparticles have been added to polymers utilizing in situ sol–gel to create nanocomposite coatings. Metal oxide nanoparticles are often created in polymer matrixes using TEOS [27], tetrabutyl titanate (TBOT) [28], and other organometallic precursors.

6.12.2 Polymer-metal oxide nanocomposite coating system

An exciting new technique for many uses is polymer–metal oxide nanocomposite varnish, which is strengthened by nanosized SiO_2, TiO_2, Al_2O_3, ZnO, and iron oxide.

6.12.2.1 Nano-SiO_2

SiO_2 nanoparticles have been extensively employed in a number of applications containing paints and polymers, due to their exceptional properties of high rigidity, lower refractive index, facile functionalization, and remarkable chemical and thermal steadiness. Production of highly reliable silica nanoparticle is possible using the flame spray pyrolysis process, the reverse microemulsion method, and the sol–gel approach. SiO_2 nanoparticles can recover the tensile strength, anti-corrosion capabilities, and mechanical properties of polymeric varnishes. SiO_2 nanoparticles in nanocomposite coatings therefore show great potential. It has been extensively researched on the method to improve anti-corrosion properties by adding SiO_2 nanoparticles to polymer matrix. SiO_2 can significantly reduce the amount of total free volume and disaggregation during curing when added to resin coatings. The use of porous SiO_2 nanoparticles as a reservoir for inhibitors to slow the corrosion process is another possible application for SiO_2.

6.12.2.2 Nano-TiO$_2$

It has been determined that the continually increasing tortuosity of the diffusion channel of chemically hostile substances is the mechanism underlying polymer/TiO$_2$ nanocomposite coatings. A coating protecting outdoor cultural treasures was created by L. D'Orazio et al. [29] using a TiO$_2$/poly(carbonate urethane) nanocomposite. The varnish with 1 wt% TiO$_2$ nanoparticles has outstanding self-cleaning and toughness characteristics. Conferring to Khademian et al. [30], poly (vinyl acetate) varnishes properties for heat steadiness, dispersion, and anti-corrosion are considerably improved when they include 1.5 wt% PANI-TiO$_2$ nanohybrid. In order to dramatically improve the anti-corrosion presentation of cold-rolled steel electrodes, Weng et al. [31] created well-dispersed electroactive polyimide-TiO$_2$ (EPTs) hybrid nanocomposite varnishes.

6.12.2.3 Nano-Al$_2$O$_3$

By incorporating nanosized Al$_2$O$_3$ into a polymer to form a polymer nanocomposite system, the varnish may acquire outstanding chemical and automated confrontation, mechanical power, UV, and temperature steadiness [32], among other features. Mechanical synthesis [33], converse microemulsion [34], spluttering [35], hydrothermal [36], combustion [37], sol–gel, and other procedures for making nanoscale alumina are discussed in the literature. Al$_2$O$_3$ nanoparticles are another favourable material that is extensively used to alter coats for corrosion conflict [38]. When stiff spherical Al$_2$O$_3$ nanoparticles [39] are present in a polymer background at a high enough attentiveness, coating pores may be filled, particle-matrix contacts are reinforced, and coating stiffness and corrosion conflict are greatly improved. Al$_2$O$_3$, which has a low surface energy and strong chemical activity, has a restricted potential for strengthening due to the nanofillers' reduced ability to dispersal and NP's inclination to aggregate. Al$_2$O$_3$ NPs can have their surfaces modified, for example by chemical surface functionalization by means of silane coupling agents and surfactant grafting, as a workable solution to this problem.

6.12.2.4 Nano-ZnO

ZnO could augment the thermal conductivity, dielectric properties, mechanical characteristics, thermal constancy, and antibacterial properties of the polymer matrix. High-temperature spluttering, molecular beam epitaxy, chemical vapour deposition, electrophoretic deposition, microwave-assisted approach, solvothermal, hydrothermal processes, and chemical bath deposition are some of the approaches that may be utilized to produce numerous morphologies of nano-ZnO. Pinholes and voids in polymeric coatings might be blocked by nanoscale ZnO with a high aspect ratio and tiny size, keeping corrosion specimens from entering the coating film's pores. In addition, ZnO nanoparticles can enhance corrosion resistance by zig-zagging the diffusion paths of the corrosive species. Numerous studies have demonstrated the important impact ZnO nanoparticles play in enhancing the anti-corrosion capabilities of predictable polymer or CP coatings. 6 wt% ZnO was added. The addition of nanoparticles to the coating's polymer boosted its super hydrophobicity, which assisted in corrosion control.

6.12.2.5 Nano-iron oxide

There are many iron oxide nanocrystals in nature, and they frequently originate in the forms of FeO, Fe_2O_3, and Fe_3O_4. Iron oxide nanoparticles can be added to PANI coatings, which combine the mechanical and conductive properties of nanocomposites to increase corrosion and abrasion resistance. Additionally, since they may potentially form a p–n type junction at the Fe_3O_4/PANI interface, Fe_3O_4/PANI nanocomposite display effective corrosion prevention when applied as a Fe_3O_4/PANI (1:1) nanocomposite coating to carbon steel in 3.5% NaCl for 60 min. The corrosion confrontation of epoxy or PU is expected to be greatly increased by the insertion of extremely modest amounts of modified Fe_3O_4 nanoparticles. The ability of Fe_3O_4 nanoparticles to increase adhesion, barrier properties, and the ability to suppress anodic corrosion are all related to coatings.

6.13.3 Carbon-based polymer nanocomposite coating system

6.13.3.1 Carbon black (CB)

A typical dye used in ink and painting is carbon black (CB) [40], which is inexpensive. High electrical and thermal conductivity, a huge superficial extent, and the capability to be easily functionalized are all features of CB nanoparticles. CB nanoparticles have a propensity to aggregate and, like other nanoparticles, have a high external action and large exact surface part. This may be avoided with appropriate polymer content control and surface functionalization. Additionally, CB nanoparticles have strong UV stabilizing qualities, extending the lifespan of epoxy.

6.13.3.2 Carbon nanotubes (CNTs)

Many research works has revealed that combining CNTs with polymers result in a nanocomposite with exceptional features. The methods that are most often utilized to create CNTs/polymer nanocomposites include solution mingling, melt involvement, and in situ polymerization. High-performance coatings are dependent on the uniform dispersal of CNTs into the polymer matrix, according to several studies. CNTs clump together into enormous bundles because of their strong surface repulsion, intrinsic electrical organization, and incompatibility with the polymer substrate. The most often used techniques for making them compatible include non-covalent functionalization with compatible additives, covalent functionalization, and fault functionalization using concentrated acids. Souto and colleagues demonstrated how the distinctive electrical properties of CNTs and CPs, such as CNTs/PANI, may be coupled to produce improved epoxy coating corrosion resistance. Due to the interfacial interactions between PANI and CNT, which improve electrochemical activity and passivation security, PANI/CNT nanocomposites exhibit excellent redox capacities, according to Rui et al. [41] By providing effective anodic shelter and a physical obstacle, PANI/CNT nanocomposites improved the protective effect of acrylate-amino resin (AA). Deyab et al. [42] claim that aluminum bipolar plates were shielded from the acidic fuel cell fluid using an anti-corrosive covering constructed of PANI nanocomposite with 0.8% CNT. Polydopamine/CNT nanocomposite coatings were suggested by Hong et al. [43] as a way to slow down carbon steel corrosion. The physical wall on the exterior and the CNT's unique electrical properties that inhibit cathodic partial reaction are both important factors in the coating's capacity to decrease corrosion.

6.13.3.3 Graphene (GN) and its byproducts

GN has exposed significant potential in a variety of applications. Graphene is most often produced using the redox [41], CVD [44], and micromechanical stripping processes. The features of graphene-based nanocomposites are significantly dependent on the nanofiller's degree of dispersion in the polymer. To produce evenly distributed graphene-based nanocomposites, many methods have been used to produce graphene byproducts such as graphene oxide (GO), reduced graphene oxide (rGO), and electrochemically exfoliated graphene (ECG). For the most part, in situ polymerization, solution mixing, and melt intercalation techniques [45] have been used to physically or chemically combine graphene nanosheets or its byproducts with polymer to create GN/polymer nanocomposites. Several investigations have identified GN and its unique lamellar-structured derivatives as superior nanofillers for high-performance anti-corrosive polymer coatings. The improved anti-corrosion property of graphene-based polymer nanocomposite coatings is most likely the result of increased contact area between the modified graphene or GO nanosheets and polymers, which are beneficial in lengthening the diffusion pathway of corrosive molecules in the matrix and provide an excellent physical barrier.

6.13.4 Hybrid polymer nanocomposite coatings

Research interests have been greatly aroused by nanohybrid architectures, which integrate several functional constituents into a distinct fundamental entity. It is anticipated that the performance of coatings would improve when different inorganic nanoparticles are combined in nanohybrids and then mixed into polymers. PANI-CeO$_2$-GO nanohybrid particles exhibit extremely active corrosion inhibition. The potential for corrosion resistance is quite significant when CPs and CNTs are added to conventional coatings.

6.14 Aqueous polymer inhibitor

6.14.1 Chitosan-based corrosion inhibitor

Chitosan (Cht), a linear polysaccharide, has a variety of polar functional groups on its linear polymer chain, including amino, hydroxyl, acetyl, and hydroxymethyl (Figure 6.8) [46]. These functional groups enable chitosan to be easily adsorbed on metal surfaces and contribute to its improved solubility in polar solutions. Additionally, because chitosan's amino group has a large degree of protonation in neutral and acidic environments due to its high acidity coefficient (pKa) value of 6.5. Chitosan is water soluble due to this characteristic, which also allows it to adhere well to negatively charged metal surfaces. Chitosan can also be changed to increase its resistance to corrosion and biological toleration. Chitosan and its derivatives are widely used in the food, biomedical, agricultural, environmental, pharmaceutical, and other sectors because they are non-toxic, protective of the environment, non-sensitizing, biocompatible, and degradable.

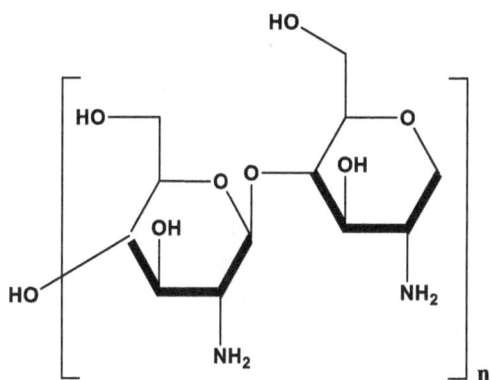

Figure 6.8: Structure of chitosan.

The primary chain of chitosan can be revised by the accumulation of polar functional groups, which improves the material's ability to inhibit corrosion and improve its adherence to steel surfaces. For the first time, 4-amino-5-methyl-1,2,4-triazole-3-mercaptan was used by Chauhan et al. to chemically adapt chitosan corrosion inhibitor. By assessing the improved corrosion inhibitor's ability to effectively block corrosion on the exterior of carbon steel in a solution of 1 M HCl, it was noted that its adsorption conformed to the Langmuir isotherm, showing both physical and chemical adsorption.

6.14.2 Corrosion inhibitor of cellulose and its products

It is common to find cellulose, a naturally occurring polymer high in carbs, in raw materials such as hemp, wood, and cotton (Figure 6.9). It is now one of the primary bases of composites as eco-friendly metal corrosion inhibitors due to the environmental preservation of its source.

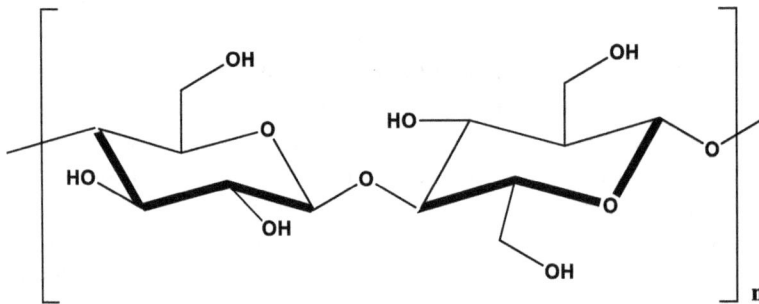

Figure 6.9: Structure of cellulose.

Nwanonenyi et al. [47] looked into the HPC's ability to adsorb aluminum and prevent corrosion in a 0.5 M HCl and 2 M H_2SO_4 medium. The findings demonstrate that HPC is a varied inhibitor and has a substantial anodic impact in both corrosion solutions. The sustained release rate increases as concentration rises, but it becomes less sustainable as temperature rises. Hassan and Ibrahim [48] investigated how methylcellulose (MC) prevented corrosion on magnesium (Mg) metal in hydrochloric acid. The results show that the rate of corrosion inhibition on Mg metal surfaces increases with increasing inhibitor concentration and decreases with increasing temperature. The isotherm models proposed by Langmuir and Freundlich are consistent with its adsorption. The MC inhibitor is able to delay the corrosion of the metal surface, which can lead to OH linking on the metal surface and adsorbence on the Mg metal exterior. As cellulose possesses a variety of polar functional groups with high electron densities but is only moderately soluble in polar solvents, polar substituents are frequently needed to modify the cellulose molecular structure. These polar functional groups can restore the effectiveness of corrosion inhibitors to stop degradation by increasing the molecular size and solubility of cellulose.

6.14.3 Starch corrosion inhibitor

Inexpensive, non-toxic, biodegradable, and with other advantages, starch is a natural, renewable polymer carbohydrate. Among other industries, it is extensively used in the food, paper, and pharmaceutical industries. It is mostly present in grains such as wheat, potatoes, and maize, and is made up of multiple glucose components. Nevertheless, natural starch is easily bound together; it has concerns with ageing and biological degradation, and it is infrequently used directly as a corrosion inhibitor due to its weak ability to resist corrosion and because of its low water solubility. The presence of amylose and amylopectin, however, leaves a lot of potential for modification. Nwanonenyi et al. [49] tested millet starch (MS) for its capacity to prevent corrosion in a 0.5 M HCl solution at 30 1 °C. According to the results, MS definitely prevents low-carbon steel from corroding in acidic solutions, and the degree to which it does so is related to the MS concentration. Wu et al. [50] calculated the efficacy of carboxymethyl starch (CMS) to prevent corrosion on low-carbon steel in a solution of hydrochloric acid. The outcomes demonstrated that in a 5% HCl solution at 24–60 °C, CMS successfully prevents corrosion. As the concentration of CMS approaches 2 g/L, the low-carbon steel corrosion rate reduces to 11.19 mm/year and the corrosion inhibition rate increases to 80.61%. Because of its poor ability to resist corrosion and low water solubility, conventional starch must be replaced regularly and cannot be used as a direct inhibitor of metal corrosion.

6.14.4 Arabic gum inhibitor

Arabic gum (GA) is a communal type of natural gum (Figure 6.10). Owing to its cheap price, non-toxicity, and high biocompatibility, it is commonly utilized as a stabilizing agent, film-forming agent, an emulsifier, and a viscosity control agent in the food and coating industries. GA is a complex combination of glycoproteins, oligosaccharides, and arabinogalactan. Its highly branched complex polysaccharide structure results in a hydrodynamic volume that is very small, which enables more functional groups to be consumed on the metal exterior and exhibits some corrosion inhibition potential [51].

Shen et al. [52] investigated the effectiveness of GA's corrosion prevention on 1,018 carbon steel submerged in brine. The upshots display that the capacity of metals to resist corrosion via chemical adsorption is mostly enhanced by the GA's surface, and that the corrosion inhibition rate grows with increasing GA concentration. Its adsorption tracks the Langmuir isotherm and can reduce corrosion by up to 94% when exposed to ambient conditions. In a high-pressure CO_2 salt water environment, Pal-

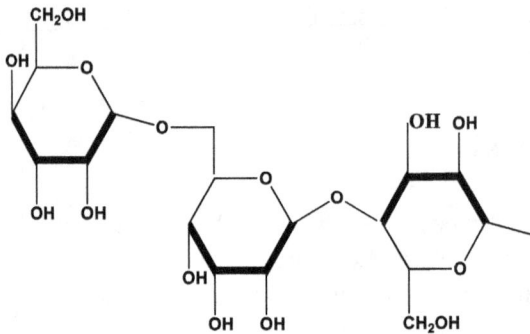

Figure 6.10: Structure of Arabic gum.

umbo et al. [53] examined the effect of GA on N80 carbon steel in terms of corrosion prevention. The results show that N80 carbon steel's corrosion inhibition rate increases as the partial pressure and concentrating of CO_2 grow. With $PCO_2 = 4$ MPa, the highest rate of corrosion inhibition is 84.53% after 24 h. GA's rate of preventing corrosion was 74.41% when the partial pressure was sustained for 168 h. In addition to study into how GA's corrosion inhibition affects different kinds of corrosive environments and materials, it is also understood that halide ions are one of the key factors influencing the presentation of GA's corrosion inhibition. Arabic gum also performs only marginally well as a biological corrosion inhibitor. Nevertheless, recent studies on Arabic gum inhibitor have mostly focused on supplemental corrosion inhibition.

6.14.5 Synthetic polymer corrosion inhibitor

The term "synthetic polymers" often refers to polymers created in a lab. A certain anti-corrosion action is provided by their large molecular weight, which allows them to retain a good coating effect even at less concentrations. As a result, the current application status of two commonly used corrosion inhibitors: resin-based corrosion inhibitor and polyethylene glycol corrosion inhibitor.

6.14.6 Resin-based corrosion inhibitor

Epoxy resin, which is also referred to as epoxy polymer, comes in a variety of forms, has a sizeable curing/amendment space, is resilient to marginal polar substituent adhesion, has short curative contraction, high mechanical strength, is resilient to acid and alkali conflict, etc. Heteroatoms like O, N, and s are easily protonated in epoxy resin, which makes it easier for epoxy polymer to dissolve in polar electrolyte [54].

Epoxy resin's polymer features also allow it to create a significant amount of surface exposure and a strong adsorption capability on the surface of metals.

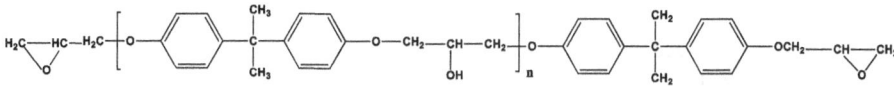

Figure 6.11: Structure of epoxy resin.

Moreover, an epoxy ring may experience a ring opening reaction, following protonation in a strong acidic solution, to produce –OH substituents with considerable polarization that interact with water (Figure 6.11). Due to their tall molar mass, which prevents them from dissolving in polar electrolytes, waterborne epoxy resins are currently primarily used as inhibitors of metal corrosion. Hsissou et al. [55], in 2020, examined the efficacy of diglycidyl aminobenzene (DGAB) epoxy prepolymer to inhibit carbon steel corrosion in acid. The outcomes show that DGAB has strong metal adsorption and corrosion blocking qualities. The Langmuir adsorption isotherm is followed by the adsorption process, and the corrosion inhibition rate may go up to 95.9% at concentrations of 10^{-3} M. A macromolecular epoxy coating (DGEDDS-MDA), based on bisphenol s diglycidyl ether, was made by Dagdag et al. using methylene diphenylamine (MDA) (DGEDDS). The core of the DGEDDS-electron-rich MDA was shown to have a high capacity for adsorption to the surface of Fe (110) and an adsorption energy of 300.67 kJ/mol. The most popular type of metal corrosion inhibitor nowadays is resin-based, which has a high resistance to salt, alkali, and acids. Moreover, the epoxy resin has a significant alteration area that may be utilized to dope various functional nanomaterials to increase its capacity for preventing corrosion.

6.14.7 Polyethylene glycol corrosion inhibitor

A high-molecular weight polymer having a molecular weight of 200 to 10,000 is polyethylene glycol (PEG). (Figure 6.12). PEG has several uses in the chemical, pharmaceutical, food, and other sectors because of its excellent dispersion, solubility, adhesion, and low cost of raw materials. PEG and its derivatives have recently been used as metal corrosion inhibitors because of their outstanding thermal, chemical, and recyclable stabilities.

Figure 6.12: Structure of polyethylene glycol (PEG).

Using PEG with unlike average molar masses (mw = 1,200, 4,000, and 6,000), Deyab and Abd El-Rehim [56] examined the corrosion protection effect of PEG on carbon steel in 1.0 M butyric acid condition. The outcomes display that the degree of corrosion of carbon steel in butyric acid shrinks with increasing PEG concentration and molar mass, but increases with increasing temperature. PEG employs physical adsorption, which trails the Temkin adsorption isotherm, to prevent carbon steel from corroding. Awad [57] looked at how the PEG molecular structure and its corresponding corrosion inhibition rate are related to one another. The rate at which the tested compound's corrosion process is slowed down is strongly correlated with its quantum chemical characteristics, according to quantum chemical analysis. There is a relationship between the inhibitor's degree of inhibiting corrosion and the reaction activity's function, E. As E decreases, the inhibitor's reactivity to the metal surface increases. In a manner similar to this, rising adsorption energy increases the steadiness of the generated composite and the pace at which corrosion is inhibited. Polyethylene glycol is not a very effective metal corrosion inhibitor. It is widely used in combination with additional chemical reagents, and reduces the hydrogen reduction reaction during the corrosive process of metals, increasing the corrosion confrontation of the coat.

6.15 Drawbacks and future perspectives

– The potential harm that nanoparticles might do to the environment and to people's health has recently received a lot of attention. The worry stems from the idea that artificial nanoparticles are quite unstable and readily change into a fairly stable state. This is a result of their tall external energy, which renders them very unbalanced and out of balance. Because of this, they could experience changes when exposed to the atmosphere, or interact with another molecule to create stable combinations that might be bad for the environment and people.
– Nanoparticles discharged into the atmosphere can be absorbed by plant roots and transferred to the stems and leaves through the vascular systems. The ingested nanoparticles may interact with plant chemicals and undergo transformation. In recent years, authors have made an effort to show how nanoparticles may change. It has been demonstrated that CuO may partially change into Cu^{2+} and Cu^+ species, and that ZnO nanoparticles can partially transform into Zn^{2+} species such as Zn-citrate, Zn-phosphates, Zn-nitrate, Zn-histidine, and Zn-phytate. The main problem with the advancement of polymer nanocomposites for corrosion guard is these recent discoveries.
– It is not advised to use polymer amalgams and nanocomposites as inhibitory films or defensive coverings in hostile conditions that contain CO_2 and H_2S as well as at extremely high temperature. High-pressure gases can easily infiltrate polymer coatings and films, posing a danger to interface corrosion damage. Polymers

disintegrate at very high temperatures. In actuality, they are used in low-corrosion situations when the temperature is below 120 °C.

– Despite the aforementioned concerns, polymer composites and nanocomposites continue to provide an accessible way to combat the threat of metal corrosion. Future research should concentrate on creating technologies that will expand the areas in which polymer composites and nanocomposites may be used to defend against corrosion. The current chapter suggests that naturally occurring polymers are not being used to their full potential. This class of polymers should be further investigated since they are more lucrative and environmentally beneficial. Rarely is the durability and reactivity of nanoparticles on integration into polymer matrices taken into account. It would be crucial to fully comprehend this element, especially during processing.

6.16 Conclusions

There are many solutions for utilizing CP to protect metals and composites against corrosion. It provides effective adhesion to the metal substrate and offers long-lasting functionality while taking the environment into account. The findings of the analysis of the shielding effectiveness of various CP-based coverings in several configurations are quite encouraging, recognizing the limitations on the usage of heavy metals and chromatin processes as dangerous to human health and the environment. The science of intelligent coatings, the terminology lately used to describe multifunctional coatings that may provide the metal substrate with more than one function, is the foundation for recent improvements in CP-based coatings. In this chapter, we followed techniques and tactics while focusing on applicational examinations of several conducting polymers.

Nanotechnology is developing quickly and as a result, PNC methods are becoming better, more inexpensive, cleverer, and more effective. PNC approaches are projected to be beneficial for self-healing, self-repair, and as anti-corrosion coatings. To protect metals from corrosion, some coating methods have lately been used. Corrosion prevention is easier when PNC procedures are used. Certain corrosion inhibition methods have countless inhibitory properties, but they are nevertheless subject to limitations due to time, temperature, and other factors. When the layer of metals covered by the polymer-based nanocomposite is ripped by an intensely corrosive atmosphere, the nanofillers/nanomatrix included in the nanocomposite discharge active components, mending the damaged surface, and postponing the corrosion procedure.

For electrochemical sensing applications, conducting polymers – polyaniline and polypyrrole – have been employed successfully. Due to their higher surface area or higher surface-to-volume ratio for the passage of analyte gas molecules into and out of the polymer matrix when compared to their bulk form, conducting polymer nanostructures offer excellent possibilities in sensing actions. Due to the inter-domain distance, the conducting polymer's shape and film thickness are crucial to the sensing action and will lessen the analyte gas contact with the polymer. Due to its reversible doping method, polyaniline exhibits superior sensing behavior compared to others.

Most coatings have a physical barrier, electrochemical protection, or a mix of the two as part of their defence system. The monomer-conducting polymer coatings prevent corrosion on steel by forming physical barriers and having great electrochemical behaviour, but they are not particularly good at adhesion, mechanical strength, or durability behavior. Conducting polymers that have nanoparticles added to them often have improved electrochemical anti-corrosion properties as well as adhesion, thermal stability, mechanical strength, and hydrophobic qualities, in addition to filling the holes to increase the barrier effect. The nanopolymer composite coatings exhibit outstanding anti-corrosion performance and extended endurance when compared to other coatings.

References

[1]   Sharma S, Sudhakara P, Omran AAB, Singh J, Ilyas RA. Recent trends and developments in conducting polymer nanocomposites for multifunctional applications. Polymers (Basel) 2021;13:1–31. https://doi.org/10.3390/polym13172898.

[2]   Zarras P, Anderson N, Webber C, Irvin DJ, Irvin JA, Guenthner A, et al. Progress in using conductive polymers as corrosion-inhibiting coatings. Radiat Phys Chem 2003;68:387–94. https://doi.org/10.1016/S0969-806X(03)00189-0.

[3]   Engineering G, Kozhikode C The international conference on emerging trends in engineering corrosion protection by conducting polymers : a review. Dept chem Eng 2021:24–6.

[4]   Deshpande PP, Jadhav NG, Gelling VJ, Sazou D. Conducting polymers for corrosion protection: a review. J Coatings Technol Res 2014;11:473–94. https://doi.org/10.1007/s11998-014-9586-7.

[5]   Nezakati T, Seifalian A, Tan A, Seifalian AM. Conductive polymers: opportunities and challenges in biomedical applications. Chem Rev 2018;118:6766–843. https://doi.org/10.1021/acs.chemrev.6b00275.

[6]   Tallman DE, Spinks G, Dominis A, Wallace GG. Electroactive conducting polymers for corrosion control: part 1. general introduction and a review of non-ferrous metals. J Solid State Electrochem 2002;6:73–84. https://doi.org/10.1007/s100080100212.

[7]   Nautiyal A, Qiao M, Ren T, Huang TS, Zhang X, Cook J, et al. High-performance engineered conducting polymer film towards antimicrobial/anti-corrosion applications. Eng Sci 2018;4:70–8. https://doi.org/10.30919/es8d776.

[8]   Umoren SA, Solomon MM. Protective polymeric films for industrial substrates: a critical review on past and recent applications with conducting polymers and polymer composites. Nanocomposites 2019. https://doi.org/10.1016/j.pmatsci.2019.04.002.

[9]   Trivedi DC, Dhawan SK. Grafting of electronically conducting polyaniline on insulating surfaces. J Mater Chem 1992;2:1091–6. https://doi.org/10.1039/jm9920201091.

[10]   Umoren SA, Solomon MM. Synergistic corrosion inhibition effect of metal cations and mixtures of organic compounds: a review. J Environ Chem Eng 2017;5. https://doi.org/10.1016/j.jece.2016.12.001.

[11] Jeyaprabha C, Sathiyanarayanan S, Venkatachari G. Effect of cerium ions on corrosion inhibition of PANI for iron in 0.5 m h 2 SO 4. Appl Surf Sci 2006;253:432–8. https://doi.org/10.1016/j.apsusc.2005.12.081.

[12] Ebuka Arthur D, Achika Jonathan POA, CA. A review on the assessment of polymeric materials used as corrosion inhibitor of metals and alloys. Int J Ind Chem 2013;4:1–9. https://doi.org/10.1186/2228-5547-4-2.

[13] Kumar R, Satyendra Singh BCY, R. Conducting polymers: synthesis, properties and applications. Conduct Polym Synth Prop Appl 2013;2:1–358. https://doi.org/10.17148/IARJSET.2015.21123.

[14] Wessling RA. The polymerization of xylene bisdialkyl sulfonium salts. Org Chem Sulfur 1977;66:473–525. https://doi.org/10.1002/POLC.5070720109.

[15] Nandapure BI, Kondawar SB, Salunkhe MY, Nandapure AI. Magnetic and transport properties of conducting polyaniline/nickel oxide nanocomposites. Advanced Materials Letters 2013;4:134–40. https://doi.org/10.5185/amlett.2012.5348.

[16] Yan F, Xue G, Chen J, Lu Y. Preparation of a conducting polymer / ferromagnet composite ® lm by anodic-oxidation method. 2001;123:17–20.

[17] Aphesteguy JC, Bercoff PG, Jacobo SE. Preparation of magnetic and conductive ni-gd ferrite-polyaniline composite. Phys B Condens Matter 2007;398:200–3. https://doi.org/10.1016/j.physb.2007.04.018.

[18] Sulong AB, Ramli MI, Hau SL, Sahari J, Muhamad N, Suherman H. Rheological and mechanical properties of carbon nanotube/graphite/ss316l/ polypropylene nanocomposite for a conductive polymer composite. Compos Part B: Eng 2013;50:54–61. https://doi.org/10.1016/j.compositesb.2013.01.022.

[19] Phang SW, Tadokoro M, Watanabe J, Kuramoto N. Microwave absorption behaviors of polyaniline nanocomposites containing tio2 nanoparticles. Curr Appl Phys 2008;8:391–4. https://doi.org/10.1016/j.cap.2007.10.022.

[20] Phang SW, Hino T, Abdullah MH, Kuramoto N. Applications of polyaniline doubly doped with p-toluene sulphonic acid and dichloroacetic acid as microwave absorbing and shielding materials. Mater Chem Phys 2007;104:327–35. https://doi.org/10.1016/j.matchemphys.2007.03.031.

[21] Zhu J, Wei S, Zhang L, Mao Y, Ryu J, Karki AB, et al. Polyaniline-tungsten oxide metacomposites with tunable electronic properties. J Mater Chem 2011;21:342–8. https://doi.org/10.1039/c0jm02090g.

[22] Mathai PSS S. Polymer-based nanocomposite coating methods: a review s. J Sci Res 2010;2:513–24. https://doi.org/10.3329/jsr.v14i3.58338.

[23] Coyle DJ, Macosko CW, Scriven LE. The fluid dynamics of reverse roll coating. AIChE J 1990;36:161–74. https://doi.org/10.1002/aic.690360202.

[24] Guo Y, Bao C, Song L, Yuan B, Hu Y. In situ polymerization of graphene, graphite oxide, and functionalized graphite oxide into epoxy resin and comparison study of on-the-flame behavior. Ind Eng Chem Res 2011;50:7772–83. https://doi.org/10.1021/ie200152x.

[25] Bian J, Wei XW, Lin HL, Gong SJ, Zhang H, Guan ZP. Preparation and characterization of modified graphite oxide/poly(propylene carbonate) composites by solution intercalation. Polym Degrad Stab 2011;96:1833–40. https://doi.org/10.1016/j.polymdegradstab.2011.07.013.

[26] Feng L, He L, Ma Y, Wang Y. Grafting poly(methyl methacrylate) onto silica nanoparticle surfaces via a facile esterification reaction. Mater Chem Phys 2009;116:158–63. https://doi.org/10.1016/j.matchemphys.2009.03.007.

[27] Pulcinelli SH, Santilli CV, Hammer P. Structure and Properties Of Epoxy-Siloxane-Silica Nanocomposite Coatings for Corrosion Protection, 2017. https://doi.org/10.1016/j.jcis.2017.11.069.

[28] Yan F, Jiang J, Chen X, Tian S, Li K. Synthesis and characterization of silica nanoparticles preparing by low-temperature vapor-phase hydrolysis of sicl4. Ind Eng Chem Res 2014;53:11884–90. https://doi.org/10.1021/ie501759w.

[29] Abbate M, D'Orazio L. Water diffusion through a titanium dioxide/poly(carbonate urethane) nanocomposite for protecting cultural heritage: interactions and viscoelastic behavior. Nanomaterials 2017;7:https://doi.org/10.3390/nano7090271.

[30] Khademian M, Eisazadeh H. Preparation and characterization emulsion of pani-tio2 nanocomposite and its application as anti-corrosive coating. J Polym Eng 2015;35:597–603. https://doi.org/10.1515/polyeng-2014-0272.

[31] Weng CJ, Huang JY, Huang KY, Jhuo YS, Tsai MH, Yeh JM. Advanced anti-corrosive coatings prepared from electroactive polyimide-tio2 hybrid nanocomposite materials. Electrochim Acta 2010;55:8430–8. https://doi.org/10.1016/j.electacta.2010.07.063.

[32] Wang K, Ogier P, Tjiu CW, He C. Morphology and mechanical properties of epoxy/alumina nanocomposites. Key Eng Mater 2006;312:233–6. https://doi.org/10.4028/www.scientific.net/kem.312.233.

[33] Ardeshir khazaei 1*, sabereh nazari2, gholamreza karimi2, esmaeil ghaderi1, khadijeh mansouri moradian1 ZB and SN, 1department. Synthesis and Characterization of γ-Alumina.pdf n.d.

[34] Ke-long HUANG, Liang-guo YIN, Su-qin Lc LIU. Preparation and formation mechanism. 2007;2003:1–5. https://doi.org/10.1016/S1003-6326(07)60147-2.

[35] Wu S, Han H, Tai Q, Zhang J, Xu S, Zhou C, et al. Improvement in dye-sensitized solar cells employing tio2 electrodes coated with al2o3 by reactive direct current magnetron sputtering. J Power Sources 2008;182:119–23. https://doi.org/10.1016/j.jpowsour.2008.03.054.

[36] Patra AK, Arghya Dutta ABD. Self-assembled mesoporous γ-al2o3 spherical nanoparticles and their efficiency for the removal of arsenic from water.pdf. n.d.

[37] Laishram K, Mann R, Malhan N. A novel microwave combustion approach for single step synthesis of α-al 2O 3 nanopowders. Ceram Int 2012;38:1703–6. https://doi.org/10.1016/j.ceramint.2011.08.044.

[38] Elangovan N, Srinivasan A, Pugalmani S, Rajendiran N, Rajendran N. Development of poly (vinylcarbazole)/alumina nanocomposite coatings for corrosion protection of 316L stainless steel in 3.5% nacl medium. J Appl Polym Sci 2017;134:1–12. https://doi.org/10.1002/app.44937.

[39] Balaraju JN. Influence of particle size on the microstructure, hardness and corrosion resistance of electroless ni-p-al2o3 composite coatings. Surf Coat Technol 2006;200:3933–41. https://doi.org/10.1016/j.surfcoat.2005.03.007.

[40] Ali M, Lin L, Cartridge D. High electrical conductivity waterborne dispersions of carbon black pigment. Prog Org Coat 2019;129:199–208. https://doi.org/10.1016/j.porgcoat.2018.12.010.

[41] Rui M, Jiang Y, Zhu A. Sub-micron calcium carbonate as a template for the preparation of dendrite-like PANI/CNT nanocomposites and its corrosion protection properties. Chem Eng J 2020. https://doi.org/10.1016/j.cej.2019.123396.

[42] Deyab MA. Corrosion protection of aluminum bipolar plates with polyaniline coating containing carbon nanotubes in acidic medium inside the polymer electrolyte membrane fuel cell. J Power Sources 2014;268:50–5. https://doi.org/10.1016/j.jpowsour.2014.06.021.

[43] Hong M, Park Y, Kim T, Kim K, Kim J, J-g K. Polydopamine/carbon nanotube nanocomposite coating for corrosion resistance. 2020.

[44] Li X, Cai W, An J, Kim S, Nah J, Yang D, et al. Large-area synthesis of high-quality and uniform graphene films on copper foils. Science (80-) 2009;324:1312–4. https://doi.org/10.1126/science.1171245.

[45] Kalaitzidou K, Fukushima H, Drzal LT. A new compounding method for exfoliated graphite-polypropylene nanocomposites with enhanced flexural properties and lower percolation threshold. Compos Sci Technol 2007;67:2045–51. https://doi.org/10.1016/j.compscitech.2006.11.014.

[46] Yihang Z. Application of water-soluble polymer inhibitor in metal corrosion protection : Progress and challenges. 2022;1–13. https://doi.org/10.3389/fenrg.2022.997107.

[47] Nwanonenyi SC, Obasi HC, Eze IO. Hydroxypropyl cellulose as an efficient corrosion inhibitor for aluminium in acidic environments: experimental and theoretical approach. Chemistry Africa 2019;2:471–82. https://doi.org/10.1007/s42250-019-00062-1.

[48] Hassan RM, Ibrahim SM. Performance and efficiency of methyl-cellulose polysaccharide as a green promising inhibitor for inhibition of corrosion of magnesium in acidic solutions. J Mol Struct 2021;1246. https://doi.org/10.1016/j.molstruc.2021.131180.

[49] Nwanonenyi SC, Arukalam IO, Obasi HC, Ezeamaku UL, Eze IO, Chukwujike IC, et al. Corrosion inhibitive behavior and adsorption of millet (panicum miliaceum) starch on mild steel in hydrochloric acid environment. BioTriboCorrosion 2017;3. https://doi.org/10.1007/s40735-017-0115-y.

[50] Wu W, Chen T, Du H, Li D, Liu J. Carboxymethyl starch as corrosion inhibitor for mild steel. Emerg Mater Res 2016;5:277–83. https://doi.org/10.1680/jemmr.15.00037.

[51] Ali BH, Ziada A, Blunden G. Biological effects of gum arabic: a review of some recent research. Food Chem Toxicol 2009;47:1–8. https://doi.org/10.1016/j.fct.2008.07.001.

[52] Shen C, Alvarez V, Koenig JDB, Luo JL. Gum arabic as corrosion inhibitor in the oil industry: experimental and theoretical studies. Corros Eng Sci Technol 2019;54:444–54. https://doi.org/10.1080/1478422X.2019.1613780.

[53] Palumbo G, Kollbek K, Wirecka R, Bernasik A, Górny M. Effect of CO2 partial pressure on the corrosion inhibition of N80 carbon steel by gum arabic in a CO2-water saline environment for shale oil and gas industry. Materials (Basel) 2020;13:1–24. https://doi.org/10.3390/MA13194245.

[54] Jiang Y, Xing Z, Wang X, Huang S, Liu Q, Yang J. Synthesis and application of epoxy resins: a review. 2015;29:43–7. https://doi.org/10.1016/j.jiec.2015.03.026.

[55] Hsissou R, Benhiba F, Dagdag O, El Bouchti M, Nouneh K, Assouag M, et al. Development and potential performance of prepolymer in corrosion inhibition for carbon steel in 1.0 m hcl: outlooks from experimental and computational investigations. J Colloid Interface Sci 2020;574:43–60. https://doi.org/10.1016/j.jcis.2020.04.022.

[56] Deyab MA, Abd El-Rehim SS. Influence of polyethylene glycols on the corrosion inhibition of carbon steel in butyric acid solution: weight loss, EIS and theoretical studies. Int J Electrochem Sci 2013;8:12613–27.

[57] Awad MK. Quantum chemical studies and molecular modeling of the effect of polyethylene glycol as corrosion inhibitors of an aluminum surface. Can J Chem 2013;91:283–91. https://doi.org/10.1139/cjc-2012-0354.

Elyor Berdimurodov, Ilyos Eliboev, Abduvali Kholikov,
Khamdam Akbarov, Dakeshwar Kumar Verma, Mohamed Rbaa,
Omar Dagdag, Khasan Berdimuradov

7 Functionalized nanomaterials/ nanocomposites-based thin film coatings as corrosion inhibitors

Abstract: In modern times, functionalized nanomaterials/nanocomposites-based thin film coatings are employed as protective materials for steel, copper, aluminum, magnesium, and zinc-based metal materials. The reason for this is that these materials are multifunctional, super-hydrophilic, cost-effective, mechanically easy to use, non-volatile, and environmentally friendly materials. In this book chapter, the main characteristics, corrosion efficiency performance, modern trends, and corrosion protection mechanisms have been discussed and reviewed with explanations. The inorganic/polymer-based nanomaterials and nanocomposites have good polymer matrices. Their mechanical properties are unique to the various inorganic nanomaterials such as nano-calcium carbonates, metal carbides, glass flakes, and metal nanopowders. The functionalizing of nanomaterials enhances the following important properties: corrosion inhibition efficiency, chemical stability, and lifetime. The corrosion protection properties of inorganic/polymer-based nanomaterials and nanocomposites depend on the following properties: the interaction between the nanoparticles and the polymer matrix, the functional groups, the chemical structure, the fraction of nanofillers, their specific surface area, morphology, size, and the type of nanoparticles.

Keywords: Nanomaterials, Nanocomposites, Protective Coatings, Corrosion Inhibitors

7.1 Introduction

7.1.1 Corrosion and corrosion protection by a protective coating

The corrosion of metallic materials is an economic and ecological problem because, bathrooms flood, chemical plants leak, oil pipelines break, and buildings and bridges can collapse [1–3]. The corrosion may have occurred in the air, in a solution, or in an underground medium. For example, the corrosion problem of air causes serious pollution to the surrounding environment [4–8]. Medical implant corrosion may cause blood poisoning. Fires may have occurred after the corrosion of electrical contacts. The metallic corrosion of radioactive reactors causes the expansion and disposal of radioactive compounds in the environment [9–11].

https://doi.org/10.1515/9783111016160-007

In corrosion processes, electrochemical reactions occur in the cathodic and anodic regions. In the anodic regions, the metal surface is oxidized while the corrosive ions are reduced to form the gas in the cathodic region [12–15].

In the present times, various corrosion protective methods are applied in the chemical, petrochemical, crude oil, and automobile industries. For example, the functionalized nanomaterials/nanocomposites-based thin film coatings are interesting and very important materials as corrosion inhibitors. Thin film coatings are very effective ways of corrosion protection [16, 17]. In this method, the metal surface is maximally insulated from contact with the aquatic solution. The active functional groups are effectively linked with the metal surface to form rigid chemical bonds. The protective coatings may be polymeric materials. However, these polymer materials are not stable over a long time [18, 19]. The functionalization of polymeric compounds with nanomaterials and nanocomposites is a more effective way to enhance the lifetime of a protective coating. The optical, mechanical, and chemical properties of polymeric materials are enhanced with the addition of nanoparticles. The size of the nanoparticles used is under 100 nm. The corrosion effect of anodic and cathodic ions is maximally depleted by the introduction of nanomaterials and nanocomposites because of their various good properties such as tribological properties, high-temperature corrosion opposition, long haul, adhesive quality, and surface hardness [20–23]. In addition to this, the nanomaterial and nanocomposite protective thin films are smoother in thickness and more slender; these properties are more useful in reducing the operating expenses, upkeep, and enhancing adaptability in equipment design [24–26]. In modern times, functionalized nanomaterials/nanocomposites-based thin film coatings are employed as protective materials for steel [27, 28], copper, aluminum, magnesium, and zinc-based metal materials. The reason for this is that these materials are multifunctional, super-hydrophilic, cost-effective, mechanically easy to use, non-volatile, and environmentally friendly materials [29–31].

7.2 Basics of inorganic/polymer-based nanomaterials and nanocomposites as a protective coating

In corrosion protection, epoxy-based coating materials are widely used because of their low cost and efficiency. Nanomaterials and nanocomposites are added to epoxy protective coatings. As a result, the lifetime of the protective coating is increased [32–34]. The corrosion protection efficiency of epoxy-based nanomaterials and nanocomposites was studied by various methods, such as surface morphology, spectroscopy, and electrochemical and theoretical methods [35–38].

The inorganic/polymer-based nanomaterials and nanocomposites have good polymer matrices. Their mechanical properties are unique to the various inorganic nanomaterials such as nano-calcium carbonates, metal carbides, nitrides, oxides and nano-glass flakes, and metal nanopowders (Figure 7.1). The functionalizing of nanomaterials enhances the following important properties: corrosion inhibition efficiency, chemical stability, and lifetime [39–42]. The corrosion protection properties of inorganic/polymer-based nanomaterials and nanocomposites depend on the following properties: the interaction between the nanoparticles and the polymer matrix, the functional groups, the chemical structure, the fraction of nanofillers, specific surface area, their morphology, size, and the type of nanoparticles [43–45].

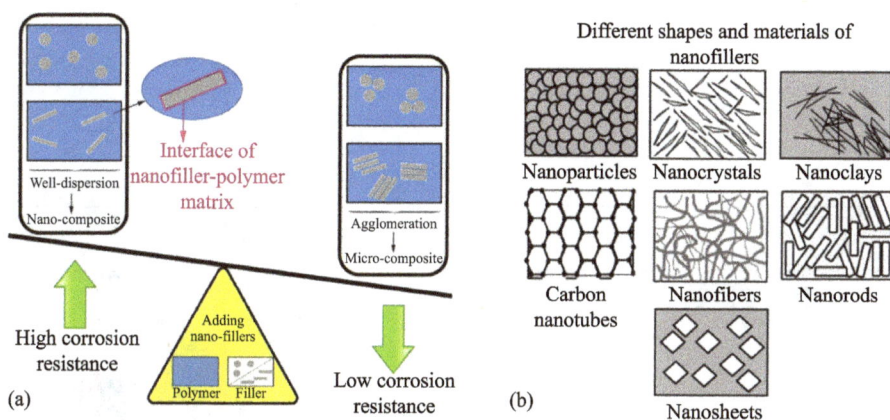

Figure 7.1: Pictorial distribution of nanofillers in the polymer matrix [43].

7.3 Metal oxide/polymer-based nanomaterials and nanocomposites as a protective coating

Khodair et al. [46] prepared epoxy-based nanomaterials and nanocomposites with the addition of magnesium oxide (MgO) nanoparticles. The obtained nanomaterial-based coating was used for protecting steel from corrosion destruction. Consequently, the corrosion resistance of the metal sample was increased by the formation of a protective Nanocomposition. The formed thin film is more stable in acidic, alkaline, and saline mediums [47–51]. The epoxy coating offers poor protection in a saline medium because of some structural defects on the surface of the protective coating [52, 53]. The introduction of magnesium oxide (MgO) nanoparticles enhanced the protective properties of the epoxy coating. Figure 7.2 shows the effect of magnesium oxide (MgO) nanoparticles on epoxy coating: (a) change in corrosion rate after the addition of various percentages of nanoparticles; (b) change in corrosion rate in various pH; (c) corrosion rate after the coating; (d) correlation between the salt concentration and coating efficiency; (e) Arrhenius plots; (f) corrosion rate before coating. It is clear from the obtained results that the corrosion rate in surface oxidation was depleted by the presence of nanoparticles. The nanomaterial- and nanocomposite-based protective thin films are more stable to change in the pH of the solution [54–56].

The protection efficiency of epoxy was 29.8% in the saline medium whilst this value was enhanced by the introduction of magnesium oxide (MgO) nanoparticles. In

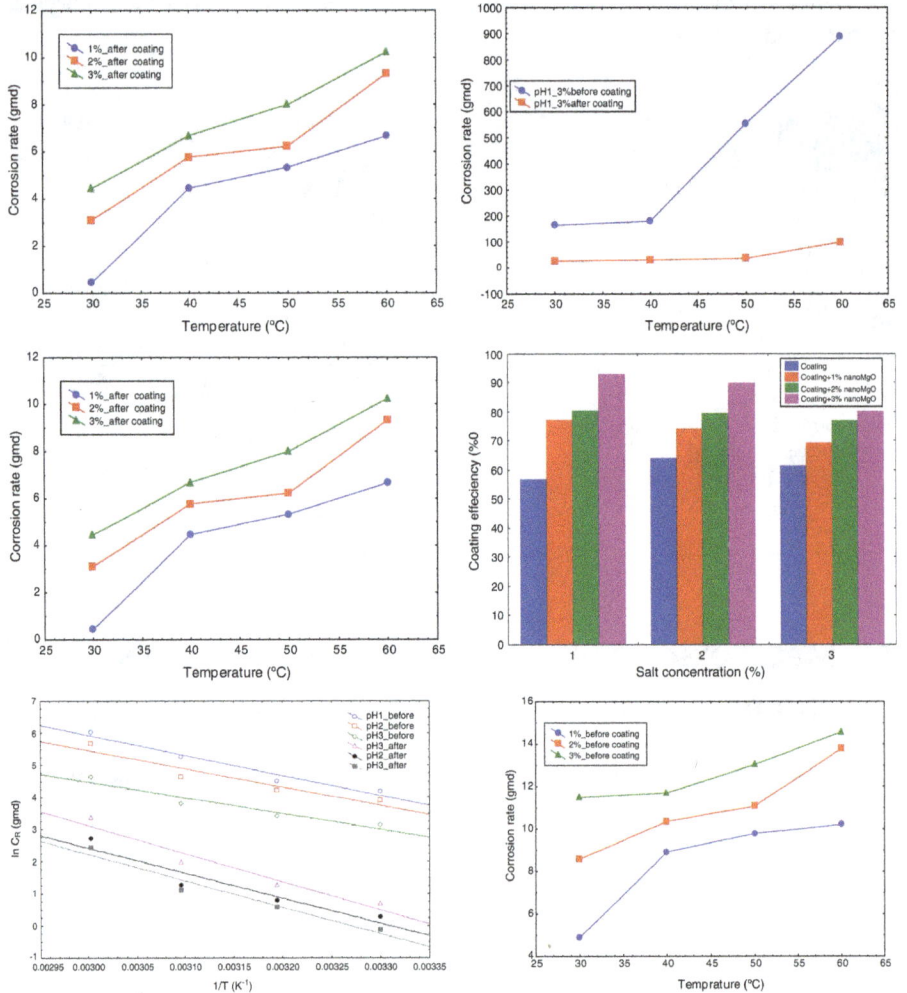

Figure 7.2: Effect of magnesium oxide (MgO) nanoparticle on epoxy coating: (a) change in corrosion rate after the addition of various percentages of nanoparticles; (b) change in corrosion rate in various pH; (c) corrosion rate after the coating; (d) correlation between the salt concentration and coating efficiency; (e) Arrhenius plots; (f) corrosion rate before coating [46].

comparison, the corrosion protection of epoxy-based nanomaterials and nanocomposites was over 93.7% while this value was very low in the absence of nanoparticles. The reason for these improvements after the introduction of a nanocompound to the epoxy coating is that:

(i) The surface morphology of the epoxy coating was enhanced with the introduction of nanoparticles, which is confirmed by the obtained results of scanning electron microscopy. The nanoparticles filled the crystallized defect of the protective thin film [57, 58].

(ii) The corrosion rate was maximally depleted by the correlation effect of epoxy-based nanomaterials and nanocomposites. The high correlation coefficient of these protective films is estimated using the polynomial interaction-effect model.

(iii) The rise in temperature does not affect the corrosion inhibition of the defender nanolayer at the interface [59, 60].

Li et al. [61] used CeO_2 nanometal oxide to prepare the nanocomposites for protective coating. The size of CeO_2 was around 3 to 5 μm. It was modified with an acrylic coating to enhance its corrosion-protective performance. In this research work, a graphene-based coating was also modified with the CeO_2 nanometal oxide to create a super protective coating. The obtained nanomaterial-based protective coating had good hydrophobicity. This property is very important in corrosion protection. This is due to the metal surface being maximally insulated by the high hydrophobicity of the material. It was found that (Figure 7.3):

(i) the corrosion rate was 53 and 48 times lower in the CeO_2 nanometal oxide and graphite(Gr)-CeO_2-modified coating nanocomposite, respectively.

(ii) the polarizing resistance was enhanced 5 times higher by metal nanooxide filler.

(iii) the correlation between the corrosion rate and the immersion time depends on the composite nature. It means that the corrosion rate is much lower in the metal oxide-functionalized nanocomposites [62, 63].

7.4 Metal/polymer-based nanomaterials and nanocomposites as a protective coating

Metal/polymer-based nanomaterial and nanocomposite protective coating were employed as effective protective materials for copper, aluminum, steel, and zinc [64–66]. The addition of nanometals (zinc, titanium, and aluminum) in polymer coatings is the main reason for the large corrosion resistance by the protective thin film. The metal-contained nanocomposite was more negative than the corroded metal surface [4, 67, 68]. This is attributed to an increase in the corrosion resistance by the formed protective thin film. Schaefer and Miszczyk [69] prepared a nanocomposite by functionalizing a protective coating with nanozinc. The corrosion inhibition of this material was

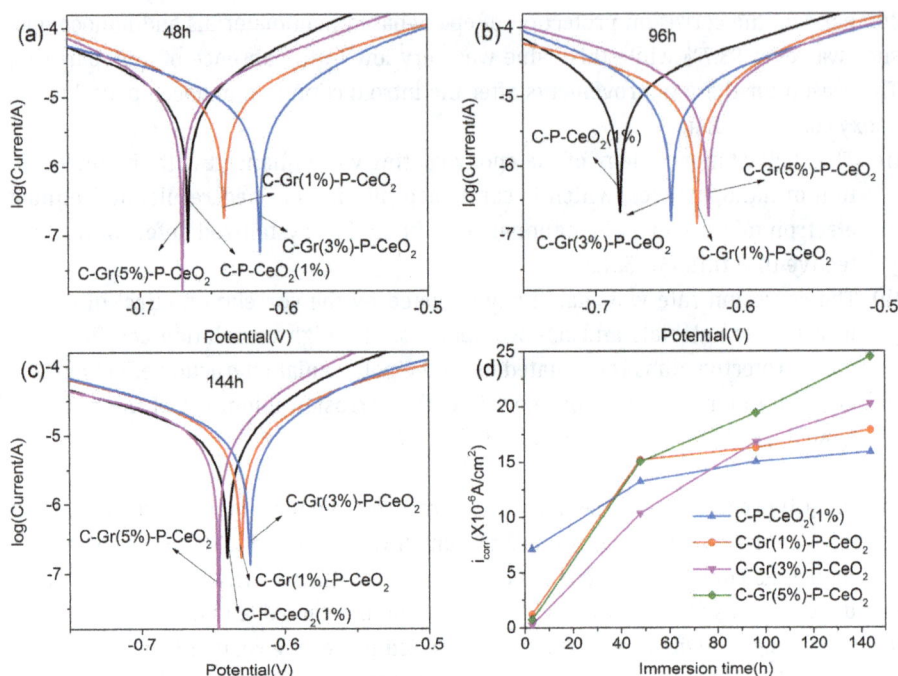

Figure 7.3: Tafel plots of various nanocomposite materials and the correlation of corrosion rate with immersed time [61].

investigated by the electrochemical impedance and potentiodynamic polarizing resistance methods. It was found that the:

(i) crystallized surface defects were covered by the nanozinc metals in the protective coating.

(ii) efficiency of the protective coating depends on the electrical connection between the spherical microparticles themselves, the zinc particles, and the steel substrate (Figure 7.4).

(iii) corrosion protection efficiency was over 90% with the introduction of metal nanoparticles in polymer-based coatings.

(iv) metal interface was insulated the Zn-contained protective coatings; as a result, the corrosion rate was depleted.

(v) introduction of small amounts of zinc nanoparticles can increase the corrosion protection of nanocomposites.

(vi) steel corrosion was maximally depleted in 3 wt% NaCl by using a protective coating of metal/polymer-based nanomaterials and nanocomposites protective coating.

Arianpouya et al. [70] prepared the polyurethane/zinc/clay nanocomposite-based protective thin film coating as a corrosion inhibitor. It was confirmed that the corrosion protection was enhanced with the addition of zinc to the polyurethane coating. The polyurethane/zinc/clay nanocomposite-based thin film protective coating was pre-pared by the ultrasonication process. Salt spray test, water absorption, DC polariza-tion techniques, and electrochemical impedance spectroscopy (EIS) were used to analyze the protection properties of the selected coating nanomaterials. The H_2O pen-etration and exfoliation adjacent to the metal surface were blocked with the presence of a corrosion inhibitor. Preparing a nanocomposite-based thin film coating depends on the polymer structural matrix and the functionalizing processes. The introduction of metal nano particles in the protective coating is the reason for promoting the filling of the voids, crevices, and pinholes of the polymer and in enhancing the adhesion ca-pabilities of the coating to the substrate and obviously increasing the barrier proper-ties of the coating.

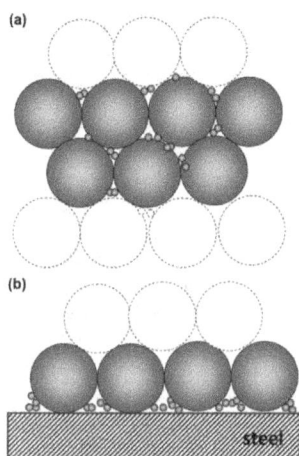

Figure 7.4: (a) The formation of an additional electrical connection between the zinc nanoparticles of the polymer coated nanocomposite and (b) coating of the steel surface [69].

7.5 Nitride/polymer-based nanomaterials and nanocomposites as a protective coating

In the corrosion protection, the nitride/polymer-based nanomaterials and nanocom-posites were used as a protective coating. The reason for this is that nitride can en-hance the stability and effectiveness of the thin film to act as a corrosion inhibitor on the metal surface. For example, hexagonal boron nitride (h-BN) was used to prepare the nanocomposite protective coating. h-BN offers effective impermeability to gas and liquid. h-BN was mixed with the water-borne epoxy (WBE) matrix to enhance the hy-drophobicity of this epoxy coating. In the preparation of nitride/polymer-based nano-materials and nanocomposites, the h-BN was functionalized to graphene oxide (GO) by the non-covalent interactions. Then, the formed GO/h-BN nanomaterials were mod-ified with waterborne epoxy (WBE) to obtain the GO/h-BN/WBE. The structural properties of the GO/h-BN/WBE were investigated by X-ray photoelectron spectroscopy (XPS), UV-vis absorbance spectroscopy, Raman spectroscopy, and transmittance electron microscopy (TEM).

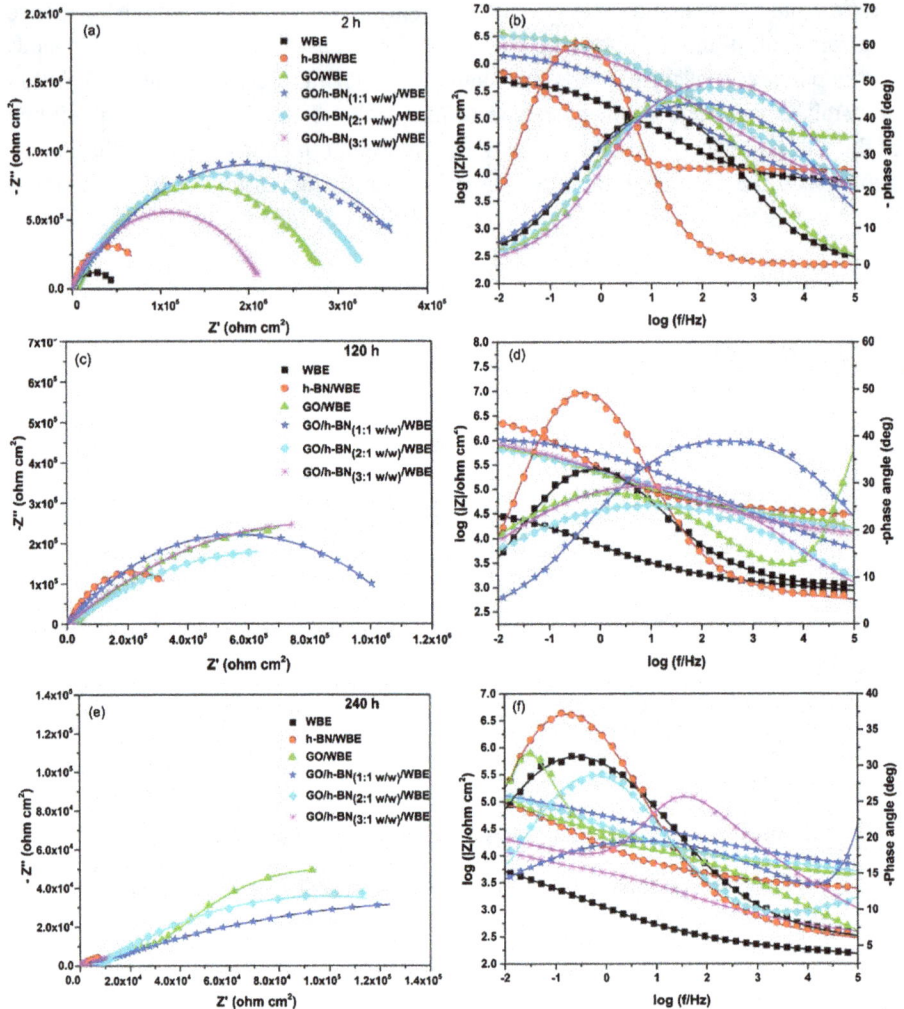

Figure 7.5: Electrochemical impedance spectroscopic results (3.5 wt% NaCl) at different times: (a, b) 2 h, (c, d) 120 h, and (e, f) 240 h [71].

The obtained results of the salt spray test and electrochemical impedance spectroscopy (EIS) (Figure 7.5) [71] indicate that the corrosion protection of GO/h-BN/WBE was higher at lower concentration levels (0.3 wt%) than without modification of nitride. This is due to the synergistic impermeable performance of h-BN and GO. In the corrosion protection of metal surfaces by the nanocomposite coating, hydrophobicity is an important factor. The corrosion resistance of the coating and the energetic barrier were enhanced by the high hydrophobicity level [71].

Cui et al. [72] prepared the nanocomposite epoxy coating from the hexagonal boron nitride nanohybrids and poly(2-butylaniline) as shown in Figure 7.6 [72]. These nanocomposites were used as synergetic anti-corrosive reinforcements of epoxy coating. The protective thin film of these nanocomposites was more stable in aggressive acidic environments. The water adsorption ability of hexagonal boron nitride nanohybrids and poly(2-butylaniline)-based nanocomposite was very low, which means that the formed thin film is more hydrophobic and insulates the water from the metal surface.

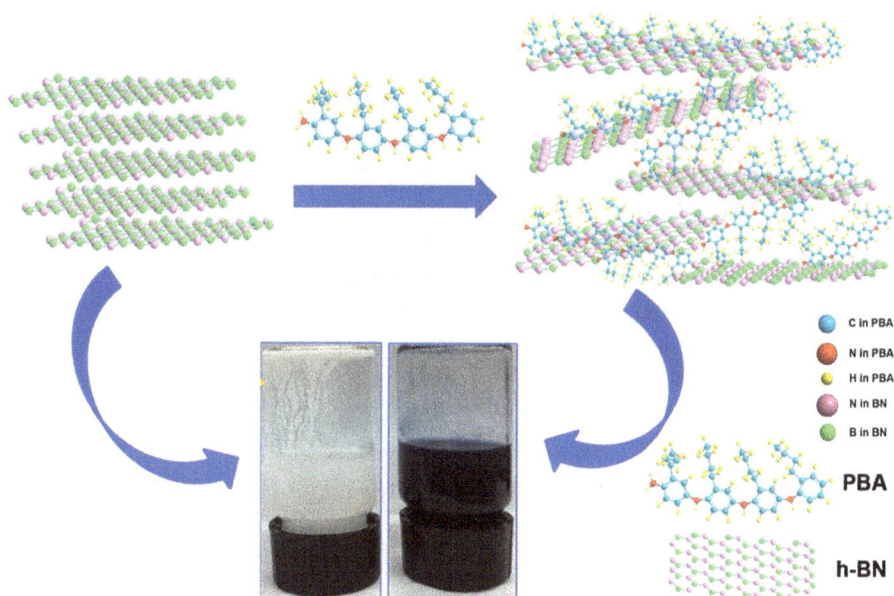

Figure 7.6: Preparation procedure of boron nitride nanohybrids and poly(2-butylaniline) nanocomposite as a protective coating film [72].

The impedance properties of the thin film are very high than those without the coating, confirming that the polarization resistance was dramatically increased with the formation of a protective thin film on the metal surface. The cathodic and anodic electrochemical actions on the metal surface were blocked by the formation of the thin film. The passivation of the metal surface was increased with the "labyrinth effect" of hexagonal boron nitride nanohybrids and poly(2-butylaniline)-based nanocomposite.

The BNxEC systems were long-term protective films on the metal surface. The synergetic effect of nanosheets promotes the BNxEC systems for corrosion protection. The passivation metal surface depends on the structural properties of the obtained nanocomposites. The solution conductivity was increased with the protective thin film on the metal surface. Corrosion was measured in a 3.5 wt% NaCl solution using the potentiodynamic and electrochemical impedance spectroscopy methods. It indicated that the delocalized electrons and aromatic rings are mainly responsible for the covalent interactions between the metal surface and the nanomaterials. The nitride and boron atoms are mainly responsible for the high protection performance of hexagonal boron nitride nanohybrids and poly(2-butylaniline)-based nanocomposite. In corrosion protection, the polymer-based nanocomposite chemically interacted with the metal surface due to the rigid covalent bonds between the metal surface and nanocomposite.

7.6 Conclusion

In modern times, functionalized nanomaterials/nanocomposite-based thin film coatings are used as corrosion Inhibitors for steel, copper, aluminum, magnesium, and zinc-based metal materials. The reason for this is that these materials are multifunctional, super-hydrophilic, cost-effective, mechanically easy to use, non-volatile, and environmentally friendly materials. In this book chapter, the main characteristics, corrosion efficiency, modern trends, and corrosion protection mechanisms were discussed and reviewed with explanations. In corrosion protection, epoxy-based coating materials are widely used because of their low cost and efficiency. Nanomaterials and nanocomposites are added to epoxy protective coatings. As a result, the lifetime of the protective coating was importantly increased. The inorganic/polymer-based nanomaterials and nanocomposites have good polymer matrices. Their mechanical properties are unique with the various inorganic nanomaterials such as nano-calcium carbonates, metal carbides, nitrides, oxides and nano-glass flakes, and metal nanopowders.

Reference

[1] Wang D, Bierwagen GP Sol-gel coatings on metals for corrosion protection. Prog Org Coat 2009;64(4):327–38.
[2] Lyon SB, Bingham R, Mills DJ Advances in corrosion protection by organic coatings: What we know and what we would like to know. Prog Org Coat 2017;102:2–7.
[3] Nazeer AA, Madkour M Potential use of smart coatings for corrosion protection of metals and alloys: A review. J Mol Liq 2018;253:11–22.
[4] Zhu M, Guo L, He Z, Marzouki R, Zhang R, Berdimurodov E Insights into the newly synthesized N-doped carbon dots for Q235 steel corrosion retardation in acidizing media: A detailed multidimensional study. J Colloid Interface Sci 2022;608:2039–49.
[5] Berdimurodov E, Eliboyev I, Berdimuradov K, Kholikov A, Akbarov K, Dagdag O, et al. Green β-cyclodextrin-based corrosion inhibitors: Recent developments, innovations and future opportunities. Carbohydr Polym 2022;119719.

[6]   Berdimurodov E, Kholikov A, Akbarov K, Guo L Inhibition properties of 4,5-dihydroxy-4,5-di-p-tolylimidazolidine-2-thione for use on carbon steel in an aggressive alkaline medium with chloride ions: Thermodynamic, electrochemical, surface and theoretical analyses. J Mol Liq 2021;327:114813.

[7]   Berdimurodov E, Guo L, Kholikov A, Akbarov K, Zhu M MOFs-based corrosion inhibitors. Supramolecular chemistry in corrosion and biofouling protection. CRC Press;2021, pp. 287–305.

[8]   Bahgat Radwan A, Mannah CA, Sliem MH, Al-Qahtani NHS, Okonkwo PC, Berdimurodov E, et al. Electrospun highly corrosion-resistant polystyrene-nickel oxide superhydrophobic nanocomposite coating. J Appl Electrochem 2021;51:1605–18.

[9]   Singh Raman RK, Tiwari A Graphene: The thinnest known coating for corrosion protection. Jom 2014;66(4):637–42.

[10]   Carneiro J, Tedim J, Fernandes SCM, Freire CSR, Silvestre AJD, Gandini A, et al. Chitosan-based self-healing protective coatings doped with cerium nitrate for corrosion protection of aluminum alloy 2024. Prog Org Coat 2012;75(1–2):8–13.

[11]   Dagdag O, Berisha A, Mehmeti V, Haldhar R, Berdimurodov E, Hamed O, et al. Epoxy coating as effective anti-corrosive polymeric material for aluminum alloys: Formulation, electrochemical and computational approaches. J Mol Liq 2021; 346:117886.

[12]   Asmatulu R Nanocoatings for corrosion protection of aerospace alloys. Corrosion protection and control using nanomaterials. Elsevier;2012, pp. 357–74.

[13]   Khramov AN, Voevodin NN, Balbyshev VN, Mantz RA Sol-gel-derived corrosion-protective coatings with controllable release of incorporated organic corrosion inhibitors. Thin Solid Films 2005;483(1–2):191–96.

[14]   Berdimurodov E, Kholikov A, Akbarov K, Guo L Experimental and theoretical assessment of new and eco-friendly thioglycoluril derivative as an effective corrosion inhibitor of St2 steel in the aggressive hydrochloric acid with sulfate ions. J Mol Liq 2021;335:116168.

[15]   Dagdag O, Haldhar R, Kim S-C, Berdimurodov E, Jodeh S, Verma C, Akpan ED, Ebenso EE. Functionalized Nanomaterials for Corrosion Mitigation: Synthesis, Characterization & Applications. InFunctionalized Nanomaterials for Corrosion Mitigation: Synthesis, Characterization, and Applications 2022 (pp. 67–85). American Chemical Society.

[16]   Ates M A review on conducting polymer coatings for corrosion protection. J Adhes Sci Technol 2016;30(14):1510–36.

[17]   Prasai D, Tuberquia JC, Harl RR, Jennings GK, Bolotin KI Graphene: corrosion-inhibiting coating. ACS Nano 2012;6(2):1102–08.

[18]   Berdimurodov E, Kholikov A, Akbarov K, Guo L, Abdullah AM, Elik M A gossypol derivative as an efficient corrosion inhibitor for St2 steel in 1 M HCl + 1 M KCl: An experimental and theoretical investigation. J Mol Liq 2021;328:115475.

[19]   Berdimurodov E, Kholikov A, Akbarov K, Guo L, Kaya S, Katin KP, et al. Novel bromide-cucurbit[7]uril supramolecular ionic liquid as a green corrosion inhibitor for the oil and gas industry. J Electroanal Chem 2021;901:115794.

[20]   Berdimurodov E, Kholikov A, Akbarov K, Guo L, Kaya S, Katin KP, et al. Novel cucurbit[6]uril-based [3]rotaxane supramolecular ionic liquid as a green and excellent corrosion inhibitor for the chemical industry. Colloids Surf A Physicochem Eng Asp 2022;633:127837.

[21]   Dagdag O, Haldhar R, Kim S-C, Berdimurodov E, Verma C, Akpan ED, Ebenso EE. Graphene and graphene oxide as nanostructured corrosion inhibitors. Carbon Allotropes: Nanostructured Anti-Corrosive Materials. 2022 Oct 3:133

[22]   Dagdag O, Haldhar R, Kim S-C, Guo L, Gouri ME, Berdimurodov E, et al. Recent progress in epoxy resins as corrosion inhibitors: design and performance. J Adhes Sci Technol 2022;37(6):1–22.

[23]   Haldhar R, Kim S-C, Berdimurodov E, Verma DK, Hussain CM (2021). Corrosion Inhibitors: Industrial Applications and Commercialization. In: Sustainable corrosion inhibitors II: Synthesis, design, and practical applications. American Chemical Society, pp. 10–219.

[24] Farag AA Applications of nanomaterials in corrosion protection coatings and inhibitors. Corros Rev 2020;38(1):67–86.
[25] Vijayan PP, Al-Maadeed M Self-repairing composites for corrosion protection: A review on recent strategies and evaluation methods. Materials 2019;12(17):2754.
[26] Deshpande PP, Jadhav NG, Gelling VJ, Sazou D Conducting polymers for corrosion protection: A review. J Coat Technol Res 2014;11(4):473–94.
[27] Berdimurodov E, Kholikov A, Akbarov K, Guo L, Kaya S, Katin KP, et al. Novel gossypol-indole modification as a green corrosion inhibitor for low-carbon steel in aggressive alkaline-saline solution. Colloids Surfaces A: Physicochem Eng Aspects 2022;637:128207.
[28] Berdimurodov E, Kholikov A, Akbarov K, Guo L, Kaya S, Kumar Verma D, et al. Novel glycoluril pharmaceutically active compound as a green corrosion inhibitor for the oil and gas industry. J Electroanal Chem 2022;907:116055.
[29] Voevodin NN, Balbyshev VN, Khobaib M, Donley MS Nanostructured coatings approach for corrosion protection. Prog Org Coat 2003;47(3–4):416–23.
[30] Maia F, Tedim J, Lisenkov AD, Salak AN, Zheludkevich ML, Ferreira MGS Silica nanocontainers for active corrosion protection. Nanoscale 2012;4(4):1287–98.
[31] Schmidt H, Langenfeld S, Nass R A new corrosion protection coating system for pressure-cast aluminium automotive parts. Mater Des 1997;18(4–6):309–13.
[32] Kaur J, Saxena A, Berdimurodov E, Verma DK Euphorbia prostrata as an eco-friendly corrosion inhibitor for steel: Electrochemical and DFT studies. Chem Pap 2022;77(2):957–76.
[33] Dagdag O, Haldhar R, Kim S-C, Safi ZS, Wazzan N, Mkadmh AM, et al. Synthesis, physicochemical properties, theoretical and electrochemical studies of Tetraglycidyl methylenedianiline. J Mol Struct 2022;5:133508.
[34] Dagdag O, Hsissou R, Safi Z, Haldhar R, Berdimurodov E, Bouchti ME, et al. Rheological and simulation for macromolecular matrix epoxy bi-functional aromatic amines. Polym Bull 2021;79(9):7571–87.
[35] Tan L, Sun Y, Wei C, Tao Y, Tian Y, An Y, et al. Design of robust, lithiophilic, and flexible inorganic-polymer protective layer by separator engineering enables dendrite-free lithium metal batteries with LiNi0. 8Mn0. 1Co0. 1O2 cathode. Small 2021;17(13):2007717.
[36] Teijido R, Ruiz-Rubio L, Echaide AG, Vilas-Vilela JL, Lanceros-Mendez S, Zhang Q State of the art and current trends on layered inorganic-polymer nanocomposite coatings for anticorrosion and multi-functional applications. Prog Org Coat 2022;163:106684.
[37] Li H-Y, Huang D-N, Ren K-F, Ji J Inorganic-polymer composite coatings for biomedical devices. Smart Mater Med 2021;2:1–14.
[38] Hu Y, Zhong Y, Qi L, Wang H Inorganic/polymer hybrid layer stabilizing anode/electrolyte interfaces in solid-state Li metal batteries. Nano Res 2020;13(12):3230–34.
[39] Wan S, Tieu AK, Xia Y, Zhu H, Tran BH, Cui S An overview of inorganic polymer as potential lubricant additive for high temperature tribology. Trib Int 2016;102:620–35.
[40] Zhou G, Liu K, Fan Y, Yuan M, Liu B, Liu W, et al. An aqueous inorganic polymer binder for high performance lithium-sulfur batteries with flame-retardant properties. ACS Cent Sci 2018;4(2):260–67.
[41] Du H, Ng SH, Neo KT, Ng M, Altman IS, Chiruvolu S, et al. (eds) Inorganic-Polymer Nanocomposites for Optical Applications, 2006.
[42] Dewangan Y, Verma DK, Berdimurodov E, Haldhar R, Dagdag O, Tripathi M, et al. N-hydroxypyrazine -2-carboxamide as a new and green corrosion inhibitor for mild steel in acidic medium: experimental, surface morphological and theoretical approach. J Adhes Sci Technol 2022;36:1–21.
[43] Pourhashem S, Saba F, Duan J, Rashidi A, Guan F, Nezhad EG, et al. Polymer/Inorganic nanocomposite coatings with superior corrosion protection performance: A review. J Ind Eng Chem 2020;88:29–57.
[44] Lloyd RR, Provis JL, Van Deventer JSJ Pore solution composition and alkali diffusion in inorganic polymer cement. Cement Concr Res 2010;40(9):1386–92.

[45] Papakonstantinou C, Balaguru P. Bond characteristics and structural behavior of inorganic polymer FRP. In Measuring, monitoring and modeling concrete properties: An International Symposium dedicated to Professor Surendra P. Shah, Northwestern University, Springer Netherlands, USA;2006, pp. 735–741.

[46] Khodair ZT, Khadom AA, Jasim HA Corrosion protection of mild steel in different aqueous media via epoxy/nanomaterial coating: preparation, characterization and mathematical views. J Mater Res Technol 2019;8(1):424–35.

[47] Prasanna SS, Balaji K, Pandey S, Rana S. Metal oxide based nanomaterials and their polymer nanocomposites. InNanomaterials and polymer nanocomposites. Elsevier;2019, pp. 123–144.

[48] Karak N Nanocomposites of epoxy and metal oxide nanoparticles. Sustainable epoxy thermosets and nanocomposites. ACS Publications;2021, pp. 299–330.

[49] Sadek RF, Farrag HA, Abdelsalam SM, Keiralla ZMH, Raafat AI, Araby E A powerful nanocomposite polymer prepared from metal oxide nanoparticles synthesized via brown algae as anti-corrosion and anti-biofilm. Front Mater 2019;6:140.

[50] Haldorai Y, Shim J-J Fabrication of metal oxide-polymer hybrid nanocomposites. Org-Inorg Hybrid Nanomater 2014;249–81.

[51] Pasha A, Khasim S, Darwish AAA, Hamdalla TA, Al-Ghamdi SA High performance organic coatings of polypyrrole embedded with manganese iron oxide nanoparticles for corrosion protection of conductive copper surface. J Inorg Organomet Polym Mater 2022;32(2):499–512.

[52] Berdimurodov E, Kholikov A, Akbarov K, Guo L, Kaya S, Verma DK, et al. New and green corrosion inhibitor based on new imidazole derivate for carbon steel in 1 M HCl medium: Experimental and theoretical analyses. Int J Eng Res Afr 2022;58:11–44.

[53] Berdimurodov E, Kholikov A, Akbarov K, Guo L, Umirov N, Verma DK, et al. (2022). Chapter19 – Ionic Liquids as Green and Sustainable Corrosion Inhibitors I. In: Guo, L, Verma, C, Zhang, DBT-E-FCI (eds). Elsevier, pp. 331–90.

[54] Pachaiappan R, Rajendran S, Show PL, Manavalan K, Naushad M Metal/metal oxide nanocomposites for bactericidal effect: A review. Chemosphere 2021;272:128607.

[55] Pal A, Prabhakar S, Bajpai J, Bajpai AK Antiviral behavior of metal oxide-reinforced polymer nanocomposites. Elsevier;2022, pp. 439–67.

[56] Kumar AA, Kumar VD, Berdimurodov E Recent trends in noble-metals based composite materials for supercapacitors: A comprehensive and development review. J Indian Chem Soc 2022;100817.

[57] Berdimurodov E, Kholikov A, Akbarov K, Obot IB, Guo L Thioglycoluril derivative as a new and effective corrosion inhibitor for low carbon steel in a 1 M HCl medium: Experimental and theoretical investigation. J Mol Struct 2021;1234:130165.

[58] Berdimurodov E, Kholikov A, Akbarov K, Xu G, Abdullah AM, Hosseini M New anti-corrosion inhibitor (3ar,6ar)-3a,6a-di-p-tolyltetrahydroimidazo[4,5-d]imidazole-2,5(1 h,3h)-dithione for carbon steel in 1 M HCl medium: gravimetric, electrochemical, surface and quantum chemical analyses. Arab J Chem 2020;13:7504–23.

[59] Rbaa M, Galai M, Dagdag O, Guo L, Tüzün B, Berdimurodov E, Zarrouk A, Lakhrissi B. Development process for eco-friendly corrosion inhibitors. InEco-Friendly Corrosion Inhibitors. Elsevier;2022, pp. 27–42.

[60] Shahmoradi AR, Ranjbarghanei M, Javidparvar AA, Guo L, Berdimurodov E, Ramezanzadeh B Theoretical and surface/electrochemical investigations of walnut fruit green husk extract as effective inhibitor for mild-steel corrosion in 1M HCl electrolyte. J Mol Liq 2021;338:116550.

[61] Li H, Wang J, Yang J, Zhang J, Ding H Large CeO2 nanoflakes modified by graphene as barriers in waterborne acrylic coatings and the improved anticorrosion performance. Prog Org Coat 2020;143:105607.

[62] Berdimurodov E, Verma C, Berdimuradov K, Quraishi MA, Kholikov A, Akbarov K, et al. 8-Hydroxyquinoline is key to the development of corrosion inhibitors: An advanced review. Inorg Chem Commun 2022;144:109839.

[63] Berdimurodov E, Verma DK, Kholikov A, Akbarov K, Guo L The recent development of carbon dots as powerful green corrosion inhibitors: A prospective review. J Mol Liq 2021;349:118124.

[64] Rbaa M, Galai M, Ouakki M, Hsissou R, Berisha A, Kaya S, et al. Synthesis of new halogenated compounds based on 8-hydroxyquinoline derivatives for the inhibition of acid corrosion: Theoretical and experimental investigations. Mater Today Commun 2022;33:104654.

[65] Rbaa M, Haida S, Tuzun B, El Hassane A, Kribii A, Lakhrissi Y, et al. Synthesis, characterization and bioactivity of novel 8-hydroxyquinoline derivatives: Experimental, molecular docking, DFT and POM analyses. J Mol Struct 2022;1258:132688.

[66] Verma DK, Dewangan Y, Singh AK, Mishra R, Susan MA, Salim R, Taleb M, El Hajjaji F, Berdimurodov E. Ionic liquids as green and smart lubricant application: an overview. Ionics. 2022 Nov;28(11): 4923–32.

[67] Rbaa M, Oubihi A, Hajji H, Tüzün B, Hichar A, Anouar EH, et al. Synthesis, bioinformatics and biological evaluation of novel pyridine based on 8-hydroxyquinoline derivatives as antibacterial agents: DFT, molecular docking and ADME/T studies. J Mol Struct 2021;1244:130934.

[68] Verma DK, Kazi M, Alqahtani MS, Syed R, Berdimurodov E, Kaya S, et al. N-hydroxybenzothioamide derivatives as green and efficient corrosion inhibitors for mild steel: Experimental, DFT and MC simulation approach. J Mol Struct 2021;1241:130648.

[69] Schaefer K, Miszczyk A Improvement of electrochemical action of zinc-rich paints by addition of nanoparticulate zinc. Corros Sci 2013;66:380–91.

[70] Arianpouya N, Shishesaz M, Arianpouya M, Nematollahi M Evaluation of synergistic effect of nanozinc/nanoclay additives on the corrosion performance of zinc-rich polyurethane nanocomposite coatings using electrochemical properties and salt spray testing. Surf Coat Technol 2013;216:199–206.

[71] Wu Y, He Y, Chen C, Zhong F, Li H, Chen J, et al. Non-covalently functionalized boron nitride by graphene oxide for anticorrosive reinforcement of water-borne epoxy coating. Colloids Surfaces A: Physicochem Eng Aspects 2020;587:124337.

[72] Cui M, Ren S, Qin S, Xue Q, Zhao H, Wang L Processable poly(2-butylaniline)/hexagonal boron nitride nanohybrids for synergetic anticorrosive reinforcement of epoxy coating. Corros Sci 2018;131:187–98.

Deepak Sharma, Hari om, Abhinay Thakur, Ashish Kumar
8 Functionalized carbon allotropes-based thin film coatings as corrosion inhibitors

Abstract: Corrosion is a major issue faced by various industries, resulting in economic losses and safety hazards. Therefore, there is an urgent need for the development of efficient and sustainable corrosion inhibitors. Carbon allotropes, such as graphene, carbon nanotubes, and carbon dots, have unique structural, electronic, and mechanical properties that make them excellent candidates for corrosion inhibition applications. The chapter begins by providing an overview of the mechanisms of corrosion and the challenges in developing effective corrosion inhibitors. It then discusses the properties and functionalization methods of various carbon allotropes, and their potential for corrosion inhibition. The discussion also covers the various techniques used for the deposition of thin film coatings, such as chemical vapor deposition, electrochemical deposition, and spray coating. Furthermore, the chapter presents recent research studies that have investigated the use of functionalized carbon allotropes as thin film coatings for corrosion inhibition. Finally, the chapter concludes by summarizing the potential of functionalized carbon allotropes-based thin film coatings as sustainable and efficient corrosion inhibitors.

Keywords: Corrosion Inhibition, Carbon Allotropes, Thin Film Coatings, Functionalization Methods, Deposition Techniques, Sustainable Corrosion Inhibitors

8.1 Introduction

Corrosion is the process of deterioration and destruction of materials, commonly metals, due to chemical reactions with the surrounding environment. It is a major issue faced by various industries, including construction, transportation, energy, and manufacturing, resulting in economic losses and safety hazards. Corrosion can lead to the failure of critical components, reduced efficiency of equipment, and increased maintenance and replacement costs. For example, in the oil and gas industry, corrosion can cause leaks, pipeline failures, and accidents, which can have serious environmental and human consequences [1–4]. In the transportation industry, corrosion of vehicles and infrastructure can compromise their structural integrity and pose risks to public safety. The cost of corrosion in the United States is estimated to be around $276 billion annually, which accounts for about 3.1% of the country's Gross Domestic Product (GDP) (NACE International, 2020). Moreover, corrosion is a global issue, affecting infrastructure and industries worldwide. Therefore,

https://doi.org/10.1515/9783111016160-008

there is an urgent need for the development of efficient and sustainable corrosion inhibitors to mitigate the adverse effects of corrosion [5–7].

The traditional approach to corrosion prevention is the use of protective coatings or sacrificial materials that can delay or prevent corrosion. However, these methods have limitations, such as environmental and health hazards, high costs, and short lifetimes. Therefore, there is a growing interest in developing sustainable and eco-friendly corrosion inhibitors that are efficient, cost-effective, and have a long-lasting effect [8–12]. Sustainable corrosion inhibitors should meet certain criteria, such as being non-toxic, biodegradable, and renewable, and have a low environmental impact. In addition, they should be compatible with existing infrastructure, easily applicable, and have a long-lasting effect. The development of such inhibitors is essential to reduce the environmental impact of corrosion prevention and ensure sustainable economic growth. Carbon allotropes, such as graphene, carbon nanotubes, and carbon dots, have gained attention in recent years as promising candidates for corrosion inhibition applications. The unique structural, electronic, and mechanical properties of these materials make them excellent candidates for developing effective and sustainable corrosion inhibitors. Below, we will discuss the properties and functionalization methods of various carbon allotropes and their potential for corrosion inhibition [13–25].

8.1.1 Graphene as a corrosion inhibitor

Graphene is a promising corrosion inhibitor due to its unique properties. It has a high surface area-to-volume ratio, which allows it to cover a larger surface area of the metal substrate, leading to better protection against corrosion [26, 27]. Graphene also has exceptional mechanical strength and stability, allowing it to withstand harsh environmental conditions. Functionalization of graphene with various molecules can further improve its corrosion inhibition properties. For example, the introduction of nitrogen or oxygen groups onto the graphene surface can increase its adsorption onto the metal surface, leading to improved corrosion inhibition. These functional groups can be introduced through different methods such as chemical vapor deposition, chemical reduction, and electrochemical deposition.

In recent years, various studies have investigated the use of graphene-based coatings as corrosion inhibitors for different metallic substrates. One study found that a graphene oxide coating exhibited excellent corrosion resistance on mild steel in a saline solution. Researchers attributed this result to the strong interaction between graphene oxide and the steel surface, which led to the formation of a protective barrier layer. Another study demonstrated the effectiveness of a reduced graphene oxide coating on aluminum alloy in a seawater environment. The coating was synthesized using a facile electrochemical deposition method and exhibited superior corrosion resistance, compared to other coatings such as polyurethane and epoxy coatings. Researchers suggested that the high surface area and the barrier properties of the reduced graphene oxide coating contributed to its excellent corrosion inhibition properties [28, 29]. Furthermore, graphene-based coatings can be combined with other corrosion inhibitors to enhance their effectiveness. For example, a study investigated the use of a graphene oxide and cerium oxide composite coating on mild steel in a corrosive environment. The composite coating exhibited superior corrosion resistance compared to the individual coatings, demonstrating the synergistic effect of combining graphene oxide and cerium oxide.

8.1.2 Carbon nanotubes as corrosion inhibitors

CNTs can act as physical barriers to protect the metallic substrate from the corrosive environment. The high aspect ratio of CNTs results in a high surface area, which can effectively cover the metallic substrate, thus providing a high level of protection. Furthermore, CNTs can also act as electron donors or acceptors, forming a charge transfer complex with the metallic substrate, which inhibits the corrosion process [30, 31]. Functionalized CNTs have been shown to exhibit enhanced corrosion inhibition properties compared to pristine CNTs. Functionalization of CNTs can be achieved by introducing functional groups such as carboxylic acid, amine, and thiol groups onto the surface of the CNTs. The functional groups can interact with the metallic substrate and form a protective layer, inhibiting the corrosion process.

Several research studies have investigated the use of CNT-based coatings as corrosion inhibitors. For example, Chen et al. (2015) synthesized CNT-based coatings using an electrochemical deposition method and investigated their corrosion inhibition properties on aluminum alloy. The results showed that the CNT-based coatings effectively inhibited the corrosion process and the corrosion rate decreased by approximately 93% [32, 33]. Similarly, Li et al. (2016) synthesized CNT-based coatings using the chemical vapor deposition method and investigated their corrosion inhibition properties on carbon steel. The results showed that the CNT-based coatings effectively inhibited the corrosion process and the corrosion rate decreased by approximately 90%. Overall, CNTs have shown great potential as corrosion inhibitors and their unique properties such as high aspect ratio, large surface area, and electron transfer capabilities make them ideal candidates for corrosion inhibition applications [34, 35]. Functionalized CNTs have been shown to exhibit enhanced corrosion inhibition properties and their synthesis using various methods such as chemical vapor deposition and electrochemical deposition makes them versatile materials for use in different industrial settings.

8.1.3 Carbon dots as corrosion inhibitors

Carbon dots (CDs) have been increasingly recognized as promising candidates for various applications, including corrosion inhibition. CDs are nanoscale carbon-based particles with unique optical, electronic, and chemical properties that make them highly desirable for corrosion inhibition applications. CDs can be synthesized using various methods such as solvothermal synthesis, hydrothermal synthesis, and microwave-assisted synthesis. CDs can also be functionalized with different molecules such as nitrogen and sulfur to enhance their corrosion inhibition properties [36, 37]. The mechanism of CDs as corrosion inhibitors is not well understood, but it is believed to be related to their unique surface properties. CDs have a high surface area and can interact with metal surfaces through π-π stacking and other van der Waals interactions. The functional groups on the surface of CDs can also form chemical bonds with metal surfaces, which can inhibit the corrosion process.

Recent research studies have shown that CDs-based coatings exhibit excellent corrosion inhibition properties on different metallic substrates such as steel and aluminum. For example, Liu et al. (2021) synthesized CDs by a one-pot solvothermal method and found that CDs-based coatings can significantly reduce the corrosion rate of carbon steel in a corrosive environment. The CDs-based coatings showed high stability and adhesion to the metal surface, indicating their potential as effective and durable corrosion inhibitors. In another study, Wang et al. (2018) synthesized sulfur-doped CDs and investigated their corrosion inhibition properties on aluminum alloy. The results showed that the CDs-based coatings can effectively reduce the corrosion rate of aluminum alloy in a corrosive environment [38, 39]. The CDs-based coatings also exhibited good adhesion to the metal surface and showed high stability during the corrosion test. Overall, CDs have shown great potential as effective and sustainable corrosion inhibitors. The synthesis of CDs is relatively simple and can be achieved using various methods. CDs can also be functionalized with different molecules to enhance their corrosion inhibition properties. Further research is needed to fully understand the mechanism of CDs as a corrosion inhibitor and to optimize their properties for specific corrosion inhibition applications [40, 41].

8.2 Functionalization of carbon allotropes for corrosion inhibition

Functionalization of carbon allotropes involves the introduction of different functional groups on their surface. These functional groups can interact with metallic substrates to form a protective layer, which inhibits the corrosion process [42, 43]. Different functionalization methods can be used to functionalize carbon allotropes for corrosion inhibition applications, including chemical vapor deposition, chemical reduction, and electrochemical deposition.

8.2.1 Chemical vapor deposition

Chemical vapor deposition (CVD) is a widely used method for the functionalization of carbon allotropes for corrosion inhibition applications. CVD involves the decomposition of a precursor gas on a substrate to form a thin film coating. The precursor gas can be functionalized with different molecules such as nitrogen, oxygen, and sulfur to enhance the corrosion inhibition properties of the carbon allotropes [44]. In CVD, the substrate is placed in a reactor and the precursor gas is introduced into the reactor chamber. The precursor gas decomposes on the substrate, leading to the formation of a thin film coating. The reaction can be carried out under different conditions such as temperature, pressure, and gas flow rate to control the thickness and morphology of the thin film coating. Graphene can be functionalized using CVD by introducing a precursor gas containing a functional group such as oxygen or nitrogen during the decomposition process. The functional groups can react with metallic substrates to form a protective layer, which inhibits the corrosion process. For example, oxygen-functionalized graphene has been shown to exhibit excellent corrosion inhibition properties on steel and aluminum substrates. Nitrogen-functionalized graphene has also been shown to exhibit good corrosion inhibition properties on aluminum substrates.

Carbon nanotubes can also be functionalized using CVD. For example, nitrogen-functionalized carbon nanotubes have been shown to exhibit excellent corrosion inhibition properties on aluminum substrates. Sulfur-functionalized carbon nanotubes have also been shown to exhibit good corrosion inhibition properties on copper substrates. Carbon dots can also be functionalized using CVD. For example, nitrogen-functionalized carbon dots have been shown to exhibit excellent corrosion inhibition properties on aluminum substrates. CVD has several advantages over other functionalization methods [45, 46]. It allows for the precise control of the thickness and morphology of the thin film coating and it can be used to functionalize different types of carbon allotropes. Additionally, CVD can be easily scaled up for industrial applications.

However, CVD also has some limitations. It requires specialized equipment and expertise, which can make it expensive compared to other functionalization methods. Additionally, the precursor gas used in CVD can be expensive and may require special handling procedures. In summary, CVD is an effective method for functionalizing carbon allotropes for corrosion inhibition applications. It allows for the precise control of the thickness and morphology of the thin film coating and it can be used to functionalize different types of carbon allotropes. However, it also has some limitations such as specialized equipment requirements and potential cost issues.

8.2.2 Chemical reduction

Chemical reduction is a widely used method for the functionalization of carbon allotropes such as graphene and carbon dots. The method involves the reduction of metal salt precursors using a reducing agent in the presence of a carbon allotrope. The reducing agent used can be functionalized with different molecules to enhance the corrosion inhibition properties of the carbon allotropes. One of the most commonly used reducing agents in chemical reduction is hydrazine, which is a colorless and highly reactive compound that acts as a reducing agent. The reduction process involves the formation of a hydrazine molecule, which donates electrons to the metal salt precursor, leading to the formation of a metal nanoparticle. The carbon allotrope, such as graphene or carbon dots, then binds to the metal nanoparticle, leading to functionalization of the carbon allotrope [47, 48].

Several studies have reported the successful functionalization of graphene and carbon dots using chemical reduction for corrosion inhibition applications. For example, Li et al. (2015) reported the functionalization of graphene oxide (GO) with nitrogen-containing molecules using chemical reduction. The resulting nitrogen-doped graphene exhibited excellent corrosion inhibition properties on a copper substrate. Similarly, Wang et al. (2017) reported the synthesis of sulfur-doped graphene using chemical reduction of graphene oxide, with thiourea as the sulfur source. The resulting sulfur-doped graphene exhibited excellent corrosion inhibition properties on a copper substrate [49].

In addition to hydrazine, other reducing agents such as sodium borohydride have also been used for the functionalization of carbon allotropes. For example, Wang et al. (2019) reported the synthesis of nitrogen-doped carbon dots using sodium borohydride as the reducing agent. The nitrogen-doped carbon dots exhibited excellent corrosion inhibition properties on a steel substrate. Overall, chemical reduction is a simple and cost-effective method for the functionalization of carbon allotropes for corrosion inhibition applications. The method can be easily scaled up for industrial applications, making it an attractive option for the development of sustainable and efficient corrosion inhibitors.

8.2.3 Electrochemical deposition

Electrochemical deposition (ECD) is a method used to functionalize carbon allotropes by applying an electric current to an electrochemical cell. The electrochemical cell consists of an anode, a cathode, and an electrolyte. The carbon allotrope is used as the cathode, and the anode is made of a metal that can be functionalized with different molecules such as nitrogen, oxygen, and sulfur to enhance the corrosion inhibition properties of the carbon allotropes. During the ECD process, the anode is oxidized, releasing metal ions into the electrolyte, while the cathode (carbon allotrope) is reduced, attracting metal ions to its surface. The metal ions react with functional groups on the carbon allotrope surface, forming a metal-functionalized coating that exhibits excellent corrosion inhibition properties. The functionalization of carbon allotropes by ECD has several advantages over other methods. One advantage is that the ECD process is relatively simple and cost-effective, compared to other methods. The ECD process can be performed using a simple electrochemical cell, which does not require complex equipment or specialized training. Another advantage of ECD is that it allows for precise control over the thickness and composition of the functionalized coating. The thickness and composition of the coating can be controlled by adjusting the duration and magnitude of the applied electric current, the composition of the electrolyte, and the functionalizing agent used.

The ECD method has been used to functionalize various carbon allotropes for corrosion inhibition applications. For example, researchers have used ECD to functionalize graphene with metal oxides such as titanium oxide and zinc oxide to enhance its corrosion inhibition properties on metallic substrates such as steel and aluminum [50, 51]. The metal oxide-functionalized graphene coating showed superior corrosion inhibition performance compared to graphene alone. ECD has also been used to functionalize carbon nanotubes with metal nanoparticles such as silver and copper to enhance their corrosion inhibition properties on metallic substrates. The metal nanoparticle-functionalized carbon nanotubes showed superior corrosion inhibition performance compared to pristine carbon nanotubes. The ECD method has also been used to functionalize carbon dots for corrosion inhibition applications. Researchers have used ECD to functionalize carbon dots with metal ions such as copper ions and iron ions to enhance their corrosion inhibition properties on metallic substrates such as steel and aluminum. The metal ion-functionalized carbon dots showed excellent corrosion inhibition performance compared to pristine carbon dots.

8.3 Mechanisms of corrosion and challenges in developing corrosion inhibitors

8.3.1 Types of corrosion

Corrosion is the degradation of materials due to chemical or electrochemical reactions with the surrounding environment. Different types of corrosion can occur, including:
- Uniform corrosion: Occurs when the entire surface of the material is corroded at a relatively uniform rate.
- Galvanic corrosion: Occurs when two dissimilar metals are in contact with each other in the presence of an electrolyte, leading to the more active metal corroding faster than the less active metal.

– Pitting corrosion: Occurs when localized areas of the material are corroded, resulting in small pits or holes.
– Crevice corrosion: Occurs in narrow crevices or gaps between two surfaces in the presence of an electrolyte.
– Stress corrosion cracking: Occurs due to the combined effect of stress and corrosion on a material, resulting in cracking or fracture of the material.

8.3.2 Factors affecting corrosion

Several factors can affect the corrosion of materials, including:
– Environment: The type and concentration of the surrounding electrolyte, humidity, temperature, and the presence of pollutants
– Material composition: The chemical composition, microstructure, and surface finish of the material
– Mechanical stress: The application of mechanical stress on the material
– Electrochemical factors: The presence of an electrolyte, potential difference, and pH

8.3.3 Challenges in developing corrosion inhibitors

Developing effective corrosion inhibitors is challenging due to several factors, including:
– Compatibility: The corrosion inhibitor must be compatible with the material being protected, the environment it is exposed to, and other components in the system.
– Durability: The corrosion inhibitor must be durable and provide long-term protection against corrosion.
– Effectiveness: The corrosion inhibitor must be effective in preventing corrosion and reducing corrosion rates.
– Cost: The cost of the corrosion inhibitor must be reasonable and economical for the application.
– Environmental impact: The corrosion inhibitor should not have a significant environmental impact and should be safe for use.

Overcoming these challenges requires a multidisciplinary approach, including materials science, chemistry, and engineering, to develop corrosion inhibitors that meet the requirements of specific applications.

8.4 Properties and functionalization methods of carbon allotropes

8.4.1 Graphene

Graphene is a two-dimensional carbon allotrope that consists of a single layer of carbon atoms arranged in a hexagonal lattice. Graphene has excellent mechanical, electrical, and thermal properties, making it a potential candidate for various applications, including corrosion inhibition. Functionalization of graphene involves the introduction of different functional groups on its surface, such as nitrogen, oxygen, and sulfur, which can interact with metallic substrates to form a protective layer, inhibiting the corrosion process [52, 53]. Graphene can be functionalized using different methods such as chemical vapor deposition, chemical reduction, and electrochemical deposition.

8.4.2 Carbon nanotubes

Carbon nanotubes (CNTs) are cylindrical structures consisting of a single layer or multiple layers of graphene rolled into a tube. CNTs have unique mechanical, electrical, and thermal properties, making them potential candidates for various applications, including corrosion inhibition. Functionalization of CNTs involves the introduction of different functional groups on their surface, such as nitrogen, oxygen, and sulfur, which can interact with metallic substrates to form a protective layer, inhibiting the corrosion process. CNTs can be functionalized using different methods such as chemical vapor deposition, arc discharge, and laser ablation.

8.4.3 Carbon dots

Carbon dots (CDs) are small carbon–based nanoparticles with a size range of 1–10 nm. CDs have excellent optical and electronic properties and can be functionalized with different molecules such as nitrogen and sulfur to enhance their corrosion inhibition properties. CDs can be synthesized using various methods such as solvothermal synthesis, hydrothermal synthesis, and microwave-assisted synthesis. CDs-based coatings exhibit excellent corrosion inhibition properties on different metallic substrates such as steel and aluminum.

8.5 Functionalization methods

The functionalization of carbon allotropes can be achieved using various methods, including:

8.5.1 Chemical vapor deposition (CVD)

Chemical Vapor Deposition (CVD) is a popular method for functionalizing carbon allotropes. This technique involves the deposition of a thin film coating onto a substrate by the decomposition of a precursor gas. The precursor gas can be functionalized with different molecules such as nitrogen, oxygen, and sulfur to enhance the corrosion inhibition properties of the carbon allotropes. CVD can be used to functionalize graphene, carbon nanotubes, and carbon dots for corrosion inhibition applications. In CVD, the precursor gas is introduced into a chamber containing the substrate, which is heated to a high temperature. The gas decomposes on the surface of the substrate, leading to the formation of a thin film coating. The reaction can be carried out under different conditions such as pressure, temperature, and gas flow rate to control the properties of the resulting coating.

The functionalization of carbon allotropes using CVD involves the use of functionalized precursor gases. These gases can be functionalized with different molecules such as nitrogen, oxygen, and sulfur to enhance the corrosion inhibition properties of the resulting coating. For example, the introduction of nitrogen into the precursor gas can lead to the formation of nitrogen-doped graphene, which has been shown to exhibit excellent corrosion inhibition properties on metallic substrates. The properties of the resulting coating can also be controlled by adjusting the deposition parameters such as the temperature and pressure. For example, increasing the temperature can lead to the formation of a more dense and uniform coating, which can enhance its corrosion inhibition properties. The gas flow rate can also be adjusted to control the thickness and composition of the coating [54–56].

One advantage of CVD is its ability to deposit coatings with high purity and uniformity. The high temperature used in CVD can lead to the removal of impurities and defects from the precursor gas, resulting in a high-purity coating. The uniformity of the coating can also be controlled by adjusting the deposition parameters such as the gas flow rate and substrate temperature. CVD has been used to functionalize different carbon allotropes for corrosion inhibition applications. Graphene, for example, has been functionalized using CVD with different precursor gases such as methane, ethylene, and acetylene to form graphene films with different properties. These functionalized graphene films have been shown to exhibit excellent corrosion inhibition properties on metallic substrates.

Carbon nanotubes have also been functionalized using CVD with different precursor gases such as acetylene and ethanol to form carbon nanotube coatings with enhanced corrosion inhibition properties. The functionalization of carbon nanotubes using CVD can lead to the formation of nanotubes with different properties such as diameter and length, which can affect their corrosion inhibition properties. Carbon dots, which are small carbon-based nanoparticles, have also been functionalized using CVD with different precursor gases such as ethylene and methane to enhance their corrosion inhibition properties. The functionalization of carbon dots using CVD can lead to the formation of dots with different properties, such as size and surface functional groups, which can affect their corrosion inhibition properties [57, 58]. One of the challenges in using CVD for functionalizing carbon allotropes for corrosion inhibition applications is the optimization of the deposition parameters. The deposition parameters such as the temperature, pressure, and gas flow rate can affect the properties of the resulting coating, including its thickness, composition, and uniformity. Therefore, the optimization of these parameters is critical for the development of functionalized carbon allotropes with optimal corrosion inhibition properties. Another challenge is the scalability of the process. CVD is a time-consuming process that requires high temperatures and specialized equipment. Therefore, the development of large-scale CVD systems for the functionalization of carbon allotropes is still a challenge.

8.5.2 Chemical reduction

Chemical reduction is a method of synthesizing materials by reducing metal salt precursors in the presence of a reducing agent and a carbon allotrope. The reducing agent is typically a small organic molecule that can donate electrons to the metal salt precursor, resulting in the formation of metal nanoparticles. The metal nanoparticles are then deposited onto the surface of the carbon allotrope, resulting in functionalized carbon materials with enhanced properties. In chemical reduction, the reducing agent plays a crucial role in determining the size and shape of the metal nanoparticles formed. Different reducing agents can lead to the formation of metal nanoparticles with different sizes and shapes. For example, sodium borohydride ($NaBH_4$) is commonly used as a reducing agent in the synthesis of metal nanoparticles due to its high reducing power. $NaBH_4$ can donate four electrons to the metal ion, leading to the formation of small and uniform metal nanoparticles. On the other hand, citric acid is a weak reducing agent that can donate one electron per molecule, leading to the formation of large and irregular-shaped metal nanoparticles.

The performance of chemical reduction can be affected by various factors, including the type of reducing agent, the concentration of the metal salt precursor, the reaction temperature, and the reaction time. The choice of reducing agent is critical in determining the size and shape of the metal nanoparticles formed. A reducing agent with high reducing power can lead to the formation of small and uniform metal nanoparticles. However, the concentration of the reducing agent must be carefully controlled to prevent the formation of agglomerated nanoparticles. The concentration of the metal salt precursor also plays a significant role in the performance of chemical reduction. A higher concentration of the metal salt precursor can lead to the formation of larger metal nanoparticles. However, too high a concentration can lead to the formation of agglomerated nanoparticles or the precipitation of metal salts, resulting in a reduced yield. The reaction temperature and time are also important factors in chemical reduction. Higher reaction temperatures can accelerate the reduction process, leading to the formation of smaller metal nanoparticles. However, high temperatures can also lead to the aggregation of metal nanoparticles or the degradation of the carbon allotrope. The reaction time also affects the size and shape of the metal nanoparticles formed. A longer reaction time can lead to the formation of larger metal nanoparticles. Chemical reduction has been used to functionalize various carbon allotropes, including graphene, carbon nanotubes, and carbon dots. The functionalization of carbon allotropes using chemical reduction involves the reduction of a metal salt precursor in the presence of the carbon allotrope and a reducing agent. The metal nanoparticles formed are deposited onto the surface of the carbon allotrope, resulting in functionalized carbon materials with enhanced properties.

8.5.3 Electrochemical deposition

Electrochemical deposition is a process that involves the transfer of metal ions from an electrolyte solution to a metal surface under the influence of an electric field. In this process, the metal ion is reduced at the cathode to form a metal atom, which is then deposited on the surface of the carbon allotrope. The carbon allotrope serves as the cathode and the metal that is to be deposited on the surface of the carbon allotrope serves as the anode. The process of electrochemical deposition can be divided into several stages, including nucleation, growth, and coalescence. In the nucleation stage, metal ions are reduced at the cathode to form metal atoms, which start to form a thin layer on the surface of the carbon allotrope. The growth stage follows the nucleation stage, where the deposited metal layer starts to grow and thicken. In the coalescence stage, adjacent metal atoms start to merge, forming a continuous and homogeneous layer.

Electrochemical deposition can be used to functionalize carbon allotropes with a wide range of metals, including zinc, copper, nickel, and silver, to enhance their corrosion inhibition properties. The metal ions that are to be deposited on the surface of the carbon allotrope are chosen, based on their ability to form a protective layer on the surface of the carbon allotrope. To enhance the corrosion inhibition properties of the carbon allotrope, the metal ion can be functionalized with different molecules such as nitrogen, oxygen, and sulfur. For example, by using a precursor solution that contains a metal ion that has been functionalized with nitrogen, a nitrogen-doped metal layer can be deposited on the surface of the carbon allotrope. This nitrogen-doped metal layer has been shown to exhibit enhanced corrosion inhibition properties. Electrochemical deposition has several advantages over other methods used to functionalize carbon allotropes for corrosion inhibition applications. One of the major advantages of electrochemical deposition is that it allows for the deposition of a thin and uniform layer of metal on the surface of the carbon allotrope. This thin and uniform layer of metal can provide excellent corrosion protection to the carbon allotrope.

Another advantage of electrochemical deposition is that it allows for the deposition of a metal layer on the surface of the carbon allotrope in a controlled manner. This controlled deposition of the metal layer allows for the precise tuning of the properties of the carbon allotrope for specific corrosion inhibition applications. Additionally, electrochemical deposition is a relatively simple and cost-effective method compared to other methods used to functionalize carbon allotropes for corrosion inhibition applications. This method does not require complex equipment or high temperatures and it can be performed under ambient conditions.

8.5.4 Microwave-assisted synthesis

Microwave-assisted synthesis involves the use of microwave radiation to initiate and control chemical reactions. Microwave radiation is a form of electromagnetic radiation that has a frequency range of 300 MHz to 300 GHz, which corresponds to a wavelength range of approximately 1 mm to 1 m. The energy carried by microwave radiation can be absorbed by polar or ionic molecules, leading to the excitation of molecular vibrations and rotations. This excitation can increase the rate of chemical reactions, resulting in faster synthesis times and higher yields. The principles of microwave-assisted synthesis are based on the ability of microwave radiation to heat materials non-uniformly. In conventional heating methods such as heating by convection or conduction, the heat is transferred from the outside to the inside of the material, leading to a temperature gradient across the material. This temperature gradient can cause thermal stress, which can lead to cracking or other forms of damage to the material. In contrast, microwave radiation can penetrate the material and heat it from the inside, resulting in a more uniform temperature distribution and reduced thermal stress [59, 60].

Microwave-assisted synthesis can be used to synthesize various carbon allotropes, including graphene, carbon nanotubes, and carbon dots. The synthesis process involves the use of a precursor material, which can be functionalized with different molecules such as nitrogen, oxygen, and sulfur to enhance the corrosion inhibition properties of the carbon allotrope. The precursor material is mixed with a reducing agent and a solvent, and the mixture is placed in a microwave reactor. Microwave radiation is then applied to the mixture, leading to the formation of the carbon allotrope. Microwave-assisted synthesis offers several advantages over conventional synthesis methods. First, it can significantly reduce the synthesis time, leading to faster and more efficient production of carbon allotropes. Second, it can lead to higher yields as the non-uniform heating provided by microwave radiation can result in more uniform temperature distributions, leading to more consistent reaction rates. Third, it can be easily scaled up for industrial production as the method is amenable to continuous-flow processing.

Despite its advantages, microwave-assisted synthesis also has some limitations. One limitation is the potential for overheating or thermal runaway, which can occur if the reaction is not monitored carefully. This can lead to the formation of unwanted byproducts or the degradation of the carbon allotrope. Additionally, the use of microwave radiation can lead to non-selective heating, which can result in the formation of multiple products. This can be mitigated by optimizing the reaction conditions such as the power and duration of the microwave radiation, and the concentration and ratio of the precursor material and the reducing agent.

8.6 Potential of carbon allotropes as corrosion inhibitors

8.6.1 Mechanisms of corrosion inhibition

Carbon allotropes have shown great potential as corrosion inhibitors due to their unique properties and ability to interact with the metal surface in a variety of ways. The mechanisms of corrosion inhibition by carbon allotropes can be broadly classified into two categories: physical and chemical. Physical mechanisms of corrosion inhibition involve the formation of a protective barrier on the metal surface that prevents contact between the metal and the corrosive environment. Carbon allotropes can act as physical inhibitors by forming a layer on the metal surface, which can act as a barrier to prevent the penetration of corrosive species. The high surface area of carbon allotropes such as graphene and carbon nanotubes allows for an effective coverage of the metal surface, thereby reducing the contact between the metal and the corrosive environment [61, 62].

Chemical mechanisms of corrosion inhibition involve the chemical interaction between the carbon allotrope and the metal surface, resulting in the formation of a passive layer on the metal surface. This passive layer prevents the penetration of corrosive species, thereby reducing the rate of corrosion. Carbon allotropes can act as chemical inhibitors by adsorbing onto the metal surface and forming a protective layer, which can interact with the corrosive species to form stable compounds.

8.6.2 Inhibitory properties of carbon allotropes

Graphene, carbon nanotubes, and carbon dots have shown significant potential as corrosion inhibitors due to their unique properties. Graphene, being a single layer of carbon atoms, has a high surface area-to-volume ratio, which allows it to effectively cover the metal surface and act as a physical barrier to prevent the penetration of corrosive species. Graphene oxide has also been shown to have inhibitory properties, possibly due to its functional groups that can interact chemically with the metal surface. Carbon nanotubes have a high aspect ratio and can align themselves in a perpendicular direction to the metal surface, allowing for an efficient coverage of the metal surface. The hollow structure of carbon nanotubes also allows for the adsorption of corrosive species, reducing their concentration and preventing them from interacting with the metal surface. Functionalization of carbon nanotubes with different molecules such as nitrogen, oxygen, and sulfur can enhance their inhibitory properties by increasing their adsorption capacity and modifying their surface chemistry. Carbon dots, being a type of carbon nanoparticle, also have a high surface area-to-volume ratio, allowing for an efficient coverage of the metal surface. The unique electronic properties of carbon dots also allow them to interact with corrosive species and form stable compounds, thereby preventing the penetration of corrosive species and reducing the rate of corrosion.

8.6.3 A comparison with traditional corrosion inhibitors

Carbon allotropes have several advantages over traditional corrosion inhibitors. Traditional corrosion inhibitors such as chromates and phosphates have been used for decades and have proven to be effective in reducing the rate of corrosion. However, they have several drawbacks, including toxicity, environmental concerns, and limited effectiveness at high temperatures and in harsh environments. Carbon allotropes, on the other hand, are non-toxic and environmentally friendly, making them an attractive alternative to traditional inhibitors. They are also effective at high temperatures and in harsh environments, making them suitable for a wide range of applications. Additionally, carbon allotropes can be functionalized with different molecules, allowing for the modification of their properties and the tailoring of their inhibitory properties to specific applications. In terms of performance, carbon allotropes have shown comparable or even superior inhibitory properties to traditional corrosion inhibitors. For example, graphene has been shown to have comparable inhibitory properties to chromates, while carbon nanotubes and carbon dots have shown superior inhibitory properties to phosphates.

8.7 Recent research studies on carbon allotropes-based thin film coatings for corrosion inhibition

In recent years, carbon allotropes-based thin film coatings have gained significant attention as promising candidates for corrosion inhibition. Among the various carbon allotropes, graphene, carbon nanotubes, and carbon dots have been extensively studied for their potential in developing thin film coatings with enhanced corrosion inhibition properties. In this discussion, we will review the recent research studies on graphene-based, carbon nanotube-based, and carbon dots-based thin film coatings for corrosion inhibition.

8.7.1 Graphene-based coatings

Graphene is a two-dimensional carbon allotrope with remarkable mechanical, electrical, and thermal properties. Due to its high surface area and unique electronic properties, graphene-based coatings have shown great potential in corrosion inhibition applications. In recent years, various studies have reported the synthesis of graphene-based coatings and their corrosion inhibition properties. One study by Xu et al. (2020) reported the synthesis of reduced graphene oxide (rGO)-based coatings on copper substrates for corrosion inhibition. The rGO was synthesized by the reduction of graphene oxide using hydrazine hydrate as a reducing agent. The rGO-coated copper substrates showed improved corrosion resistance properties compared to the uncoated substrates. The rGO coating provided a barrier against the diffusion of corrosive species, leading to a decrease in the corrosion rate.

Another study by Akhavan et al. (2020) reported the synthesis of graphene oxide (GO)-based coatings on aluminum substrates for corrosion inhibition. The GO was synthesized by the oxidation of graphite using the Hummers' method. The GO-coated aluminum substrates showed improved corrosion resistance properties compared to the uncoated substrates. The GO coating provided a barrier against the diffusion of corrosive species, leading to a decrease in the corrosion rate. In addition, the GO coating also acted as a sacrificial layer, leading to the formation of a protective oxide layer on the aluminum substrate [63, 64]. In a recent study by Zhang et al. (2021), a graphene oxide/chitosan (GO/CS) composite coating was synthesized on magnesium alloys for corrosion inhibition. The GO/CS composite coating showed improved corrosion resistance properties, compared to the uncoated substrate. The GO/CS composite coating provided a barrier against the diffusion of corrosive species and also acted as a reservoir for corrosion inhibitors, leading to a decrease in the corrosion rate. One study published in Materials Science and Engineering: A in 2021 investigated the use of graphene oxide (GO) coatings for corrosion protection of low-carbon steel. The researchers used a simple dip-coating method to deposit the GO coating on the steel surface and then performed various electrochemical tests to evaluate the corrosion inhibition properties of the coating. The results showed that the GO coating effectively inhibited the corrosion of the low-carbon steel and that the inhibition efficiency increased with the thickness of the coating.

Another study published in Carbon in 2020 reported the use of reduced graphene oxide (rGO) coatings for corrosion protection of magnesium alloys. Researchers used a spray coating method to deposit the rGO coating on the magnesium alloy surface, and then evaluated the corrosion resistance of the coating using electrochemical impedance spectroscopy and potentiodynamic polarization techniques. The results showed that the rGO coating significantly improved the corrosion resistance of the magnesium alloy, with an inhibition efficiency of up to 90%. A review article published in the Journal of Coatings Technology and Research in 2020 summarized the recent advances in graphene-based coatings for corrosion inhibition. The article highlighted the various functionalization methods used to enhance the corrosion inhibition properties of graphene-based coatings, including nitrogen doping, metal oxide deposition, and polymer grafting. The review also discussed the challenges in developing practical graphene-based coatings, including issues with scalability and cost-effectiveness.

8.7.2 Carbon nanotube-based coatings

Carbon nanotubes (CNTs) are one-dimensional carbon allotropes with unique mechanical and electrical properties. Due to their high aspect ratio, high surface area, and unique electronic properties, CNTs have shown great potential for corrosion inhibition applications. In recent years, various studies have reported the synthesis of CNT-based coatings and their corrosion inhibition properties. One study by Liu et al. (2020) reported the synthesis of CNT-based coatings on carbon steel substrates for corrosion inhibition. The CNTs were synthesized by chemical vapor deposition (CVD) and functionalized with amino groups using 3-aminopropyltriethoxysilane (APTES). The APTES-functionalized CNT-coated carbon steel substrates showed improved corrosion resistance properties compared to the uncoated substrates. The APTES-functionalized CNT coating provided a barrier against the diffusion of corrosive species and also acted as a reservoir for corrosion inhibitors, leading to a decrease in the corrosion rate. Recent studies have explored the potential of carbon nanotubes (CNTs) as a material for thin film coatings with corrosion inhibition properties. One study by Liu et al. (2021) investigated the use of functionalized CNTs as a coating for copper surfaces. The CNTs were functionalized with polyethyleneimine (PEI) and polyacrylic acid (PAA) to enhance their adhesion to the copper surface and provide corrosion inhibition. The coated copper samples showed significantly reduced corrosion rates compared to uncoated samples, with the PEI-functionalized CNT coating performing the best.

Another study by Zhang et al. (2021) explored the use of CNTs as a corrosion inhibitor for steel. The CNTs were functionalized with amines and coated onto the steel surface using an electrophoretic deposition method. The coated steel samples showed a significant decrease in corrosion rate compared to uncoated samples, with the amine-functionalized CNT coating performing the best. The study also found that the coating remained effective even after 168 h of salt spray testing. A study by Navarro et al. (2020) investigated the use of CNTs as a coating for aluminum alloys. The CNTs were functionalized with different nitrogen-containing molecules and deposited onto the aluminum surface using a spray coating method. The coated aluminum samples showed a significant decrease in corrosion rate compared to uncoated samples, with the nitrogen-functionalized CNT coatings performing the best. The study also found that the coatings had good adhesion to the aluminum surface and remained effective after exposure to a corrosive environment.

In another study by Al-Juaid et al. (2020), the potential of CNTs as a corrosion inhibitor for mild steel was investigated. The CNTs were functionalized with sulfonic acid and coated onto the mild steel surface using a dip-coating method. The coated steel samples showed a significant decrease in corrosion rate compared to uncoated samples, with the sulfonic acid-functionalized CNT coating performing the best. The study also found that the coating remained effective even after exposure to a corrosive environment for up to 21 days.

8.7.3 Carbon dots-based coatings

Carbon dots (CDs) are small carbon nanoparticles, typically less than 10 nm in size, with unique optical and electronic properties, making them suitable for various applications, including corrosion inhibition. CDs can be synthesized from various carbon sources such as citric acid, glucose, and ethylene glycol using various methods such as hydrothermal synthesis, microwave synthesis, and pyrolysis. The resulting CDs can be functionalized with various molecules to enhance their corrosion inhibition properties. Several recent studies have investigated the potential of CDs as corrosion inhibitors in various metallic systems. For example, Wang et al. (2021) synthesized CDs from citric acid and used them as a corrosion inhibitor for copper in 3.5% NaCl solution. They found that the CDs effectively reduced the corrosion rate of copper by forming a protective film on the surface of the metal. The CDs also exhibited good stability and were found to be non-toxic. In another study, Li et al. (2021) synthesized CDs from glucose and used them as a corrosion inhibitor for steel in simulated concrete pore solution. They found that the CDs effectively reduced the corrosion rate of steel by adsorbing onto the surface of the metal and forming a protective film. The CDs also exhibited good stability and were found to be non-toxic. Recent research studies have focused on the use of CDs as effective corrosion inhibitors for various metallic substrates. For instance, Wang et al. (2020) reported the synthesis of amine-functionalized CDs for the corrosion protection of carbon steel in 3.5% NaCl solution. The authors found that the CDs coating reduced the corrosion rate of carbon steel by 70% compared to the uncoated substrate. Another study by Xie et al. (2021) demonstrated the use of carboxyl-functionalized CDs for the corrosion protection of aluminum alloy in 3.5% NaCl solution. The authors found that the CDs coating significantly reduced the corrosion rate of the aluminum alloy and improved its mechanical properties [65–69].

Moreover, Li et al. (2021) reported the use of multi-functionalized CDs for the corrosion protection of copper in 0.5 M NaCl solution. The authors found that the CDs coating exhibited superior corrosion inhibition properties compared to traditional corrosion inhibitors such as benzotriazole and imidazole. In addition, other researchers have explored the use of CDs in combination with other corrosion inhibitors to enhance their corrosion inhibition properties. For example, Chen et al. (2020) reported the use of CDs, in combination with molybdate ions, for the corrosion protection of copper in 0.5 M NaCl solution. The authors found that the CDs/molybdate coating exhibited superior corrosion inhibition properties compared to the individual CDs or molybdate coatings. In a recent study by Yan et al. (2020), CDs were synthesized from glucose and functionalized with amino groups to enhance their corrosion inhibition properties. The resulting CDs were used as a corrosion inhibitor for copper in 3.5% NaCl solution. They found that the amino-functionalized CDs effectively reduced the corrosion rate of copper by forming a protective film on the surface of the metal. The CDs also exhibited good stability and were found to be non-toxic [70–74].

8.8 Conclusion and future prospects

The use of functionalized carbon allotropes such as graphene, carbon nanotubes, and carbon dots as thin film coatings for corrosion inhibition has shown great potential in recent research studies. These materials offer several advantages over traditional corrosion inhibitors, including high mechanical strength, chemical stability, and excellent electrical conductivity. Additionally, their functionalization with various molecules such as nitrogen, oxygen, and sulfur has been shown to enhance their corrosion inhibition properties. Graphene-based coatings have been extensively studied and have shown promising results in inhibiting corrosion in different environments, including acidic and alkaline solutions, as well as in the presence of salt and organic compounds. The unique properties of graphene such as its large surface area and high electrical conductivity make it an excellent candidate for use as a corrosion inhibitor. Carbon nanotubes have also been investigated as a potential corrosion inhibitor due to their high aspect ratio, excellent mechanical strength, and good electrical conductivity. The functionalization of carbon nanotubes with various molecules has been shown to enhance their corrosion inhibition properties. Furthermore, the use of carbon nanotube-based coatings has been demonstrated to improve the corrosion resistance of various materials, including aluminum, steel, and copper. Carbon dots, a relatively new class of carbon allotropes, have shown great potential as corrosion inhibitors due to their excellent photoluminescence and biocompatibility. Carbon dots have been synthesized using various methods and have been functionalized with different molecules to enhance their corrosion inhibition properties. Recent research studies have shown that carbon dots-based coatings can effectively inhibit corrosion in acidic and alkaline solutions.

Overall, the use of functionalized carbon allotropes-based thin film coatings for corrosion inhibition offers a sustainable and efficient alternative to traditional corrosion inhibitors. These materials have shown excellent corrosion inhibition properties and can be tailored to suit specific applications through functionalization.

8.9 Future directions for research and development

There is still much research to be carried out to fully understand the potential of functionalized carbon allotropes-based thin film coatings as corrosion inhibitors. Some of the key areas that require further investigation include:

– Optimization of synthesis methods: Although various synthesis methods have been developed for the production of functionalized carbon allotropes, there is still a need for optimization of these methods to improve the quality and reproducibility of the coatings.
– Understanding the underlying mechanisms of corrosion inhibition: While the use of functionalized carbon allotropes as corrosion inhibitors has shown promising results, the mechanisms underlying their inhibition properties are not yet fully understood. Further research is needed to investigate the interaction between the coatings and the corrosive environment.
– Development of scale-up techniques: Currently, most research studies on functionalized carbon allotropes-based thin film coatings for corrosion inhibition have been conducted at the laboratory scale. There is a need to develop scalable techniques for the production of these coatings to enable their practical use in industrial settings.
– Assessment of long-term stability and durability: While functionalized carbon allotropes-based thin film coatings have shown excellent corrosion inhibition properties in short-term studies, there is a need to assess their long-term stability and durability. The coatings need to be tested under various conditions to determine their resistance to degradation over time.
– Exploration of new applications: Although functionalized carbon allotropes-based thin film coatings have shown great potential for corrosion inhibition, there is a need to explore their potential in other applications such as in the biomedical and energy sectors.

In conclusion, the use of functionalized carbon allotropes-based thin film coatings for corrosion inhibition offers a sustainable and efficient alternative to traditional corrosion inhibitors. Further research is needed to optimize the synthesis methods, understand the underlying mechanisms of corrosion inhibition, develop scalable techniques for production, and assess long term.

References

[1] Saleh TA, Ibrahim MA. Advances in functionalized Nanoparticles based drilling inhibitors for oil production. Energy Reports [Internet] 2019;5:1293–304. Available from: https://doi.org/10.1016/j.egyr.2019.06.002.

[2] Bawazeer TM, El-Ghamry HA, Farghaly TA, Fawzy A. Novel 1,3,4-Thiadiazolethiosemicarbazones derivatives and their divalent cobalt-complexes: Synthesis, characterization and their efficiencies for acidic corrosion inhibition of carbon steel. J Inorg Organomet Polym Mater [Internet] 2020;30 (5):1609–20. Available from: https://doi.org/10.1007/s10904-019-01308-8.

[3] Li P, He L, Li X, Liu X, Sun M. Corrosion Inhibition Effect of N-(4-diethylaminobenzyl) Quaternary Ammonium Chitosan for X80 Pipeline Steel in Hydrochloric Acid Solution. Int J Electrochem Sci 2021;16(1):1–21.

[4] Cui M, Qiang Y, Wang W, Zhao H, Ren S. Microwave Synthesis of Eco-friendly Nitrogen Doped Carbon Dots for the Corrosion Inhibition of Q235 Carbon Steel in 0.1 M HCl. Int J Electrochem Sci 2021;16(1):1–23.

[5] Das M, Biswas A, Kumar Kundu B, Adilia Januário Charmier M, Mukherjee A, Mobin SM, et al.. Enhanced pseudo-halide promoted corrosion inhibition by biologically active zinc(II) Schiff base complexes. Chem Eng J 2019;357(October):447–57.

[6] Solomon MM, Umoren SA, Quraishi MA, Tripathy DB, Abai EJ. Effect of akyl chain length, flow, and temperature on the corrosion inhibition of carbon steel in a simulated acidizing environment by an imidazoline-based inhibitor. J Pet Sci Eng [Internet] 2020;187:106801. Available from: https://doi.org/10.1016/j.petrol.2019.106801.

[7] Loto CA, Loto RT. Corrosion inhibition effect of Allium Cepa extracts on mild steel in H 2 SO 4. Der Pharma Chem 2016;8(20):272–81.

[8] Thakur A, Kaya S, Kumar A. Recent Innovations in Nano Container-Based Self-Healing Coatings in the Construction Industry. Curr Nanosci 2021;18(2):203–16.

[9] Thakur A, Kaya S, Abousalem AS, Experimental KA. DFT and MC simulation analysis of Vicia Sativa weed aerial extract as sustainable and eco-benign corrosion inhibitor for mild steel in acidic environment. Sustain Chem Pharm [Internet] 2022;29(July):100785. Available from: https://doi.org/10.1016/j.scp.2022.100785.

[10] Thakur A, Kumar A. Recent advances on rapid detection and remediation of environmental pollutants utilizing nanomaterials-based (bio)sensors. Sci Total Environ [Internet] 2022;834 (January):155219. Available from: https://doi.org/10.1016/j.scitotenv.2022.155219.

[11] Thakur A, Sharma S, Ganjoo R, Assad H, Kumar A. Anti-Corrosive Potential of the Sustainable Corrosion Inhibitors Based on Biomass Waste: A Review on Preceding and Perspective Research. J Phys Conf Ser 2022;2267(1):012079.

[12] Thakur A, Kumar A, Sharma S, Ganjoo R, Assad H Materials Today : Proceedings Computational and experimental studies on the efficiency of Sonchus arvensis as green corrosion inhibitor for mild steel in 0 . 5 M HCl solution. Mater Today Proc [Internet]. 2022;66:609–21. Available from: https://doi.org/10.1016/j.matpr.2022.06.479

[13]   Thakur A, Kumar A. Sustainable Inhibitors for Corrosion Mitigation in Aggressive Corrosive Media: A Comprehensive Study. BioTriboCorros Internet2021;7(2):1–48. Available from. https://doi.org/10.1007/s40735-021-00501-y.

[14]   Thakur A, Kumar A, Kaya S, Marzouki R, Zhang F, Guo L. Recent Advancements in Surface Modification, Characterization and Functionalization for Enhancing the Biocompatibility and Corrosion Resistance of Biomedical Implants. Coatings 2022;12:1459.

[15]   Thakur A, Kaya S, Abousalem AS, Sharma S, Ganjoo R, Assad H, et al.. Computational and experimental studies on the corrosion inhibition performance of an aerial extract of Cnicus Benedictus weed on the acidic corrosion of mild steel. Process Saf Environ Prot [Internet] 2022;161:801–18. Available from: https://doi.org/10.1016/j.psep.2022.03.082.

[16]   Verma C, Thakur A, Ganjoo R, Sharma S, Assad H. Coordination bonding and corrosion inhibition potential of nitrogen-rich heterocycles : Azoles and triazines as specific examples. Coord Chem Rev [Internet] 2023;488(November 2022):215177. Available from: https://doi.org/10.1016/j.ccr.2023.215177.

[17]   SHARMA D. ABHINAY THAKUR 2 MKS, 1 KJ, * and HO 1 AK 3 AKS 1. Synthesis, Electrochemical, Morphological, Computational and Corrosion Inhibition Studies of 3-(5-Naphthalen-2-yl-[1,3,4] oxadiazol-2-yl)-pyridine against Mild Steel in 1 M HCl. Asian J Chem 2014;35(5):1079–88.

[18]   Bashir S, Thakur A, Lgaz H, Chung I-M, Kumar A. Computational and experimental studies on Phenylephrine as anti-corrosion substance of mild steel in acidic medium. J Mol Liq 2019;293:111539.

[19]   Bashir S, Thakur A, Lgaz H, Chung I-M, Kumar A. Corrosion Inhibition Performance of Acarbose on Mild Steel Corrosion in Acidic Medium: An Experimental and Computational Study. Arab J Sci Eng [Internet] 2020;45(6):4773–83. Available from: https://doi.org/10.1007/s13369-020-04514-6.

[20]   Thakur A, Savaş K, Kumar A. Recent Trends in the Characterization and Application Progress of Nano-Modified Coatings in Corrosion Mitigation of Metals and Alloys. Appl Sci 2023;13:730.

[21]   Bashir S, Thakur A, Lgaz H, Chung IM, Kumar A. Corrosion inhibition efficiency of bronopol on aluminium in 0.5 M HCl solution: Insights from experimental and quantum chemical studies. Surfaces Interfaces [Internet] 2020;20(April):100542. Available from: https://doi.org/10.1016/j.surfin.2020.100542.

[22]   Thakur A, Kumar A. Recent trends in nanostructured carbon-based electrochemical sensors for the detection and remediation of persistent toxic substances in real- time analysis. Mater Res Express 2023;10:034001.

[23]   Parveen G, Bashir S, Thakur A, Saha SK, Banerjee P, Kumar A. Experimental and computational studies of imidazolium based ionic liquid 1-methyl- 3-propylimidazolium iodide on mild steel corrosion in acidic solution Experimental and computational studies of imidazolium based ionic liquid 1-methyl- 3-propylimidazolium. Mater Res Express 2020;7(1):016510.

[24]   Kumar A, Thakur A. Encapsulated nanoparticles in organic polymers for corrosion inhibition [Internet]. Corrosion Protection at the Nanoscale. Elsevier Inc., pp. 345–362; 2020, Available from: http://dx.doi.org/10.1016/B978-0-12-819359-4.00018-0.

[25]   Thakur A, Kumar A, Kaya S, Vo DVN, Sharma A. Suppressing inhibitory compounds by nanomaterials for highly efficient biofuel production: A review. Fuel [Internet] 2022;312(September 2021):122934. Available from. https://doi.org/10.1016/j.fuel.2021.122934.

[26]   Baig N, Chauhan DS, Saleh TA, Quraishi MA. Diethylenetriamine functionalized graphene oxide as a novel corrosion inhibitor for mild steel in hydrochloric acid solutions. New J Chem 2019;43 (5):2328–37.

[27]   Verma C, Quraishi MA. Chelation capability of chitosan and chitosan derivatives: Recent developments in sustainable corrosion inhibition and metal decontamination applications. Curr Res Green Sustain Chem [Internet] 2021;4(September):100184. Available from: https://doi.org/10.1016/j.crgsc.2021.100184.

[28] Wang C, Chen J, Han J, Wang C, Hu B. Enhanced corrosion inhibition performance of novel modified polyaspartic acid on carbon steel in HCl solution. J Alloys Compd [Internet] 2019;771:736–46. Available from: https://doi.org/10.1016/j.jallcom.2018.08.031.

[29] Xu H, Zhang Y. A review on conducting polymers and nanopolymer composite coatings for steel corrosion protection. Coatings 2019;9(12).

[30] Lee M-H. Cobalt nanoparticles immobilized in 3D mesoporous silica KIT-6 for excellent catalytic hydrogen generation from ammonia borane. J Mater Sci Eng 2017;06(03):4172.

[31] Sadek RF, Farrag HA, Abdelsalam SM, Keiralla ZMH, Raafat AI, Araby E. A powerful nanocomposite polymer prepared from metal oxide nanoparticles synthesized Via brown algae as anti-corrosion and anti-biofilm. Front Mater 2019;6(July):1–17.

[32] Thomas NC, Beaumont OA, Deacon GB, Gaertner C, Forsyth CM, Somers AE, et al.. Preparation and Structures of Rare Earth 3-Benzoylpropanoates and 3-Phenylpropanoates. Aust J Chem 2020;73 (12):1250–9.

[33] Fouda AEAS, El-bendary MM, Din H ESE, Structure MMM. Characterizations and Corrosion Inhibition of New Coordination Polymer Based on Cadmium Azide and Nicotinate Ligand. Prot Met Phys Chem Surfaces 2018;54(4):689–99.

[34] Hegazy MA, Badawi AM, El Rehim SS A, Kamel WM. Corrosion inhibition of carbon steel using novel N-(2-(2-mercaptoacetoxy)ethyl)-N,N-dimethyl dodecan-1-aminium bromide during acid pickling. Corros Sci [Internet] 2013;69:110–22. Available from: http://dx.doi.org/10.1016/j.corsci.2012.11.031.

[35] Mishra M, Tiwari K, Singh AK, Singh VP. Versatile coordination behaviour of a multi-dentate Schiff base with manganese(II), copper(II) and zinc(II) ions and their corrosion inhibition study. Inorganica Chim Acta [Internet] 2015;425(Ii):36–45. Available from: http://dx.doi.org/10.1016/j.ica.2014.10.026.

[36] Liu JJ, Li ZY, Yuan X, Wang Y, Huang CC. A copper(I) coordination polymer incorporation the corrosion inhibitor 1H-benzotriazole: Poly[μ3-benzotriazolato-κ3 N1:N2:N3-copper(I)]. Acta Crystallogr Sect C: Struct Chem 2014;70(6):599–602.

[37] Umoren SA, Solomon MM. Effect of halide ions on the corrosion inhibition efficiency of different organic species – A review. J Ind Eng Chem 2015;21:81–100.

[38] Petkova G, Sokolova E, Raicheva S, Ivanov P. Corrosion inhibition of copper in near neutral aqueous solutions by gamma-pyrodiphenone. J Appl Electrochem 1998;28(10):1067–71.

[39] Sesmero E, Calatayud DG, Perles J, López-Torres E, Mendiola MA. The Reactivity of Diphenyllead(IV) Dichloride with Dissymmetric Thiosemicarbazone Ligands: Obtaining Monomers, Coordination Polymers, and an Organoplumboxane. Eur J Inorg Chem 2016;2016(7):1044–53.

[40] Sekine I, Sanbongi M, Hagiuda H, Oshibe T, Yuasa M, Imahama T, et al.. Corrosion Inhibition of Mild Steel by Cationic and Anionic Polymers in Cooling Water System. J Electrochem Soc 1992;139 (11):3167–73.

[41] Ahoulou S, Vilà N, Pillet S, Schaniel D, Walcarius A. Coordination Polymers as Template for Mesoporous Silica Films: A Novel Composite Material Fe(Htrz)3@SiO2 with Remarkable Electrochemical Properties. Chem, Mater 2019;31(15):5796–807.

[42] Othman NK, Yahya S, Ismail MC. Corrosion inhibition of steel in 3.5% NaCl by rice straw extract. J Ind Eng Chem [Internet] 2019;70(March):299–310. Available from: https://doi.org/10.1016/j. jiec.2018.10.030.

[43] Papadaki M, Demadis KD. Structural mapping of hybrid metal phosphonate corrosion inhibiting thin films. Comments Inorg Chem 2009;30(3–4):89–118.

[44] Shokry H, Yuasa M, Sekine I, Issa RM, El-Baradii HY, Gomma GK. Corrosion inhibition of mild steel by Schiff base compounds in various aqueous solutions: Part I. Corros Sci 1998;40(12):2173–86.

[45] Qin L, Xiao SL, Ma PJ, Cui GH. Synthesis, crystal structures and catalytic properties of Ag(I) and Co(II) 1D coordination polymers constructed from bis(benzimidazolyl)butane. Trans Met Chem 2013;38 (6):627–33.

[46]   Iannuzzi M, Young T, Frankel GS. Aluminum Alloy Corrosion Inhibition by Vanadates. J Electrochem Soc 2006;153(12):B533.

[47]   Kuznetsov YI, Andreev NN, Marshakov AI. Physicochemical Aspects of Metal Corrosion Inhibition. Russ J Phys Chem A 2020;94(3):505–15.

[48]   Kharitonov DS, Kurilo WA, II, Zharskii IM. Korrosionsinhibierung der Aluminiumlegierung AA6063 in alkalischen Medien durch Vanadate. Materwiss Werksttech 2017;48(7):646–60.

[49]   Kim YS, Lee SH, Son MY, Jung YM, Song HK, Lee H. Succinonitrile as a corrosion inhibitor of copper current collectors for overdischarge protection of lithium ion batteries. ACS Appl Mater Interfaces 2014;6(3):2039–43.

[50]   Li X, Deng S, Xie X. Experimental and theoretical study on corrosion inhibition of oxime compounds for aluminium in HCl solution. Corros Sci [Internet] 2014;81:162–75. Available from: http://dx.doi.org/10.1016/j.corsci.2013.12.021.

[51]   Yang Z, Peng H, Wang W, Liu T. Crystallization behavior of poly(ε-caprolactone)/layered double hydroxide nanocomposites. J Appl Polym Sci 2010;116(5):2658–67.

[52]   Da Silva JEP, De Torresi SIC, Torresi RM. Polyaniline acrylic coatings for corrosion inhibition: The role played by counter-ions. Corros Sci 2005;47(3 SPEC. ISS.). 811–22.

[53]   Tao Z, Zhang S, Li W, Hou B. Adsorption and corrosion inhibition behavior of mild steel by one derivative of benzoic-triazole in acidic solution. Ind Eng Chem Res 2010;49(6):2593–9.

[54]   Singh R, Pratibha Chaudhary NKK. A Review: Organotin Compounds in Corrosion Inhibition. Rev Inorg Chem [Internet] 2010;30(4):275–95. Available from: http://www.reference-global.com/doi/abs/10.1515/REVIC.2010.30.4.275.

[55]   Negm N, Yousef M, Tawfik S. Impact of Synthesized and Natural Compounds in Corrosion Inhibition of Carbon Steel and Aluminium in Acidic Media. Recent Patents Corros Sci 2013;3(1):58–68.

[56]   Schultz ZD, Biggin ME, White JO, Gewirth AA. Infrared-Visible Sum Frequency Generation Investigation of Cu Corrosion Inhibition with Benzotriazole. Anal Chem 2004;76(3):604–9.

[57]   Zhong WY, Zheng ZH. Spectro-electrochemical studies of the corrosion inhibition of copper by mercaptobenzothiazole. J Inorg Chem (Chinese) 1999;15(6):759–60.

[58]   Radovanović MB, Tasić ŽZ, Mihajlović MBP, Simonović AT, Antonijević MM. Electrochemical and DFT studies of brass corrosion inhibition in 3% NaCl in the presence of environmentally friendly compounds. Sci Rep 2019;9(1):1–16.

[59]   Hassan R M, Ibrahim S M, Khairou K S. Novel Synthesis of Diketopectate Coordination Biopolymer Derivatives as Alternative Promising In Biomedicine, Pharmaceutics and Food Industrial Applications. Nutr Food Process 2019;2(2):01–5.

[60]   Solomon MM, Umoren SA. Enhanced corrosion inhibition effect of polypropylene glycol in the presence of iodide ions at mild steel/sulphuric acid interface. J Environ Chem Eng [Internet] 2015;3(3):1812–26. Available from: http://dx.doi.org/10.1016/j.jece.2015.05.018.

[61]   Qian B, Wang J, Zheng M, Hou B. Synergistic effect of polyaspartic acid and iodide ion on corrosion inhibition of mild steel in H2SO4. Corros Sci Internet2013;75:184–92. Available from: http://dx.doi.org/10.1016/j.corsci.2013.06.001.

[62]   Karthikaiselvi R, Subhashini S. Study of adsorption properties and inhibition of mild steel corrosion in hydrochloric acid media by water soluble composite poly (vinyl alcohol-o-methoxy aniline). J Assoc Arab Univ Basic Appl Sci [Internet] 2014;16:74–82. Available from. http://dx.doi.org/10.1016/j.jaubas.2013.06.002.

[63]   Atta AM, El-Azabawy OE, Ismail HS, Hegazy MA. Novel dispersed magnetite core-shell nanogel polymers as corrosion inhibitors for carbon steel in acidic medium. Corros Sci [Internet] 2011;53(5):1680–9. Available from: http://dx.doi.org/10.1016/j.corsci.2011.01.019.

[64]   Rakha TH, Bekheit MM, Ibrahim KM. Ligational, thermal, corrosion inhibition and antimicrobial properties of salicylidinebenzenesulphonylhydrazone. Trans Met Chem 1989;14(5):371–4.

[65] Manoj A, Ramachandran R, Menezes PL. Self-healing and superhydrophobic coatings for corrosion inhibition and protection. Int J Adv Manuf Technol 2020;106(5–6):2119–31.

[66] Raj R, Taryba MG, Morozov Y, Kahraman R, Shakoor RA, Montemor MF. On the synergistic corrosion inhibition and polymer healing effects of polyolefin coatings modified with Ce-loaded hydroxyapatite particles applied on steel. Electrochim Acta [Internet] 2021;388:138648. Available from. https://doi.org/10.1016/j.electacta.2021.138648.

[67] Tao Z, Zhang S, Li W, Hou B. Corrosion inhibition of mild steel in acidic solution by some oxo-triazole derivatives. Corros Sci Internet 2009;51(11):2588–95. Available from. DOI: http://dx.doi.org/10.1016/j.corsci.2009.06.042.

[68] Shainy KM, Rugmini Ammal P, Unni KN, Benjamin S, Joseph A. Surface interaction and corrosion inhibition of mild steel in hydrochloric acid using pyoverdine, an eco-friendly bio-molecule. BioTriboCorros 2016;2(3):1–12.

[69] Singh P, Singh DP, Tiwari K, Mishra M, Singh AK, Singh VP. Synthesis, structural investigations and corrosion inhibition studies on Mn(ii), Co(ii), Ni(ii), Cu(ii) and Zn(ii) complexes with 2-amino-benzoic acid (phenyl-pyridin-2-yl-methylene)-hydrazide. RSC Adv [Internet]2015;5(56):45217–30. Available from: http://dx.doi.org/10.1039/C4RA11929K.

[70] Mo W, Xiong H, Li T, Guo X, Li G. The catalytic performance and corrosion inhibition of CuCl/Schiff base system in homogeneous oxidative carbonylation of methanol. J Mol Catal A Chem 2006;247 (1–2):227–32.

[71] Umoren SA, Solomon MM, Udosoro II, Udoh AP. Synergistic and antagonistic effects between halide ions and carboxymethyl cellulose for the corrosion inhibition of mild steel in sulphuric acid solution. Cellulose 2010;17(3):635–48.

[72] Kloet JV, Schmidt W, Hassel A, Stratmann M. The role of chromate in filiform corrosion inhibition. Electrochim Acta 2004;49(9–10):1675–85.

[73] Xue YN, Xue XZ, Miao M, Liu JK. Mass preparation and anticorrosion mechanism of highly triple-effective corrosion inhibition performance for co-modified zinc phosphate-based pigments. Dye Pigment Internet2019;161(October 2018):489–99. Available from: https://doi.org/10.1016/j.dyepig.2018.10.001.

[74] Tian YQ, Xu HJ, Weng LH, Chen ZX, Zhao DY, You XZ. CuI(im)]∞: Is this air-stable copper(I) imidazolate (8210)-net polymer the species responsible for the corrosion-inhibiting properties of imidazole with copper metal?. Eur J Inorg Chem 2004;9:1813–6.

P. R. Rajimol, Sarah Bill Ulaeto, Ben John, T. P. D. Rajan

9 Metal-organic frameworks (MOFs)-based functionalized thin film coating for corrosion protection

Abstract: Among the various corrosion protection methods, the use of corrosion inhibitors in coatings is the most effective and commonly used method. MOFs are a relatively new class of hybrid systems containing metal ions and organic ligands with exceptional surface area and porous nature. Many factors like the hetero atom density, high crystallinity, versatile and tunable porous nature, wide range of compositions, high surface area, stability, etc., make MOFs ideal for corrosion inhibition applications. They are potential candidates to be used as corrosion inhibitors in different ways, like dispersed in corrosive media, as a container for loading other inhibitors, as thin films deposited on the metal surface, and incorporated in the polymer matrix. Among these methods, incorporation in the polymer matrix is commonly used and the most effective as it enhances the potential of the respective barrier coatings and mechanical properties. Various corrosion analysis methods like EIS and potentiodynamic polarization studies in thin film coatings are discussed in this chapter. Epoxy medium is generally reinforced with MOFs for corrosion protection applications, improving the coating's mechanical strength, adhesion property, barrier protection, and stability. All these factors make MOFs ideal for both real-time applications and research purpose. The different methods of preparation of the MOFs, their application methods and examples, limitations and possible rectifications, and corrosion inhibition mechanisms are discussed in this chapter.

Keywords: Corrosion Inhibition, MOFs, Polymer Coatings, EIS and Potentiodynamic Polarization, Corrosion Inhibition Mechanism

9.1 Introduction

Corrosion is the deterioration of the metal surface due to chemical or environmental factors. It is a spontaneous and unavoidable process that affects all industries severely. The annual cost of corrosion maintenance and loss due to accidents is almost 3–4% of a country's GDP [1]. Many methods like cathodic protection, anodic protection, thin film polymer coatings, etc. are used to reduce the corrosion rate. Among all the methods reported, a barrier coating with a thin film containing some corrosion inhibitors is the most commonly used method.

Many organic, inorganic, and bio-based pigments or fillers are used as corrosion inhibitors. Previously, chromates and phosphates were used widely, but they are now banned due to their toxic, carcinogenic, and endocrine-disrupting nature. Many bio-

https://doi.org/10.1515/9783111016160-009

based products, coordination compounds, and inorganic salts have been used recently in corrosion inhibition coatings [2, 3]. Materials with greater surface area and hetero atom density are the ideal choices for selection as corrosion inhibitors. Metal-organic frameworks (MOFs) have recently been recognized as potent candidates for metallic corrosion prevention.

MOFs are synthesized by the reaction between metal cations and organic ligands. They are also called porous polymers or coordination polymers. Versatile and tunable pore structure, large surface area, flexible structure, variety of compositions, and high crystallinity make them suitable for many applications [4]. Some MOFs have about 6,200 m^2/g of interior surface area, and more than 90% of the internal volume is free. This free space allows the loading of other compounds and makes MOFs an ideal choice for catalysis, corrosion inhibition, water treatment, etc. Different synthesis methods can prepare a large variety of MOFs from coordination metal cations and heteroatom-bearing organic ligands [5].

Due to their flexible and highly tuneable geometry, dynamic behavior to external stimuli, or their sensitivity to stimuli such as temperature, pH change, or pressure, MOFs stand apart from other inorganic mesoporous, nanoporous, or microporous materials. Along with the donor-acceptor interaction in coordination bonding, weak coordinative interactions, including H-bonding, pi-pi stacking, and van der Waals interactions, support the structure of MOFs. Due to these weak forces of attraction, structural flexibility can be attained [6].

Because they can be utilized for catalysis, drug delivery, gas storage or separation, and biological imaging, MOFs have attracted much attention. In the realm of electrochemistry, MOFs have been utilized to create materials for supercapacitors, fuel cells, rechargeable batteries, electrocatalysis, and corrosion prevention [7]. Because of their loading and releasing characteristics, MOFs are excellent candidates for catalysis; they are effective adsorbents for selective separation and for storage of gases like H_2 and CH_4. Some biomedical uses include pharmaceutical loading, anti-cancer use, anti-bacterial applications, and wound healing [5].

9.1.1 Method of synthesis of MOFs

Compared to other organic corrosion inhibitors, MOFs have some advantages in the synthesis steps involved regarding the precursors and yield. The precursor compound for MOFs synthesis are generally affordable inorganic salts like nitrates and chlorides; and ligands are multidentate ligands like nitriles, carboxylates, etc. The yield is usually high compared to organic synthetic routes. Many low-temperature synthesis procedures are also reported [7]. MOFs can be prepared by different methods like diffusion, solvo-thermal, isothermal, microwave-assisted synthesis, electrochemical synthesis, etc. The various methods have diverse requirements and thus, different limitations and advantages [5]. The conditions for each method are given below in Figure 9.1.

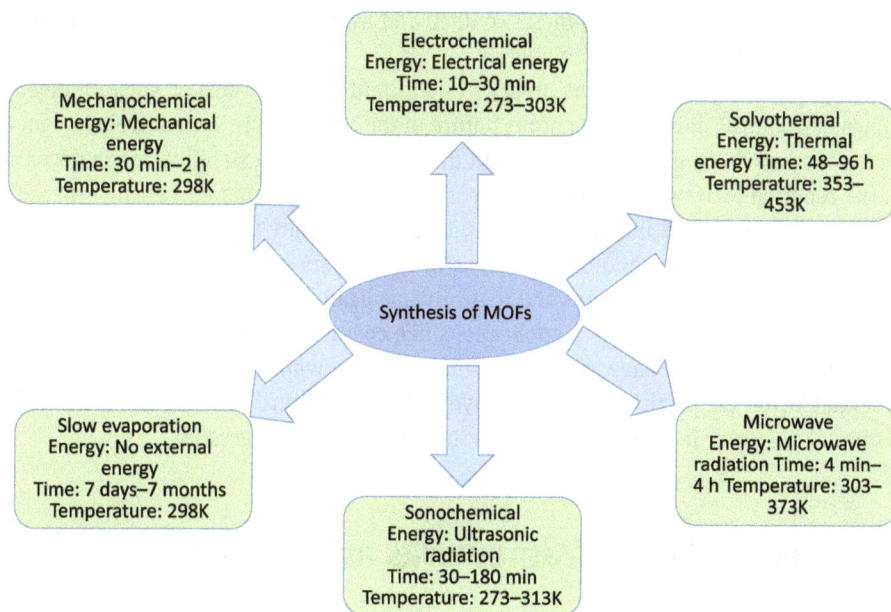

Figure 9.1: Methods for the preparation of metal-organic frameworks (figure adapted from ref. [5]).

9.1.2 Stability of MOFs

"Stability" or "robustness" is more of a collection of parameter sets than a specific quality. The stability of the inhibitors is a crucial consideration when adding them to coatings that prevent corrosion. There are many genres in which stability can be studied, including mechanical, thermal, chemical, and water stability. MOF stability describes how resistant the framework is to structural degradation in operating circumstances. Recent research has produced a wide variety of thermally and chemically withstanding MOFs, with potential applications [5].

In any industrial sector, components with high thermal stability are essential. Due to structural degradation, heat treatment operations result in metal-oxo-cluster dehydration, linker dehydrogenation, melting, graphitization, or amorphization. These effects are more apparent as the temperature rises. Additionally, they only occur at temperatures higher than the breakdown point. In addition to the breakdown of the ligand-metal linkages and the subsequent burning of the organic spacer, this scenario typically happens with the combustion and/or release of the guest molecules. The factors affecting the thermal stability of the MOFs are the coordination number, spatial arrangements, nature of ligand (aromatic/ aliphatic), the oxidation state of the central metal ion, ion size, etc. A framework with a compact packing, metallic ions with a stable oxidation state, a strong ligand-metal contact, and a tiny aromatic linker are the best options for excellent thermal stability [5].

In practical applications, the ability to withstand high pressure or vacuum determines mechanical stability. The porous nature of the MOFs results in relatively low mechanical strength, compared to zeolites. In the case of chemical stability, only a portion of the reported MOFs is found to be stable in non-inert environments. Reactive species' stability can be divided into several areas, including stability in a range of pH values, stability under extreme conditions (such as in the presence of NH_3 or H_2S), and moisture stability. Divalent metal ions and their organic ligands are frequently unstable and have brittle structures. In many applications like corrosion inhibition, catalysis, water purification, etc. MOFs have to interact with water; so water/ chemical stability is an essential factor. ZIF particles are usually hydrophobic in nature. The possible competing nature of the water in ligand replacement should also be considered [5].

9.2 Metal-organic frameworks as corrosion inhibitors

A particle's ability to suppress corrosion depends on various physicochemical factors, including its size, shape, active chemical groups, aromaticity, and ability to donate electrons. Organic molecules having heteroatoms like N, P, O, or S are discovered to be highly effective inhibitors of corrosion for many metals. Most MOFs contain heteroatoms with a lone pair of electrons or the pi electron cloud, which are suitable corrosion inhibitors, similar to coordination compounds [6].

Like the applications of all other corrosion inhibitors, MOFs also can be employed in different ways to prevent or reduce the rate of metallic corrosion. Most MOFs are highly porous materials with enormous surface area; they can be used as containers for loading other corrosion inhibitors. The combined effect may result in a synergistic effect in corrosion inhibition with exceptional performance. The loading of the inhibitor in the container allows the controlled release of the compound and enhanced stability of the inhibitor in aggressive mediums. They can also be used as a corrosion inhibitor in picking solutions by simply dispersing them in the corrosive medium. In the case of thin-layer corrosion protection, MOFs are most widely used in two methods. The first is direct in situ deposition of the MOFs on the surface to protect, and the second is the incorporation of the MOFs in the polymer matrix. Epoxy coatings are really compatible with the MOFs, and the combined results of barrier protection and inhibitor action are outstanding.

9.2.1 MOFs forming a thin-layer protection in corrosive solutions

Huiwen Tian et al. [8] inhibited mild steel corrosion by incorporating triazole derivatives in a ZIF-8 metal-organic framework. Ball-milled ZIF-8 was introduced into the inhibitor solution and was stirred for 24 h. The inhibitor-loaded MOFs were then collected
by centrifugation and dried for solvent removal. Electrochemical impedance spectroscopy (Figure 9.2) and potentiodynamic polarization studies were performed in 0.5M
NaCl solution to evaluate the efficiency of the modified corrosion inhibitor. It was found
that the MOF-inhibitor complex forms a thin layer of protection on the metal surface
(Figure 9.3) by physisorption using the lone pair of electrons from the heteroatom. The
loading of the inhibitor in the ZIF-8 reduced the optimal dosage by a factor of four.
The effective loading and the thin film formation on the surface of the mild steel are
the combined result of intermolecular hydrogen bonding and coordination bonding.

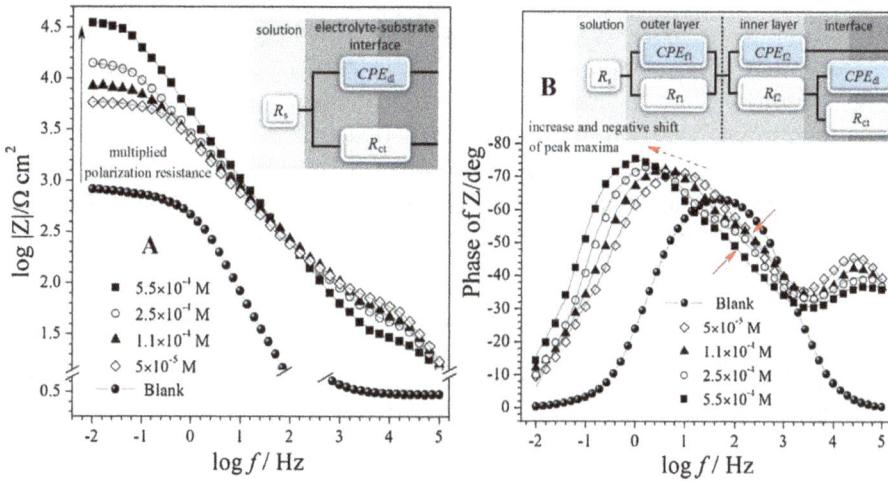

Figure 9.2: The equivalent circuit and Bode plot of M S samples after 12 h immersion in NaCl (0.5M) with
various amounts of the inhibitor (figure adapted with permission from ref. [8]. Distributed under a
Creative Commons Attribution License 4.0 (CC BY) https://creativecommons.org/licenses/by/4.0/).

Even though the grooves remain more or less visible, the contact angle of the covered
surface was 110°, indicating hydrophobicity. But, the property of the layers in the logarithmic stage was found to be less hydrophobic, which indicates a cracked structure,
possibly due to a heavily disordered molecular orientation on the surface.

By the reaction between AgNO$_3$, quinoxaline (qox), and 4-amino benzoic acid (4-
aba), Etaiw et al. synthesized yellow crystals of MOF [Ag(qox)(4-ab)] at room temperature conditions [9]. The Nyquist plot obtained from the EIS measurements in 1M HCl
fitted to an electrochemical equivalent circuit suggests the formation of a thin layer
by the adsorption of the inhibitor molecule on the metal surface. The pi electrons

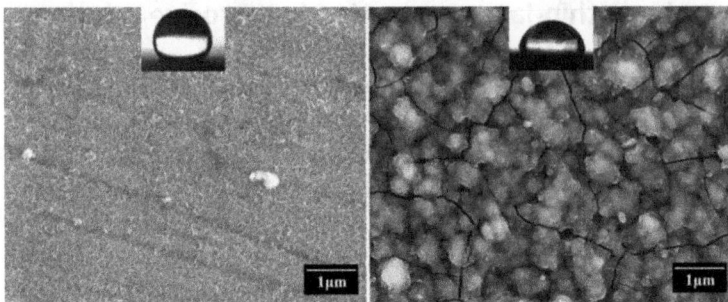

Figure 9.3: At an inhibitor concentration of 5.5 × 10^{-4}M, iron-deposited electrode in the parabolic (left) state, showing a hydrophobic surface, and the logarithmic (right) layer showing a hydrophilic surface nature from SEM and contact angle studies (figure adapted with permission from ref. [8]. Distributed under a Creative Commons Attribution License 4.0 (CC BY) https://creativecommons.org/licenses/by/4.0/).

from the inhibitor orbital are transferred to the metal atom's vacant d-orbital, resulting in a strong adsorption layer on the metal surface. The protective layer of MOF prevents corrosive ions and water molecules from interacting with metal atoms.

The corrosion inhibition efficiency of the MOFs containing the same organic ligand and different metal ions may vary. Kumaraguru et al. [10] analyzed the effect of metal atoms on corrosion inhibition by evaluating three metal-organic frameworks of the trimesic acid ligand with cobalt, copper, and nickel. The three nanoparticles in the morphology seem different. Ni-containing MOF was a one-dimensional nanorod structure and the Cu-MOF was a foam-like structure. At the same time, the Co-bearing particles were solid chunky pieces with polydispersed nature particles, with etched edges.

A small amount of acrylic resin was added to each of the nanoparticles for coating purposes. The amount of resin was meagre compared to the substance under investigation. The coating thickness was approximately 120 μm; the drying process was carried out at room temperature for seven days. Coated mild steel samples were evaluated with the help of EIS measurements in 0.1 M HCl aqueous solution as the electrolyte solution. The observed corrosion prevention order is Ni-MOF > Co-MOF > Cu-MOF. In terms of thermal stability, the performance of Cu-MOF stand out when compared to the other two MOFs. The results revealed that, like all other properties, the corrosion inhibition property not only depends on the ligand present but also on the metal atom present.

9.2.2 MOFs incorporated in polymer composite thin film coatings

Among all the corrosion prevention methods, barrier coatings with organic polymers are the most accepted. The significant drawbacks of this method are the relatively low mechanical strength, scratch and other defects formed, microcracks formed during curing, etc.; all these factors eventually affect corrosion [5]. Recently, many micro and nanofillers have been studied and incorporated into the polymer matrix to improve its mechanical properties. The self-healing efficiency of many fillers helps improve the polymer's performance.

While incorporating the inorganic/ hybrid fillers in the polymer matrix, the central aspect being considered is the compatibility of the organic and inorganic components. To enhance the organic polymers' physical, mechanical, thermal, and corrosion-inhibiting properties, numerous nano- and microparticles have been developed. As a hybrid entity of organic and inorganic components, the compatibility between MOFs and organic polymers is highly appreciable. Adding micro/nano-MOFs to organic films is an intriguing and practical way to enhance the coating's thermomechanical properties. Additionally, it has been noted that these structures efficiently disperse the stress-energy imparted to the surfaces [11, 12].

Tang and Tanase explored the manufacturing processes of MOFs and their composites and assessed their ability to suppress corrosion using various methods. When compared to pure polymer membranes, they found that MOFs added to polymer matrices improved the blending efficiency and permeability [13]. Also, incorporating the inhibitors in the containers and then dispersing them in the polymer matrix can induce the self-healing efficiency and controlled release of the inhibitor. It can avoid unwanted interactions with the polymer, leaching of inhibitors, etc. [14] In this approach, the MOFs can either be a container or act as an inhibitor in the epoxy medium.

For the best performance of functional polymers, the compatibility between the filler and the organic part plays an important role. Among all the MOFs-polymer combinations, the epoxy combination of zeolitic imidazolate framework-8 (ZIF-8) is the most successful and frequently reported modification, particularly graphene oxide-modified ZIF-8. For the pH-dependent smart delivery of drugs, ZIF-8 is widely used [15]; its property can be used for the controlled release of the inhibitor. More studies are based on ZIF-8 because of this pH-dependent release. Studies on another polymeric systems and polyvinyl butyral resin are also reported.

Song Duan et al. [16] investigated the effect of ZIF-8 MOFs, synthesized by the micro emulsion method, in corrosion inhibition when incorporated in the epoxy matrix using EIS and salt spray analysis. They found that the chemical reaction between ZIF-8 and the epoxy resolves the filler/ resin interphase problem. Epoxy coatings containing different mass fractions of uniform-sized ZIF-8 particles were prepared. The corrosion inhibition potential and other mechanical properties were studied. Both, corrosion resistance and mechanical strength were found to be improved by the addition of ZIF-8 particles as crosslinkers and, thus, the barrier properties improved. Impedance measurements showed that the corrosion inhibition efficiency increases as the matrix's mass fraction of the ZIF-8 increases. The results were also confirmed with the help of salt spray test results (Figure 9.4b) and adhesion test. Incorporation of the filler also improved the tensile strength and friction properties.

a)

b)

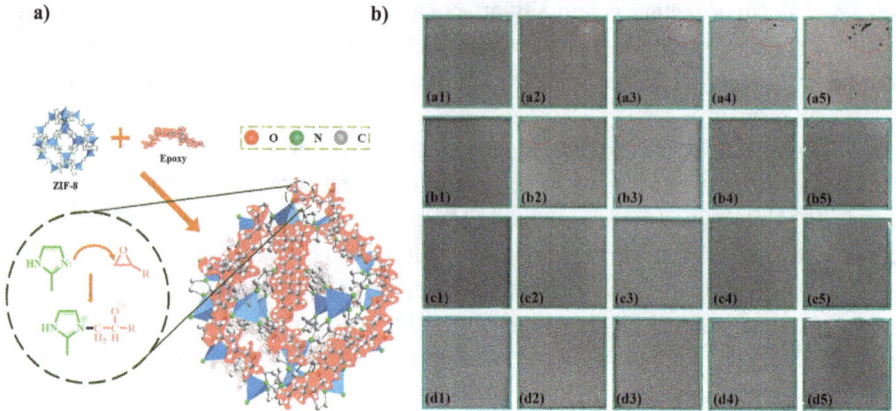

Figure 9.4: (a) Reaction between ZIF-8 and epoxy. (b) Visual appearance of different coatings after salt spray exposure of 0, 168, 720, 1,440, and 2,000 h (a-d) in increasing order of ZIF-8 concentration (figure adapted with permission from ref. [16]. Distributed under a Creative Commons Attribution License 4.0 (CC BY) https://creativecommons.org/licenses/by/4.0/).

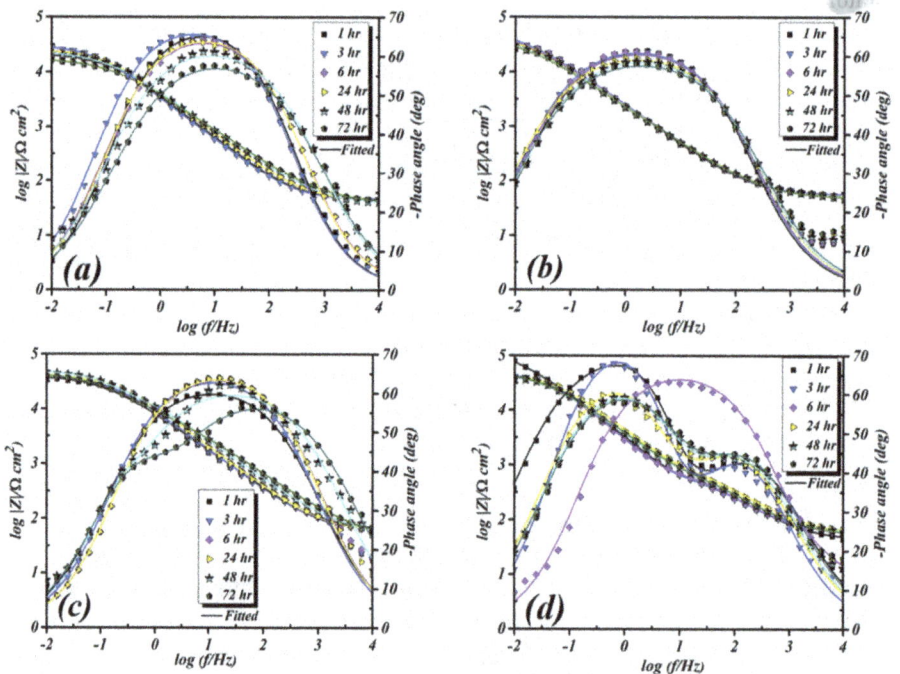

Figure 9.5: Bode plots for the (a) pure epoxy coating, (b) GO /Epoxy coating, (c) ZIF-8/ Epoxy coating, and (d) GO@ZIF-8/Epoxy coated samples with an artificial scratch (1 cm in length) immersed in saline solution (3.5 wt%) for different durations (figure adapted with permission from ref. [17]. Distributed under a Creative Commons Attribution License 4.0 (CC BY) https://creativecommons.org/licenses/by/4.0/).

Ramezanzadeh et al. [17] developed ZIF-8 particles in GO sheets through a one-pot synthesis strategy. GO@ZIF-8 exhibited a 79% improved surface area compared to graphene oxide. NaCl solution containing GO@ZIF-8 particles demonstrated a 70% improvement in corrosion resistance of mild steel, evaluated by polarization studies. The GO@ZIF-8/epoxy coating showed low-frequency impedance ($|Z|10$ mHz) values of greater than 1,010 $\Omega.cm^2$, revealing enhanced barrier protection (Figure 9.5). Also, the cathodic delamination strength and the wet adhesion strength of the epoxy improved by incorporating these nanoparticles in the epoxy matrix.

In another study, Li et al. [18] synthesized a hybrid corrosion inhibitor, M-ZIF-8/GO, by loading 2-Mercaptobenzimidazole (M) in ZIF-8 particles deposited on GO (Figure 9.6). They incorporated this hybrid inhibitor in an epoxy matrix and the resultant coating, M-ZIF-8/GO/EP, exhibited excellent corrosion inhibition, pH-triggered performance, and self-healing activity. It was found from electrochemical studies that the released inhibitor from the MOFs at the site of damage forms a protective layer with the metal. Salt spray analysis and EDS mapping (Figure 9.7) revealed a large reduction in the corrosion formation in the sample coated with M-ZIF-8/GO/EP. The reaction between epoxy and M-ZIF-8/GO/EP increased the compatibility of M-ZIF-8/GO/EP and the crosslinking density of the epoxy coating.

Baohui Ren [19] incorporated green corrosion inhibitor zinc gluconate in ZIF-8 to form ZnG@ZIF-8, which dispersed in the epoxy matrix to create a hybrid polymer composite with enhanced properties. The resultant ZnG@ZIF-8/Epoxy coatings displayed good anti-corrosion performance. The composite layer was particularly studied for Mg alloys by electrochemical impedance and potentiodynamic polarization studies (Figure 9.8). The performance of ZnG@ZIF-8/EP coating was better than pure Epoxy coatings, ZIF-8/Epoxy coatings, and ZnG + ZIF-8/Epoxy coatings due to uniform distribution, reduced defects, controlled release, and enhanced compatibility.

To explore the combined properties of hollow mesoporous silica nanoparticles (HMSN) and ZIF-8 particles, Chengliang Zhou et al. [20] incorporated conventional corrosion inhibitor benzotriazole (BTA) inside HMSN and subsequently loaded it into the sacrificial template, ZIF-8. The pH-responsive ZIF-8 acted as a smart gatekeeper to obtain novel intelligent nanocomposite HMSN-BTA@ZIF-8. The ZIF-8 and the tertiary amine group present in the ZIF-8 ensured the compatibility of the inhibitor with the polymer matrix and also controlled the release of the inhibitor. Electrochemical studies revealed superior corrosion inhibition and excellent self-healing property of the modified epoxy coatings in both acidic and alkaline conditions. The uniform distribution of HMSN-BTA@ZIF-8 closed the micropores formed during the curing and prolonged the path of the corrosive ions to the metal. The curing between the epoxy and the imidazole groups in the capped ZIF-8 resulted in increased cross-linking density. ZIF-8 is basically hydrophobic in nature, and the release and adsorption of the inhibitor from the damaged area resulted in the self-healing action. All these factors are the reasons for the excellent corrosion inhibition of this modified epoxy coating. The Tafel polarization curve in Figure 9.9 demonstrates the difference in the efficiency of pure and modified epoxy coating.

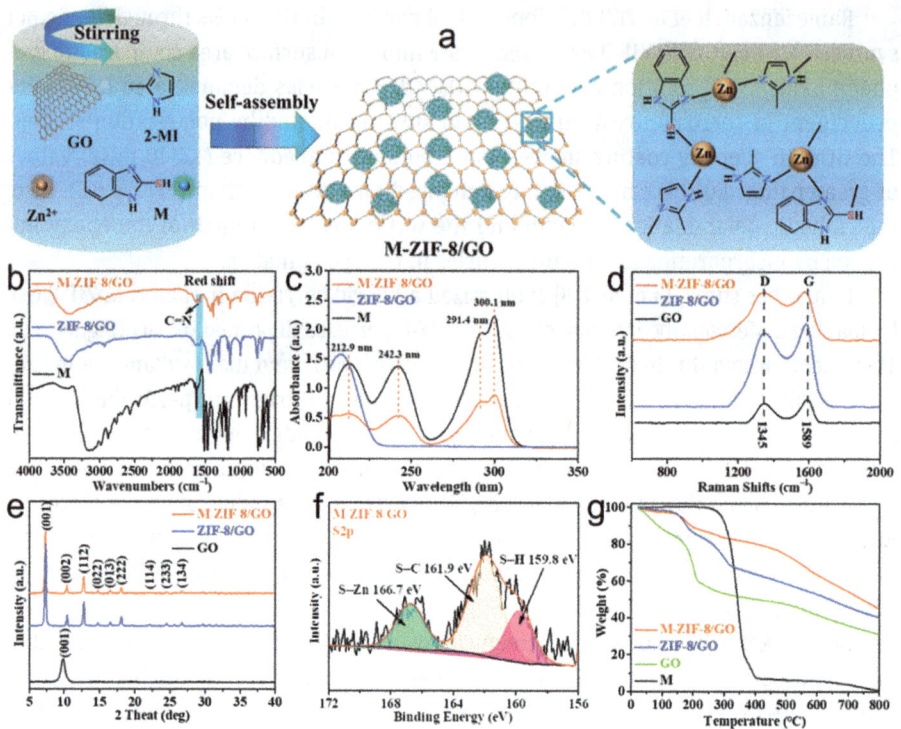

Figure 9.6: (a) Design of synthesis of M-ZIF-8/GO composite nanomaterial, (b) FT-IR, (c) UV-vis spectra, (d) Raman spectra, (e) XRD spectra, (f) S2p high-resolution XPS spectra of M-ZIF-8/GO, and (g) TGA (figure adapted with permission from ref. [18]. Distributed under a Creative Commons Attribution License 4.0 (CC BY) https://creativecommons.org/licenses/by/4.0/).

Guo et al. [21] explored the idea of active corrosion protection by incorporating self-healing efficiency in epoxy polymer coatings by impregnating MOFs. The primary requirements for a self-healing coating are the on-demand supply of the inhibitor or curing agent and the ability to prevent the unwanted leakage of the active component before any damage occurs. Here, benzotriazole (BTA) was used as the inhibitor and self-healing agent. The ligand approach method (Figure 9.10) was utilized for the successful synthesis of ZIF-7@BTA nanoparticles with 30% BTA concentration. It was observed that the loaded compound could be released into the scratched area of the polymer matrix with a corrosion inhibition efficiency of 99.4% in an acidic environment. Only a limited amount of BTA was leached out in a neutral environment.

ZIF-7@BTA nanoparticles were added to the matrix at a rate of 1 wt% with respect to the total weight of epoxy and hardener. Then, the organic primer layer was coated on the steel plates with a rod applicator. Finally, the coated samples were cured at 80 °C for 24 h in an oven. The thickness of the dry coating was 70 ± 10 μm. The compatibility and uniform distribution of ZIF-7@BTA in the polymer matrix was confirmed

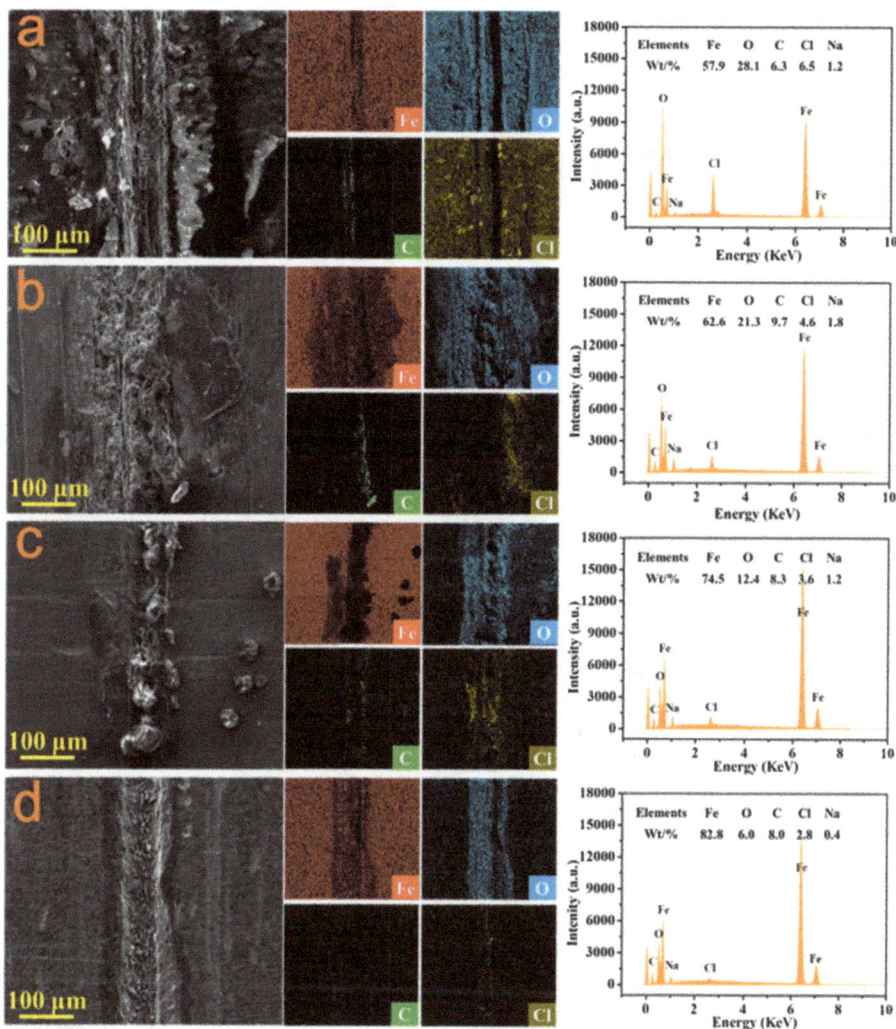

Figure 9.7: SEM images and elemental analysis of corroded layer under the scratch (a) Epoxy, (b) GO/EP, (c) ZIF-8/GO/EP, and (d) M-ZIF-8/GO/EP coatings after 40 days salt spray analysis (figure adapted with permission from ref. [18]. Distributed under a Creative Commons Attribution License 4.0 (CC BY) https://creativecommons.org/licenses/by/4.0/).

with the help of SEM-EDX mapping. The adsorptive films formed by the inhibitor on the surface of the metal in acidic conditions are the reason for the self-healing action of the composite coating in acidic conditions. The pH-responsive nature of ZIF-7@inhibitor can be utilized for the manufacturing of coatings with smart properties [21].

Another graphene oxide-based metal-organic framework system was reported by Lashgari et al. [22] by developing ZIF-67, a new family of MOF synthesized on a graphene oxide platform. ZIF-67/GO NPs were developed from Co^{2+} central atom

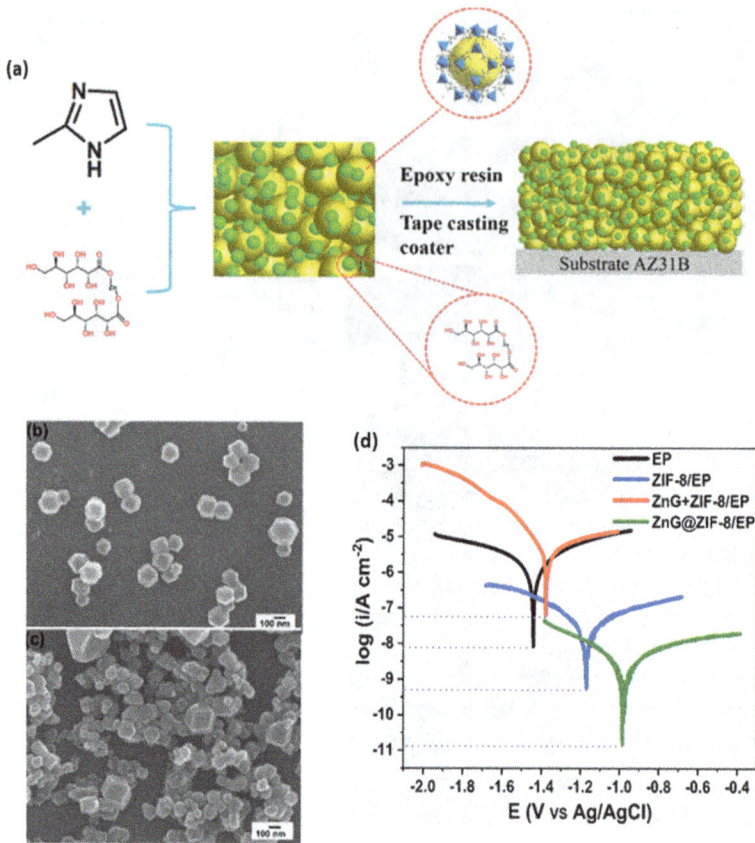

Figure 9.8: (a) The design strategy for the preparation of ZnG@ZIF-8/EP coatings; (b, c) SEM of ZIF-8, ZnG@ZIF-8; (d) polarization curves of the EP coatings in a 3.5 wt% NaCl solution (figure adapted with permission from ref. [19]. Distributed under a Creative Commons Attribution License 4.0 (CC BY) https:// creativecommons.org/licenses/by/4.0/).

with 2-methylimidazole as the ligands. Incorporating ZIF-67/GO NPs in the epoxy matrix resulted in passive and active anti-corrosion protection by barrier protection and self-healing, respectively. To improve the hydrophobicity and thereby control the solubility in the saline media, the nanoparticles were modified with 3-Aminopropyl triethoxysilane (APS) to synthesize ZIF-67/GO@APS NPs. This modification also enhanced the fillers' compatibility with the epoxy matrix. The potentiodynamic polarization studies clearly showed that the inhibitor was a mixed-type inhibitor and worked by the O_2 reduction/ Fe oxidation mechanism. The *icorr* value was reduced to 1.41 μA/cm. EIS and salt spray studies confirmed the anti-corrosion and self-healing potential of the coating. Among the two nanoparticles, ZIF-67/GO@APS NPs showed better performance. The modification with APS led to better dispersion and lower adsorption of water; DFT calculations also supported the results.

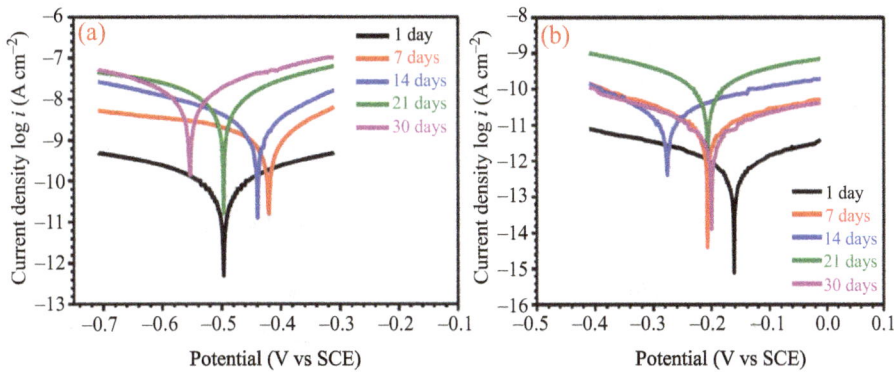

Figure 9.9: Time-dependent Tafel plots of (a) neat epoxy coating and (b) HMSN-BTA@ZIF-8-2% throughout 30 days in 3.5 wt% NaCl solution (figure adapted with permission from ref. [20]. Distributed under a Creative Commons Attribution License 4.0 (CC BY) https://creativecommons.org/licenses/by/4.0/).

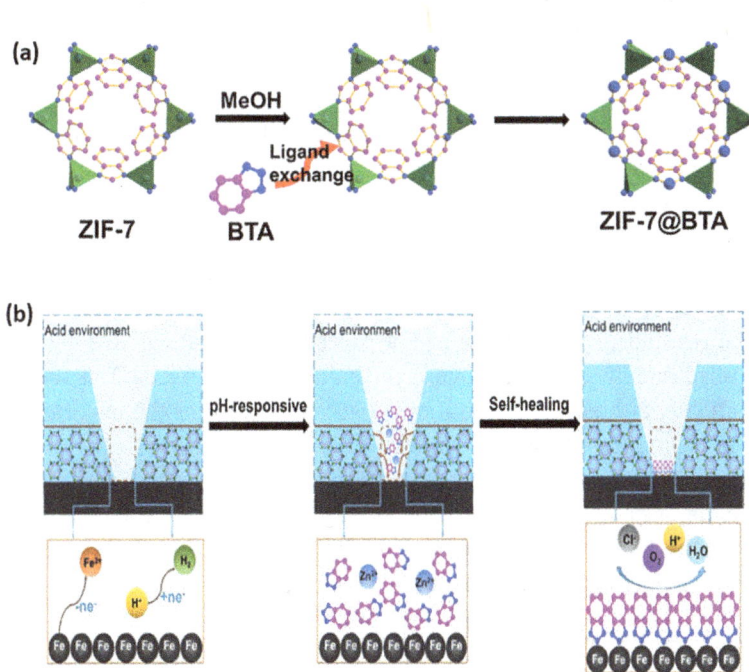

Figure 9.10: (a) Ligand exchange method for synthesizing ZIF-7@BTA. (b) Self-healing mechanism of ZIF-7@BTA-EP coatings on carbon steel (figure adapted with permission from ref. [21]. Distributed under a Creative Commons Attribution License 4.0 (CC BY) https://creativecommons.org/licenses/by/4.0/).

A number of Zr-based MOFs were developed by Ramezanzadeh et al. [23] to evaluate their corrosion inhibition efficiency in saline medium. Zr-UIO-66 MOF, Zr-UIO-NH$_2$-MOF, and Zr-NH$_2$-UIO-MOF were covalently functionalized with Glycidyl Methacrylate (GMA) to prepare GMA@NH$_2$-UIO as the filler to disperse in the epoxy medium. The controlled release of the inhibitor, zirconium ions and organic ligands from the MOFs, was evaluated and proved by a water stability test in acidic, neutral, and basic solutions of 3.5 wt% NaCl. The solution phase test and scratched epoxy coatings' Tafel polarization and EIS measurements proved the increased corrosion inhibition efficiency of the modified epoxy coatings. The composite film filled with GMA@NH$_2$-UIO particles showed the highest impedance at the low-frequency region in the Nyquist plot and the highest coating undamaged index (85.93%) after 120 h of immersion. The accelerated corrosion tests – salt spray, cathodic disbanding and pull-off test, etc. – showed excellent adhesion property of the modified coating. The hybrid nanofillers improved the scratch resistance and hardness of the polymer matrix.

In another investigation, Wang et al. [24] grafted dopamine (DA) onto the top of MOFs and then incorporated DA-MOFs within waterborne epoxy resins. Tests such as salt spray analysis, adhesion analysis, and EIS analysis were used to investigate the coating's potential. Compared to pure epoxy coatings, the coatings that incorporated DA-MOFs showed superior corrosion prevention and hydrophobicity. The EIS investigation of dopamine-modified MOFs with coating revealed resistance exceeding 3.18×10^8 Ω cm^2. The film containing 0.5 wt% DA-MOFs showed the best corrosion inhibition. MOFs' functionality reacts chemically with epoxy (Figure 9.11) to form a uniformly distributed, close-packed structure.

The best performance was exhibited by a coating containing 0.5 wt% filler; on either side of this concentration, the effect was lower. This coating exhibited the best compatibility, cross-linking density, compactness, better shielding property, etc. Although the initial resistance value for the higher concentration coatings was higher, the value diminished rapidly after some time as the immersion progressed. This shows the lower compatibility at higher concentrations. Figure 9.12 explains the importance of the optimum concentration of the inhibitor filler in the epoxy matrix. At very low or zero concentrations of the filler, the defects or pores that originated at the time of curing or during the lifetime allows the penetration of the water molecules and other corrosive ions, directly leading to the metal surface. At very high concentrations, there is an agglomeration of the particles in the epoxy medium, resulting in a non-uniform distribution of the particles in the epoxy matrix and open pores. This directs the corrosive ions and water toward the metal surface. At the filler's optimum concentration, the particles' uniform distribution is achieved, and all the pores formed during the curing are closed at a height above the metal surface [24].

Nanoceria-decorated Cerium (III)-Imidazole network (NC/CIN)-based nanopigment was synthesized by Motamedi et al. [25] by a one-pot coprecipitation synthetic strategy (Figure 9.13). The modified epoxy composite coating exhibited excellent anticorrosion and thermomechanical properties. The coating exhibited tremendous corrosion inhibition potential in the EIS and salt spray test; the impedance measurements showed high impedance even after immersion for 7 weeks. After the immersion test,

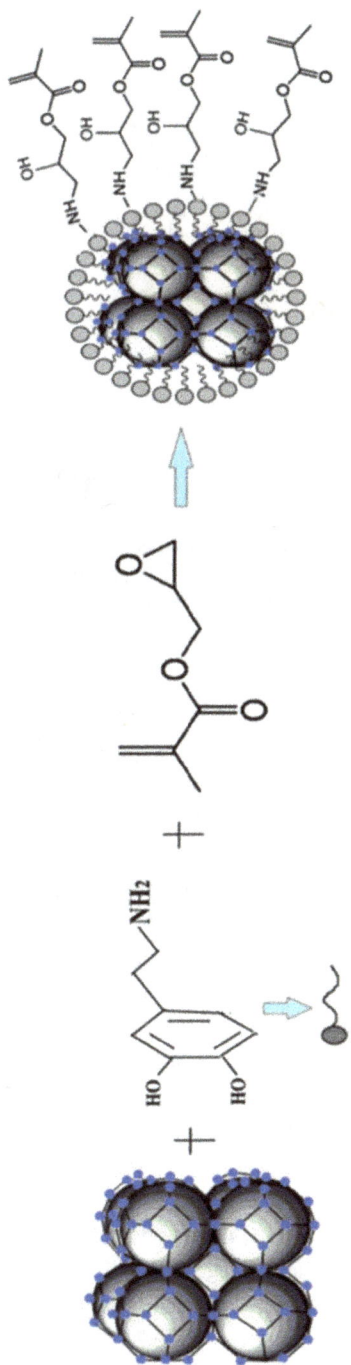

Figure 9.11: DA-MOFs reaction with waterborne epoxy (figure adapted with permission from ref. [24]. Distributed under a Creative Commons Attribution License 4.0 (CC BY) https://creativecommons.org/licenses/by/4.0/).

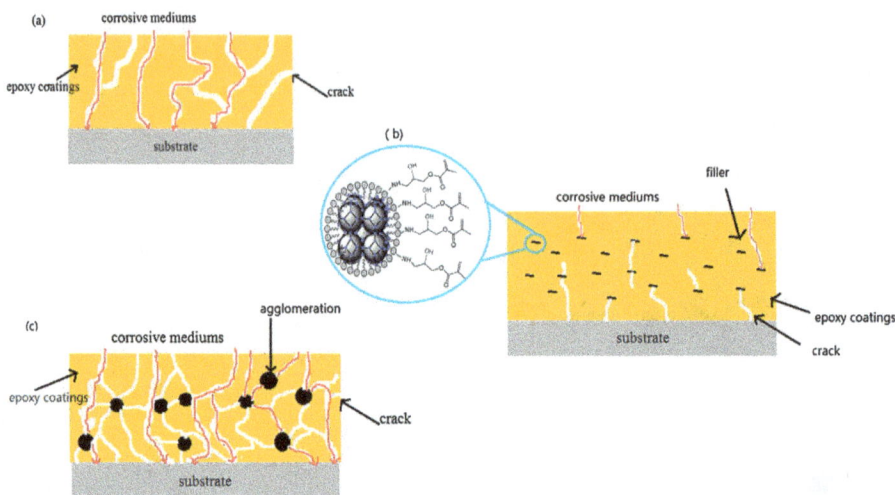

Figure 9.12: The schematic diagram represents the anti-corrosion activity of fillers at various concentrations, explaining the importance of optimum concentrations (figure adapted with permission from ref. [24]. Distributed under a Creative Commons Attribution License 4.0 (CC BY) https://creativecom mons.org/licenses/by/4.0/).

the self-repairing property was also observed in the case of artificially damaged coatings. The cross-linking density, toughness, and thermomechanical properties were also improved by incorporating the nanopigment in the epoxy coating.

Yue Zhao et al. [26] incorporated benzotriazole corrosion inhibitor in a nanocontainer derived from graphene oxide (GO) and MOFs in a simple one-step assembly and one-pot synthesis route (Figure 9.14). The resultant composite coating of polyvinyl butyral (PVB) resin exhibited both active and passive corrosion protection. Active corrosion protection was performed with the help of the excellent barrier properties of GO, and the controlled releasing mechanism of the nanocontainer was responsible for the active protection. ZIF-8 and UiO-66 were chosen as two examples for this study to form GO/ZIF-8@BTA or GO/UiO-66@BTA. The corrosion inhibition efficiency was evaluated by the immersion test and EIS analysis of various samples with and without an artificial defect on the surface of the copper alloy and archaized bronze. Both GO/ZIF-8@BTA and GO/UiO-66@BTA have corrosion inhibition efficiency over pure inhibitors.

For an effective protection of Mg alloys, Zheng et al. [27] developed a modified caprolactone polymer matrix by incorporating a copper-based MOF, HKUST-1 modified with folic acid (Figure 9.15a). The resultant polymer film exhibited excellent corrosion inhibition and enhanced bio-compatibility, thus showing promise to be used in living systems. The modified HKUST-1 dispersed in caprolactam and the hydrogen bonding ensured the uniform distribution of the inhibitor. In the polarization test (Figure 9.15 b), it was found that the corrosion current density decreased drastically from $7.18 \pm 3.243 \times 10^{-7}$ to

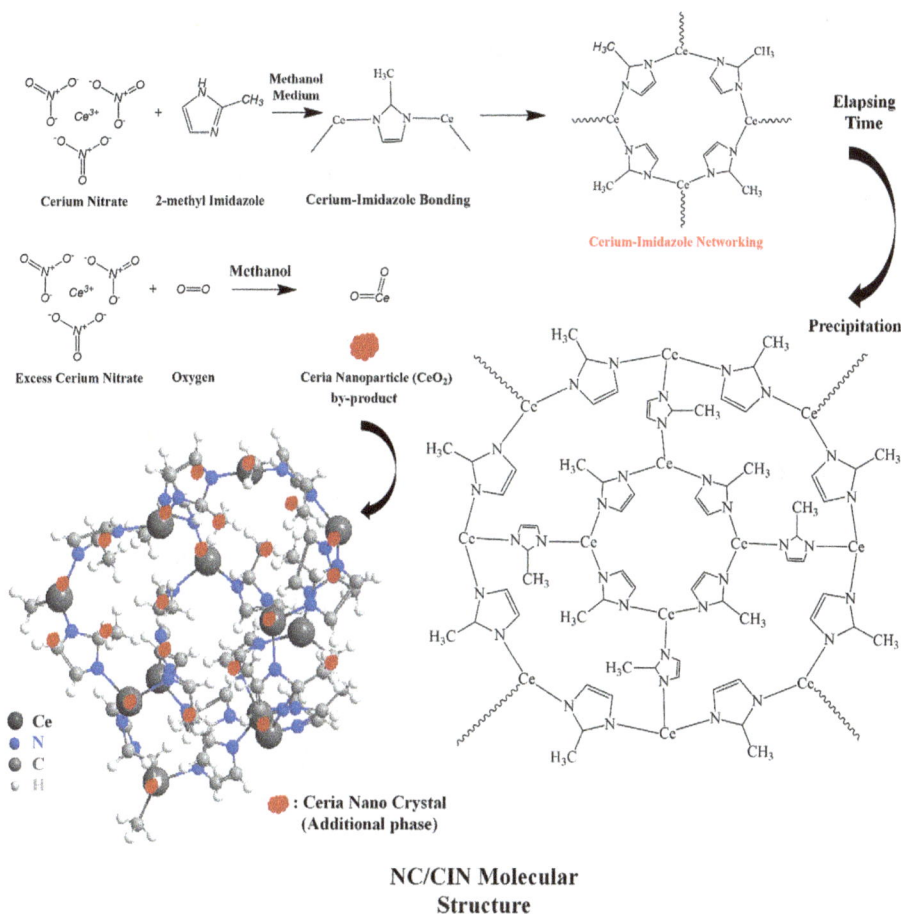

Figure 9.13: Steps involved in the one-pot co-precipitation preparation method and chemical structure of the NC/CIN nanopigment (figure adapted with permission from ref. [25]. Distributed under a Creative Commons Attribution License 4.0 (CC BY) https://creativecommons.org/licenses/by/4.0/).

$1.10 \pm 0.937 \times 10^{-10}$ A/cm^2. An additional benefit of the hybrid polymer is that the released copper ions helped in the proliferation and differentiation of osteoblastic cells. The system is a promising multifunctional coating system for future developments of biological implants.

In another investigation, Keshmiri et al. [28] synthesized an eco-friendly lanthanide-based MOF loaded into GO and subsequently incorporated it into the epoxy matrix to study its corrosion inhibition potential. A cerium-based MOF with 1,3,5-benzene tricarboxylic acid as the organic linker was the particle loaded into the graphene oxide. GO@Ce-MOF/EP was found to have an excellent corrosion inhibition potential by barrier protection, and the active corrosion inhibition was ensured by the self-healing effect (Figure 9.16). The results were confirmed by EIS and salt spray analysis. In the EIS

Example 1

Example 2

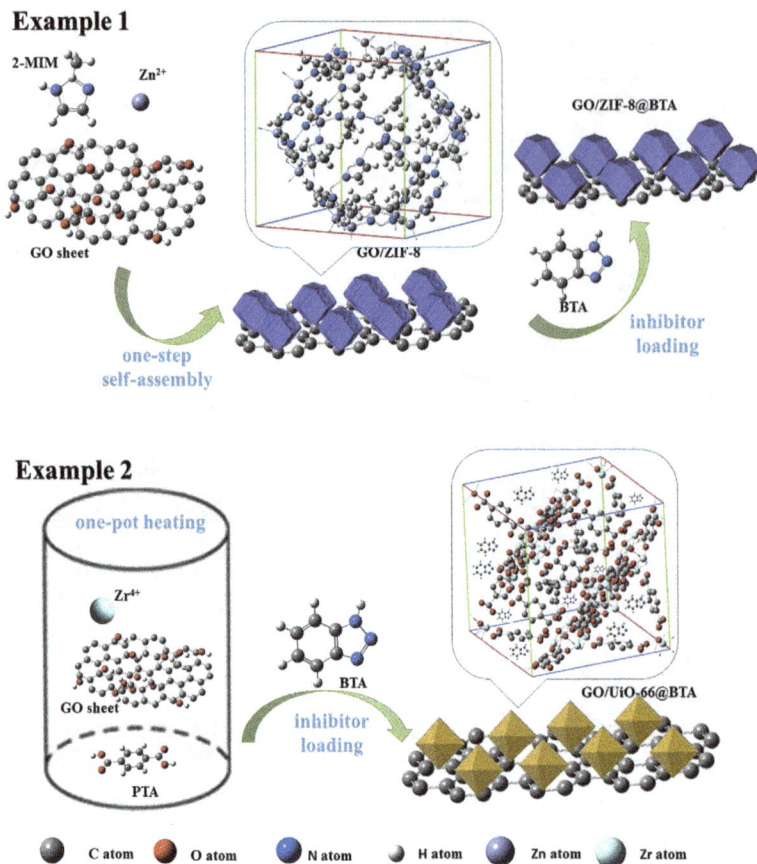

Figure 9.14: Synthesis approach of containers and the method of incorporation of inhibitors in the containers (figure adapted with permission from ref. [26]. Distributed under a Creative Commons Attribution License 4.0 (CC BY) https://creativecommons.org/licenses/by/4.0/).

measurements of the scratched coatings (Figure 9.17), it was found that the epoxy coating containing the MOFs-loaded graphene oxide showed enhanced corrosion inhibition potential than the pure epoxy coatings.

9.2.3 MOFs deposited as corrosion inhibition film

MOFs can be deposited directly on the metal surface to be protected without incorporating into any polymeric matrix. MOFs or a combination of MOFs with suitable inhibitors are the perfect choices for corrosion inhibition if compatibility issues are solved and proper application methods are developed. However, finding appropriate materials and adaptable pathways to construct composites based on MOF is necessary. It is helpful for

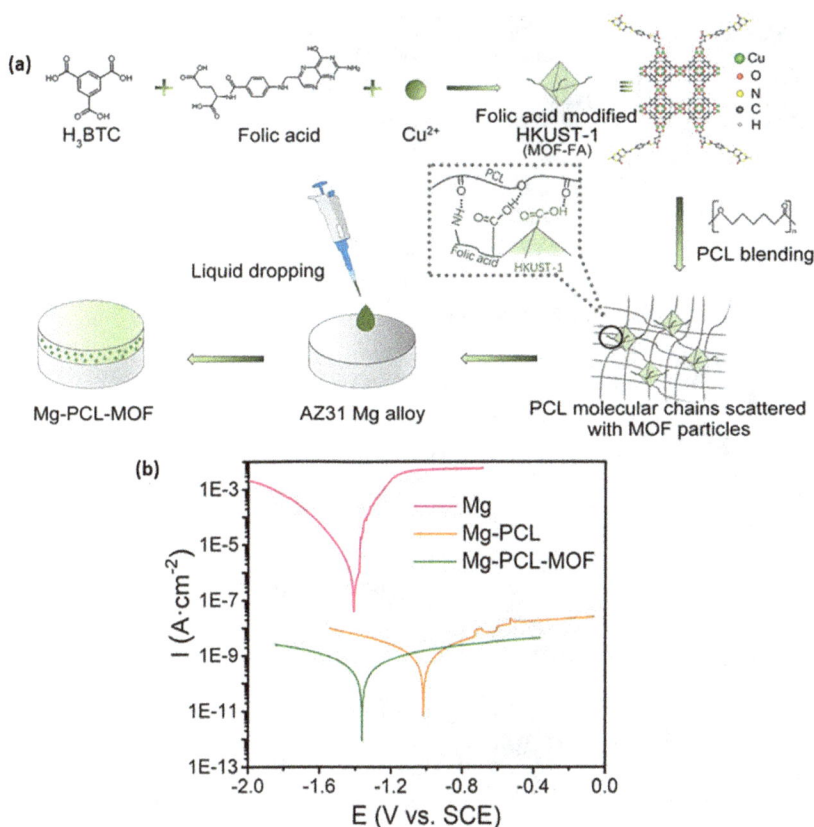

Figure 9.15: (a) Schematic illustration of the fabrication of the composite coating. (b) Tafel plots for the samples (figure adapted with permission from ref. [27]. Distributed under a Creative Commons Attribution License 4.0 (CC BY) https://creativecommons.org/licenses/by/4.0/).

composite manufacture, due to its flexible crystalline structures, neatly ordered metal ions, and organic linkers. Due to the invention of metal polymers, silica, graphene, nanoparticles, proteins, carbon nanotubes, quantum dots, metal oxides, organic compounds, polyoxometalates, and other materials, a number of MOF composites have been produced. The combination of different nanomaterials gives advantages over many functional properties [29].

Weijin Li et al. [30] deposited a thin film of the MOF HKUST-1 on the surface of the bronze alloy; the film was used as a container for the inhibitor CTAB (Cetyltrimethyl ammonium bromide). The pores created by the MOFs on the surface of the metal helped in the selective positioning of the inhibitor. CTAB@HKUST-1 films deposited by the electrophoretic method (Figure 9.18) exhibited excellent corrosion inhibition for bronze without affecting its adhesion and passive protection properties. In this case, the modified thin film coating particularly exhibited its corrosion inhibition efficiency

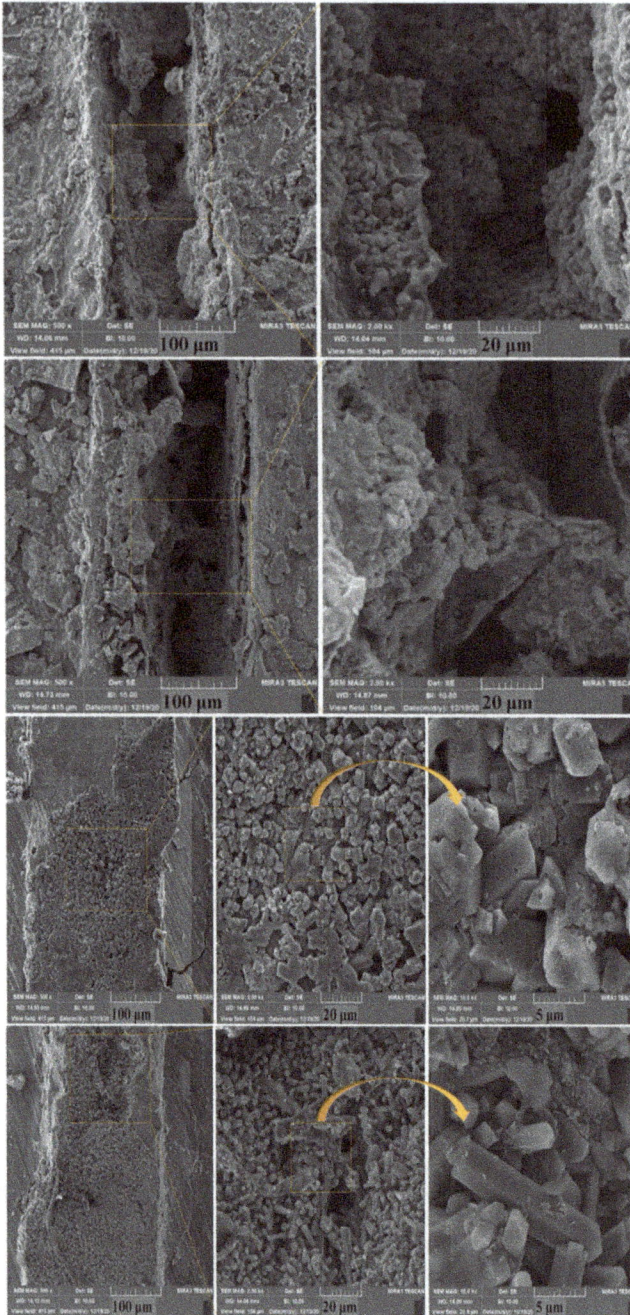

Figure 9.16: FE-SEM micrographs of the scratched zones in various magnifications. (A1, A2) Blank/EP, (B1, B2) GO/EP, (C1, C2, C3) Ce-MOF/EP (D1, D2, D3) GO@Ce-MOF/EP coatings immersed in saline solution (figure adapted with permission from ref. [28]).

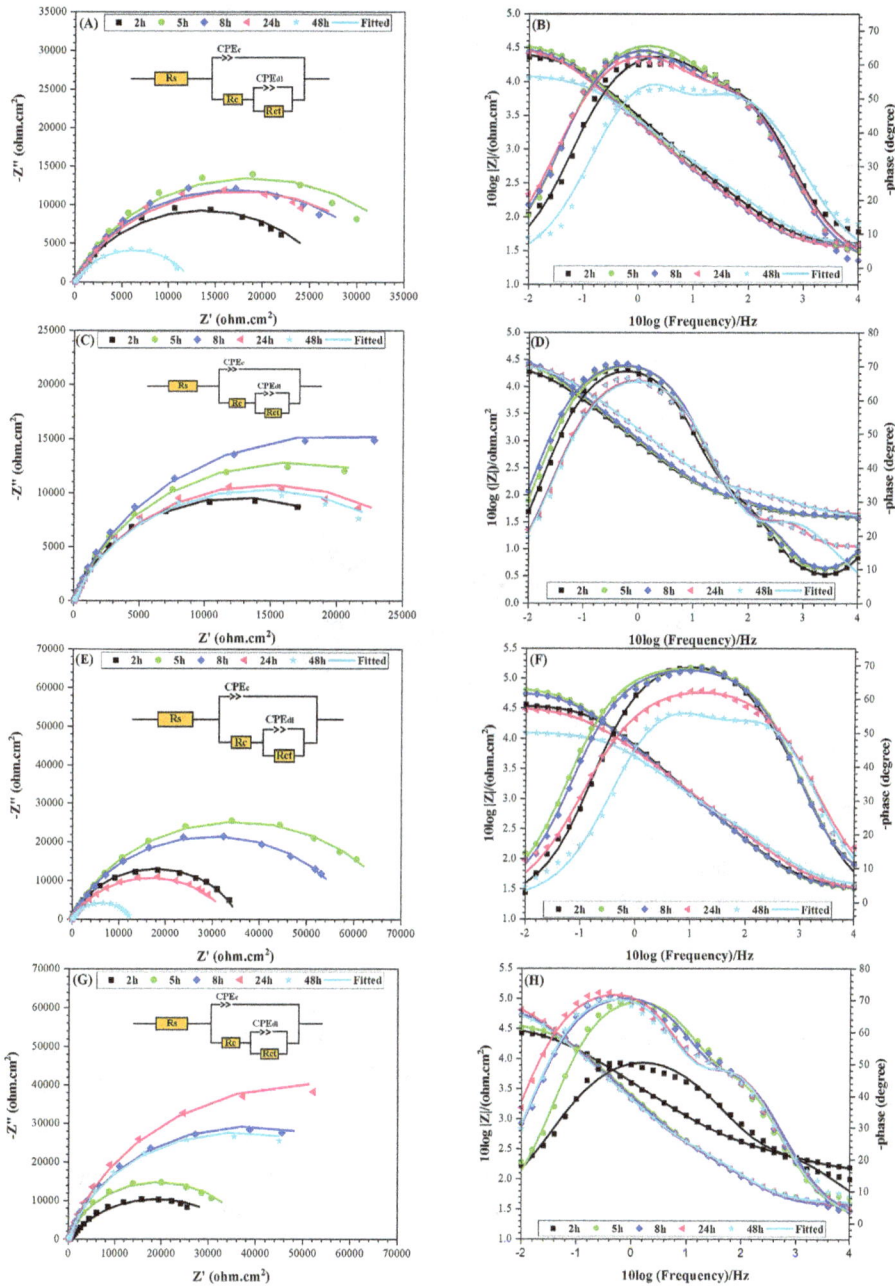

Figure 9.17: Nyquist, Bode diagrams, and equivalent electrical circuit of artificially scratched: (a, b) Blank/EP, (c, d) GO/EP, (e, f) Ce-MOF/EP, and (g, h) GO@Ce-MOF/ EP coatings kept in NaCl solution. (figure adapted with permission from ref. [28]. Distributed under a Creative Commons Attribution License 4.0 (CC BY) https://creativecommons.org/licenses/by/4.0/).

at the defect sites. They also suggested that the duration of the anti-corrosion activity can be enhanced by improving the stability of the host MOF. This method is particularly useful for protecting artworks and statues of historical value.

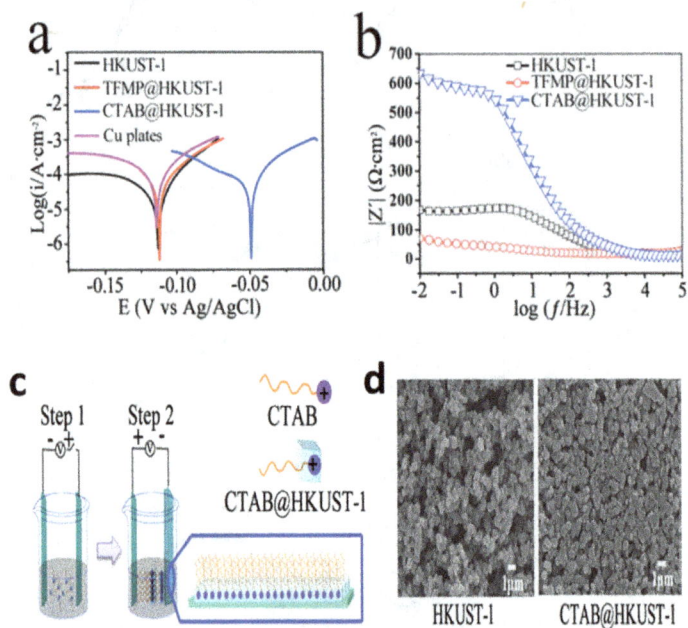

Figure 9.18: (a) Tafel plots of the CTAB@HKUST-1/Cu, TFMP@HKUST-1/Cu, and HKUST-1/Cu coatings and Cu plates. (b) Bode plots of the as-prepared MOF films after immersion in 1 M NaCl. (c) Schematic diagram for the synthesis of thin film CTAB@HKUST-1 on bronze surface. (d) SEM micrographs of the film deposited on the bronze surface (figure adapted with permission from ref. [30]. Distributed under a Creative Commons Attribution License 4.0 (CC BY) https://creativecommons.org/licenses/by/4.0/).

Among all the materials used for biological implants, Mg is the most widely used due to its bio-compatibility, mechanical strength, and biodegradability. But the higher corrosion rate is the major limitation faced by the researchers. Phosphonic acid and its derivatives are the commonly used and ideal corrosion inhibitors for Mg implants due to their similarity toward many bone and cell membranes components. Pu et al. [31] created a metal-organic complex system of 1-hydroxyethylidene-1,1-diphosphonic acid (HEDP) and Zr on the Mg surface by immobilization and self-assembly (Figure 9.19). Chelator was first adsorbed on alkaline-pre-treated Mg, and then Zr ions and HEDP were alternately coated using a dip-coating process. This coating was discovered to be uniform, compact, and secured to Mg by a framework that resembled metal-organic complexes. This layer of MOF ensured a reduced corrosion rate and long-term in vitro degradation of magnesium in a saline solution of phosphate buffer.

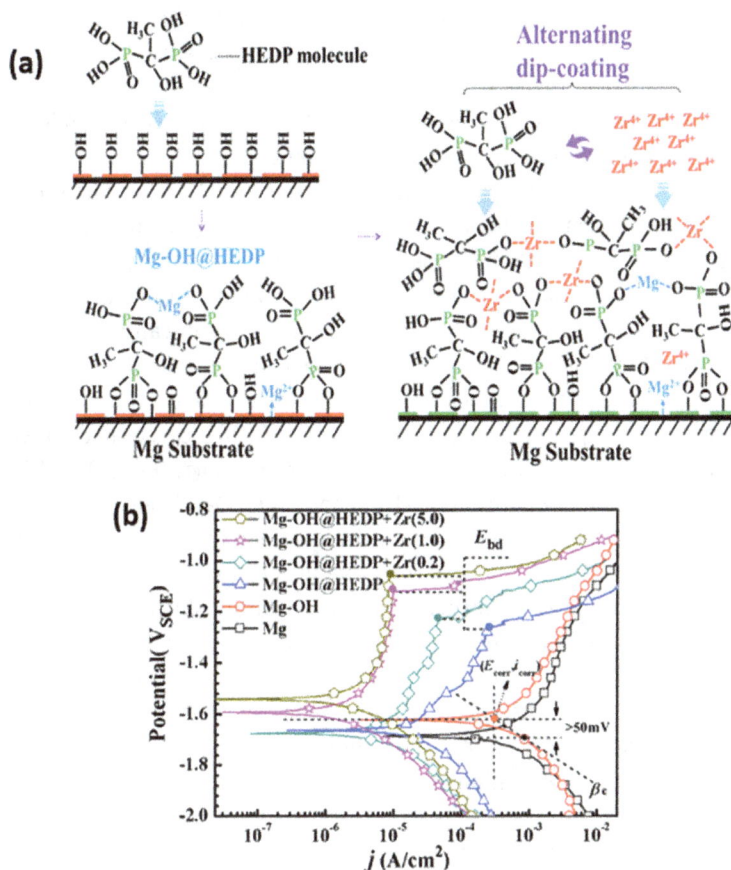

Figure 9.19: (a) Method of synthesis of Zr ions-incorporated HEDP layer on the surface of Mg by chemical immobilization, self-assembly deposition, and concurrent chelation; (b) polarization study results (figure adapted with permission from ref. [31]. Distributed under a Creative Commons Attribution License 4.0 (CC BY) https://creativecommons.org/licenses/by/4.0/).

9.3 Corrosion inhibition mechanism of MOFs

The corrosion inhibition potential of a MOF and its mechanism of action depend on many factors, including the nature of the metal, the nature of the ligand present in it, and particularly, on the nature of ligands, as it can contribute to inhibition in different ways. The presence of heteroatoms bearing the lone pair of electrons, the presence of pi-electron cloud, the steric factors and the spatial arrangement of the ligands, the Lewis acid-base nature, etc. are the significant factors to be considered while selecting the ligands for corrosion inhibition purposes. The lone pair of electrons and the pi electrons can form co-ordinate bonds with the metal ions on the alloy surface, thus preventing its further oxidation or reaction with other corrosive ions.

The hydrophobic nature of the MOF and pH-sensitive release ability are the factors of the whole MOF structure that determine its potential in passive corrosion inhibition. The release of the inhibitor in the damaged area and the resulting self-healing ability are the features contributing to active corrosion protection [21, 22]. When triamine functionalities are present in the MOFs structure, the epoxy medium's cross-slinking density and dispersion ability increases, thereby reducing the diffusion paths for corrosive species [20]. Additionally, nanofillers have the ability to seal off micropores that are formed during the curing process of the coating, significantly reducing the pathways through which corrosive electrolyte solution can penetrate [16, 20].

The mechanism is also explained by Ren et al. using the schematic model shown in Figure 9.20. Comparing the protective effectiveness of the MOF-incorporated epoxy (ZIF-8/EP), zinc gluconate-loaded MOFs dispersed in epoxy (ZnGu@ZIF-8/EP) and in pure epoxy coating (EP). From the figure, it is clear that water, O_2, Cl^-, etc. can easily attack the metal surface by penetrating through the neat epoxy matrix's pores. In coatings containing ZIF-8 particles alone, the corrosive ions can be avoided to an extent due to the hindrance of the ZIF-8 particles and their hydrophobicity. But, due to the absence of strong corrosion inhibitors, the ZIF-8 alone cannot provide long-lasting protection. However, ZnG@ZIF-8/EP coatings exhibited the largest contact angle of 96°, higher than EP and ZIF-8-EP. The ZnG@ZIF-8/EP coating showed better corrosion resistance because of the following reasons. (i) the inhibitor adsorbs on the surface of the metal and forms a stable coordination complex; this prevents the reaction of the corrosive ions with the metal atom. (ii) During the immersion period, the hydrophobicity of the coatings enhances the barrier protection and thus prevents water and other corrosive ions from entering the polymer matrix. (iii) The in situ-loaded inhibitors and the uniform distribution of ZnG@ZIF-8 enhanced the corrosion protection. It was observed that direct mixing of ZnG or ZIF-8 decreases EP's efficiency [19].

As explained earlier by Wang et al. [24], the pores formed during the curing are sealed by fillers, which block the penetration of the water molecules and other corrosive ions in the polymer matrix and prevent them from reaching the metal surface. But, here, the concentration of the fillers plays an important role. There is a decrease in corrosion inhibition efficiency at concentrations above and below the optimum concentration. The optimum concentration depends on many factors like particle size, charge distribution, loading percentage, the viscosity of the polymer matrix, etc.

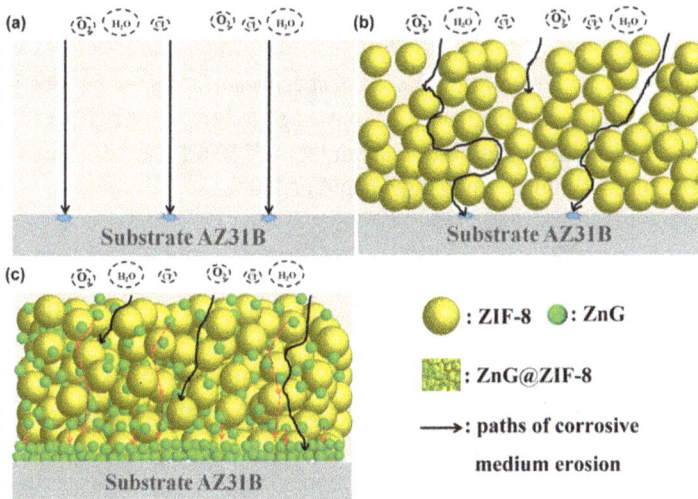

Figure 9.20: Schematic representation of the possible anti-corrosion mechanism. (a) EP coatings, (b) ZIF-8/EP coatings, and (c) ZnG@ZIF-8/EP coatings [19]. Distributed under a Creative Commons Attribution License 4.0 (CC BY) https://creativecommons.org/licenses/by/4.0/.

9.4 Limitations of MOFs in thin film coatings and possible remedies

Apart from the highlighted advantages of MOFs, there are some limitations in the applications of MOFs in thin film coatings. The stability in water and the dispersion ability in the polymer matrix are significant factors. Even though some MOF structures are very compatible with the polymer matrix, some of them have poor dispersion in the polymer matrix and lower stability in water. Excellent water stability is crucial for applications like corrosion inhibition, water filtration, and solution phase catalysis. However, some MOFs are said to be stable in water, and many MOF materials are even employed for the long-term removal of pollutants from aqueous waste. For this application, the University of Oslo (UiO), Material of Institute Lavoisier (MIL), and Zeolitic Imidazolate Framework (ZIF) series substrates are often utilized MOFs [32].

Generally, nanoscale materials face difficulty in dispersion in the epoxy matrix, and MOFs being nanoscale particles, also experience the same. But due to the hybrid nature of the MOFs, the functional groups in the organic ligands interact with the epoxy and enable uniform distribution of the particles, compared to other nanoparticles of mineral character. Also, the large surface area of the MOFs can ensure the interfacial interaction between the metal and the polymer [33]. The agglomeration of the particles at a higher concentration is another limitation that requires serious attention. Salman Shahid et al. suggested the in situ growth of the MOFs particle and avoiding simple physical mixing of the MOFs and the polymer. This results in uniform dispersion of the nanoparticles and better compatibility [34].

Outstanding surface-active sites, a large amount of specific surface area, the availability of more straightforward preparation techniques, a range of pore sizes, a variety of topologies, and other corrosion inhibition advantages of MOFs are sufficient to rule out all the aforementioned shortcomings that could be fixed.

9.5 Conclusion

Corrosion is a spontaneous and unavoidable problem faced by all industries and hence corrosion prevention and protection of structures and systems are very crucial. Among all the corrosion protection methods, thin film coatings are the most widely used and convenient method. Many of the commonly used inhibitors create many health issues and environmental problems, and some are really costly. MOFs are an excellent choice for the replacement of toxic chromates, phosphates, heavy metals, and costly organic inhibitors. Outstanding surface-active sites, a large amount of specific surface area, the availability of simpler preparation techniques, a range of pore sizes, and a variety of topologies are the advantages of MOFs, seen as good corrosion inhibitors. They can be used as (i) containers for loading other corrosion inhibitors, (ii) they can be dispersed in the corrosive medium, thus forming a layer of protection on the surface of the metal to be protected. In the thin film coating procedure (i) they can be embedded in the polymer matrix used for barrier coating for enhanced passive protection, inducing active protection like self-healing properties, (ii) can be directly deposited on the metal surface using various techniques. Among all the reported works, incorporating the MOFs in the epoxy medium is the most effective way and is frequently studied. ZIF-type MOFs are ideal for incorporation into the epoxy matrix. Detailed studies on other combinations of different polymers and MOFs are an unexplored area.

References

[1]   Nash W, Zheng L, Birbilis N Deep learning corrosion detection with confidence. Npj Mater Degrad 2022;6:26. https://doi.org/10.1038/s41529-022-00232-6.
[2]   Rajimol PR, Ulaeto SB, Puthiyamadam A, Neethu S, Rajan TPD, Radhakrishnan K V, et al. Smart anticorrosive and antimicrobial multifunctional epoxy coating using bergenin and malabaricone C bio-nanocomposite dispersoids on mild steel and aluminium-6,061 alloy. Prog Org Coat 2022;169:106924. https://doi.org/10.1016/j.porgcoat.2022.106924.
[3]   Ravi RP, Ulaeto SB, Deva Rajan TP, Radhakrishnan KV. 8 Natural product-based multifunctional corrosion inhibitors for smart coatings. Corros Mitigation De Gruyter 2022;139–74. https://doi.org/10.1515/9783110760583-008.

[4]   Ramesh M, Deepa C Metal-organic frameworks and their composites. In: Khan, A, Verpoort, F, Asiri, AM, Hoque, ME, Bilgrami, AL, Azam, M, et al. (eds). Metal-organic frameworks for chemical reactions. Elsevier, pp. 1–18; 2021. https://doi.org/10.1016/B978-0-12-822099-3.00001-0.

[5]   Ramezanzadeh M, Ramezanzadeh B. Thermomechanical and anticorrosion characteristics of metal-organic frameworks. In: Khan, A, Verpoort, F, Asiri, AM, Hoque, ME, Bilgrami, AL, Azam, M, et al. (eds). Metal-Organic Frameworks for Chemical Reactions. Elsevier, pp. 295–330; 2021. https://doi.org/10.1016/B978-0-12-822099-3.00012-5.

[6]   Morozan A, Jaouen F Metal organic frameworks for electrochemical applications. Energy Environ Sci 2012;5:9269. https://doi.org/10.1039/c2ee22989g.

[7]   Morozan A, Jaouen F Metal organic frameworks for electrochemical applications. Energy Environ Sci 2012;5:9269. https://doi.org/10.1039/c2ee22989g.

[8]   Tian H, Li W, Liu A, Gao X, Han P, Ding R, et al. Controlled delivery of multi-substituted triazole by metal-organic framework for efficient inhibition of mild steel corrosion in neutral chloride solution. Corros Sci 2018;131:1–16. https://doi.org/10.1016/j.corsci.2017.11.010.

[9]   Etaiw SEH, Fouda -AE-AS, Amer SA, El-bendary MM Structure, characterization and anti-corrosion activity of the new metal-organic framework [Ag(qox)(4-ab)]. J Inorg Organomet Polym Mater 2011;21:327–35. https://doi.org/10.1007/s10904-011-9467-9.

[10]   Kumaraguru S, Pavulraj R, Mohan S Influence of cobalt, nickel and copper-based metal-organic frameworks on the corrosion protection of mild steel. Trans IMF 2017;95:131–36. https://doi.org/10.1080/00202967.2017.1283898.

[11]   Lin Y-T, Don T-M, Wong C-J, Meng F-C, Lin Y-J, Lee S-Y, et al. Improvement of mechanical properties and anticorrosion performance of epoxy coatings by the introduction of polyaniline/graphene composite. Surf Coat Technol 2019;374:1128–38. https://doi.org/10.1016/j.surfcoat.2018.01.050.

[12]   Shi X, Nguyen TA, Suo Z, Liu Y, Avci R Effect of nanoparticles on the anticorrosion and mechanical properties of epoxy coating. Surf Coat Technol 2009;204:237–45. https://doi.org/10.1016/j.surfcoat.2009.06.048.

[13]   Tang Y, Tanase S Water-alcohol adsorptive separations using metal-organic frameworks and their composites as adsorbents. Microporous Mesoporous Mater 2020;295:109946. https://doi.org/10.1016/j.micromeso.2019.109946.

[14]   Yeganeh M, Keyvani A The effect of mesoporous silica nanocontainers incorporation on the corrosion behavior of scratched polymer coatings. Prog Org Coat 2016;90:296–303. https://doi.org/10.1016/j.porgcoat.2015.11.006.

[15]   Su L, Li Y, Liu Y, Ma R, Huang F, et al. Antifungal-inbuilt metal-organic-frameworks eradicate *Candida albicans* biofilms. Adv Funct Mater 2020;30:2000537. https://doi.org/10.1002/adfm.202000537.

[16]   Duan S, Dou B, Lin X, Zhao S, Emori W, Pan J, et al. Influence of active nanofiller ZIF-8 metal-organic framework (MOF) by microemulsion method on anticorrosion of epoxy coatings. Colloids Surf A Physicochem Eng Asp 2021;624:126836. https://doi.org/10.1016/j.colsurfa.2021.126836.

[17]   Ramezanzadeh M, Ramezanzadeh B, Mahdavian M, Bahlakeh G Development of metal-organic framework (MOF) decorated graphene oxide nanoplatforms for anti-corrosion epoxy coatings. Carbon N Y 2020;161:231–51. https://doi.org/10.1016/j.carbon.2020.01.082.

[18]   Li H, Qiang Y, Zhao W, Zhang S 2-Mercaptobenzimidazole-inbuilt metal-organic-frameworks modified graphene oxide towards intelligent and excellent anti-corrosion coating. Corros Sci 2021;191:109715. https://doi.org/10.1016/j.corsci.2021.109715.

[19]   Ren B, Chen Y, Li Y, Li W, Gao S, Li H, et al. Rational design of metallic anti-corrosion coatings based on zinc gluconate@ZIF-8. Chem Eng J 2020;384:123389. https://doi.org/10.1016/j.cej.2019.123389.

[20]   Zhou C, Li Z, Li J, Yuan T, Chen B, Ma X, et al. Epoxy composite coating with excellent anticorrosion and self-healing performances based on multifunctional zeolitic imidazolate framework derived nanocontainers. Chem Eng J 2020;385:123835. https://doi.org/10.1016/j.cej.2019.123835.

[21] Guo Y, Wang J, Zhang D, Qi T, Li GL pH-responsive self-healing anticorrosion coatings based on benzotriazole-containing zeolitic imidazole framework. Colloids Surf A Physicochem Eng Asp 2019;561:1–8. https://doi.org/10.1016/j.colsurfa.2018.10.044.

[22] Lashgari SM, Yari H, Mahdavian M, Ramezanzadeh B, Bahlakeh G, Ramezanzadeh M Synthesis of graphene oxide nanosheets decorated by nanoporous zeolite-imidazole (ZIF-67) based metal-organic framework with controlled-release corrosion inhibitor performance: Experimental and detailed DFT-D theoretical explorations. J Hazard Mater 2021;404:124068. https://doi.org/10.1016/j.jhazmat.2020.124068.

[23] Ramezanzadeh M, Ramezanzadeh B, Bahlakeh G, Tati A, Mahdavian M Development of an active/barrier bi-functional anti-corrosion system based on the epoxy nanocomposite loaded with highly-coordinated functionalized zirconium-based nanoporous metal-organic framework (Zr-MOF). Chem Eng J 2021;408:127361. https://doi.org/10.1016/j.cej.2020.127361.

[24] Wang N, Zhang Y, Chen J, Zhang J, Fang Q Dopamine modified metal-organic frameworks on anti-corrosion properties of waterborne epoxy coatings. Prog Org Coat 2017;109:126–34. https://doi.org/https://doi.org/10.1016/j.porgcoat.2017.04.024.

[25] Motamedi M, Ramezanzadeh M, Ramezanzadeh B, Mahdavian M One-pot synthesis and construction of a high performance metal-organic structured nano pigment based on nanoceria decorated cerium (III)-imidazole network (NC/CIN) for effective epoxy composite coating anti-corrosion and thermo-mechanical properties improvement. Chem Eng J 2020;382:122820. https://doi.org/10.1016/j.cej.2019.122820.

[26] Zhao Y, Jiang F, Chen Y-Q, Hu J-M Coatings embedded with GO/MOFs nanocontainers having both active and passive protecting properties. Corros Sci 2020;168:108563. https://doi.org/10.1016/j.corsci.2020.108563.

[27] Zheng Q, Li J, Yuan W, Liu X, Tan L, Zheng Y, et al. Metal-Organic Frameworks Incorporated Polycaprolactone Film for Enhanced Corrosion Resistance and Biocompatibility of Mg Alloy. ACS Sustain Chem Eng 2019;7(18):114–24. https://doi.org/10.1021/acssuschemeng.9b05196.

[28] Keshmiri N, Najmi P, Ramezanzadeh M, Ramezanzadeh B Designing an eco-friendly lanthanide-based metal organic framework (MOF) assembled graphene-oxide with superior active anti-corrosion performance in epoxy composite. J Clean Prod 2021;319:128732. https://doi.org/10.1016/j.jclepro.2021.128732.

[29] Ramesh M, Deepa C. Metal-organic frameworks and their composites. In: Khan, A, Verpoort, F, Asiri, AM, Hoque, ME, Bilgrami, AL, Azam, M, et al. (eds) Metal-Organic Frameworks for Chemical Reactions. Elsevier, pp. 1–18; 2021. https://doi.org/10.1016/B978-0-12-822099-3.00001-0.

[30] Li W, Ren B, Chen Y, Wang X, Cao R Excellent Efficacy of MOF Films for Bronze Artwork Conservation: The Key Role of HKUST-1 Film Nanocontainers in Selectively Positioning and Protecting Inhibitors. ACS Appl Mater Interfaces 2018;10(37):529–34. https://doi.org/10.1021/acsami.8b13602.

[31] Pu S, Chen M, Chen Y, Zhang W, Soliman H, Qu A, et al. Zirconium ions integrated in 1-hydroxyethylidene-1,1-diphosphonic acid (HEDP) as a metalorganic-like complex coating on biodegradable magnesium for corrosion control. Corros Sci 2018;144:277–87. https://doi.org/10.1016/j.corsci.2018.09.003.

[32] Zhang S, Wang J, Zhang Y, Ma J, Huang L, Yu S, et al. Applications of water-stable metal-organic frameworks in the removal of water pollutants: A review. Environ Pollut 2021;291:118076. https://doi.org/10.1016/j.envpol.2021.118076.

[33] Seidi F, Jouyandeh M, Taghizadeh M, Taghizadeh A, Vahabi H, Habibzadeh S, et al. Metal-organic framework (MOF)/epoxy coatings: A review. Materials 2020;13:2881. https://doi.org/10.3390/ma13122881.

[34] Shahid S, Nijmeijer K, Nehache S, Vankelecom I, Deratani A, Quemener D MOF-mixed matrix membranes: Precise dispersion of MOF particles with better compatibility via a particle fusion approach for enhanced gas separation properties. J Memb Sci 2015;492:21–31. https://doi.org/10.1016/j.memsci.2015.05.015.

Manilal Murmu, Naresh Chandra Murmu, Priyabrata Banerjee

10 Functionalized quantum dots/carbonaceous quantum dots-based thin film coating as corrosion inhibitor

Abstract: With the progress of nanoscience and nanotechnology, carbonaceous quantum dots or simply quantum dots (CDs) have opened new frontiers in materials science with their diverse application prospects. The CDs as well as the functionalized CDs-based thin film coatings have also emerged as ecological, environment-friendly, and efficient corrosion inhibitors in the domain of metal protection from the adverse environments. Chemically functionalized CDs possessing conjugated π-bonds extended in their molecular skeleton along with the existence of heteroatoms facilitate them to adsorb and impart barrier property to metallic surfaces owing to their aqueous solubility, bio-compatibility, chemical stability, and high thermal stability as well as their non-flammability. Researches pertaining to chemical functionalization of CDs as well as heteroatoms-doped CDs – new materials for highly efficient corrosion inhibitors for several metallic materials and used in various aggressive electrolytic solutions such as saline water – have been overviewed and discussed comprehensively. Additionally, the various adsorption mechanisms as well as the corrosion inhibition mechanism of CDs on metallic surface have been explored. Finally, the future scope for the targeted functionalization of CDs to be used as efficient anti-corrosive additives for adverse environmental conditions have been discussed.

Keywords: Carbon Quantum Dots, Chemical Functionalization, Corrosion Inhibition, Thin Film Coating, Adsorption, Self-Healing Coating, Super Hydrophobic Coating, Self-Cleansing Coating

10.1 Introduction

Corrosion of metals or alloy-made materials and several metallic structures exposed to humid or corrosive environment lead to severe damages, causing a huge economic drain owing to their maintenance and replacement. Sometimes, hazardous and life-threatening

Acknowledgments: MM is very thankful to the Ministry of Tribal Affairs, New Delhi, India for his National Fellowship for Higher Education of Scheduled Tribes candidates, NFST (vide award letter no. F1-17.1/2014-15/RGNF-2014-15-ST-JHA-71,559). PB and NCM would like to acknowledge the Department of Higher Education, Science and Technology and Biotechnology, Govt. of West Bengal, India for providing financial assistance to carry out this work [vide sanction order no. 78(Sanc.)/ST/P/S&T/6 G-1/2018 dated 31.01.2019 and project no. GAP-225612].

https://doi.org/10.1515/9783111016160-010

accidents are also attributed to damages of metallic equipment or structures due to corrosion, which leads to loss of lives and adverse affect on the surroundings [1, 2]. Several safety hazards and accidents have been attributed to the adverse effect of corrosion on metallic structures of various equipment in several industries and sectors across the globe [3]. There have been great economic losses incurred by industrialists and manufacturers worldwide. Based on reports of the National Association of Corrosion Engineers (NACE), the global cost due to the consequences of corrosion is approximately US$2.5 trillion, which is equivalent to 3.4% of the gross domestic product of the world [4, 5]. Such economic loss primarily occurs in service and maintenance in the industry and agricultural sectors. In the United States and United Kingdom, the corrosion costs in the services, industries, and agriculture sectors are 79%, 20%, and 1%, respectively. In Japan, the costs of corrosion in these sectors are 73%, 26%, and 1%, respectively. In India, it is 57%, 26%, and 17%, respectively. In this way, the annual cost of corrosion in the developed and the developing countries are estimated and reported to lie within 1 to 5% of the global gross domestic product. This is a huge economic drain for a nation [3, 6, 7]. In this regard, keeping in mind the environment friendly nature of the corrosion inhibiting organic molecules [3, 8–10], several corrosion inhibitors with drugs [11, 12], plant extracts [12–14], biomolecules [15], and ionic liquids [12, 16, 17] as additives are in progress. There is recent progress in the development of micro or nanocontainer-based intelligent coatings [18, 19], bioactive surface coatings [20], etc. to protect metallic materials.

Recently, Long et al. reviewed the application of CDs for controlling environmental pollution, environment pollution sensing, contaminant adsorption, pollutant degradation, membrane separation, and anti-microbial activities [21]. There are scanty articles in which the corrosion inhibitory action of carbon dots as corrosion inhibiting additives in corrosive electrolytes has been reviewed. Pourhashem et al. briefly reviewed the application of carbon nanostructures such as carbon black, carbon dot, carbon nanotube, graphene, graphene oxide, fullerene, and functionalized graphene oxide as additives for corrosive electrolytes and as nanofillers in corrosion resistant coatings [22]. Herein, there is a scanty focus on the corrosion inhibitory action of carbon dots. Recently, Verma et al. shed more light on the applications of CDs synthesized from ammonium citrate, citric acid, L-histidine, thiourea, and urea that exhibit remarkable corrosion inhibition when compared with traditional corrosion inhibitors for carbon steel, copper, and mild steel. The heteroatoms-doped CDs have been found to show enhanced anti-corrosive properties. Mostly, the summarized CDs exhibited mixed types of corrosion inhibitory actions along with the Langmuir type of adsorption during their adsorption on the metal substrates. It has been found that their high solubility and excellent stability make the functionalized CDs preferential materials for use as corrosion inhibiting additives [23]. Berdimurodov et al. systematically overviewed the use of CDs as anti-corrosive additives in corrosive electrolytes such as acidic, saline, carbon dioxide saturated, and microbial solutions for the protection of metal substrates. The CDs are highly water soluble, biocompatible, less toxic, chemically and thermally stable, non-flammable, and also minimize bacterial growth. These properties make them a good choice for suitable use as corrosion inhibitors. The reported corrosion inhibition efficiency was found to be greater than 95%, with minute addition at moderate temperatures [24].

With this aim, the major works pertaining to chemical functionalization of CDs and heteroatoms-doped CDs as new materials for highly efficient corrosion inhibitors for several metallic materials used in various aggressive electrolytic solution such as saline water have been overviewed and discussed comprehensively, including the interpretation on the current challenging fields of corrosion inhibition. Additionally, the adsorption mechanisms as well as the corrosion inhibition mechanism of CDs on metallic surface have been explored. Finally, the future scope for targeted functionalization of CDs to be used as efficient anti-corrosive additives in adverse environmental conditions has been discussed.

10.2 Carbon dots: synthesis, functionalization, and properties

Carbon quantum dots (CQDs) or simply carbon dots (CDs) are zero-dimensional carbon nanomaterials that are less than 10 nm in size and possess surface passivation properties [25–27]. The CDs consist of an sp^2 hybridized carbon center with the coexistence of an sp^3 hybridized carbon center as also conjugated aromatic carbon centers in its core structure [28]. CDs can be doped with heteroatoms to achieve tuneable properties. This is accomplished by doping heteroatoms like nitrogen (N), sulphur (P), and boron (B). The CDs possess pyrrolic N, pyridinic N, graphitic N, amino N, and O atoms (vide Figure 10.1 (a, b)) having lone electron pairs, making them fascinating nanomaterials [24, 29]. Some literatures also report CDs or CQDs as quasi-spherical carbon nanoparticles having size < 100 nm; they possesses quantum confinement as well as edge defects [22, 30].

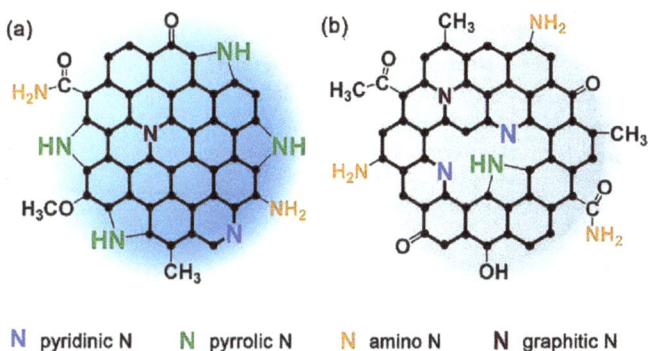

N pyridinic N   **N** pyrrolic N   **N** amino N   **N** graphitic N

Figure 10.1: Schematic for synthetic techniques of CQDs (reproduced with permission from ref. [29]).

10.2.1 Synthesis approaches for CDs or CQDs

Primarily, the synthesis approaches of CDs are generally categorized into two classes, based on the sources of bulk materials or the small molecules: (a) top-down and (b) bottom-up procedures. The first procedure includes the disintegration of a large carbon source from bulk substances like graphene, carbon nanotubes, carbon nanofiber, and so on; while the bottom-up process involves the polymerization and carbonization of small molecules or polymers as precursor [31]. Both these synthetic procedures are completed through chemical oxidation, electrochemical reaction, hydrothermal/solvo-thermal method, laser-ablation approach, microwave treatment, and pyrolysis reactions [31]. Some of the other synthesis approaches for CDs include plasma-induced approach [32], solvent-free gaseous deprotonation approach [33], facile ultrasonic approach [34, 35], etc., Some of the most imported approaches have been discussed briefly in the subsequent subsection. A pictorial representation of the synthetic techniques of CDs or CQDs is shown in Figure 10.2.

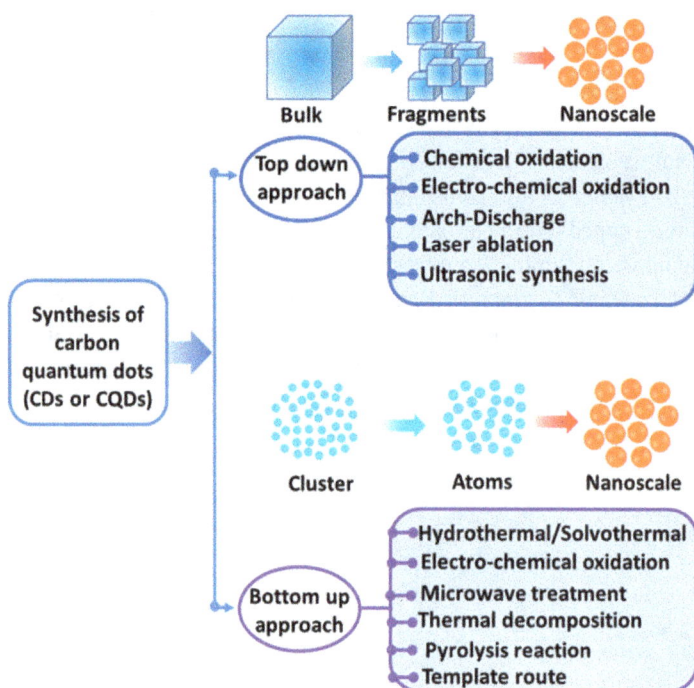

Figure 10.2: Pictorial representation of the synthetic techniques of CDs or CQDs.

10.2.1.1 Chemical oxidation methods for the synthesis of CDs

The chemical oxidation methods, also known as acidic oxidation, for the synthesis of CDs requires strong solutions of acids, for instance concentrated nitric acid, in combination with strong sulfuric acid, or hydrogen peroxide, and ferric chlorides like facile oxidants for oxidizing the reactants to get CDs. In 2004, Xu et al. reported the pioneering work in carbon nanomaterials – synthesis of CDs through chemical oxidation methods by using concentrated nitric acid to oxidize arc-discharged soot, followed by employing gel electrophoresis methods to isolate the suspension in order to accidently get blue-, green-, yellow- and orange-emitting CDs, though with a maximum photoluminescence quantum yield (PLQY) of only 1.6% [36].

10.2.1.2 Electrochemical oxidation methods for the synthesis of CDs

In the electrochemical oxidation methods, a certain voltage is applied to initiate the oxidation reaction at the anode, which facilitates the exfoliation of CDs from the carbon sources, based on electrochemistry. Thus, it is mandatory to use carbon materials as the working electrode, which must possess good conductivity and target-specific carbon sources. The electrochemical synthesis of CDs was reported by Sham et al. in 2007 for the first time, though having a PLQY of only 6.4% [37]. In this work, a three-electrode system was chosen with multiwalled carbon nanotubes (MWCNTs)-covered carbon paper as the working electrode and carbon source, a platinum as the counter electrode and $Ag/AgClO_4$ as the reference electrode along with electrolyte solution made of tetrabutylammonium perchlorate and acetonitrile. As a result, water soluble and blue-emitting CDs were obtained. The authors claimed that during the electrochemical reactions cycles, the structural skeleton of tetrabutylammonium cations disintegrated near the defects and were intercalated in the gaps of MWCNTs. Subsequently, the CDs got exfoliated into the electrolyte [37].

10.2.1.3 Hydrothermal/solvothermal approach for the synthesis of CDs

In this approach, initially, the reactant, based on small molecules or bulk materials, is well dissolved/dispersed in appropriate solvents. Subsequently, the reactants solution mixture is placed in high-pressure autoclave, and then subjected to high temperature and pressure, where these are polymerized and carbonized or broken down to form the desired CDs. The hydrothermal/solvothermal approach for CDs synthesis was first reported by Wu's group, and showed the preparation of graphene quantum dots (GQDs) (which are highly soluble in water) and cutting down of oxidized graphene sheets. The obtained GQDs were reported to have a PLQY of 5%, and these GQDs exhibited bright blue photoluminescence [38].

10.2.1.4 Laser-ablation approach for the synthesis of CDs

In general, a high-energy laser beam is used for bombarding carbon targets, making the carbon nanoparticle to peel using the laser ablation technique. These peeled of carbon are then passivated in order to obtain the CDs. Furthermore, if the most specific carbon target or its medium is favorably chosen, then the desired CDs can be obtained without further passivation.

10.2.1.5 Microwave treatment approach for the synthesis of CDs

The microwave treatment approach is a fast synthetic approach for preparing CDs. Here, microwave energy is applied on the reactant molecules that are polar in nature. The energy is absorbed by these molecules and they convert the electromagnetic energy to internal energy. This internal energy generated from inside of the materials increases the temperature of the reactants rapidly. The microwave treatment approach for CDs synthesis was initially reported by Yang et al. in 2009. These researchers mixed PEG200N and glucose, and then used it as a precursor. It was revealed within minutes that the size and photoluminescence of the CDs changed, leading to the formation of a stable photoluminescence with excellent water dispersibility due to the abundant traps and functional groups [39].

10.2.1.6 Pyrolysis reaction for the synthesis of CDs

The pyrolysis route, which employs the one-pot synthetic approach for the synthesis of oil-soluble CDs in non-coordinating solvent, was first reported by Liu's group in 2010. Highly luminescent CDs having 53% PLQY was obtained by carbonizing anhydrous citric acid as the carbon source in a hot mixture of a non-coordinating solvent, namely octadecene, and the surface passivating agent, 1-hexadecylamine, at 300 °C. This was optimized through changing the reaction solvent, the capping agent, the reaction time to control the depolymerization, decomposition, and pyrolysis process, with the aim of obtaining CDs with various structural and optical properties [40].

10.2.2 Functionalization of CDs

The intrinsic structure and surface states of CDs can be tuned by the functionalization approaches to improve the photophysical performance, broadening the applications in diverse fields. It can be achieved primarily by employing two broadly categorized approaches, viz., surface modification and heteroatom doping. Surface modification is a procedure for the functionalization of CDs that are target-specific since there are abundant functional groups present at the periphery of CDs. These groups can easily be functionalized by surface modification as these functional groups are available for various interactions such as coordination and electrostatic interaction. Such surface modification can be achieved by ionic modification approaches, based on metal and lanthanide ion, small organic and biological molecules (such as proteins and DNA), and polymer, providing unique and target-specific properties as well as improving the quantum yield. Heteroatom doping is a highly efficient method to tune the intrinsic structures as well as the electron distribution in CDs. These changes are reflected from the changes in energy differences in HOMO as well as LUMO of CDs, affecting the fluorescent properties. Heteroatom doping can be broadly categorized into non-metallic doping such as nitrogen, sulfur, phosphorous, boron, fluorine, silicon, iodine, chloride, tellurium, *etc.*, which also improves the quantum yield, and metallic doping using metallic dopants such as copper, gadolinium, manganese, zinc, germanium, magnesium, cobalt, ruthenium, iron, lanthanide, etc. that modulate the band structures of CDs. Additionally, the introduction of metallic dopants prevents the overconsumption of carboxyl as well as amino groups used as precursors at the time of dehydration and carbonization through chelate formation by chemical entities with metal ions [41–43].

10.2.3 Properties of CDs and functionalized CDs

Several functional groups or polymer chains impart unique characteristic properties in CDs, such as easy synthesis, excellent water solubility, superior biocompatibility, biological activities, low toxicity, environment friendly, photoluminescence, interchangeable optoelectronic properties, chemical stability, high thermal stability, and non-flammability. There is still scope for incorporating tuneable functionality by extending the π-conjugation and heteroatoms in CDs [22, 28, 43–45]. This multifunctional carbon nanostructure has found its applications in several fields such as biomedical purposes, traceable drug delivery, therapeutic applications, sensing and transfecting properties, bone tissue engineering, thin film coating for substrate protection, and so on [46, 47]. Recently, these CDs have emerged as a novel family of multifunctional carbon nanoparticles in research and development and have tremendously attracted researchers and scientists as well as industrialists [48].

10.3 CDs-based thin film as corrosion inhibitors

CDs are also deposited on metal substrates' surfaces to obtain a thin protective film for protection from adverse and corrosive environments. Saikia and Karak reported a chemically stable cellulose nanofiber (CNF)-functionalized, CDs-polyaniline nanofiber (PN), abbreviated as CCPN [49]. The CCPN was then sonicated in tetrahydrofuran (THF) for 30 min and added to sorbitol-based hyperbranched epoxy (SMD). Under continuous stirring for 3 h at 45–50 °C, polyamidoamine as a curing agent was added to it to get PCCN. It has been found that ultrasonication facilitates the interaction of CNF as well as CDs with aniline during the fabrication process. This is because, there is a possibility of the formation of strong inter- or intramolecular H-bond, leading to undue agglomerated accumulation. Thus, in order to get rid of this anomaly, the ultrasonication procedure is necessary to breakdown the agglomeration before adding the oxidant with the aim of promoting the even growth of PN over the CNF surface. Furthermore, the existing strong secondary interaction amid amine (from aniline) and hydroxyl or other functional groups that are polar in nature and present in CNF and CDs facilitate the uniform polymerization of aniline on CNF-peripheral reactive sites. Hence, the instant addition of the oxidant initiates instant polymerization reactions on the CNF surface. Furthermore, the addition of sodium hypochlorite solution (6 wt%) leads to the formation of long nanofibers possessing a large aspect ratio. Next, the targeted CCPN nanocomposite was prepared by properly dispersing it in tetrahydrofuran (THF) through ultrasonication and by the subsequent addition of hyper-branched epoxy as shown in the schematic diagram in Figure 10.3. Here, the hyperbranched epoxy was initially prepared by taking aliphatic polyols. Sorbitol functioned as branch-extending moiety by participating in the epoxidation reaction occurring between the bisphenol A-based epoxy and the epichlorohydrin. Subsequently, it formed globular-shaped hyperbranched structure, which comprised several functional groups. This reduced the viscosity of the epoxy and simultaneously made the matrix more flexible, which is attributed to its distribution, aspect ratio, and also its interactions with the matrix.

The curing process was accomplished by poly(amidoamine) at 100 °C for 30 min. The curing process has been found to increase with the addition of a high percentage of CCPN, which is attributed to the free -OH functional groups present in SMD, which interact with the polar functional groups of CCPN through polar–polar interactions. Furthermore, the presence of PN causes the enhancement of the basicity of the medium and increases the curing rate of the nanocomposite. It has also been found that when the percentage of the nanohybrid in the matrix is increased, the curing time is decreased [49]. It has also been reported that the nanocomposite exhibited good mechanical properties such as confirmed from Figure 10.4(a–c) showing stress-strain profile, elongation and tensile strength, as well as thermogram of SMD, and different percentage compositions (0.1, 0.25 and 0.5 weight percentage of PCCN as PCCN0.1, PCCN0.25, PCCN0.5, respectively) of the nanocomposite as well as thermal stability.

The nanocomposite showed good mechanical properties, evidenced by 82% elongation at break, 39 MPa tensile strength, greater than 10 kg scratch resistance, and 16.5 kJ/m impact resistance, along with good thermal stability up to 300 °C. The reported nanocomposite also possesses better biodegradability of pristine epoxy; this property is attributed to the accelerated degradation capability due to the presence of CD and CNF within the nanocomposite. Furthermore, the weight loss study also showed high chemical stability in aqueous HCl (10%) solution, aqueous NaOH (15%)

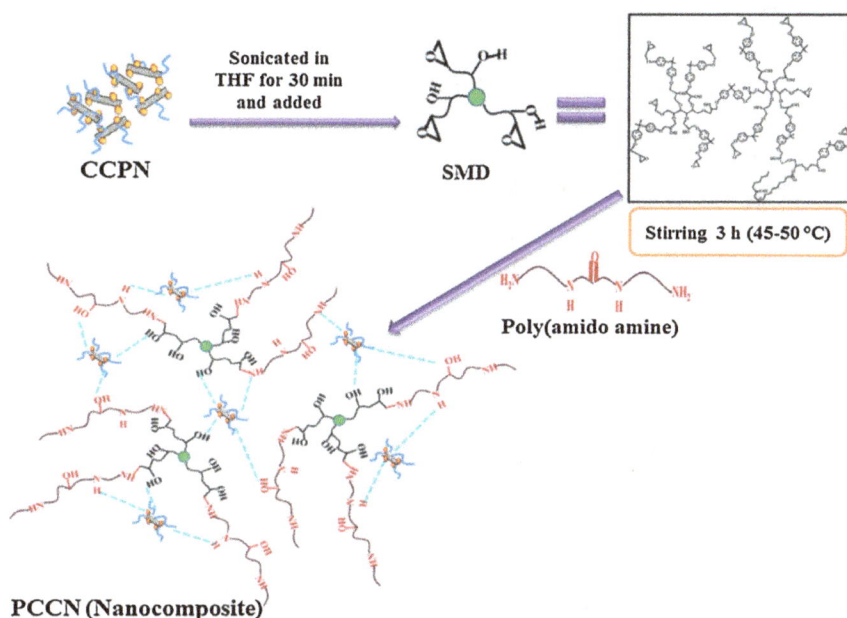

Figure 10.3: Fabrication procedure of the PCCN nanocomposite form, CCPN (reproduced with permission from ref. [49]).

solution, aqueous NaCl (15%) solution as well as in water, as tabulated in Table 10.1. It revealed that the order of the chemical resistance exhibited by the nanocomposite is in the order: NaCl > HCl > NaOH > Water.

Furthermore, the same group reported the fabrication of multiwalled carbon nanotube (MWCNT)-polyaniline nanofiber-CD nanohybrid. This fabricated nanohybrid was used in synthesized hyperbranched epoxy. This fabricated nanocomposite acts as a thermosetting nanocomposite having uniform as well as even dispersion of the prepared nanohybrid, and exhibited admirable mechanical properties, viz., 69 MPa tensile strength, 45% elongation at break, and greater than 10 kg scratch resistance, along with 16.7 kJ/m impact resistance. It was also thermally stable above 264 °C and showed high chemical resistance. Additionally, the cured nanocomposite applied on mild steels in saline media exhibited excellent corrosion inhibition capability, showing 4.62×10^{-4} mpy corrosion rate, suggesting that this nanocomposite can be effectively applied for protecting materials from degradation [50].

Recently, Peng et al., reported that organic quantum dots (OCDs) can be used as anti-corrosive coatings on aluminum alloy for its protection, when exposed in 3.5 wt% NaCl solution [51]. The OCDs were prepared from cetylpyridinium chloride monohydrate after treating them with sodium hydroxide at room temperature. The prepared homogeneous solutions of OCDs in petroleum ether were used for the immersion of an aluminum substrate to obtain the coating of OCDs. This coating was dried at room

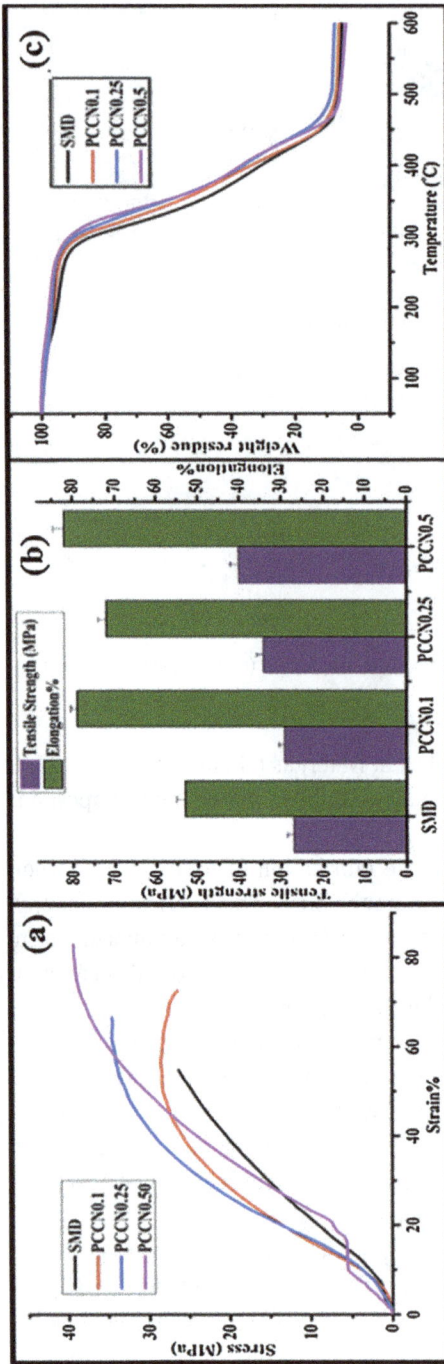

Figure 10.4: Mechanical properties: (a) stress–strain profile, (b) elongation and tensile strength, (c) thermogram of SMD, along with different percentage compositions of the nanocomposite (reproduced with permission from ref. [49]).

Table 10.1: Weight loss (%) data determined for SMD, along with the different percentage compositions of the nanocomposite (i.e., PCCN0.1, PCCN0.25, and PCCN0.5) in different media (reproduced with permission from ref. [49]).

Chemical medium	SMD	PCCN0.1	PCCN0.25	PCCN0.5
Aqueous HCl (10%)	0.61	0.54	0.45	0.38
Aqueous NaOH (5%)	0.82	0.76	0.66	0.43
Aqueous NaCl (15%)	0.55	0.48	0.41	0.36
Water	0	0	0	0

temperature to obtain a superhydrophobic aluminum alloy surface, which can resist water contact up to a temperature of 161.4° with 0.5 mg/mL OCDs in the coating formulation. It suggests that ~ 4.1% of the water droplets came in contact with the aluminum alloy, while the residual water droplets remained suspended in the air, resulting it super hydrophobicity. The fabricated OCDs coating on the aluminum surface also exhibited self-cleaning property, which is reflected from the observation that the contaminants were encapsulated or diffused into water droplets, and rolled off from the surface. The anti-corrosion properties of the OCD coating on the aluminum substrate were ascertained from electrochemical techniques. Potentiodynamic polarization showed a drastic reduction in the corrosion current density, the lowest being 0.5 mg/mL of OCDs formulation fabricated coating, when exposed to 3.5% NaCl media. Similarly, another electrochemical technique, namely electrochemical impedance spectroscopy, also confirmed the high charge transfer resistance exhibited by the 0.5 mg/mL OCDs formulation fabricated coating, exposed to corrosive media, in comparison to other composition containing coatings [51].

10.4 Functionalization of thin film CDs encapsulated as corrosion inhibitors

10.4.1 Functionalization of CDs epoxy thin film as corrosion inhibitors

Liu et al., revealed in situ bonding modification of zinc molybdenum oxide (ZM) using CDs, which synergistically improved the anti-corrosion property, exhibiting photoelectron suppression effect by CDs-modified ZM (CDs@ZM) [52]. Here, the CDs were synthesized by adopting the bottom-up process, exploiting citric acid and urea, by employing the hydrothermal procedure. The obtained product from sodium molybdate and zinc nitrate was termed as ZM; the ZM was dried at 60 °C to obtain zinc molybdenum oxide hydrate (ZM1), while the calcined ZM1 at 400 and 600 °C for 2 h gave trizinc dimolybdate, ZM2 and ZM3. The CDs@ZM1 and CDs@ZM1 were prepared using the in situ bond formation procedure. It has been reported that the complete surface coverage and optimum protection from corrosion exhibited by the CDs@ZM2 coating were evidenced microscopically. Through electrochemical study using electrochemical impedance spectroscopy, results also revealed that epoxy coating of CDs@ZM2 pigment exhibited high charge transfer resistance, compared to other coating compositions such as CDs@ZM2, CDs@ZM3, CDs@ZM1, ZM2, ZM3, and ZM1 pigment-infused epoxy coating in sequence, upon exposure in saline media (vide Figure 10.5). Based on the observation of Figure 10.5, it can be said that the ZM as well as the CDs@ZM coating results in a capacitive loop. High-frequency regions correspond to reactions occurring at the coating/electrolyte interface, while low-frequency regions correspond to results of reactions at the metal/coating interface. After the modification of ZM with CDs, the radius of the Nyquist plot showed an increment, as shown in Figure 10.5, suggesting a significant increase in the shielding effect of the applied coating. The CDs@ZM2 pigment containing the coating showed a higher charge transfer resistance, which is an indication of the greater reduction in the electron transference from the metal to the applied coating that is exposed to corrosive electrolytes for 72 h. It is also said that the photoelectron production efficiency increased remarkably. These photoelectrons in the coating retarded the electron loss from the metal substrate during the corrosion processes and ultimately increased the charge transfer resistances. Deposition of molybdate on metal surfaces also hinders penetration of the corrosive species. In addition, the presence of electron-withdrawing carbonyl group ($>C = O$) in the CDs-modified materials are also attributed for retarding the rate of charge transfer. Furthermore, the formation of a covalent bond by the peptide bond (-CONH) in CDs with polymer chain aids the self-healing property as well as increases the compactness of the coating, leading to inhibition of the penetration of the corrosive species. The bode plot presented in Figure 10.6(a) showed that CDs-modified ZM pigments infused epoxy coating offer high impedance modulus at low frequencies because of the electron suppression and the shielding effect of oxidation, minimizing the media indicating and improving the corrosion inhibition efficiency. Again, the higher phase angles at the low-frequency regions indicate better integrity of coating for the protective layer, inhibiting penetration by the corrosive species. Furthermore, based on the reported Tafel polarization curve (vide Figure 10.6(b)), it is obvious that the applied CDs@ZM coating exhibited corrosion inhibitory action on both the anode and the cathode region. The attainment of more positive values of corrosion potential (−889 mV) and lower current density (23.67 $\mu A/cm^2$) confirmed the improvement in corrosion inhibition performances, which is attributed to the anode inhibition effect of CDs as well as the deposited molybdate barrier layer.

When the steel substrates are immersed in thee corrosive electrolytes, anodic and cathodic reactions take place, as depicted in eq. (10.1–10.2) and eq. (10.3), respectively.

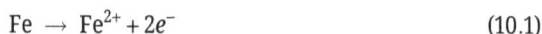

$$Fe \rightarrow Fe^{2+} + 2e^- \tag{10.1}$$

$$Fe^{2+} \rightarrow Fe^{3+} + e^- \tag{10.2}$$

$$O_2 + 2H_2O + 4e^- \rightarrow 4OH^- \tag{10.3}$$

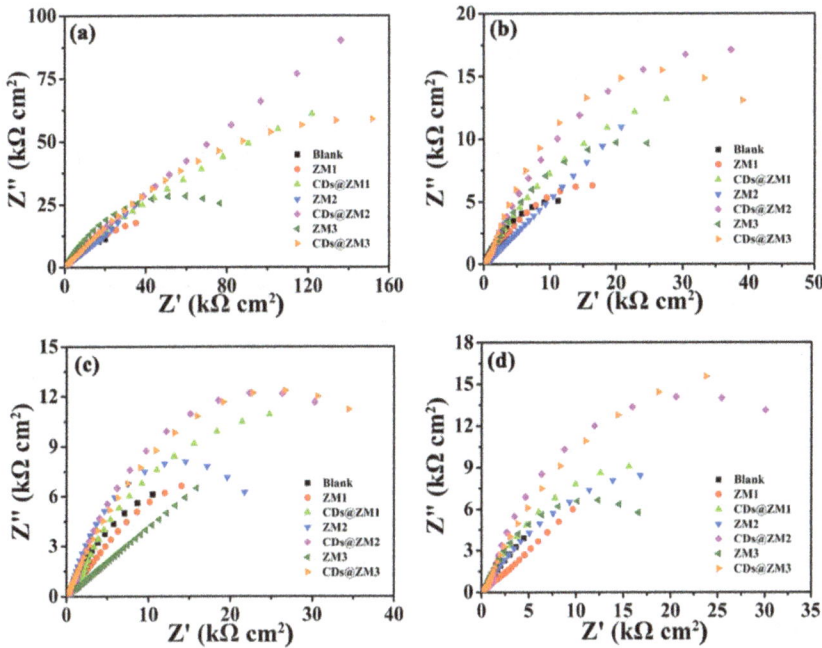

Figure 10.5: Nyquist plots for uncoated (i.e., blank) and coated steel substrates, with different coating compositions, explored for different immersion times (a) 6 h, (b) 24 h, (c) 48 h, and (d) 72 h in 3.5 wt% NaCl solution (reproduced with permission from ref. [52]. Copyright, 2020 Elsevier Ltd.).

It is known that the molybdenum zinc oxides facilitate dissolution of molybdate ions in the electrolyte, subsequently competing with the chloride anions and lessening the contact of the corrosive ions with the iron substrates. In addition to it, the zinc as well as the molybdate anions are also capable of interacting with the charged iron atoms and the hydroxyl (-OH$^-$) ions formed at the cathode to produce an insoluble passivation film. This insoluble film gets deposited at the sites where corrosion occurs and minimizes the impending corrosive dissolution, as depicted schematically in Figure 10.7 – shown by Stage A exhibiting the layer shielding effect of the applied coating. Simultaneously, the suppression of the electron loss on the steel surface can also be attributed to the applied CDs@ZM coating, as depicted schematically in Figure 10.7 – shown by Stage B. The suppression of the electron loss minimizes the oxidation reaction occurring at the anodic part of the steel substrate. ZM and CDs@ZM pigments generate photoelectrons, and their accumulation on the steel substrates significantly retards the initiation of metallic dissolution. Subsequently, it leads to an inference that the corrosion inhibition performances increase in the sequence from ZM1, ZM3, ZM2, CDs@ZM1, CDs@ZM3, and CDs@ZM2; where CDs@ZM2-pigmented epoxy exhibited the highest corrosion inhibition performance.

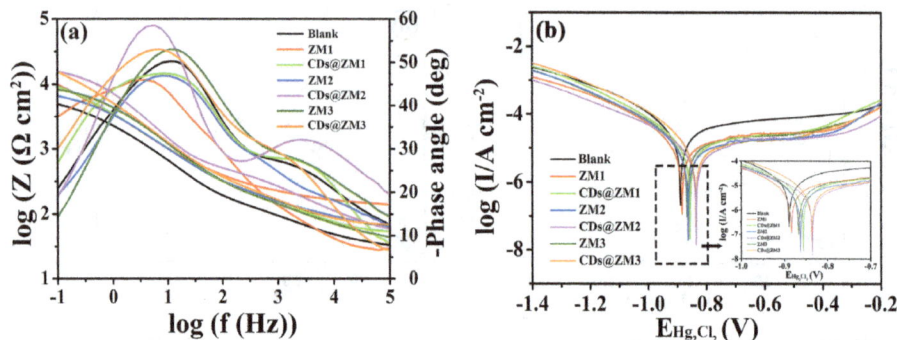

Figure 10.6: (a) Bode plots and (b) Tafel polarization plots for uncoated (i.e., blank) and coated steel substrates with different coating composition explored for 72 h immersion in 3.5 wt% NaCl solution (reproduced with permission from ref. [52]. Copyright, 2020 Elsevier Ltd.).

Recently, Ye et al. revealed a functionalization strategy of CDs obtained from citric acid derivatives, which were subsequently used in modifying graphene. The modified CDs acted as intercalators, which were dispersed in the epoxy matrix to get CDs-modified graphene/epoxy (CDs-G/EP) coating. Microstructural analysis revealed the compactness of the CDs-G/EP, which might have come due to the π–π interactions between the functionalized CDs and graphene. The functionalized CDs enhance the dispersion as well as interface compatibility of graphene. The coating compositions are: EP, 0.5 wt% graphene in epoxy (G0.5%/EP), CDs-modified G0.5%/EP, designated as CDs-G0.5%/EP, and CDs-modified 1 wt% graphene in epoxy as CDs-G1%/EP.

The results obtained from electrochemical studies revealed that G0.5%/EP revealed better protection ability for steel. The Nyquist plot revealed capacitive property, with increase in charge transfer resistance (R_{ct}) and resistance due to coating (R_c) indicating the barrier property of coating. The appearance of Q_c and Q_{dl} are due to the fitting the electrochemical parameters in the Nyquist plot due to the phase components of the coating as well as the electrical double layer capacitances. The appearance of R_{ct} and Q_{dl} suggested diffusion of the corrosive species toward the substrate's surface through coating. The plots of R_{ct}, R_c, Q_{dl}, and Q_c are presented in Figure 10.8. The figure shows that in the beginning of the immersion process, the R_c values reduce and the Q_c values increase. It is attributed to the penetration of the corrosive chloride toward the coating at the coating–steel interface and the initiation of corrosion of the steel surface (vide Figure 10.8 (a) and (c)).

It implies that with the addition of graphene, the variation of the trend of coating retarded corrosion to some degree. The excellent compactness of the coating composite,

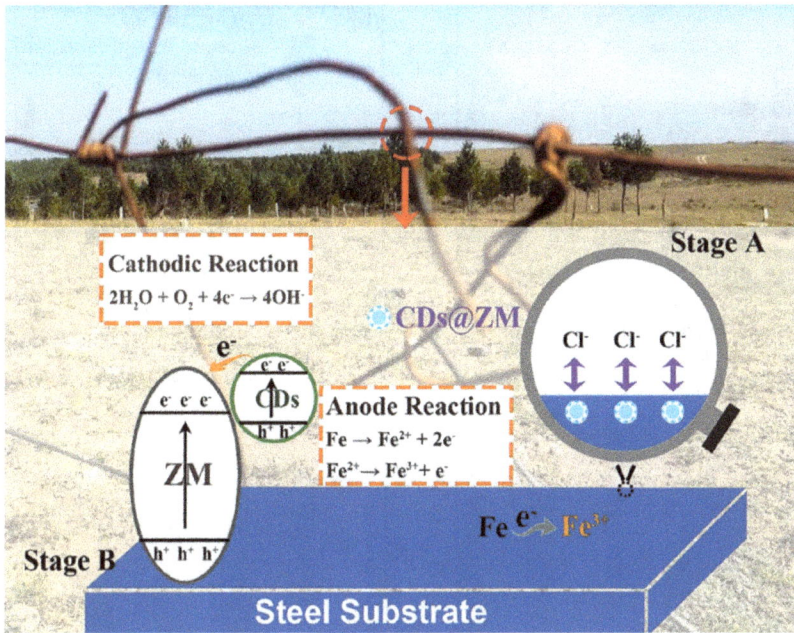

Figure 10.7: Mechanism of corrosion inhibition exhibited by CDs@ZM (reproduced with permission from ref. [52]. Copyright, 2020 Elsevier Ltd.).

CDs-G0.5%EP, is main reason for exhibiting superior coating resistance as well as lower coating capacitance throughout the whole period of immersion. But after 5 days of immersion, it can be found that there may be micropores that allows the passage of corrosive electrolytes through the pure EP coating (vide Figure 10.8 (b) and (d)). Thus, due to the excellent barrier property of CDs-G0.5%EP, it slowed down the diffusion of the corrosive electrolytes toward the steel substrates for up to 10 days. But with the incorporation of CDs into the coating, there is further delay in the diffusion of the corrosive electrolytes toward the substrates, which is attributed to the highly dispersed graphene that acts as a filler to fill up the micropores in the coating, enhancing its compactness and enabling it to withstand the corrosive electrolytes till 20 days of immersion. Afterward, the R_{ct} declined and Q_{dl} increased, owing to the dissolution of the metal surface's atoms at the interface. The CDs-modified coating also exhibited remarkable reduction in oxygen permeability and water absorption, as evidenced from Figure 10.9 [53].

It has been found that there is an uptrend increase in the radius of the capacitive resistance property with the addition of CDs-G. This might be arising due to the barrier properties exhibited by the highly dispersed graphene. Furthermore, the compactness helps in suppressing the impact of the corrosive substances. Owing to the high compactness, it exhibited superior coating resistance and lowest capacitance throughout the immersion experiment of the coating substrate in the corrosive medium. A schematic diagram considering experimental finding shows that there are

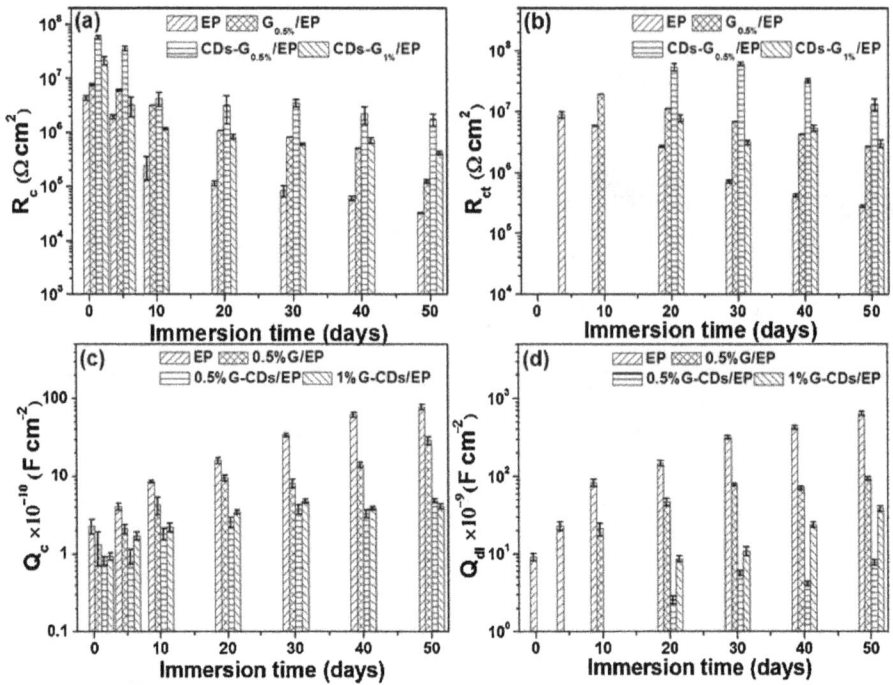

Figure 10.8: Plot of (a) Rc; (b) Rct; (c) Qc; (d) Q_{dl} obtained for coating, upon exposure to 3.5 wt% NaCl solution (reproduced with permission from ref. [53]. Copyright © 2019 Elsevier Ltd.).

micro pores or holes present in the pristine epoxy coating (vide Figure 10.9(a)), it can easily be understood that the corrosive ingredients dissolved in the electrolytes can easily penetrate the coating and lead to corrosion at the epoxy–metal substrate interface. Next, Figure 10.9(b–d) lead us to interpret that when the corrosive medium passes through the nanocomposite coating, it comes across graphene-filled pores that are dispersed in the composite formulation, and this barrier property delays the corrosive electrolytes from penetrating the metal substrates. Upon incorporation of CDs-modified graphene, they were well dispersed and were capable to fill the micropores in the coating to a greater extent, leading to the formation of a highly impermeable barrier coating that is effective till 20 days when exposed to a corrosive solution. Furthermore, it has been observed that the corrosion dissolution at the interface becomes intense after a prolonged immersion period. This observation is well confirmed from the increased R_{ct} and Q_{dl} values till 20 days of immersion and the subsequent decrease of Rct and increase of Qdl with time, as determined from electrochemical studies.

Similarly, Yu et al., have also reported the anti-corrosive property of hexagonal boronnitride (h-BN), which is structurally analogues to graphene [28]. The h-BN increased the rescuer's consideration because of its electrical insulating as well as barrier properties, which are prerequisite properties for the exhibition of anti-corrosive

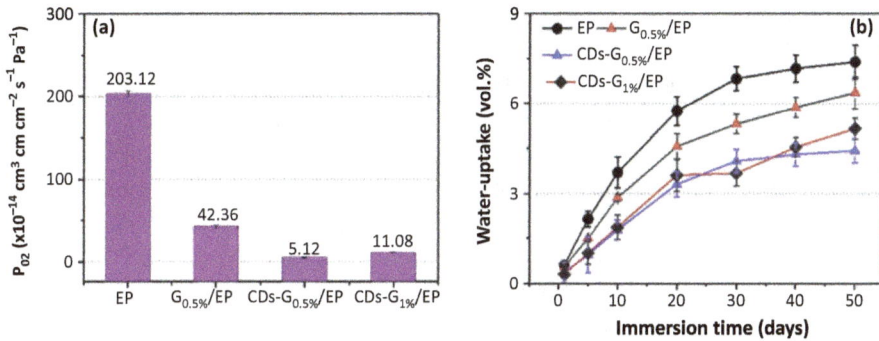

Figure 10.9: Plot of (a) oxygen permeability coefficient and (b) water absorption of coating (reproduced with permission from ref. [53]. Copyright © 2019 Elsevier Ltd.).

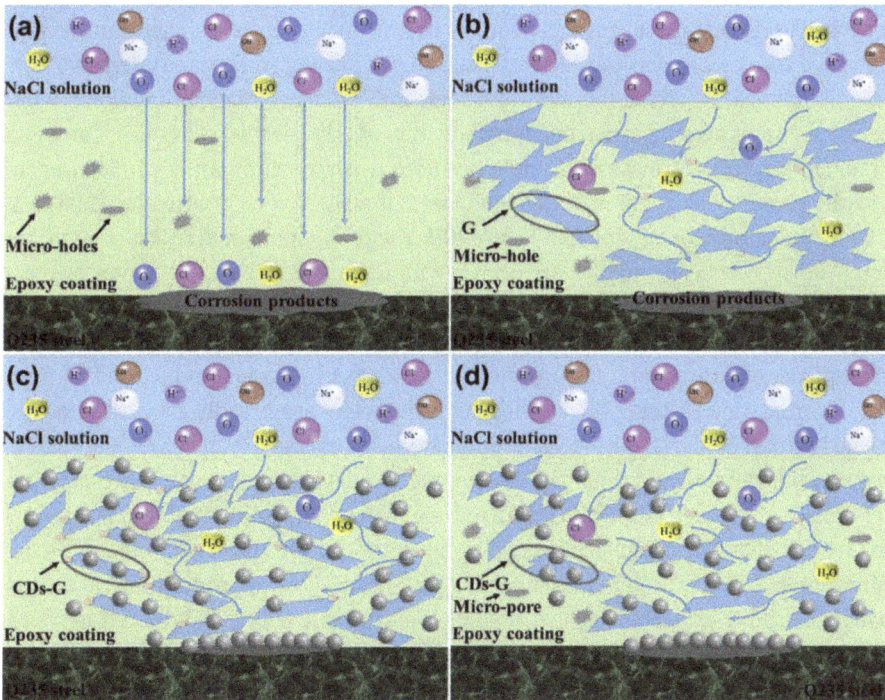

Figure 10.10: Schematic diagram showing the corrosion inhibition mechanism by coatings: (a) EP; (b) G0.5%/EP; (c) CDs-G0.5%/EP; (d) CDs-G1%/EP (reproduced with permission from ref. [53]. Copyright © 2019 Elsevier Ltd.).

behavior. In this work Yu et al., have attempted to reduce the formation of easy aggregates of h-BN in water by improving its hydrophilicity through modification with CDs. The synthesized CDs assisted in dispersing the h-BN nanosheets in waterborne epoxy

(WEP), resulting h-BN@CDs/WEP. This is achieved by the π–π interactions occurring between the h-BN and CDs. In addition to it, the most plausible reaction of oxygen functional groups present in CDs with the unreacted epoxy and hardener enhances the compatibility as well as crosslinking density between h-BN and the WEP matrix. Owing to this, the attained barrier property of nanosheet-infused nanocomposite is reinforced, which can be confirmed from the electrochemically determined impedance results showing an impedance modulus of 10^8 Ω cm^2 for h-BN@CDs/WEP – higher than that of pristine WEP coating of 10^5 Ω cm^2, which is higher than that of h-BN/WEP, determined by electrochemical impedance spectroscopy explored a in saline environment. Additionally, the measured local electrochemical impedance spectroscopy as well as salt spray also revealed superior anti-corrosive property even in a damaged condition [28].

The metal substrate with carbon dots in a polymer matrix acts as a self-healing anti-corrosive coating. There are several innovations for the exploitation of CDs for self-healing coating for imparting a stimulus responsive and protective coating on metallic materials to protect these from adverse environments. For decades, there are several reports of the use of microcapsule-encapsulated self-healing composite systems for corrosion inhibition [54]. Next, there are also some innovatively designed corrosion inhibiting coatings that become active upon the initiation of corrosion reactions [55]. Corrosion inhibition through extrinsic self-healing has been overviewed by An et al. [56]. In general, the protective coating might be composed of three layers: pre-treatment layer, a primer, and a top coat as shown in Figure 10.7. The pre-treatment layer acts as an adhesion promotor, which combines the primer with the metal substrate, whereas primers usually comprise chromated pigments acting as anti-corrosive agents. The top coat above the primer plays a crucial role in protecting the other layers as well as the metal substrates from environmental factors such as ultraviolet (UV) radiation as well as from high external temperature. The self-healing phenomena are broadly grouped into two: (a) intrinsic self-healing, which deals with the intrinsic reversibility of constituents participating in chemical bond formations that are rearranged in response to an external stimulus; and (b) extrinsic self-healing, which deals with activation through the intentional release of self-healing agents embedded in the matrix or encapsulated in the nanocontainer, upon damage. The multilayer coating on a metal substrate with a crack at the top layer penetrating toward the primer can be visualized as shown in Figure 10.11 in which the primer is applied usually on the pre-treatment layer and the top coat in order to protect them from ultra-violet (UV) irradiation and crack due to external environmental or other effects. This crack may penetrate and allow corrosive ingredients to reach the substrate and initiate corrosive dissolution of the metallic substrates [56, 57]. The mechanism of (a) self-healing through capsule and (b) fiber-based approaches that contain encapsulated corrosion inhibitors that are released at the time of crack has been depicted in Figure 10.12(a, b) [56, 57].

The mechanism of self-healing through capsule and fiber-based approaches containing encapsulated corrosion inhibitors that are released at the time of crack have been shown in Figure 10.12. Nano- or micro-sized healing agents are usually embed-

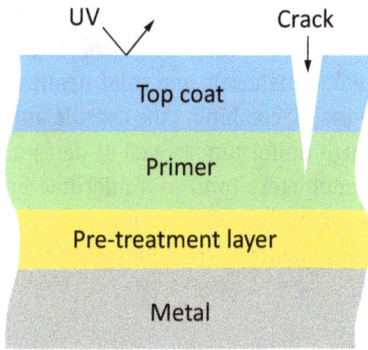

Figure 10.11: Multilayer coating on metal substrate with crack at the top layer penetrating toward the primer (reproduced with permission from ref. [56].).

ded in the structure of the composite matrix that contains an incorporated catalyst capable of reacting with the healing agent, leading to the formation of a polymeric network, upon polymerizing it. Such polymerization reactions are initiated whenever cracks are formed due to the damage of the matrix. Cracks break nano/micro capsule, facilitating healing agents to ooze out onto the surface of cracks through capillary action. Subsequently, the healing agents approach the catalyst, which catalyze the polymerization reaction and help in bond formation between the adjacent cracked planes.

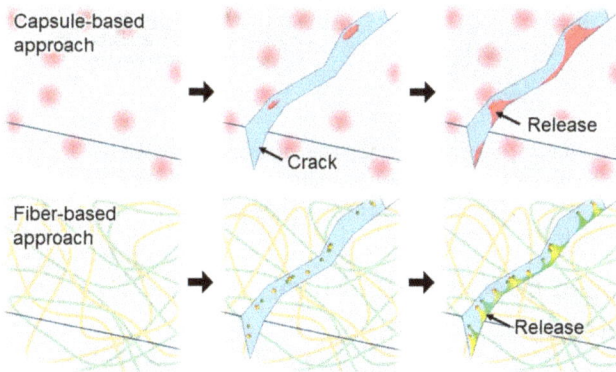

Figure 10.12: Schematic diagram showing the mechanism of self-healing through capsule and fiber-based approaches, containing encapsulated corrosion inhibitors that are released at the time of the crack (reproduced with permission from ref. [56]).

Despite this, there are several self-healing agents that are easily available. They might possess fast healing nature that have been developed so far; however, there are also some flaws when the economic feasibility and scalability of the self-healing coatings are considered. As reported by Yu et al., it has been found that there may be inhomogeneity of the dispersed capsules, negative impact on the mechanical properties, or they may be obtained using complicated fabrication procedures, as depicted in Figure 10.13. In this regard, some fiber-based self-healing coating strategies are developed with the intension of addressing some of the disadvantages as shown in Figure 10.13 [28]. These may be with some bottlenecks such as limited available materials and mild healing-ability. But, a combination of these might address the issues; combining the capsule and fiber-based coating strategies might introduce ultra-high uniformity as well as density, where widespread recovered cracks, from a few nanometers (nm) to a micrometer (μm), are introduced. Rapid healing ability may be accomplished within a few minutes, along with the enhancement of mechanical properties, owing to the fiber-reinforced composite nature.

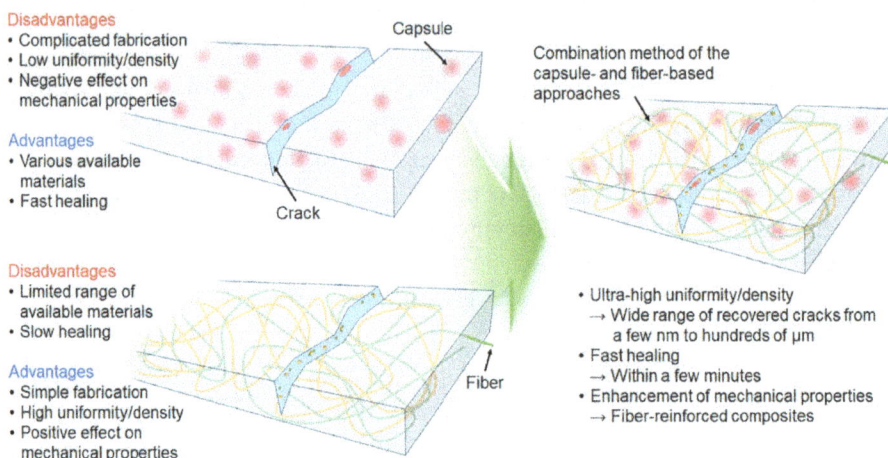

Figure 10.13: Pictorial representation showing the advantages and disadvantages of capsule and fiber-based carbon dots' self-healing coatings (reproduced with permission from ref. [56].).

10.4.2 Heteroatom doped CDs-based thin film to minimize corrosion

Recently, Wang et al., synthesized nitrogen-doped carbond dots (CNDs or NCDs) using 4-aminosalicyclic acid through the solvothermal process [58]. The diameter and the height of the NCDs were reported to be 3–5 nm; this was established by the results of scanning probe microsopy and transmission electron microscopy. They revealed that 0.5 wt% NCDs embedded into epoxy-forming NCDs-nanocomposite act as anti-corrosive waterborne epoxy coating for metal substrate, protecting it from corrosion in a 3.wt% NaCl solution for a 70-day immersion period. The NCDs act as nanofillers, which upon addition of suitable amount fill up defects as well as tiny holes formed aat the time of the curing process when water evaporates. These NCDs nanofillers that are in contact with the metal substrate surface (herein Q235 carbon steel) are adsorbed through physisorption as well as chemisorption, forming a dense passivation film of Fe_3O_4. The impermeable barrier film is strengthened by the bond formation between the NCDs and the waterborne epoxy hydrogen bonding, covalenyt bond fornmation as well as the van der Waals interaction occurring between the NCDs and the waterborne epoxy. This dense film acts as a barrier, which isolates the susceptible metal from corroive attack by the electrolyte and protects it from corrosion damages [58]. Ren et al. reported NCDs preparation using the hydrothermal method to obtain 4–6 nm sized nanomaterials [59]. The incorporation of 2.0% of synthesized small-sized NCDs into the WEP coating exhibited 364 times higher impedance modulus after 800 h immersion, compared to pristine WEP coating, as determined from electrochemical impedance spectroscopy, suggesting superior corrosion inhibition efficacy. The high corrosion inhibition efficacy is attributed to surface functional groups that are capable of interacting with the functional group of WEP, remarkably suppressing the pore size. Additionally, the trapped NCDs also act as a barrier, which isolates the contact of the corrosive electrolytes with the metal substrates apart from these NCDs nanomaterials hindering the dissolution of oxygen toward the susceptible metal surfcace.

Again, a deeper investigation into the corrosion inhibition mechanism by the NCDs was revealed by Li et al. [60]. They synthesized nitrogen-doped carbon dots (NCDs) having ~ 100 nm size through the solvothermal technique, taking cost-effective citric acid and ethylene diamine as carbon and nitrogen sources, respectively. It resulted in a high yield, and was easy to fabricate. In addition to these properties, the product obtained after reactions is enriched with acylamino, carboxyl, and hydroxyl groups, which offer advantageous functionalities for obtaining improved corrosion inhibition efficiency [60], These synthesized CNDs showed best performance in the decrease of the water absorption property measured using gravimetric measurement with the optimum addition (~ 0.05–0.15 weight %) into epoxy (E-44) coating. The anti-corrosion capability and service life are also well predicted by the water absorption property study by other researchers [53, 61, 62]. The effect of CNDs on the anti-corrosive property in waterborne epoxy coating is revealed through electrochemical measurements [59]. The results deduced from the electrochemical measurements have been presented in Figure 10.14, which shows the plot of (a) charge transfer resistance (R_{ct}) and the exposure time for neat, 0.05% and 0.20% CNDs-incorporated epoxy nanocomposites and (b) charge transfer resistance (R_{coat}) and the exposure time for neat epoxy 0.05%, 0.15% and 0.20% CNDs/epoxies. Figure 10.15 shows the porosity of different coatings determined for neat epoxy, 0.05, 0.15 and 0.20 CNDs/epoxies. The increase in exposure period causes a significant drop of Rct for both pure epoxy as well as 0.05CNDs/epoxy, while for 0.20CNDs/epoxy, it increases till 240 h and then drops afterward. It suggests that CNDs are interfering with the electron transfer, which is reflected from the Rct values. But, for 0.15CNDs/epoxy, the non-appearance of R_{ct} suggests the non-occurrence of corrosion degradation during the entire immersion period, thus explaining superior inhibiting capability. Similarly, the higher values of R_{coat} revealed the barrier property of the coating, which decreases with prolong immersion.

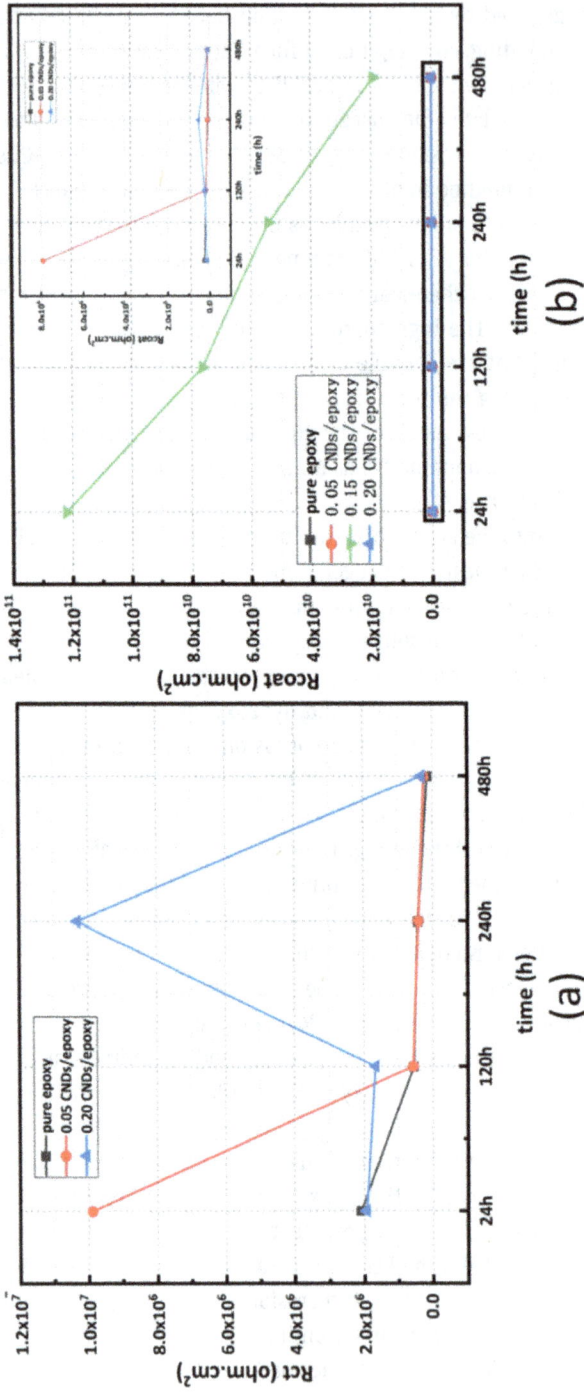

Figure 10.14: Plot of (a) charge transfer resistance (R_{ct}) versus exposure time for pure epoxy coating, 0.05 and 0.20 CNDs/epoxies and (b) charge transfer resistance of coating (R_{coat}) versus exposure time for pure epoxy coating, 0.05, 0.15 and 0.20 CNDs/epoxies (reproduced with permission from ref. [60]. Copyright © 2019 Elsevier Ltd.).

Porosity of coating, determined from R_{coat} and coating thickness values, as shown in Figure 10.15, revealed that 0.15CND/epoxy is least porous (5–6 order of magnitude), and also remains almost constant, implying that this coating is highly impermeable by corrosive electrolytes till 480 h immersion in corrosive electrolytes. It has been also seen that the porosity of pure epoxy coating increases with exposure soaking time. The porosity of 0.15CNDs/epoxy decreased up to 240 h immersion in corrosive electrolytes and then increased thereafter, suggesting the diffusion of corrosive electrolytes, while the low content of CNDs in the epoxy showed an almost similar trend of porosity increase. Hence, 0.15 wt % CNDs in epoxy matrix can exhibit best corrosion inhibition performance.

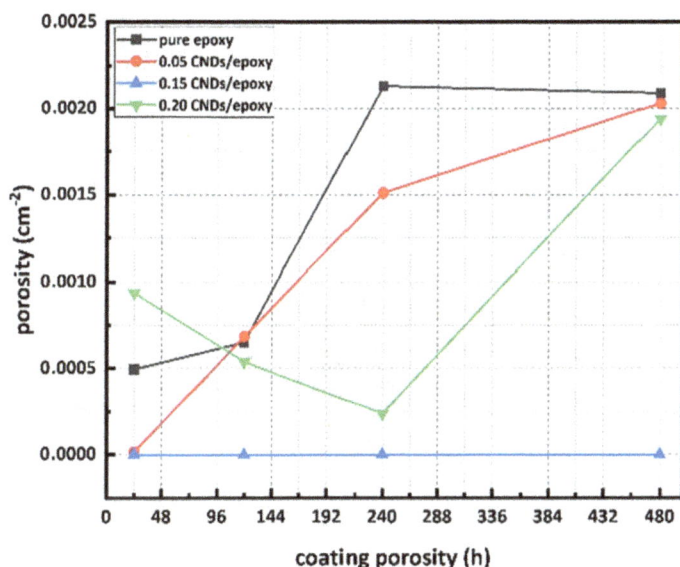

Figure 10.15: Porosity of different coatings, determined for pure epoxy coating, 0.05, 0.15 and 0.20 CNDs/epoxies (reproduced with permission from ref. [60]. Copyright © 2019 Elsevier Ltd.).

The oxidation and reduction of metallic materials leads to its degradation in a corrosive media containing water and oxygen. The water, oxygen, and the corrosive chloride (Cl⁻) ions present in the media can easily penetrate through the inherent micro-pore channels within the pure epoxy coating on the metal substrates (vide Figure 10.16 (a)). It has been revealed that the measured porosity of the CNDs/epoxy coating is reduced with the addition of NCDs into the pure epoxy. The porosity at the 0.15 wt% CNDs into epoxy is extremely reduced to 6.23×10^{-9} cm^{-2} after 480 h soaking, which is 5–6 times lower compared to other concentrations and pure epoxy coating. The porosity decrease is the consequence of H-bond as well as covalent bond formation between the CNDs and the epoxy (*vide* Figure 10.16 (b)). Here, the dispersibility and bonding of CNDs within the epoxy gives rise to a stable nanocomposite coating. The CNDs possess functionalities like amides, amino, and carboxyl groups. Here, amide groups act as a curing agent that cures the used epoxy by forming covalent bonds with the epoxy groups. Furthermore, the amides, amino, and carboxyl groups might also form hydrogen and covalent bonds between the CNDs nanoparticles (vide Figure 10.16(c)). Again, the nanoparticles of CNDs present in both sides of the inherent micropore channel of the coating interact with each other, reducing the porosity. In these circumstances, the bond formation within the interior wall of the micropore channel inhibits the penetration of corrosive electrolytes toward the metal substrates and retards the deteriorating epoxy coating, subsequently enhancing the corrosion inhibition capability of the CNDs/epoxy nanocomposite coating [59].

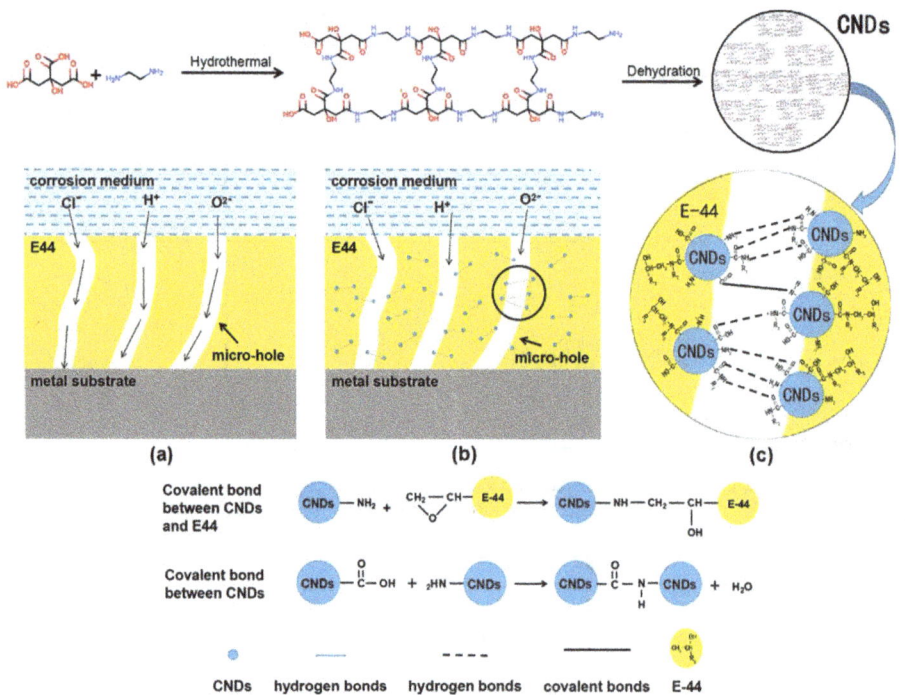

Figure 10.16: Mechanism of protection of a metal substrate using (a) pristine epoxy, and (b) CNDs/epoxy coatings, while (c) shows magnified interpretation of the CNDs/epoxy interaction at the micro hole within the nanocomposite coating. The inset at the top shows the illustrative synthetic procedure of CNDs, the inset at the bottom shows the covalent bond formation between CNDs and E44 epoxy; and between CNDs (reproduced with permission from ref. [60]. Copyright © 2019 Elsevier Ltd.).

Research and developments for capsules having impregnated corrosion inhibitors have attracted significant number of researchers for developing self-healing corrosion inhibition of metallic materials exposed to various corrosive environment. In this regard, Bao et al. have recently reported the salt-responsive zinc oxide (ZnO) microcapsules encapsulating nitrogen-doped carbon dots (NCDs) for increased longevity of the tin plate having an applied coating thickness of ~20 μm [63]. The NCDs are synthesized from o-phenylenediamine through the solvothermal process as reported by Cui et al. [64]. A three-layered coating, i.e., waterborne polyacrylate (WPA)-NCD@ZnO, was applied to attain saturation in a short span of time, as confirmed from the modelled equivalent electric circuits. This is because of the dense and impermeable coating of WPA/NCD@ZnO applied onto tin plate. The waterborne polyacrylate coating limits the shielding behavior of coating. The aggressive ingredients may penetrate toward the coating metal interface through defects and pores present in the WPA coating layer. Diffused corrosive electrolytes at the WPA metal interface initiates the corrosive dissolution. The ferrous (Fe^{2+}) and ferric (Fe^{3+}) ions formed due to the dissolution of metals could be converted to Fe_2O_3 and Fe_3O_4 in the presence of oxygen. Both of these oxides get deposited where the metallic dissolution takes place, forming a dense oxide layer that acts as a barrier, which plays a positive role in protecting the metal substrates. The addition of anti-corrosive additives imparts protective properties to the coating layer. Here, the WPA/ZnO composite coating increases the density of the coating with ZnO addition, which fills the pores in the coating. This happens because the Zn^{2+} ions on the surface of ZnO at the coating substrate interface interacts chemically with the hydroxyl groups of the WPA through hydrogen bond formation and electrostatic interaction [65], leading to the formation of a dense film. But, in WPA/NCD@ZnO composite coating, initially, this coating inhibited the penetration of the corrosive ingredients along with the electrolytes since, the added filler fills the pores and chemically interacts with the WPA, forming a dense and impermeable layer. When the corrosive species penetrates, upon damage, the NCDs get released gradually from the microcapsule and moves toward the metal surface along with the electrolytes and gets adsorbed onto the metal surface there. These released NCDs, which possessing lone pairs of electrons on the nitrogen heteroatoms, interact with the unoccupied orbitals of the metal atoms at the surfaces, thereby giving rise to a protective film. This protective film helps in resisting it from further dissolution due to the penetrated corrosive electrolytes. In this way, this composite coating acts as a self-healing anti-corrosive coating. Furthermore, it has been predicted that the developed microcapsule might be applied to other polymeric materials such as polyurethane, alkyd, and epoxy resins with the aim of obtaining self-healing anti-corrosive coating with high performance [63].

Graphene exhibits ultra-high anti-corrosive property; it is also highly impermeable. But it unfortunately possesses an aggregation tendency and aligns in waterborne epoxy in a random way, forming a conducting framework that accelerates corrosive dissolution. By the way, Wang et al. reported NCDs-decorated graphene (i.e., N-CDs@Gr), stabilized by π–π interactions, making the composite stable and with water dispersibility greater than 90 days, and appreciably compatible in waterborne epoxy [66]. The applied N-CDs@Gr waterborne epoxy (WEP) coating (N-CDs@Gr/WEP) applied on steel plate through spin coating technique demonstrated prolonged corrosion inhibition. Impedance modulus was reported to be 200 times higher when compared to WEP coating even after immersing for 90 days in saline water. Furthermore, it showed greater than 10^9 Ω cm^2 after 260 days in harsh oxygen environment, as verified by electrochemical measurement. Furthermore, the salt spray study results explored for 13 days of immersion also revealed an effective corrosion inhibitory action of the coating as shown in Figure 10.17. After 13 days, huge corrosion was noticed at the scratch of the WEP coating along with numerous blisters and corrosion spots as shown in Figure 10.17(a2). The Gr/WEP coating partially acted as an anti-corrosive coating. A large amount of rust was also reported to be agglomerated at scratches in the Gr/WEP coating, comparatively higher when compared to the WEP coating (*vide* Figure 10.17(b1, b2). It is attributed to corrosion initiation due to the aggregated and randomly aligned graphene. Furthermore, as shown in Figure 10.17(c2), there are a smaller number of blisters as well as corrosion spots, but a severe corrosion has been noticed in the surroundings of the scratches. It suggests that the applied N-CQDs@WEP coating acts as protective film in which the N-QCDs help in enhancing the crosslinking density of the N-QCDs/WEP, leading to the formation of a N-QCDs@Fe^{3+} layer as a corrosion inhibiting layer at the coating–steel interface, upon interaction with iron atoms. The N-QCDs@Fe^{3+} layer formation might be accomplished through interactions between the heteroatoms having lone pairs of electrons (herein, N and O) and the carboxyl groups present in the N-QCDs with unoccupied 3d orbitals of Fe^{3+} cations. This N-QCDs@Fe^{3+} layer formation limits the mass transfer of the corrosive species like H$_2$O, O$_2$, Cl$^-$, etc. toward the cathode regions of the steel plates under investigation, thereby reducing the rate of corrosion on the steel substrates. It implies that the N-QCDs/WEP acts as a better corrosion inhibiting coating compared to Gr/WEP or WEP coating. Further, corrosion reduction can easily be visualized with the incorporation of graphene along with N-QCDs designated as N-CQDs@Gr/WEP coating (vide Figure 10.17(d1)). Figure 10.17(d2) shows the superior corrosion inhibition property of N-CQDs@Gr/WEP coating on a steel substrate upon exposure to corrosive electrolytes for 13 days of immersion. This finding was in corroboration with the electrochemically determined corrosion inhibition capability of N-CQDs@Gr/WEP coating on steel substrates. High corrosion inhibition capability is caused by the formation of a synergistic shielding layer as well as enhanced crosslinking capability of N-QCDs as well as the orientation of graphene in the WEP matrix applied as an anti-corrosive coating on the metal substrates. The excellent anti-corrosive property is attributed to the compatibility as well as high dispersibility of the added nanofiller. The densely packed barrier layer of coating retards the diffusion of the corrosive species toward the metal substrates, thereby minimizing the corrosive dissolution effect of the aggressive electrolytes.

The compatibility as well as high dispersibility of the added nanofiller, N-QCDs, as well as the orientation of the homogeneously dispersed graphene nano sheets into the WEP matrix also played an important role in forming a protective as well as an impermeable barrier layer. The sole addition of graphene may lead to agglomeration, creating voids due to the less compatibility and dispersibility of graphene nanosheets in the WEP matrix. Upon addition of the N-QCDs, the nanofiller plays a crucial role in the formation of a compact layer. Amino, carboxyl, epoxy, and hydroxyl functional groups in

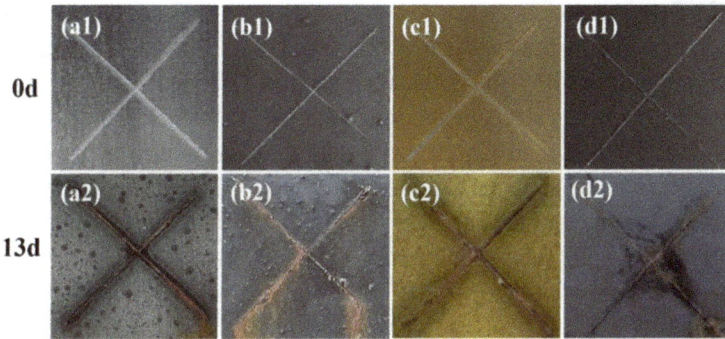

Figure 10.17: Snapshots of the steel substrates coated with WEP, Gr/WEP, N-CQDs/WEP, N-CQDs@Gr/WEP where the upper panel (a1–d1) shows the fresh coating and the lower panel (a2–d2) shows the coating after 13 days of salty spray exposure (reproduced with permission from ref. [66]. Copyright © 2019 Elsevier Ltd.).

N-QCDs might facilitate H-bond as well as covalent bond formation with the WEP as well as with the adjacent N-QCDs as shown in Figure 10.18. Additionally, the aromatic heterocyclic moieties present in the N-QCDs might also participate in the van der Waals type of interaction with the adjacent N-QCDs. Hence, it is obvious that the N-QCDs act as crosslinking agents and all such chemical bonding as well as interactions lead to the formation of denser polymeric network of WEP. It gives rise to defects- and voids-free, smoother coating of N-QCDs@Gr/WEP on the metal substrates (vide Figure 10.19 (a)). It is noteworthy to mention that the N-QCDs possesses abundant edge sites and a larger surface area.

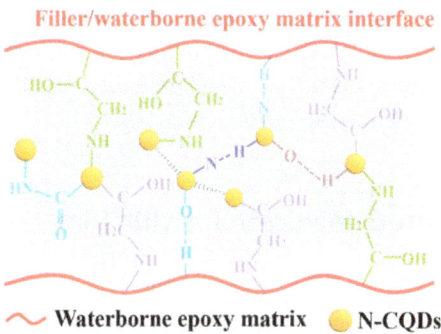

Figure 10.18: Schematic diagram showing the interactions between the adjacent N-CQDs along with the waterborne epoxy matrix with N-CQDs (reproduced with permission from ref. [66]. Copyright © 2019 Elsevier Ltd.).

These properties of N-QCDs facilitate a fast transfer of electrons, and the nitrogen atom doping impel it to lose electroneutrality, to form charged sites that are suitable to adsorb O_2. Thereby, it weakens the O-O bond and facilitates O_2 reduction. Thus, O_2 molecules diffuse through the coating, and when these O_2 molecules reach the coating–steel interface, these O_2 adsorb on the active sites of N-CQDs. As a result, the O-O bond disintegrates, generating oxygen atoms as shown in Figure 10.19 (b). The generated oxygen atoms are highly reactive in nature and these react with iron, forming dense iron oxide, Fe_3O_4, as shown in Figure 10.19 (b). In this way, the dense iron oxides are formed and uniformly distributed on the steel surface, since the N-QCDs and N-QCDs@Gr are dispersed uniformly in the WEP matrix. In 3.5 wt% NaCl solution coupled by pure O_2, the rate of oxygen diffusion is enhanced and the larger oxygen penetration facilitates the formation of a sufficiently dense oxide passive film on the entire surface that behaves like a barrier layer, saving the steel plate (vide Figure 10.19(b)).

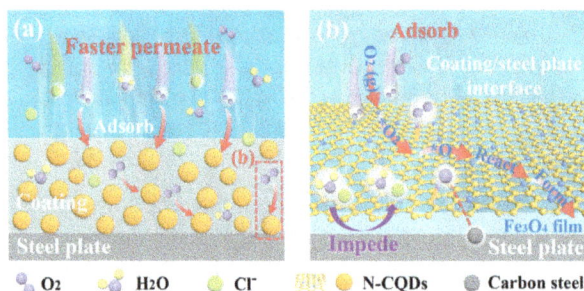

Figure 10.19: Schematic diagram showing (a) dense passive film formation on steel plate, and (b) magnified view of (a) (reproduced with permission from ref. [66]. Copyright © 2019 Elsevier Ltd.).

The authors said that only the N-QCDs as fillers in the WEP would not be able to provide corrosion inhibition in an oxygenated environment. Herein, the N-QCDs helped in improving the dispersibility as well as compatibility of the introduced graphene, rendering it to align in a horizontal fashion, despite high centrifugal as well shear forces. This make the graphene exert a barrier effect to a greater degree. In this way, the aligned graphene and N-CQDs help in synergistically improving the corrosion inhibition performance of the composite coating in the adverse environment [66].

10.4.3 Functionalization of CDs-infused polyurethane thin film as corrosion inhibitors

The influence of the incorporation of CDs into the polyurethane and its enhancement in the anti-corrosive properties have also been reported recently. Khatoon et al., synthesized CDs and functionalized it by polycarbazole (PC_Z) in order to encapsulate as naono-filler well dispersed in polyurethane (PU) matrix resulting a $CD@PC_Z/PU$ nanocomposite [67]. This $CD@PC_Z/PU$ nanocomposite coating applied on metal substrates exhibited superior anti-corrosive property compared to the bare substrate as well as the CD/PU- and PC_Z/PU-coated substrates. It suggested that the applied coating of the synthesized nanofiller-filled polyurethane coating is capable of showing barrier as well as surface passivation property, hindering the penetrating corrosive electrolytes such as H_2O, O_2, Cl^-, etc. The nanofiller imparted barrier property and the functionalized conductive nanoparticle facilitated passivation layer formation. It has also been reported that the as the corrosive ions penetrates upon prolonged immersion time, the PC_z starts to form passive layer, which is capable of enhancing anti-corrosion behavior. PCz/PU coating showed long-term protection from corrosion, which is attributed to the barrier property of PU as well as to the passivation property of the PCz nanoparticle. Increasing the immersion time led to pitting corrosion at the PCz/PU interface due to penetration of the corrosive electrolytes. The PCz/PU shows an initial increase in corrosion inhibition capability due to the formed passive layer, which inhibits the initial corrosion damages and fails after prolonged immersion time, owing to its breakdown. But in the case of the CD/PU coating, the CDs are capable of withstanding for a comparatively longer time as well as exhibiting corrosion inhibiting capability since the CD nanoparticles retard the penetration of the corrosive electrolytes. The strong and compact barrier layer formation is attributed to the fact that there might be a strong H-bond formed between the CDs and the PU matrix, as well as due to the van der Waals bond interaction between the neighboring CDs. This coating can withstand 10 days of immersion in an aggressive media. The free electrons present in CDs are capable of interacting with the unoccupied d-orbital present in the iron atoms. Additionally, there is also a possibility of simultaneous electron donations from the d-orbital of the iron atom to the available π^*-orbital in the CDs, which strengthen the interactions, leading to strong adhesion of the coating on the metal surface [67].

Recently, Abbas Madhi reported carbon quantum dots (CQDs)-reinforced epoxy/polyurethane (EP/PU) smart hybrid coating. Salt spray tests established that its anti-corrosion behavior increased with increased contents of CQDs in the composite. The EP/PU/CQDs coatings underwent fewer chemical changes than the pure coatings, which suggested better protective function against corrosion. The coatings containing CQDs exhibited higher crosslinking density along with stronger adhesive bonding at the coating–steel interface. Moreover, homogenous distribution of CQDs through the polymer matrix and the filling of empty voids with these nanoparticles enhanced the resistance to corrosion. The best performance of protection from corrosion was recorded for EP/PU/CQDs 3 wt%; one could witness the least blistering and rusting value at the scratch region versus the other samples. The fact behind the enhanced protection was the presence of sufficient chemical interactions at the coating–steel interface, by which the penetration of the corrosive factors decelerated at the scratch region. CQDs could entrap the diffused oxygen and reduce it in the damaged region, and finally disconnect the coating matrix from oxygen. Hence, one could utilize CQDs as efficient additives to improve the anti-corrosive behavior of the polymer coatings. A schematic of the anti-corrosive mechanism of EP/PU/CQDs coatings is shown in [68].

Recently, Ojo et al., synthesized novel small-sized (9 nm) carbon quantum dots (cqds) from the aqueous extract of *Phoenix datylifera* fruit pulp, employing thermal decomposition methods. The surface of the synthesized cqds were subsequently functionalized using a silanizing agent tridecafluorooctyltriethoxysilane in order to introduce hydrophobicity. This silane modified cqds (i.e., Scqds) was then introduced into the waterborne polyurethane (WPU) to make a composite WPU/Scqds. The WPU/Scqds was found to show excellent thermal stability as well as superhydrophobicity, compared to the WPU composite with cqds, i.e., WPU/cqds, established by thermogravimetry and water contact angle measurement, respectively. The WPU/Scqds coating applied on the steel substrates showed superior corrosion inhibition capability in comparison to WPU/cqds. The applied coating could withstand 96 h salt spray test showing only less than 5% corroded regions. It suggested that the hydrophobic nature of the WPU/Scqds coating does not allow corrosive ingredients to penetrate through it, thus retarding the rate of corrosion of metallic substrates. It can be predicted that this coating might be a potential coating material to introduce anti-corrosive property on metal substrates [69].

10.5 Summary and outlook

In this chapter, a recent overview on the carbonaceous quantum dots or simply quantum dots (CDs) that have opened new frontiers in materials science with their efficient applications as anti-corrosive materials when incorporated as coatings applied for the protection of metal substrates has been presented. The CDs as well as the functionalized CDs-based thin film coatings have also emerged as ecological, environment-friendly, and efficient corrosion inhibitors in the domain of metal protection from adverse environments. Chemical functionalization of CDs imparted conjugated π-bonds with their extension across the molecular skeleton; heteroatoms facilitated their adsorption and increased their barrier property or surface passivation behavior on the metallic surfaces, owing to their aqueous solubility, bio-compatibility, chemical stability, and high thermal stability as well as non-flammability. The functionalized CDs nanomaterials exhibited efficient adsorption on the metal substrates and facilitated the formation of a highly adherent thin film that acted as a barrier layer, shielding the metallic materials from the corrosive electrolytes; they also acted as barrier to the diffusion of corrosive species, thereby protecting the metal substrates from further corrosion. The major works pertaining to chemical functionalization of CDs as well as heteroatoms-doped CDs as efficient nanomaterials for superior corrosion inhibitors for several metallic materials used in various aggressive electrolytic solution, such as saline water, revealed various adsorption mechanisms as well as corrosion inhibition mechanism of CDs on the metallic surface through chemical bond formation and lone pairs of electron sharing as well as by the synergistic interaction by the functionalized CDs or heteroatom-doped CDs with the available 3d orbital of the targeted metal surface atoms, making a highly adherent and protecting film. Synergistic electron sharing from the metal surface atoms to the functionalized CDs have also been revealed. The surface passivation as well as the barrier property of the functionalized CDs reacted with the binder or the resin matrix, facilitating a highly impermeable thin film that hindered the diffusion or penetration of the corrosive electrolytes or species like H_2O, O_2, Cl^-, etc. though the nanopores present in the coating matrix or the crack appearing on the caoting. Finally, the futuristic scopes for the targeted functionalization of CDs to be used as efficient corrosion inhibitors in adverse and extreme environmental conditions need more insightful exploration. Owing to the various synthetic approaches and the strategy of purification employed, the corrosion inhibition property of the functionalized CDs might be affected by impurities as well as by inhomogeneity. Hence, efficient synthetic and purification methods need to be standardized for obtaining highly pure and uniform CDs of high quantum yield. Future research might also emphasize the exploitation of combination approaches of innovatively designed CDs-based corrosion inhibitory nanomaterials that are encapsulated as well as fiber-based self-healing coating fabrication approaches that possess ultrahigh uniformity as well as density and fast healing ability, along with the enhancement of mechanical properties of fiber reinforced anti-corrosive coating for the fabrication of self-healing thin film coating on the metal surfaces. Furthermore, an in-depth understanding of the adsorption mechanism as well as the barrier properties of the functionalized CDs, incorporated in epoxy, waterborne epoxy, polyurethane and other matrices, is highly necessary through theoretical calculations like DFT and MD simulation approach, and its subsequent validation and corroboration with the experimentally determined results. The future research should also devote on the use of functionalized CDs for improving the anti-bacterial activities, along with the reduction of the selective toxicity on the host. The prospect of tailoring the synthesis methodologies for CDs would make the nanostructured materials acquiescent to diverse applications such as anti-corrosive agents or additives in the next generation of coating materials.

References

[1] Sastri VS. Green corrosion inhibitors theory and practice. 2011;302.
[2] Quraishi MA, Chauhan DS, Saji V. Heterocyclic Organic Corrosion Inhibitors. Elsevier; 2020.
[3] Verma C, Quraishi MA. Recent progresses in Schiff bases as aqueous phase corrosion inhibitors: Design and applications. Coord Chem Rev 2021;446:214105.
[4] Koch G. Cost of corrosion. In: Trends in Oil and Gas Corrosion Research and Technologies, Elsevier, pp. 3–30; 2017.
[5] Goyal M, Kumar S, Bahadur I, Verma C, Ebenso EE. Organic corrosion inhibitors for industrial cleaning of ferrous and non-ferrous metals in acidic solutions: A review. J Mol Liq 2018;256:565–73.
[6] Verma C, Quraishi MA, Rhee KY. Electronic effect vs. Molecular size effect: Experimental and computational based designing of potential corrosion inhibitors. Chem Eng J 2022;430:132645.
[7] Aslam R, Mobin M, Zehra S, Aslam J. A comprehensive review of corrosion inhibitors employed to mitigate stainless steel corrosion in different environments. J Mol Liq 2022;364:119992.

[8]   Verma C, Haque J, Quraishi MA, Ebenso EE. Aqueous phase environmental friendly organic corrosion inhibitors derived from one step multicomponent reactions: A review. J Mol Liq 2019;275:18–40.

[9]   Olajire AA. Corrosion inhibition of offshore oil and gas production facilities using organic compound inhibitors – A review. J Mol Liq 2017;248:775–808.

[10]   Quraishi MA, Chauhan DS, Ansari FA. Development of environmentally benign corrosion inhibitors for organic acid environments for oil-gas industry. J Mol Liq 2021;329:115514.

[11]   Gece G. Drugs: A review of promising novel corrosion inhibitors. Corros Sci 2011;53(12):3873–98.

[12]   Popoola LT. Progress on pharmaceutical drugs, plant extracts and ionic liquids as corrosion inhibitors. Heliyon 2019;5(2):e01143.

[13]   Zakeri A, Bahmani E, Aghdam ASR. Plant extracts as sustainable and green corrosion inhibitors for protection of ferrous metals in corrosive media: A mini review. Corr Commun 2022;5:25–38.

[14]   Shang Z, Zhu J. Overview on plant extracts as green corrosion inhibitors in the oil and gas fields. J Mater Res Technol 2021;15:5078–94.

[15]   Quraishi MA, Chauhan DS, Saji VS. Heterocyclic biomolecules as green corrosion inhibitors. J Mol Liq 2021;341:117265.

[16]   Kobzar YL, Fatyeyeva K. Ionic liquids as green and sustainable steel corrosion inhibitors: Recent developments. Chem Eng J 2021;425:131480.

[17]   Zunita M, Kevin YJ. Ionic liquids as corrosion inhibitor: From research and development to commercialization. Results Eng 2022;15:100562.

[18]   Chen Z, Scharnagl N, Zheludkevich ML, Ying H, Yang W. Micro/nanocontainer-based intelligent coatings: Synthesis, performance and applications – A review. Chem Eng J 2023;451:138582.

[19]   Liu T, Zhang D, Ma L, Huang Y, Hao X, Terryn H, et al.. Smart protective coatings with self-sensing and active corrosion protection dual functionality from pH-sensitive calcium carbonate microcontainers. Corros Sci 2022;200:110254.

[20]   Singh N, Batra U, Kumar K, Ahuja N, Mahapatro A. Progress in bioactive surface coatings on biodegradable Mg alloys: A critical review towards clinical translation. Bioact Mater 2023;19:717–57.

[21]   Long C, Jiang Z, Shangguan J, Qing T, Zhang P, Feng B. Applications of carbon dots in environmental pollution control: A review. Chem Eng J [Internet] 2021;406:126848. Available from: https://linkinghub.elsevier.com/retrieve/pii/S1385894720329764.

[22]   Pourhashem S, Ghasemy E, Rashidi A, Vaezi MR. A review on application of carbon nanostructures as nanofiller in corrosion-resistant organic coatings. J Coat Technol Res 2020;17(1):19–55.

[23]   Verma C, Alfantazi A, Quraishi MA. Quantum dots as ecofriendly and aqueous phase substitutes of carbon family for traditional corrosion inhibitors: A perspective. J Mol Liq [Internet] 2021;343:117648. Available from: https://linkinghub.elsevier.com/retrieve/pii/S0167732221023734.

[24]   Berdimurodov E, Verma DK, Kholikov A, Akbarov K, Guo L. The recent development of carbon dots as powerful green corrosion inhibitors: A prospective review. J Mol Liq [Internet] 2022;349:118124. Available from: https://linkinghub.elsevier.com/retrieve/pii/S016773222102849X.

[25]   Wang Y, Hu A. Carbon quantum dots: synthesis, properties and applications. J Mater Chem C Mater 2014;2(34):6921.

[26]   Fernando KAS, Sahu S, Liu Y, Lewis WK, Guliants EA, Jafariyan A, et al.. Carbon Quantum Dots and Applications in Photocatalytic Energy Conversion. ACS Appl Mater Interfaces 2015;7(16):8363–76.

[27]   Gao X, Cui Y, Levenson RM, Chung LWK, Nie S. In vivo cancer targeting and imaging with semiconductor quantum dots. Nat Biotechnol 2004;22(8):969–76.

[28]   Yu Y, Cui M, Zheng W, Zhao H. Eco-friendly functionalization of hexagonal boron nitride nanosheets with carbon dots towards reinforcement of the protective performance of water-borne epoxy coatings. New J Chem 2022;46(13):6330–42.

[29] Hu R, Li L, Jin WJ. Controlling speciation of nitrogen in nitrogen-doped carbon dots by ferric ion catalysis for enhancing fluorescence. Carbon N Y 2017;111:133–41.

[30] Zhang Z, Zhang J, Chen N, Qu L. Graphene quantum dots: an emerging material for energy-related applications and beyond. Energy Environ Sci 2012;5(10):8869.

[31] He C, Xu P, Zhang X, Long W. The Synthetic Strategies, Photoluminescence Mechanisms and Promising Applications of Carbon Dots: Current State and Future Perspective. Vol. 186, Carbon. Elsevier Ltd, pp. 91–127; 2022.

[32] Wang J, Wang CF, Chen S. Amphiphilic Egg-Derived Carbon Dots: Rapid Plasma Fabrication, Pyrolysis Process, and Multicolor Printing Patterns. Angew Chem Int Ed 2012;51(37):9297–301.

[33] Yan H, He C, Li X, Zhao T. A solvent-free gaseous detonation approach for converting benzoic acid into graphene quantum dots within milliseconds. Diam Relat Mater 2018;87:233–41.

[34] Park SY, Lee HU, Park ES, Lee SC, Lee JW, Jeong SW, et al.. Photoluminescent Green Carbon Nanodots from Food-Waste-Derived Sources: Large-Scale Synthesis, Properties, and Biomedical Applications. ACS Appl Mater Interfaces 2014;6(5):3365–70.

[35] Aslandaş AM, Balcı N, Arık M, Şakiroğlu H, Onganer Y, Meral K. Liquid nitrogen-assisted synthesis of fluorescent carbon dots from Blueberry and their performance in Fe 3+ detection. Appl Surf Sci 2015;356:747–52.

[36] Xu X, Ray R, Gu Y, Ploehn HJ, Gearheart L, Raker K, et al.. Electrophoretic Analysis and Purification of Fluorescent Single-Walled Carbon Nanotube Fragments. J Am Chem Soc 2004;126(40):12736–7.

[37] Zhou J, Booker C, Li R, Zhou X, Sham TK, Sun X, et al.. An Electrochemical Avenue to Blue Luminescent Nanocrystals from Multiwalled Carbon Nanotubes (MWCNTs). J Am Chem Soc 2007;129(4):744–5.

[38] Pan D, Zhang J, Li Z, Wu M. Hydrothermal Route for Cutting Graphene Sheets into Blue-Luminescent Graphene Quantum Dots. Adv Mater 2010;22(6):734–8.

[39] Zhu H, Wang X, Li Y, Wang Z, Yang F, Yang X. Microwave synthesis of fluorescent carbon nanoparticles with electrochemiluminescence properties. Chem Commun 2009;34:5118.

[40] Wang F, Pang S, Wang L, Li Q, Kreiter M, yan LC. One-Step Synthesis of Highly Luminescent Carbon Dots in Noncoordinating Solvents. Chem Mater 2010;22(16):4528–30.

[41] bin CB, Liu ML, Li CM, Huang CZ. Fluorescent carbon dots functionalization. Adv Colloid Interface Sci 2019;270:165–90.

[42] Falara PP, Zourou A, Kordatos KV. Recent advances in Carbon Dots/2-D hybrid materials. Carbon N Y 2022;195:219–45.

[43] Wareing TC, Gentile P, Phan AN. Biomass-Based Carbon Dots: Current Development and Future Perspectives. ACS Nano 2021;15(10):15471–501.

[44] Namdari P, Negahdari B, Eatemadi A. Synthesis, properties and biomedical applications of carbon-based quantum dots: An updated review. Biomed Pharm 2017;87:209–22.

[45] Hu Y, Yang J, Tian J, Jia L, Yu JS. Waste frying oil as a precursor for one-step synthesis of sulfur-doped carbon dots with pH-sensitive photoluminescence. Carbon N Y 2014;77:775–82.

[46] Emam HE. Clustering of photoluminescent carbon quantum dots using biopolymers for biomedical applications. Biocatal Agric Biotechnol 2022;42:102382.

[47] Khan ME, Mohammad A, Yoon T. State-of-the-art developments in carbon quantum dots (CQDs): Photo-catalysis, bio-imaging, and bio-sensing applications. Chemosphere 2022;302:134815.

[48] Nasrollahzadeh M, Sajjadi M, Sajadi SM, Issaabadi Z. Green Nanotechnology. In: p. 145–98, 2019.

[49] Saikia A, Karak N. Cellulose nanofiber-polyaniline nanofiber-carbon dot nanohybrid and its nanocomposite with sorbitol based hyperbranched epoxy: Physical, thermal, biological and sensing properties. Colloids Surf A Physicochem Eng Asp 2020;584:124049.

[50]   Saikia A, Karak N. Fabrication of renewable resource based hyperbranched epoxy nanocomposites with mwcnt-polyaniline nanofiber-carbon dot nanohybrid as tough anticorrosive materials. Express Polym Lett 2019;13(11):959–73.

[51]   Peng H, Li L, Wang Q, Zhang Y, Wang T, Zheng B, et al.. Organic carbon dot coating for superhydrophobic aluminum alloy surfaces. J Coat Technol Res 2021;18(3):861–9.

[52]   Liu XR, Yuan XY, Liu JK, Sun XW, Yang XH. Research on correlation between corrosion resistance and photocatalytic activity of molybdenum zinc oxide modified by carbon quantum dots pigments. Dyes Pigm 2020;175:108148.

[53]   Ye Y, Chen H, Zou Y, Zhao H. Study on self-healing and corrosion resistance behaviors of functionalized carbon dot-intercalated graphene-based waterborne epoxy coating. J Mater Sci Technol 2021;67:226–36.

[54]   Ullah H, Azizli KA M, Man ZB, Ismail MBC, Khan MI. The Potential of Microencapsulated Self-healing Materials for Microcracks Recovery in Self-healing Composite Systems: A Review. Polym Rev 2016;56(3):429–85.

[55]   White SR, Sottos NR, Geubelle PH, Moore JS, Kessler MR, Sriram SR, et al.. Autonomic healing of polymer composites. Nature [Internet] 2001;409(6822):794–7. Available from: https://www.nature.com/articles/35057232.

[56]   An S, Lee MW, Yarin AL, Yoon SS. A review on corrosion-protective extrinsic self-healing: Comparison of microcapsule-based systems and those based on core-shell vascular networks. Chem Eng J 2018;344:206–20.

[57]   Zwaag van der S. Self-Healing Materials: An Alternative Approach to 20 Centuries of Materials Science. Sybrand Zwaag, editor. Chem Int – Newsmagazine IUPAC [Internet] 2008;30(6):Available from: https://www.degruyter.com/document/doi/10.1515/ci.2008.30.6.20/html.

[58]   Wang J, Du P, Zhao H, Pu J, Yu C. Novel nitrogen doped carbon dots enhancing the anticorrosive performance of waterborne epoxy coatings. Nanoscale Adv 2019;1(9):3443–51.

[59]   Ren S, Cui M, Zhao H, Wang L. Effect of nitrogen-doped carbon dots on the anticorrosion properties of waterborne epoxy coatings. Surf Topogr 2018;6(2).

[60]   Li S, Du F, Lin Y, Guan Y, Qu WJ, Cheng J, et al.. Excellent anti-corrosion performance of epoxy composite coatings filled with novel N-doped carbon nanodots. Eur Polym J 2022;163.

[61]   Cui M, Ren S, Chen J, Liu S, Zhang G, Zhao H, et al.. Anticorrosive performance of waterborne epoxy coatings containing water-dispersible hexagonal boron nitride (h-BN) nanosheets. Appl Surf Sci 2017;397:77–86.

[62]   Pourhashem S, Rashidi A, Vaezi MR, Bagherzadeh MR. Excellent corrosion protection performance of epoxy composite coatings filled with amino-silane functionalized graphene oxide. Surf Coat Technol 2017;317:1–9.

[63]   Bao Y, Yan Y, Wei Y, Ma J, Zhang W, Liu C. Salt-responsive ZnO microcapsules loaded with nitrogen-doped carbon dots for enhancement of corrosion durability. J Mater Sci 2021;56(8):5143–60.

[64]   Cui M, Ren S, Zhao H, Wang L, Xue Q. Novel nitrogen doped carbon dots for corrosion inhibition of carbon steel in 1 M HCl solution. Appl Surf Sci 2018;443:145–56.

[65]   Ramezanzadeh M, Bahlakeh G, Sanaei Z, Ramezanzadeh B. Interfacial adhesion and corrosion protection properties improvement of a polyester-melamine coating by deposition of a novel green praseodymium oxide nanofilm: A comprehensive experimental and computational study. J Ind Eng Chem [Internet] 2019;74:26–40. Available from: https://linkinghub.elsevier.com/retrieve/pii/S1226086X19300292.

[66]   Wang X, Li C, Zhang M, Lin D, Yuan S, Xu F, et al.. A novel waterborne epoxy coating with anti-corrosion performance under harsh oxygen environment. Chem Eng J [Internet] 2022 [cited 2022 Oct 24] 430: Available from: https://doi.org/10.1016/j.cej.2021.133156.

[67] Khatoon H, Iqbal S, Ahmad S. Influence of carbon nanodots encapsulated polycarbazole hybrid on the corrosion inhibition performance of polyurethane nanocomposite coatings. New J Chem 2019;43(26):10278–90.

[68] Madhi A. Smart epoxy/polyurethane/carbon quantum dots hybrid coatings: Synthesis and study of UV-shielding, viscoelastic, and anti-corrosive properties. Polym-Plast Technol Mater 2022.

[69] Bankole-Ojo OS, Oyedeji FO, Raju KVSN, Ramanuj N. Bankole-Ojo et al Salt spray resistance of superhydrophobic waterborne polyurethane/carbon dots nanocomposites. J Mater Environ Sci [Internet] 2022;2022(7):7. Available from: http://www.jmaterenvironsci.com.

Khalid Bouiti, Nabil Lahrache, Ichraq Bouhouche, Najoua Labjar,
Meriem Bensemlali, Souad El Hajjaji, Ghita Amine Benabdallah,
Hamid Nasrellah

11 Functionalized hybrid sol–gel thin film coatings for corrosion inhibition

Abstract: This chapter discusses the use of hybrid coatings synthesized and deposited by the sol–gel method in protection against corrosion, especially since the damage to metallic components is a significant problem contributing to the deterioration of the equipment and entails substantial costs for their recovery or prevention., This process is based on two essential steps from the sol to the gel so that the chain develops and the coating is obtained after evaporation and treatment; the sol–gel process is easily synchronized with all known deposition techniques, which indicates the reliability of this approach. Several studies demonstrate improved efficiency for corrosion prevention with hybrid layers by altering precursors and materials used, providing localized protection of the substrate, preventing the progression of damage, controlling the flow of current to the substrate, and reducing the diffusion of oxidizing agents to the material.

Keywords: Hybrid, Sol, Gel, Coating, Corrosion, Inhibition

11.1 Introduction

The world demands materials with more complicated functionality. As a result, the discovery of novel synthesis techniques has been a prominent focus of scientific research [1]. It is fairly simple to mix powdered reagents and heat them to produce a wide range of inorganic compounds [2, 3]. Even if the reaction conditions are straightforward to carry out, there are various constraints. These are mostly due to the variability of the starting materials [4].

Grinding is an effective method for resolving this issue since it both reduces particle size and increases specific surface area. Particle form control using solid-state techniques is notoriously challenging [4, 5], especially when phase-wise crushing of the material is required for prolonged heating or numerous different treatments.

The capacity to produce a material from a chemically homogenous precursor is the primary benefit of sol–gel chemistry, which is an alternative to solid-state chemistry. Sol–gel chemistry promises improved control over particle size and shape [6, 7].

To assure reaction homogeneity, several sol–gel methods have been developed to separate the phase during synthesis [4], since production of a precursor at room temperature does not guarantee it. This ability to adjust the structure of nano systems in

https://doi.org/10.1515/9783111016160-011

the early phases of processing contributed to the improvement of sol–gel processing, which enables the manufacture of pure materials with improved capabilities [7].

Due to their superior properties, these compounds may be shaped into a wide variety of forms, including fibers, films, fine powders, and even monoliths, expanding their potential range of uses.

It is common practice to apply protective films or coatings to substrates in order to generate and improve desirable substrate features such as resistance to scratching and abrasion [8, 9], adhesion [10], optical appearance [7], bioactivity [11], hydrophilic or hydrophobic properties [12], and antibacterial and antireflective properties [13].

Hybrid sol–gel coatings can offer excellent protection against corrosion for metals such as aluminum, magnesium, and steel. The sol–gel process involves the formation of a thin, dense layer on the metal surface, which acts as a barrier against corrosion agents.

11.1.1 Transition sol–gel

11.1.1.1 Sol and gel step

Sol refers to a stable suspension of solid particles in a liquid [7]. Moreover, the particles must be denser than the surrounding liquid and tiny enough that the forces holding them in suspension are not too significant [11, 12].

According to the medium, the production of aerosol solids, including smoke [13], which are solid particles suspended in a gas, or a liquid aerosol such as fog, or colloidally dispersed droplets in a liquid, producing an emulsion, may occur [11, 14].

A colloidal system, sol is mainly composed of a liquid dispersion medium and a solid dispersed phase, with a 1 μm dimensional limit [15, 16].

The Wiener process is responsible for establishing equilibrium in the system although the size of the system cannot be more than 1 μm. Forces of gravity dominate over those that cause the sedimentation of particles beyond this dimensional limit [17].

In a liquid medium, the gel's expansion is restricted solely by the volume of the container holding the liquid, since the gel is a porous, three-dimensionally interconnected solid network [18].

Colloidal gel is formed by a stable network of colloidal sol particles [19]. Gels are considered polymeric if their solid network is made up of macromolecules that were first dissolved in a polymeric solution [18]. In order to create its structure, a polymer may include the repetition of one or more of its basic components [20].

This solid phase expansion is due to the confinement of the liquid inside a solid network [21]. The inability of a material to bend elastically or under shear stress is a characteristic of the sol–gel phase transition [21, 22].

Gels are formed when a fluid medium, such as a liquid or gas, fills the pores of a very porous solid network and the two remain in equilibrium. In this equilibrium state, the liquid diffuses through the gel's solid network, which makes up most of the gel's volume. In addition, it does not move out of the system in which it may reach thermodynamic equilibrium on its own [23].

This procedure is chemically similar to an organic polycondensation reaction in which tiny molecules create polymeric structures by losing their substituents. Typically, the reaction produces a three-dimensional network of cross-links [22]. Various benefits are associated with the usage of tiny molecules as precursors for the creation of cross-linked materials., including a high degree of control over the purity and composition of the final materials and the use of solvent-based chemistry, which offers significant advantages for the processing of the formed materials [24].

11.1.1.2 Sol–gel process

Precursor selection is crucial to process chemistry. The most prevalent ingredients in sol–gel processing are inorganic salts, metal alkoxides, and organosilanes [25, 26].

To manufacture pure materials with enhanced qualities, the sol–gel approach not only provides for the customization of the structure of the nano systems from the earliest processing stages but also permits control of the nano architecture throughout the synthesis [24].

Sol–gel manufacturing is a tried and effective method for creating hybrid sol–gel materials [27]. The process involves diluting precursors of sol–gel reactive alkoxides with a suitable solvent and then subjecting them to a series of hydrolysis and condensation reactions with water in the presence of a catalyst. The inorganic basis is created when hydroxides, formed in the hydrolysis process, condense to form the skeleton of the molecule [28, 29].

11.1.2 Hydrolysis

Condensation reactions are able to take place even before the hydrolysis reactions have been completed, despite the fact that hydrolysis is the first stage in the process.

Hydrolysis processes should produce Si-OH groups when water is added to silicon alkoxide. A catalyst is required [30] due to the sluggish reaction of silicon alkoxide with water and alcohol.

It involves the hydrolysis-induced generation of reactive species. The alkoxide species is hydrolyzed, and hydroxyl groups are produced as a consequence of a nucleophilic substitution by the oxygen atom of a water molecule on the silicon atom of the alkoxide molecule [28, 31].

Through this mechanism, alcohol is liberated while metal hydroxide is produced.

$$M(OR)_n + H_2O \rightarrow M(OH)(OR)_{n-1} + ROH$$

Alcohol may react with a hydrolyzed species to produce water and an alkoxide ligand by esterification [22, 32], but the reaction can also proceed in the other direction.

This shows that alcohol is not only a passive solvent but instead plays an active role in the process. Yet, since water and alkoxysilanes are incompatible, a mutual solvent is required to achieve solution homogeneity; however, in certain instances, the alcohol produced as a by-product of hydrolysis may be adequate [31, 32].

Furthermore, it is important to note that this first phase should not be regarded as well-established; in reality, condensation begins as soon as hydrolysis commences.

Hydrolysis may be conducted under neutral circumstances, but often an acid or basic catalyst is added to the water to speed up the hydrolysis events or to modify the shape of the resultant gel [4, 33].

The use of either an acid or a basic catalyst requires a variety of distinct chemical reactions, each affecting the hydrolysis kinetics.

Hypervalent intermediate production may be attributed to the protonation of the alkoxy leaving group under acidic circumstances, which causes the ligand to become destabilized and gives way for the nucleophilic assault of a water molecule [22]. The proton transfer that takes place between the water molecule and the alkoxy group [34] is what makes this destabilization adjustable. As a consequence of the proton transfer that takes place, the σ-bond that was present in the alkoxy molecule is destroyed, and this results in the alkoxy being effectively replaced by a hydroxyl group. [22].

Under basic circumstances, the hydroxyl group launches a nucleophilic assault on the silica, causing the material to take on a partial positive charge [22]. The metal center is hydroxylated after the release of an alkoxy group stabilizes the intermediate formed during hydrolysis [35, 36].

Faster hydrolysis and more open, linked networks result from acid-catalyzed hydrolysis [10]. In contrast, base-catalyzed hydrolysis speeds up the condensation process but slows down the hydrolysis reactions [37].

Management of the reaction kinetics, which in turn affects the shape and structure of the final materials, requires study of the inverse relationship between the hydrolysis and condensation processes.

11.1.3 Condensation

Condensation occurs once the alkoxide precursors are hydrolyzed. In order to create the inorganic three-dimensional network, hydroxyl groups must react with one another or with non-hydrolyzed alkoxy groups to produce oxo-bridges while simultaneously losing a water or alcohol molecule [38].

For transition metal alkoxide precursors, the condensation reaction is irreversible. In rare instances, less condensed hydrolyzed species or alkoxide groups may be reformed from condensed species by hydrolysis or alcoholysis processes when alkoxysilane precursors are used alone or in conjunction with transition metals [39].

When the steric barrier surrounding the silicon atom is too great to permit the development of chemical structures with less mechanical stress, several reactions become conceivable. Less mechanical stress is required for the formation of chemical structures [22, 39].

Specifically, the electronic clouds for the intermediate atoms must be redistributed. Sanderson's hypothetical method is used to study electron transport [40]. When atoms in an intermediate add a partial charge of an integer, they are said to be in a "leaving group," and they may then leave the intermediate. The magnitude of the partial charge determines whether the departing group is transformed into an electrically neutral molecule, an anion, or a cation [18].

11.1.4 Gelification

These reactions result in gelation and the production of a gel composed of M-OM (or M-OH-M) links, whose viscosity increases with time [41]. This gel comprises solvents and precursors that have not yet interacted.

The "gel" phase is defined and characterized in the sol–gel technique by a 3D solid "skeleton" that is submerged in a liquid phase. Commonly, the solid phase is a polymeric sol that has been compressed to the point that the particles have become intertwined in a three-dimensional network [42].

All of the necessary processes to produce this substance can be done at room temperature. Conditions such as temperature, pH, precursor and solvent type, and reagent concentration all have a role in the outcomes of chemical reactions [4].

11.1.5 Aging

At the gel stage, the process that leads to gelatinization continues. The whole process of gel development with time is known as aging [43]. The aging of the gel is characterized by physicochemical changes that occur after gelation [22].

During the aging process, the material shrinks due to the cross-linking phenomena by releasing the solvent, also known as syneresis. The gel may be dried [44] at either atmospheric or supercritical settings, regardless of syneresis. Without the solute, the network is denser or less dense in both cases [45].

11.1.6 Evaporation

Following gelation, the material dries out because of capillary forces in the pores, which might result in a volume reduction [46].

During drying, a sol–gel material is produced, and the alcohol or water must be able to escape as the gel hardens. As the gel solidifies, water may escape. This process of evaporation occurs [47]. The evaporation reaction starts through the pores and channels of the porous sol–gel substance.

There are several ways of drying based on the materials collected. The xerogel is produced by a standard drying procedure. The evaporation of the solvent permits the creation of a xerogel that may be densified by thermal treatment at a moderate temperature [48]. Densification is dependent upon the kind of material and the desired qualities.

Gel drying is an intricate process. To prevent fragmentation of the xerogel, it is crucial that the solvent evaporates slowly [24]. Due to the internal stresses that arise during the drying process, it is difficult to create a solid material, which might result in cracking [48].

After being dried in an autoclave at very high pressure, aerogel is the result. Supercritical solvent evaporation produces a non-densified aerogel [49]. A very porous substance with excellent insulating characteristics is the end product. The change from "sol" to "gel" permits the adjustment of viscosity, which in turn enables the manufacture of fibers and layers on various supports by dipping or spraying [39, 49].

Sol–gel enables the production of high-purity, homogeneous materials with a variety of chemical compositions, and the liquid method of elaboration for very versatile shape as shown in Figure 11.1.

Supercritical drying is better suitable for inorganic sol–gels and hybrids with a low organic species concentration and results in solids with good porosity and no cracks [50]. Although the elimination of solvents results in a dry-solid phase with enhanced mechanical capabilities, a final cure is necessary to stabilize the hybrid solid phase produced [51] to its fullest extent.

11.1.7 Hybrid coating sol–gel

Depending on their intrinsic surface or mass properties, thin films may serve a variety of functions. They may be dense, porous, structured, multilayered, or composite, among other characteristics.

Sol–gel films are applied in diverse fields, such as optics, electrical, protective, and analytical applications [52]. The majority of these functionalities need uniform performance all over the surface of the material, making uniform and tightly regulated thickness a requirement [24].

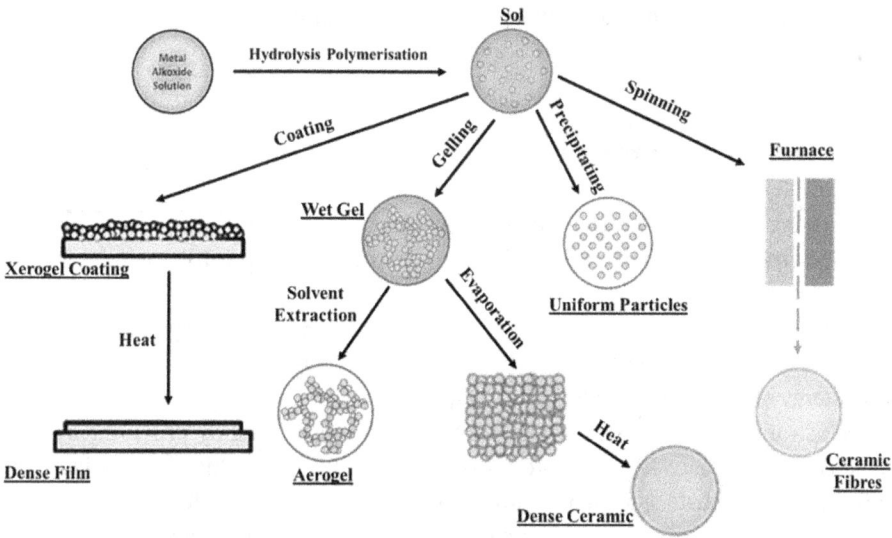

Figure 11.1: Sol–gel process.

By spreading a solution over a material and evaporating volatile chemicals, liquid deposition processes are highly adaptable ways to produce relatively uniform coatings with typical thicknesses [47].

These methods are a viable alternative to conventional dry deposition procedures including grinding, spraying, and evaporating [53]. Since the coating's form and chemistry can be precisely controlled by CVD and PECVD, these processes are favored due to their faster manufacturing and reasonable production costs [54].

Typically, the various liquid deposition processes are explained using the same progression of sequential stages: formulation of the chemical solution, distribution of the solution to the substrate surface, evaporation of the liquid layer, post-treatment of the layer, and stabilization of the layer [55, 56].

11.2 Coating sol–gel steps

11.2.1 Formulation of the chemical solution

For the deposition process to be as reproducible as possible, the formulation of the initial chemical solution ought to be as stable under room temperature and over time [55].

Precursors and other solutes may be of many sorts and forms, but they must be entirely solubilized in the liquid-phase solvent to produce homogeneous mixtures. Long-term stability is one of the original solution's most crucial features [47, 53]. In fact, very stable solutions are favored due to their ability to be held for extended durations.

11.2.2 Deposition process

The layering procedure involves applying a surface coat of solution to a specific surface. Various techniques for pouring fluids, including sol-gel mixtures, onto interfaces are used [57].

Despite the surface tension of the solution, which works to keep the solution in droplet form [58], the solution may be spread out throughout the surface if the surface energy of the material is high enough [58, 59].

Substrate surface activation may be used to avoid wetting by providing a great chemical contact between the material and the solution [58], which is especially important for high-surface-tension solutions. Roughness is a crucial feature that, when uniform and composed of suitably small peak and valley portions, may foster affinity [60].

The substrate's dimensions and geometry dictate the processing method and degree of precision required [24]. The uniformity of the solution and the amount of solution administered per unit area are the primary factors under control [4].

11.2.3 Evaporation process

Evaporation releases the solvent in liquid deposition thin film production [61]. Evaporation occurs when molecules in the liquid state obtain enough energy to cross the surface tension barrier and transition into the gaseous state.

Evaporation rates are higher at elevated temperatures and for liquids with lower surface tension. Evaporation periods during thin film manufacture may vary from a few seconds to several hours depending on the conditions. This point in the tuning process is often considered to be the most stable [9]. Evaporation takes place when a molecule's thermal energy is high enough to break through the surface tension barrier [62]. For it to penetrate the surface, its speed must be greater than the cohesive energy of the surface.

11.2.4 Patterning processes

The deposited sol–gel films may undergo further shaping procedures before they are thermally stabilized. Top-down methods may be used to selectively condense or shape the material at the nanometer dimension after liquid layering but before heat condensation of the sol–gel system [63].

Thus, we may expect significant progress toward the objective of integrating complex hierarchical structures with new capabilities into practical devices due to the combination of sol–gel-produced layers with top-down shaping [64]. After being exposed, the material that did not condense is dissolved selectively using the proper development techniques. Molding procedures may also be used to impart textures to sol–gel films [65, 66], albeit depending on the layer structure.

11.2.5 Post-deposition treatments

Surfaces may be functionalized, components removed selectively, and networks consolidated with the use of chemical processes [22]. Gradual precursor condensation is linked to the accumulation of internal failure stresses that are partially but not fully released due to the layer's robust adhesion to the substrate [34]. The breakdown of the layer is due to the stresses that develop at its boundaries as shown in Figure 11.2.

They are more severe for thicker layers and increase with density; for any given material, there is a threshold thickness beyond which fractures initiate and propagate across the layer. Stabilizing large layers often requires more careful and extended treatments than maintaining thin coatings to enable the system, the time to gradually settle, to allow the system to gradually reorganize into forms with lower breakdown energy [36].

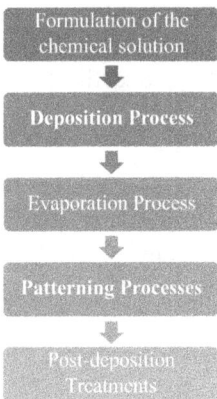

Figure 11.2: Coating sol–gel phases.

11.3 Coating techniques

11.3.1 Dip coating

For using a deposition technique of spin coating, dip coating, or spray coating, the sol–gel process may be easily included. Once the sol with the proper composition has been synthesized for the intended use, it is deposited onto the substrate using a coating method to create the layer [67]. The substrate is slowly dipped into the sol and held there for as long as is necessary to completely wet its surface and apply the protective layer [56, 68]; this allows for the creation of chemical bonding among the surface and the sol.

Following the holding period within the sol, the material is vertically removed from the solution under controlled temperature and pressure at a predetermined and regular rate [56].

The substrate's upward motion results in the evacuation of sol and the uniform layering of sol on the substratum's interface. Sol gels while its solvent evaporates, a process that occurs simultaneously [69].

The viscosity of the solvent and the rate of shrinkage are the most important factors in determining the film thickness. Specific evidence for this may be found in [70] where it is shown that a thinner coating results from a reduction in shrinkage and viscosity rates.

Coating structure and characteristics are determined by the shape and texture of the coating particles, the condensation process, and the forces generated during the film deposition [71], all of which are influenced by environmental and process conditions during the evaporation stage and, potentially, heat treatment of the coating [24].

11.3.2 Spin coating

Centrifugal action causes the liquid to migrate counterclockwise, pushing the sol to be ejected radially outside from the substrate surface [72]. Once the substrate has been spun for the allotted amount of time, the excess fluid moves to the surface and evaporates as droplets, and the sol layer gels and develops into a thin film as a result of the evaporation of volatile species [56, 73].

Other than viscosity, the concentration of the sol, the rate of dryness, and the tension of the superficial layer of the sol all play vital roles in determining the coating's qualities. Furthermore, the specified instrumental parameters, such as the end rotational speed, acceleration [69, 74], and the rate of change to attain the final rotational speed, are very important. The more polished the film is, greater the spinning angular velocity and longer the spinning step time [74].

11.3.3 Spray-coating

Since spray coating is combined with the sol–gel process, the sol is vaporized or fogged to form a thin layer of particles using airflow or by applying pressure to the sol [75]. The tiny droplets are then dispersed throughout the whole substrate using a spray mechanism. Spray coverage is achieved by using a sol with a lower viscosity that is used in the spin-and-dip coating method [47, 76]. Tiny droplets may clump together if the area is wet as shown in Figure 11.3. The method is extensively employed throughout the industry due to the higher speed of coating compared to other methods; it can be applied to a variety of substrate shapes, and it generates significantly less coating sol waste [55, 61]. However, clogging may develop over time owing to solvent evaporation and the resultant gelling of the sol, and covering hydrophobic surfaces might be a problem [77].

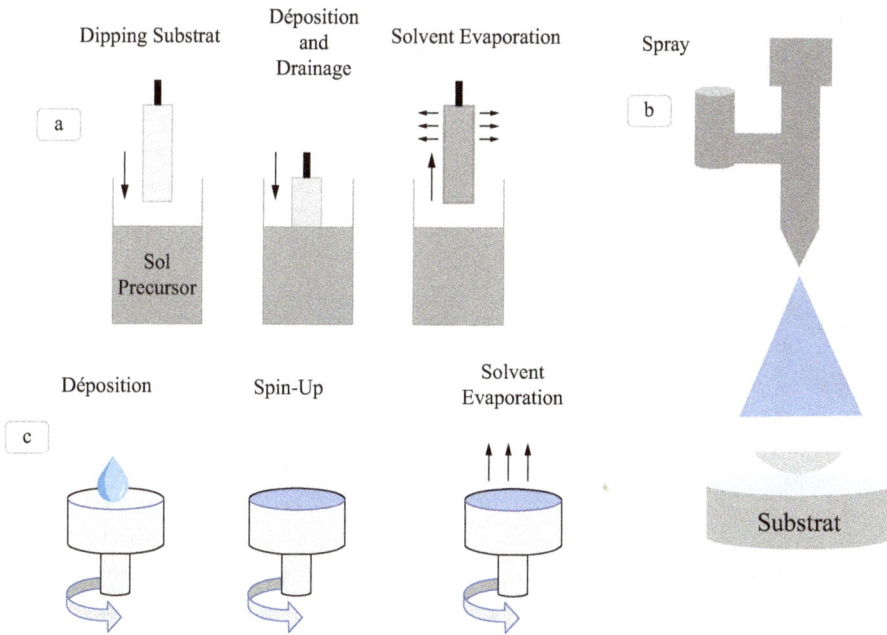

Figure 11.3: Coating techniques: (a) dip coating, (b) spray coating, and (c) spin coating.

11.4 Hybrid sol–gel protective layer from corrosion

Coatings made using the sol–gel method have improved corrosion resistance and chemical stability [25]. Furthermore, the sol–gel method has been shown to be an effective alternative to the standard, harmful pretreatments and surface treatments used to induce metal corrosion [78, 79]. Coatings and films may be deposited to protect surfaces against corrosion [80, 81]. This is a common method that involves altering the coating's chemical makeup.

Metals and alloys show improved corrosion resistance due to the formation of a thick barrier [80] that insulates the substrate from the external oxidizing environment. Coatings may be used to prevent corrosion by depositing a layer between the metal and its environment as shown in Figure 11.4. Metals and alloys prized for their mechanical, physical, or chemical qualities, but with very limited corrosion resistance in hostile conditions, may find their areas of usage broadened as a result of this [81, 82].

They allow for localized substrate protection, halting the spread of deterioration, limiting current flow on the substrate, inhibiting oxidant diffusion to the material, minimizing water and electrolyte penetration, and releasing incorporated inhibitory species that contribute to passivation of the substrate or prevent corrosion process [83, 84].

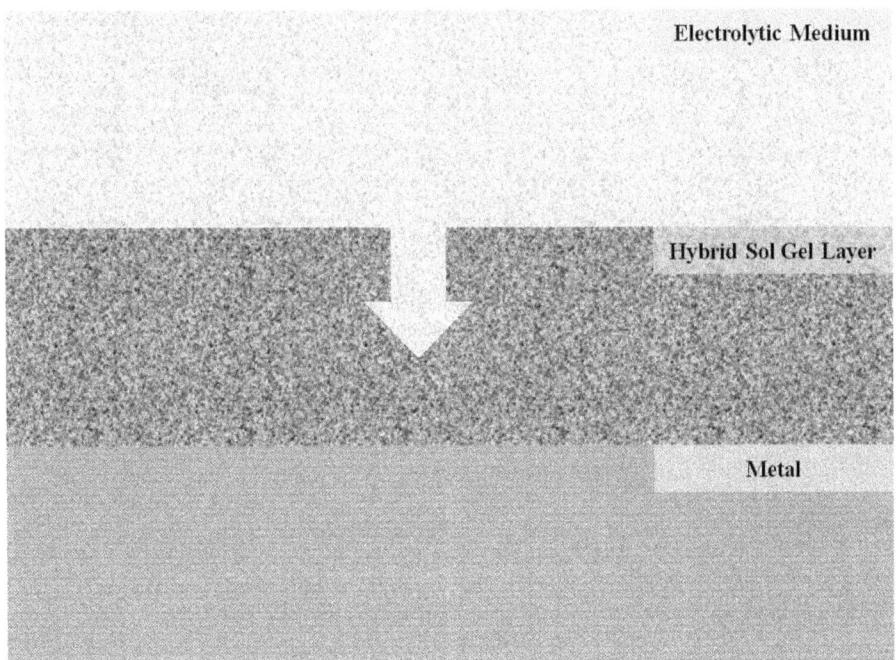

Figure 11.4: Metal protection with hybrid sol–gel layer.

11.4.1 Steel protection

Phosphating and chromating are two common conversion coatings used to passivate steel. Such treatments provide a passivation layer that is impervious to further chemical assault [85]. However, there are several efforts underway on a global scale to eliminate chromates because of the high toxicity attributed to their use [86]. Global efforts are being undertaken to discover ecologically treatment.

Sol–gel-produced hybrid coatings based on siloxanes are a good option because the metal oxides present on the treated steel surfaces form a covalent bond between the inorganic components [87]. Hybrid coatings are preferable to conventional conversion coatings because the presence of metal oxides on treated surfaces facilitates a covalent bond between the inorganic components of siloxane [88].

Coating manufacturing parameters, such as aging, effect pH, temperature, and precursor wt%, have a significant impact on corrosion protection capabilities.

11.4.2 Aluminum protection

Aluminum-based alloys find widespread use thanks to their advantageous physical and processing properties, which include low density, ease of forming, strong corrosion resistance to diverse conditions, high electrical and thermal conduction, and the relatively cheap cost. Alloys based on aluminum benefit greatly from these characteristics due to their high economic and industrial relevance [87, 89]. Due to their inherent propensity to form a passivation layer, aluminum alloys are often put to use; this layer may also be synthesized artificially by oxidizing the substrate through anodizing.

However, when exposed to aggressive conditions such as those holding chloride ions, this passivation layer deteriorates and corrosion ensues [89]. Hybrid sol–gel coatings, in addition to their coating properties, may generate a stable metal oxide by coupling their inorganic functionality to the passivation layer [90].

11.4.3 Copper protection

Copper's long-term resistance to corrosion, degeneration, degradation from high temperatures, and degradation from UV rays are all extraordinary [91].

However, corrosion processes are sped up in wet settings. There have been a few studies [88] on hybrid sol–gel coatings for copper protection. At room temperature, Cu_2O or Cu_2O and CuO make up most of the native layer, which is just a few nanometers thick and typically contaminated with carbon [92].

Table 11.1: Studies on corrosion control with hybrid sol–gel coverings.

Metals	Precursors	Results	References
AISI 304 stainless steel plates	Al_2O_3-TiO_2	A study has shown that a single layer of Al_2O_3 can increase corrosion resistance by up to 97%. Sol–gel dipping was used to deposit coatings on AISI 304 stainless steel plates.	[93]
Stainless steel	TSTE-TEOS-GPTMS	In 0.5 M HCl, (3-Glycidyloxypropyl) trimethoxysilane (GPTMS) and tetraethoxysilane (TEOS) are used to make hybrid sol–gel coatings containing tamarind tannin extracts, and 94.02% corrosion inhibition.	[94]
Steel substrate	TEOS-TMOMS	A silane hybrid coating comprising tetraethylorthosilicate and trimethoxymethylsilane applied to mild steel substrate boosted its protective efficiency, according to a recent research. Hybrid coating protection improves with a greater condensation reaction, denser network, and lower corrosion intensity.	[95]
Steel substrate	SiO2–NH2@GO	These studies used a sol–gel technique and a SiO_2–NH_2@GO hybrid material to develop a durable and environmentally friendly rosin-based sealant. The longevity of the sealant was extended by including a functional nanocomposite. The hybrid sealant has demonstrated improved adhesion, filling efficiency, and long-term performance in thermal stabilities and corrosion resistance. The new hybrid sealant is environmentally friendly, affordable, and has a unique preparation procedure, which is a new approach to sealant development.	[96]
Stainless steel	Polysiloxane hybrid	The study explored the effectiveness of hybrid polysiloxane films as a corrosion protection solution for stainless steel substrates. The films were fabricated using a sol from co-polycondensation and free radical polymerization. The levels of polymerization, polycondensation, adhesion, and thermal stability of the hybrid-coated stainless steel were high. Even after three weeks of immersion in a 3.5% NaCl solution, the layer with a ratio of 2% was the most successful at preventing corrosion and functioning as a physical barrier,, reducing current densities of magnitude related to the bare electrode.	[97]

Carbon steel	TEOS-GPTMS	The study looked at how well hybrid organic–inorganic sol–gel films loaded with silica or alumina nanoparticles improved the electrochemical corrosion resistance of low-carbon steel. The nanoparticles were injected into the sol–gel films, which were produced from 3-GPTMS and TEOS precursors. The electrochemical performance of the coated steel was evaluated using EIS and SEM, and the study discovered that sol–gel films containing silica nanoparticles improved the barrier properties of the silane coating and had the highest initial pore resistance after five days of immersion in a 0.05 M NaCl solution. According to the study, adding silica nanoparticles to silane sol–gel coatings might be a more ecologically friendly choice for boosting metal corrosion resistance.	[98]
Galvanized steel and electroplated ZnFe steel	TEOS-PDMS-OH	This study prepared organic–inorganic hybrid materials using a sol–gel process and evaluated their corrosion protection performance on galvanized steel and steel electroplated with ZnFe. Electrochemical techniques were used to evaluate the degradation behavior of the coatings. Results showed that the hybrid films exhibited good protective properties, but the overall performance of the coating systems was highly dependent on the type of metallic coating applied to the steel.	[99]
Hot-dip galvanized steel	Amino-alcohol-silicates	OIH matrices were synthesized by sol–gel method and deposited on HDGS using a dip-coating process. Electrochemical studies including electrochemical impedance spectroscopy, macrocell current density, open circuit potential monitoring, and polarization resistance were carried out in mortar to assess the performance of the coatings. The analysis revealed the OIH-coated HDGS materials had better performance against corrosion, compared to uncoated HDGS samples.	[100]
Hot-dip galvanized steel	(3-isocyanatopropyltriethoxysilane) with diamine-alkylethers	This study investigates the electrochemical behaviour as well as the interface analyses for an ecologically friendly hybrid inorganic–organic covering for the hot galvanized steel contacting cemented medium. The coating was attained through solution–gel processes and five different molecular weight diamine-alkylethers; the protective characteristics against corrosion of the coated materials was tested by EIS, PDP methods, and macrocell current density for a period of 74 days. Results showed that the coatings provide barrier properties that protect the hot-dip galvanized steel when it first contacts cementitious media.	[101]

(continued)

Table 11.1 (continued)

Metals	Precursors	Results	References
Hot-dip galvanized steel	Ureasilicates and amino-alcoholsilicates	This study synthesized two types of organic–inorganic hybrid coatings on hot-dipped steel: aminoalcohol silicates and ureasilicates. The coatings were assessed by electrochemical and SEM imaging studies and were found to have better resistance to corrosion in high alkalinity environments relative to the non-coated materials. The U(X) overlays showed superior corrosion behavior compared to the A(X) overlays, with lower and higher rates of corrosion observed for one dipping stage of U(X) and three dipping stages of Cr^{3+}-doped A(X), by employing oligopolymers at MW 400 & 900.	[102]
Galvanized steel	Ureasilicate matrix	The objective of the study was to investigate the performance of the electrochemical EC behavior of OIH coatings applied to hot-dipped galvanized steel immersed in a simulated concrete pore solution with a pH greater than 12.5. The EC results of the OIH coverings were evaluated with a variety of methods, including EIS, PDP plots, macrocellular DC, and polarization resistance. The layers were produced using the sol–gel technique and coated onto the surface HDGS in one or three dip processes.	[103]
Stainless steel	Aminopropyl silane-based coatings	This study discusses the preparation and characterization of hybrid aminopropyl silane-based coatings on steel using the sol–gel method. The coatings were made by dip coating, and various parameters were studied to understand their effect on properties like thickness, roughness, contact angle, and abrasion resistance. The coatings were found to be smooth and hydrophilic, and had good thermal stability. The abrasion resistance improved with decreasing pH and thicker coatings were delaminated, while thinner coatings were smeared and had lubricating properties.	[104]

AISI 304 stainless steel	TEOS-MTES	This research describes the preparation and characterization of hybrid sol–gel organic–inorganic coatings for protecting different metal surfaces. The coatings were prepared by incorporating non-hydrolyzable groups in combination within various mole proportions of sol–gel reactants solubilized in different solvents. The authors found that ethyl acetate embedding in the solution–gel preparation gave the highest performance, while 2-propanol was suitable for stainless steel. EIS analysis indicated a significant enhancement in resistance against corrosion with the incorporation of hydrotalcite particles in the covering. SEM indicated that the coatings were homogeneous on low-carbon steel, but showed cracks on stainless steel.	[105]
Q235 steel	TEOS-TSUPQD	Hydrolysis and condensation of tetraethoxysilane on Q235 steel developed unique electroactive sol–gel hybrid coatings. In hybrid coatings, the Q235 substrate's redox catalytic activity created passive films.	[106]
Steel substrate	ZPCC-CC	Post-treatment with a hybrid silane coating improved the barrier properties of zinc phosphate conversion coating (ZPCC) pretreated steel. SC deposition on ZPCC produced a smooth, crack-free, thick coating with increased hydrophobicity. After SC modification, corrosion resistance increased significantly.	[107]
Mild steel	GPS-TEOS-MTES	A hybrid silane coating post-treated on zinc phosphate conversion coating (ZPCC) pretreated steel improved its barrier properties, according to a research. A smooth, crack-free, thick film with increased hydrophobicity was produced via SC deposition on ZPCC. SC adjustment significantly improved corrosion resistance.	[108]
Steel sheet	TiO₂	A titanium oxide-containing hybrid organic–inorganic sol–gel coating improves steel corrosion resistance. Results demonstrated thick, homogenous, corrosion-resistant coatings.	[109]
A5083	TEOS-OTES-PMMA	Inorganic–organic hybrid films-coated Al alloy A5083 for maritime environment corrosion resistance. Sol–gel hybrid films were made by tetraethyl orthosilicate, Octyltriethoxysilane, and Poly (methyl methacrylate) (PMMA).	[110]

(continued)

Table 11.1 (continued)

Metals	Precursors	Results	References
AA7,075-T6	TEOS-ZTP	The degree of protection relied on zirconium concentration and aging duration in coatings made with zirconium propoxide, 3-(trimethoxysilyl)propyl methacrylate, and tetraethyl orthosilicate. The connection between silicon, oxygen, and zirconium protected against corrosion.	[111]
AA7,075-T6	TEOS-MMA-MAPTMS	This research investigated how polymerization time affects the structure and morphology of an acrylic monomer-siloxane network hybrid material. Polymerization time has been extensively examined using various methods.	[112]
AA7,075	TEOS-GPTMS	The combination of silica matrix barrier characteristics and cerium nitrate active protection improves corrosion prevention. Coating precursors were tetraethoxysilane and 3-glycidoxypropyl-trimethoxymethylamine.	[113]
Al AA2024	LDH film	The effective corrosion prevention provided by AA2024 was investigated using a bilayer system consisting of a laminated double hydroxide conversion layer covered by a sol–gel film. When the LDH layer was loaded with a vanadate corrosion inhibitor, the analysis proved that the system exhibited effective corrosion control and self-healing properties. During corrosion testing, microstructural and surface chemical analyses revealed an accumulation of chlorine signals in the LDH coatings, indicating that chloride scavenging and vanadate inhibitor release from the LDH layer may account for the high corrosion prevention efficiency.	[114]

Aluminum and copper	BTMSE–MPTMS	In this study, sol–gels and hybrid sol–gel coatings were prepared using bis(trimethoxysilyl)ethane and (3-mercaptopropyl)trimethoxysilane under acidic conditions. The coverings were analyzed by X-ray photoelectron spectroscopy and reflection adsorption IR spectra, which showed covalent bonding between the MPTMS sol–gel coating and aluminum but no bond interaction of the thiol group towards the copper. The BTMSE sol–gel coating was found to covalently interact with aluminum. The IR spectra showed the progression of the condensation reaction in the sol–gel BTMSE film, and the sol–gel coated films showed better corrosion protection than the uncoated metals. Corrosion protection was characterized using cyclic voltammetry, potentiodynamic polarization, and impedance spectroscopy methods.	[115]
Al alloy 2024-T3	MBT	In this study, the authors have synthesized microcapsules loaded with a corrosion inhibitor called 2-mercaptobenzothiazole using interfacial polycondensation. These microcapsules were spherical and had a diameter range of 100 nm–2 μm. The MBT loading content was 5 wt% and release studies showed that MBT is preferentially released under acidic and alkaline conditions, following the Fickian diffusion model. The microcapsules were added to a hybrid sol–gel coating, and testing showed that they did not negatively affect the coating's barrier properties and contributed to better adhesion to the metallic substrate. Moreover, the addition of these microcapsules imparted active corrosion protection, making them a promising additive for high-performance coatings in the field of corrosion protection.	[116]
Al alloy AA6061-T6	TiO2 NPs	The study involved the use of a hybrid sol–gel matrix doped with TiO2 nanoparticles to produce layers with improved anti-corrosive and hydrophobic properties on aluminum alloy. The coverings were treated using the dipping technique with the integration of two different organic modified silica alkoxides and a fluorinated metal-alkoxide precursor. Experimental results showed that the presence of TiO2 NPs played a crucial role in enhancing the hydrophobicity and mechanical properties of the coatings while also providing effective corrosion resistance. The study provides valuable information on the use of hybrid sol–gel coverings for anticorrosive purposes and could have significant implications in the development of new coatings for various industrial applications.	[117]
Al 7,075	TEOS–MTMS	The coatings of BTA are applied to aluminum alloy substrates using the sol–gel process, with a titanium oxide coating coating those functions as a corrosion inhibitor.	[118]

(continued)

Table 11.1 (continued)

Metals	Precursors	Results	References
Mg alloy AZ31	8-Hydroxyquinoline	This work used sol–gel OIH layers deposited by an inhibitor to create novel protective coatings for the AZ31 magnesium alloy. The coatings were created by mixing 3-glycidoxypropyltrimethoxysilane and zirconium tetrapropoxide, with 8-Hydroxyquinoline added at two phases of the synthesis. Tests and imaging techniques validated the coatings' efficiency, and the coatings were shown to resist corrosion for up to two weeks without degrading the sol–gel matrix. According to the findings, adding the inhibitor the magnesium.	[119]
Mg alloy AZ91D	Hybrid silica sol-gel matrix	The study examined at the usefulness of using halloysite nanotubes overloaded with cationic corrosion inhibitors to increase resistance of metallic surfaces in this study. The inhibitors were contained in halloysite clay nanotubes and disseminated in a hybrid silica sol–gel matrix, which was subsequently applied to AZ91D magnesium alloy substrates. After differing lengths of immersion in a salt solution, the coatings were evaluated using various methods. The results validated the coatings' self-healing characteristics and promising nature for corrosion prevention. Overall, the research revealed the feasibility of integrating cationic corrosion inhibitors into halloysite nanotubes for self-healing coatings in a variety of applications.	[120]
Mg–Al alloys	TEOS–MEMO	The purpose of this study was to safeguard light Mg-Al alloys used in biomedical prostheses against corrosion and mechanical damage caused by physiological human fluid. The researchers created organic–inorganic hybrid coatings doped with ecologically acceptable and non-toxic corrosion inhibitors and tested their efficiency in a typical corrosion test. The results revealed that adding an inhibitor to the sol–gel coating considerably enhanced corrosion performance, and the L-cysteine-doped hybrid sol–gel films were an outstanding barrier. Adding TiO2 nanoparticles, 2-Aminopyridine, and quinine chemical molecules further decreased the corrosion rate. Graphene oxide showed a comparable corrosion reaction as the hybrid sol–gel covering without inhibitors.	[121]

AZ31B alloys	GPTMS-TEOS And GPTMS-TEOS-GN	This work presents a study in which hybrid pure and modified graphene sol–gel silica films were compared in terms of their morphology and electrochemical properties when coated on AZ31B anodized Mg alloys. The coatings were made with precursor solution containing GPTMS and TEOS, on the added of the graphene chemically modified nanoparticles. The coatings were found to be homogeneous and adherent, with a uniform thickness of around 4 μm. The presence of graphene nanosheets was confirmed through EDS analysis, SEM images, and Raman analysis. The results of potentiodynamic polarization indicated that silica coatings containing graphene improved the resistance to corrosion of AZ31B alloys.	[122]
Mg alloy AZ31B	PHS-TEOS	The study synthesized hybrid organic–inorganic coatings with phosphonate functionalities via a sol–gel method using a mixture of diethylphosphonatoethyltriethoxysilane and tetraethoxysilane with varying molar ratios. The coatings' morphology and surface chemistry were analyzed, and their corrosion protection performance was tested. The study found that the phosphonate-containing coatings offered better corrosion protection than pure silica sol–gel coatings due to the combination of the organosilicate matrix's barrier properties and the strong chemical bonding of the phosphonate groups to the magnesium substrate.	[123]
Mg alloy AZ31B	ORMOSIL materials	The study aimed to create coatings for magnesium AZ31B alloy substrates using a sol–gel method. The coatings' corrosion protection properties were enhanced due to the protective layer characteristics of the organosilica matrix and the bond between the phosphono groups and the surface of the metal. The use of sol–gel methods to synthesize these coatings offers a cost-effective and environmentally friendly approach to corrosion protection.	[124]
Mg alloy AZ31B	Hybrid organic–inorganic sols	This study developed a coating for Mg alloy AZ31B. OIH sols were formulated by co-polymerization of epoxysiloxane and Ti or Zr alkoxides, with tris(trimethylsilyl) phosphate added for extra corrosion protection. The sol–gel coating effectively prevented corrosion attacks and demonstrated good adhesion to the metal substrate. The sol–gel film's structure and thickness were characterized using TEM and SEM, while the corrosion behavior of the substrates was tested using EIS. The silylphosphate holding sol–gel layer's chemical composition was explored using XPS with depth profiling.	[125]

(continued)

Table 11.1 (continued)

Metals	Precursors	Results	References
Ni-Co alloy	Polyurea formaldehyde/ SiO2	In this study, face polymerization and urea formaldehyde resin and TEOS sol–gel were used to create microcapsules with a hybrid shell of polyurea formaldehyde/SiO2 and a linseed oil core. Several approaches were used to describe the microcapsules, which ranged in size from 5–200 m on average and had a hybrid shell layer under 1 μm. The microcapsules were then implanted into an Ni-Co alloy film with electrodeposition process, and the corrosion behavior of the coating was studied. The results indicate that the coating exhibits potential self-repair characteristics. This research has crucial implications for the creation of new, more durable, and corrosion-resistant materials.	[126]
Zn/Mg/NC	HSG	Zinc and magnesium hybrid sol–gel materials, containing nanocarbon, were synthesized and incorporated into ecofriendly anti-corrosive water-based paint formulations to protect carbon steel from corrosion. The corrosion protection efficiency of the painted panels increased as the concentration of the HSG increased, with maximum inhibition efficiency up to 96.5%. The study reported that the featured corrosion resistances for the prepared paint formulations have not been reported yet in the open literature.	[127]
Copper substrates	TTIP-TBOT-GPTMS	Sol–gel was used to produce two titanium-containing organic–inorganic hybrid coating materials. These coatings on copper substrates resist corrosion in 3.5 wt% sodium chloride media.	[128]
Copper rod	TEOS-GPTMS	The research examines how binary coatings reduce copper corrosion in electrochemically investigated copper. In a mixed-material environment, the binary coating significantly reduced copper corrosion.	[129]

11.5 Conclusion

In conclusion, functionalized hybrid sol–gel thin film coatings have shown immense potential in the field of corrosion inhibition. These coatings combine the benefits of sol–gel technology and functionalization techniques, resulting in enhanced corrosion protection for various metallic substrates. Firstly, the sol–gel process allows for the formation of thin films with excellent adhesion to the substrate surface. This ensures a continuous and uniform coating, which effectively acts as a barrier against corrosive agents. Additionally, sol–gel coatings possess high chemical stability and resistance to environmental factors, making them suitable for long-term corrosion protection. The functionalization of sol–gel coatings involves the incorporation of specific organic or inorganic compounds to impart additional desirable properties. For instance, the addition of corrosion inhibitors, such as organic molecules or nanoparticles, enhances the protective capabilities of the coating. These inhibitors can form a passive layer on the metal surface, slowing down the corrosion process and reducing the degradation rate.

Furthermore, functionalization techniques enable the modification of surface properties, such as hydrophobicity or hydrophilicity, to control the interaction between the coating and the corrosive environment. This tailored approach allows for the development of coatings that are specifically designed for the targeted application and substrate. The versatility of functionalized hybrid sol–gel thin film coatings is evident in their successful application in various industries, including aerospace, automotive, marine, and oil and gas. These coatings have demonstrated excellent corrosion resistance in harsh environments, extending the lifespan of metallic components and structures, reducing maintenance costs, and improving overall system reliability. However, further research and development efforts are still needed to optimize the performance of functionalized hybrid sol–gel coatings. This includes exploring new functionalization strategies, investigating the long-term durability of the coatings, and understanding their behavior under different operating conditions.

In conclusion, functionalized hybrid sol–gel thin film coatings offer a promising solution for corrosion inhibition. Their ability to combine the advantages of sol–gel technology with tailored functionality makes them a valuable tool in the ongoing battle against corrosion, with the potential to revolutionize the protection of metallic substrates in various industrial sectors.

References

[1] Kharissova OV, Kharisov BI, Oliva González CM, Méndez YP, López I. Greener synthesis of chemical compounds and materials. R Soc Open Sci 2019;6(11):191378.
[2] Gonzalo-Juan I, Riedel R. Ceramic synthesis from condensed phases. Chem Texts 2016;2(2):6.
[3] Luc W, Jiao F. Synthesis of nanoporous metals, oxides, carbides, and sulfides: beyond nanocasting. Acc Chem Res 2016 19;49(7):1351–58.

[4]   Danks AE, Hall SR, Schnepp Z. The evolution of 'sol–gel' chemistry as a technique for materials synthesis. Mater Horiz 2016;3(2):91–112.

[5]   Petrakis E, Karmali V, Bartzas G, Komnitsas K. Grinding kinetics of slag and effect of final particle size on the compressive strength of alkali activated materials. Minerals 2019 19;9(11):714.

[6]   Bhushan B (ed). Encyclopedia of nanotechnology [Internet]. Dordrecht: Springer Netherlands; 2012. http://link.springer.com/10.1007/978-90-481-9751-4.

[7]   Pillai SC, Hehir S (eds). Sol-gel materials for energy, environment and electronic applications [Internet]. Cham: Springer International Publishing; 2017 [cited 2022 Sep 6]. (Advances in Sol-Gel Derived Materials and Technologies). Available from: http://link.springer.com/10.1007/978-3-319-50144-4.

[8]   Hofacker S, Mechtel M, Mager M, Kraus H. Sol–gel: a new tool for coatings chemistry. Prog Org Coat 2002;45(2–3):159–64.

[9]   Wang D, Bierwagen Gordon P. Sol–gel coatings on metals for corrosion protection. Prog Org Coat 2009;64(4):327–38.

[10] Amiri S, Rahimi A. Hybrid nanocomposite coating by sol–gel method: a review. Iran Polym J 2016;25 (6):559–77.

[11] Bolintineanu DS, Grest GS, Lechman JB, Pierce F, Plimpton SJ, Schunk PR. Particle dynamics modeling methods for colloid suspensions. Comput Part Mech 2014;1(3):321–56.

[12] Nutan MTH, Reddy IK. General principles of suspensions. In: Kulshreshtha AK, Singh ON, Wall GM (eds). Pharmaceutical Suspensions [Internet] (pp. 39–65). New York, NY: Springer New York, 2010 [cited 2022 Sep 25]. Available from: http://link.springer.com/10.1007/978-1-4419-1087-5_2.

[13] Schramm LL. Emulsions, Foams, Suspensions, and Aerosols., 503.

[14] Kingsley Ogemdi I. Properties and uses of colloids: A review. Colloid Surf Sci 2019;4(2):24.

[15] Birdi KS. Handbook of surface and colloid chemistry (p. 752). 3rd ed. Taylor & Francis Group; 2009.

[16] Chin˜as-Castillo F, Spikes HA. Mechanism of action of colloidal solid dispersions. J Tribol 2003;125 (3):552–57.

[17] Ghosh SK. Functional coatings (pp. 369). WILEY-VCH Verlag GmbH&Co. KGaA; 2006.

[18] Pierre AC. Introduction to sol-gel processing [Internet]. Cham: Springer International Publishing; 2020 [cited 2022 Oct 2]. Available from: http://link.springer.com/10.1007/978-3-030-38144-8.

[19] Hood M, Mari M, Muñoz-Espí R. Synthetic strategies in the preparation of polymer/inorganic hybrid nanoparticles. Materials 2014;7(5):4057–87.

[20] Pomogailo AD, Dzhardimalieva GI. Nanostructured materials preparation via condensation ways [Internet]. Dordrecht: Springer Netherlands; 2014 [cited 2022 Oct 2]. Available from: http://link. springer.com/10.1007/978-90-481-2567-8.

[21] Innocenzi P, Zub IL, Kessler VG North Atlantic treaty organization (eds). Sol-gel methods for materials processing: focusing on materials for pollution control, water purification, and soil remediation (pp. 505). Dordrecht, Netherlands: Springer; 2008. (NATO science for peace and security series – C: Environmental security).

[22] Owens GJ, Singh RK, Foroutan F, Alqaysi M, Han CM, Mahapatra C, et al. Sol–gel based materials for biomedical applications. Prog Mater Sci 2016;77:1–79.

[23] Dimitriyev MS, Chang YW, Goldbart PM, Fernández-Nieves A. Swelling thermodynamics and phase transitions of polymer gels. Nano Futur 2019;3(4):042001.

[24] Bokov D, Turki Jalil A, Chupradit S, Suksatan W, Javed Ansari M, Shewael IH, et al. Nanomaterial by Sol-Gel Method: Synthesis and application Wang Z, editor. Adv Mater Sci Eng 2021;1–21.

[25] Zheludkevich ML, Salvado IM, Ferreira MGS. Sol–gel coatings for corrosion protection of metals. J Mater Chem 2005;15(48):5099.

[26] Kakde V, Mannari V. Advanced chrome-free organic–inorganic hybrid pretreatments for aerospace aluminum alloy 2024-T3 – application of novel bis-ureasll sol–gel precursors. J Coat Technol Res 2009;6(2):201–11.

[27]   Kessler VG. The synthesis and solution stability of alkoxide precursors. In: Klein, L, Aparicio, M, Jitianu, A (eds). Handbook of Sol-Gel Science and Technology [Internet] (pp. 1–50). Cham: Springer International Publishing; 2016. [cited 2022 Oct 2]. Available from: http://link.springer.com/10.1007/978-3-319-19454-7_1-1.

[28]   Almeida RM, Gonçalves MC. Sol–Gel process and products. In: Richet P, Conradt R, Takada A, Dyon J (eds) Encyclopedia of Glass Science, Technology, History, and Culture [Internet] (pp. 969–79). 1st ed, Wiley; 2021 [cited 2022 Oct 14]. Available from: https://onlinelibrary.wiley.com/doi/10.1002/9781118801017.ch8.2.

[29]   Guibal E, Vincent T, Jouannin C. Immobilization of extractants in biopolymer capsules for the synthesis of new resins: a focus on the encapsulation of tetraalkyl phosphonium ionic liquids. J Mater Chem 2009;19(45):8515.

[30]   Milea CA, Bogatu C, Du A. The influence of parameters in silica sol-gel process. 2011;4(1):9.

[31]   Schubert U. Chemistry and fundamentals of the Sol-Gel process. In: Levy, D, Zayat, M (eds). The Sol-Gel Handbook [Internet] (pp. 1–28). Weinheim, Germany: Wiley-VCH Verlag GmbH & Co. KGaA; 2015 [cited 2022 Oct 14], Available from: https://onlinelibrary.wiley.com/doi/10.1002/9783527670819.ch01.

[32]   Deshmukh R, Niederberger M. Mechanistic aspects in the formation, growth and surface functionalization of metal oxide nanoparticles in organic solvents. Chem – Eur J 2017;23 (36):8542–70.

[33]   ALOthman Z. A review: fundamental aspects of silicate mesoporous materials. Materials 2012;5 (12):2874–902.

[34]   Rodič P, Iskra J, Milošev I. Study of a sol–gel process in the preparation of hybrid coatings for corrosion protection using FTIR and 1H NMR methods. J Non-Cryst Solids 2014;396–397:25–35.

[35]   Anderson R, Lockwood G. P invaluable contributions of toxicology. 608.

[36]   Yang X. Sol-Gel synthesized nanomaterials for environmental applications, 137.

[37]   Kanamori K, Nakanishi K. Controlled pore formation in organotrialkoxysilane-derived hybrids: from aerogels to hierarchically porous monoliths. Chem Soc Rev 2011;40(2):754–70.

[38]   Sanchez C, Julián B, Belleville P, Popall M. Applications of hybrid organic–inorganic nanocomposites. J Mater Chem 2005;15(35–36):3559.

[39]   Montes S, Maleki H. Aerogels and their applications. In: Colloidal Metal Oxide Nanoparticles [Internet] (pp. 337–99). Elsevier; 2020 [cited 2022 Oct 14]. Available from: https://linkinghub.elsevier.com/retrieve/pii/B9780128133576000152

[40]   Zhang J, Hu Y, Li Y. Inorganic Gels. In: Gel Chemistry (pp. 191–208). Singapore: Springer Singapore; 2018 [Internet]. [cited 2022 Oct 14]. (Lecture Notes in Chemistry; vol. 96). Available from: http://link.springer.com/10.1007/978-981-10-6881-2_6.

[41]   Serra A, Ramis X, Fernández-Francos X. Epoxy Sol-Gel hybrid thermosets. Coatings 2016;6(1):8.

[42]   Lee KY, Mahadik DB, Parale VG, Park HH. Composites of silica aerogels with organics: a review of synthesis and mechanical properties. J Korean Ceram Soc 2020;57(1):1–23.

[43]   Navas D, Fuentes S, Castro-Alvarez A, Chavez-Angel E. Review on Sol-Gel synthesis of perovskite and oxide nanomaterials. Gels 2021;7(4):275.

[44]   Estella J, Echeverría JC, Laguna M, Garrido JJ. Effects of aging and drying conditions on the structural and textural properties of silica gels. Microporous Mesoporous Mater 2007;102(1–3):274–82.

[45]   Strøm RA, Masmoudi Y, Rigacci A, Petermann G, Gullberg L, Chevalier B, et al. Strengthening and aging of wet silica gels for up-scaling of aerogel preparation. J SoL-Gel Sci Technol 2007;41 (3):291–98.

[46]   Zemke F, Scoppola E, Simon U, Bekheet MF, Wagermaier W, Gurlo A. Springback effect and structural features during the drying of silica aerogels tracked by in-situ synchrotron X-ray scattering. Sci Rep 2022;12(1):7537.

[47]   Butt MA. Thin-film coating methods: A successful marriage of high-quality and cost-effectiveness – a brief exploration. Coatings 2022;12(8):1115.

[48]   Jin C. Porous Silica Xerogel. In: Reference Module in Materials Science and Materials Engineering [Internet] (p. B9780128035818023000). Elsevier; 2016 [cited 2022 Oct 15]. Available from: https://linkinghub.elsevier.com/retrieve/pii/B9780128035818023353

[49]   Fricke J, Emmerling A. Aerogels – preparation, properties, applications. In: Reisfeld, R, Jjørgensen, CK (eds). Chemistry, Spectroscopy and Applications of Sol-Gel Glasses [Internet] (pp. 37–87). Berlin, Heidelberg: Springer Berlin Heidelberg; 1992 [*cited 2022 Oct* 15]. (Clarke MJ, Goodenough JB, Ibers JA, Jørgensen CK, Mingos DMP, Neilands JB, et al., eds. Structure and Bonding; vol. 77). Available from: http://link.springer.com/10.1007/BFb0036965.

[50]   Dervin S, Pillai SC. An introduction to Sol-Gel processing for aerogels. In: Pillai SC, Hehir S (eds) Sol-Gel Materials for Energy, Environment and Electronic Applications [Internet] (pp. 1–22). Cham: Springer International Publishing, 2017 [cited 2022 Oct 15]. Advances in Sol-Gel Derived Materials and Technologies). Available from: http://link.springer.com/10.1007/978-3-319-50144-4_1.

[51]   Oubaha M. Introduction to hybrid Sol-Gel materials. In: World Scientific Series in Nanoscience and Nanotechnology [Internet] (pp. 1–36). WSC; 2019 [cited 2022 Oct 15]. Available from: https://www. worldscientific.com/doi/abs/10.1142/9789813270565_0001.

[52]   Aegerter MA, Mennig M (eds). Sol-Gel technologies for glass producers and users [Internet]. Boston, MA: Springer US; 2004 [cited 2022 Oct 15]. Available from: http://link.springer.com/10.1007/ 978-0-387-88953-5.

[53]   Franklin T, Yang R. Vapor-deposited biointerfaces and bacteria: An evolving conversation. ACS Biomater Sci Eng 2020;6(1):182–97.

[54]   Doll GL, Mensah BA, Mohseni H, Scharf TW. Chemical vapor deposition and atomic layer deposition of coatings for mechanical applications. J Therm Spray Technol 2010;19(1–2):510–16.

[55]   Eslamian M. Inorganic and organic solution-processed thin film devices. Nano-Micro Lett 2017;9(1):3.

[56]   Brinker CJ. Dip coating. In: Schneller, T, Waser, R, Kosec, M, Payne, D (eds). Chemical Solution Deposition of Functional Oxide Thin Films (pp. 233–61). Vienna: Springer Vienna, 2013 [Internet]. [cited 2022 Oct 15]. Available from: http://link.springer.com/10.1007/978-3-211-99311-8_10.

[57]   Poddighe M, Innocenzi P. Hydrophobic thin films from sol–gel processing: A critical review. Materials 2021;14(22):6799.

[58]   Liu Y, Yan X, Wang Z. Droplet dynamics on slippery surfaces: small droplet, big impact. Biosurfe Biotribol 2019;5(2):35–45.

[59]   Sanchez C, Rozes L, Ribot F, Laberty-Robert C, Grosso D, Sassoye C, et al. "Chimie douce": A land of opportunities for the designed construction of functional inorganic and hybrid organic-inorganic nanomaterials. Comptes Rendus Chim 2010;13(1–2):3–39.

[60]   Fu Q, Wu X, Kumar D, Ho Jwc, Kanhere PD, Srikanth N, et al. Development of Sol–Gel Icephobic coatings: effect of surface roughness and surface energy. ACS Appl Mater Interfaces 2014;6 (23):20685–92.

[61]   Tranquillo E, Bollino F. Surface Modifications for Implants Lifetime extension: An Overview of Sol-Gel Coatings. Coatings 2020;10(6):589.

[62]   Mbam SO, Nwonu SE, Orelaja OA, Nwigwe US, Gou XF. Thin-film coating; historical evolution, conventional deposition technologies, stress-state micro/nano-level measurement/models and prospects projection: a critical review. Mater Res Express 2019;6(12):122001.

[63]   Gupta R, Kumar A. Bioactive materials for biomedical applications using sol–gel technology. Biomed Mater 2008;3(3):034005.

[64]   Pandey S, Mishra SB. Sol-gel derived organic–inorganic hybrid materials: synthesis, characterizations and applications. J Sol-Gel Sci Technol 2011;59(1):73–94.

[65]   Jones JR. Review of bioactive glass: From Hench to hybrids. Acta Biomater 2013;9(1):4457–86.

[66]   Bollino F, Armenia F, Tranquillo E. Zirconia/hydroxyapatite composites synthesized via Sol-Gel: Influence of hydroxyapatite content and heating on their biological properties. Materials 2017;10(7):757.

[67] Purcar V, Răditoiu V, Răditoiu A, Manea R, Raduly FM, Ispas GC, et al. Preparation and characterization of some Sol-Gel modified silica coatings deposited on polyvinyl chloride (PVC) substrates. Coatings 2020;11(1):11.

[68] Savignac P, Menu MJ, Gressier M, Denat B, Khadir Y, Manov S, et al. Improvement of adhesion properties and corrosion resistance of Sol-Gel coating on zinc. Molecules 2018;23(5):1079.

[69] Mitzi DB, Kosbar LL, Murray CE, Copel M, Afzali A. High-mobility ultrathin semiconducting films prepared by spin coating. Nature 2004;428(6980):299–303.

[70] Zhang Z, Peng F, Kornev K. The Thickness and Structure of Dip-Coated Polymer Films in the Liquid and Solid States. Micromachines 2022;13(7):982.

[71] Shanmugam N, Pugazhendhi R, Madurai Elavarasan R, Kasiviswanathan P, Das N. Anti-reflective coating materials: A holistic review from PV perspective. Energies 2020;13(10):2631.

[72] Sahu N, Parija B, Panigrahi S. Fundamental understanding and modeling of spin coating process: A review. Indian J Phys 2009;83(4):493–502.

[73] Schneller T, Waser R, Kosec M, Payne D (eds.) Chemical solution deposition of functional oxide thin films [Internet]. Vienna: Springer Vienna; 2013 [[cited 2022 Oct 16]. Available from]. http://link.springer.com/10.1007/978-3-211-99311-8.

[74] Tyona MD. A theoretical study on spin coating technique. Adv Mater Res 2013;2(4):195–208.

[75] Palumbo F, Lo Porto C, Fracassi F, Favia P. Recent advancements in the use of aerosol-assisted atmospheric pressure plasma deposition. Coatings 2020;10(5):440.

[76] Habibi M, Ahmadian-Yazdi MR, Eslamian M. Optimization of spray coating for the fabrication of sequentially deposited planar perovskite solar cells. J Photonics Energy 2017;7(2):022003.

[77] Glocker DA, Ranade S (eds.) Medical coatings and deposition technologies (p. 768). Hoboken, New Jersey : Salem, Massachusetts: John Wiley & Sons Inc.; Scrivener Publishing LLC; 2016.

[78] Yun H, Li J, Chen HB, Lin CJ. A study on the N-, S- and Cl-modified nano-TiO2 coatings for corrosion protection of stainless steel. Electrochimica Acta 2007;52(24):6679–85.

[79] Shan CX, Hou X, Choy KL, Choquet P. Improvement in corrosion resistance of CrN coated stainless steel by conformal TiO2 deposition. Surf Coat Technol 2008;202(10):2147–51.

[80] El Hajjaji S, Ben Bachir A, Aries L. Electrolytic deposits on stainless steel as high temperature coatings. Surf Eng 2001;17(3):201–04.

[81] Inamuddin, AMI, Luqman M, Altalhi T (eds.) Sustainable corrosion inhibitors (p. 225). Millersville, PA, USA: Materials Research Forum LLC; 2021. (Materials research foundations).

[82] Durán A, Conde A, Coedo AG, Dorado T, García C, Ceré S. Sol–gel coatings for protection and bioactivation of metals used in orthopaedic devices. J Mater Chem 2004;14(14):2282–90.

[83] O'Brien T, Bommaraju TV, Hine F. Handbook of Chlor-Alkali technology (pp. 5). New York: Springer; 2005.

[84] Sastri VS, Ghali E, Elboujdaini M. Corrosion Prevention and Protection, 581.

[85] Flores-Álvarez JF, Rodríguez-Gómez FJ, Onofre-Bustamante E, Genescá-Llongueras J. Study of the electrochemical behavior of a Ti6Al4V alloy modified by heat treatments and chemical conversion. Surf Coat Technol 2017;315:498–508.

[86] Osborne JH. Observations on chromate conversion coatings from a sol–gel perspective. Prog Org Coat 2001;41(4):280–86.

[87] Figueira RB, Silva CJR, Pereira EV. Organic–inorganic hybrid sol–gel coatings for metal corrosion protection: A review of recent progress. J Coat Technol Res 2015;12(1):1–35.

[88] Artesani A, Di Turo F, Zucchelli M, Traviglia A. Recent advances in protective coatings for cultural heritage–an overview. Coatings 2020;10(3):217.

[89] Boidot A, Gheno F, Bentiss F, Jama C, Vogt JB. Effect of aluminum flakes on corrosion protection behavior of water-based hybrid zinc-rich coatings for carbon steel substrate in NaCl environment. Coatings 2022;12(10):1390.

[90] Álvarez P, Collazo A, Covelo A, Nóvoa XR, Pérez C. The electrochemical behaviour of sol–gel hybrid coatings applied on AA2024-T3 alloy: Effect of the metallic surface treatment. Prog Org Coat 2010;69(2):175–83.

[91] Tuck CDS, Powell CA, Nuttall J. 3.07 Corrosion of Copper and its Alloys, 37.

[92] Honkanen M, Hoikkanen M, Vippola M, Vuorinen J, Lepistö T. Aminofunctional silane layers for improved copper–polymer interface adhesion. J Mater Sci 2011;46(20):6618–26.

[93] Gecu R, Birol B, Özcan M. Improving wear and corrosion protection of AISI 304 stainless steel by Al2O3-TiO2 hybrid coating via sol–gel process. Trans IMF 2022;100(6):324–32.

[94] Abdulmajid A, Hamidon TS, Hussin MH. Tamarind shell tannin-doped hybrid sol–gel coatings on mild steel in acidic medium toward improved corrosion protection. J Coat Technol Res 2022;19 (2):527–42.

[95] Ramezanzadeh B, Akbarian M, Ramezanzadeh M, Mahdavian M, Alibakhshi E, Kardar P. Corrosion protection of steel with zinc phosphate conversion coating and post-treatment by hybrid organic-inorganic Sol-Gel based silane film. J Electrochem Soc 2017;164(6):C224–30.

[96] Wang G, Zhou Z, Hu Q, Shi X, Zhang X, Zhang K, et al. Preparation of eco-friendly natural rosin-based SiO2–NH2@GO hybrid sealant and study on corrosion resistance of Fe-based amorphous coating for steel substrate. Carbon 2023;201:170–88.

[97] Sarmento VHV, Schiavetto MG, Hammer P, Benedetti AV, Fugivara CS, Suegama PH, et al. Corrosion protection of stainless steel by polysiloxane hybrid coatings prepared using the sol–gel process. Surf Coat Technol 2010;204(16–17):2689–701.

[98] Balan P, Ng A, Beng Siang C, Singh Raman RK, Chan ES. Effect of nanoparticle addition in hybrid sol-gel silane coating on corrosion resistance of low carbon steel. Adv Mater Res 2013;686:244–49.

[99] Souza MEPD, Ariza E, Ballester M, Yoshida IVP, Rocha LA, Freire CMDA. Characterization of organic-inorganic hybrid coatings for corrosion protection of galvanized steel and electroplated ZnFe steel. Mater Res 2006;9(1):59–64.

[100] Figueira RB, Silva CJR, Pereira EV, Salta MM. Alcohol-aminosilicate hybrid coatings for corrosion protection of galvanized steel in mortar. J Electrochem Soc 2014;161(6):C349–62.

[101] Figueira RB, Silva CJR, Pereira EV, Manuela Salta M. Ureasilicate hybrid coatings for corrosion protection of galvanized steel in cementitious media. J Electrochem Soc 2013;160(10):C467–79.

[102] Figueira RB, Silva CJR, Pereira EV. Hybrid Sol-Gel coatings for corrosion protection of hot-dip galvanized steel in alkaline medium. Surf Coat Technol 2015;265:191–204.

[103] Figueira RB, Silva CJR, Pereira EV. Hybrid sol–gel coatings for corrosion protection of galvanized steel in simulated concrete pore solution. J Coat Technol Res 2016;13(2):355–73.

[104] Hanetho SM, Kaus I, Bouzga A, Simon C, Grande T, Einarsrud MA. Synthesis and characterization of hybrid aminopropyl silane-based coatings on stainless steel substrates. Surf Coat Technol 2014 Jan;238:1–8.

[105] Hernandez M, Barba A, Genesca J, Covelo A, Bucio E, Torres V. Characterization of Hybrid Sol-Gel coatings doped with hydrotalcite-like compounds on steel and stainless steel alloys. ECS Trans 2013;47(1):195–206.

[106] Alibakhshi E, Akbarian M, Ramezanzadeh M, Ramezanzadeh B, Mahdavian M. Evaluation of the corrosion protection performance of mild steel coated with hybrid sol-gel silane coating in 3.5 wt.% NaCl solution. Prog Org Coat 2018;123:190–200.

[107] Omar SA, Ballarre J, Ceré SM. Protection and functionalization of AISI 316 L stainless steel for orthopedic implants: hybrid coating and sol gel glasses by spray to promote bioactivity. Electrochimica Acta 2016;203:309–15.

[108] Asadi N. Effect of curing condition on the protective performance of an eco-friendly hybrid silane sol-gel coating with clay nanoparticles applied on mild steel, 27.

[109] Guin AK, Nayak S, Rout TK, Bandyopadhyay N, Sengupta DK. Corrosion resistance nano-hybrid Sol-Gel coating on steel sheet. ISIJ Int 2011;51(3):435–40.

[110] Naknikham U, Youngthin I, Tapasa K. Enhanced corrosion resistance of aluminium alloy using hybrid Sol-Gel coating. 2021;6:8.

[111] Rodič P, Iskra J, Milošev I. A hybrid organic–inorganic sol-gel coating for protecting aluminium alloy 7075-T6 against corrosion in Harrison's solution. J Sol-Gel Sci Technol 2014;70 (1):90–103.

[112] Rodič P, Korošec RC, Kapun B, Mertelj A, Milošev I. Acrylate-based hybrid Sol-Gel coating for corrosion protection of AA7075-T6 in aircraft applications: The effect of copolymerization time. Polymers 2020;12(4):948.

[113] Shanaghi A, Kadkhodaie M. Investigation of high concentration of benzotriazole on corrosion behaviour of titania–benzotriazole hybrid nanostructured coating applied on Al 7075 by the sol–gel method. Corros Eng Sci Technol 2017;52(5):332–42.

[114] Yasakau KA, Kuznetsova A, Kallip S, Starykevich M, Tedim J, Ferreira MGS, et al. A novel bilayer system comprising LDH conversion layer and sol-gel coating for active corrosion protection of AA2024. Corros Sci 2018;143:299–313.

[115] Li YS, Lu W, Wang Y, Tran T. Studies of (3-mercaptopropyl)trimethoxylsilane and bis(trimethoxysilyl) ethane sol–gel coating on copper and aluminum. Spectrochim Acta A Mol Biomol Spectrosc 2009;73 (5):922–28.

[116] Maia F, Yasakau KA, Carneiro J, Kallip S, Tedim J, Henriques T, et al. Corrosion protection of AA2024 by sol–gel coatings modified with MBT-loaded polyurea microcapsules. Chem Eng J 2016;283:1108–17.

[117] Rivero P, Maeztu J, Berlanga C, Miguel A, Palacio J, Rodriguez R. Hydrophobic and corrosion behavior of sol-gel hybrid coatings based on the combination of TiO2 NPs and fluorinated chains for aluminum alloys protection. Metals 2018;8(12):1076.

[118] Tiringer U, Milošev I, Durán A, Castro Y. Hybrid sol–gel coatings based on GPTMS/TEOS containing colloidal SiO2 and cerium nitrate for increasing corrosion protection of aluminium alloy 7075-T6. J Sol-Gel Sci Technol 2018;85(3):546–57.

[119] Galio AF, Lamaka SV, Zheludkevich ML, Dick LFP, Müller IL, Ferreira MGS. Inhibitor-doped sol–gel coatings for corrosion protection of magnesium alloy AZ31. Surf Coat Technol 2010;204 (9–10):1479–86.

[120] Adsul SH, Siva T, Sathiyanarayanan S, Sonawane SH, Subasri R. Self-healing ability of nanoclay-based hybrid sol-gel coatings on magnesium alloy AZ91D. Surf Coat Technol 2017;309:609–20.

[121] Rodríguez-Alonso L, López-Sánchez J, Serrano A, Rodríguez de la Fuente O, Galván JC, Carmona N. Hybrid Sol-Gel coatings doped with non-toxic corrosion inhibitors for corrosion protection on AZ61 magnesium alloy. Gels 2022;8(1):34.

[122] Afsharimani N, Talimian A, Merino E, Durán A, Castro Y, Galusek D. Improving corrosion protection of Mg alloys (AZ31B) using graphene-based hybrid coatings. Int J Appl Comput Sci 2021;8.

[123] Khramov AN, Balbyshev VN, Kasten LS, Mantz RA. Sol–gel coatings with phosphonate functionalities for surface modification of magnesium alloys. Thin Solid Films 2006;514(1–2):174–81.

[124] Khramov AN, Johnson JA. Phosphonate-functionalized ORMOSIL coatings for magnesium alloys. Prog Org Coat 2009;65(3):381–85.

[125] Lamaka SV, Montemor MF, Galio AF, Zheludkevich ML, Trindade C, Dick LF, et al. Novel hybrid sol–gel coatings for corrosion protection of AZ31B magnesium alloy. Electrochimica Acta 2008 May;53(14):4773–83.

[126] Sadabadi H, Allahkaram SR, Ghader O, Rohatgi PK. Synthesis and characterization of hybrid shell microcapsules for anti-corrosion Ni-Co coating. J Met Mater Miner 2022;32(4):143–49.

[127] Elhalawany N, Serour J, Saleeb M, Abdlekarim A, Morsy F. Enhanced anticorrosion properties of water-borne anticorrosive paints based on Zinc/magnesium/nanocarbon (Zn/Mg/NC). Egypt J Chem 2021;0(0):0–0.

[128] Tong H, Zhao Z, Wu X, Liu W. Titanium-containing organic–inorganic hybrid coatings for the corrosion protection of copper in sodium chloride medium. Mol Cryst Liq Cryst 2021;722 (1):87–94.

[129] Karthik N, Lee YR, Sethuraman MG. Hybrid Sol-Gel/thiourea binary coating for the mitigation of copper corrosion in neutral medium. Prog Org Coat 2017;102:259–67.

Sukdeb Mandal, Sanjukta Zamindar, Manilal Murmu,
Priyabrata Banerjee

12 Incorporating microcapsules-embedded corrosion inhibitor in functionalized thin film coatings

Abstract: Recently, functionalized thin film coatings with embedded smart microcapsules have emerged as ecological, environment-friendly, and efficient corrosion inhibitors in the domain for the protection of metal substrates from adverse environments such as acidic solutions, saline environments, and so on. The utilization of microcapsules that have the capability to release encapsulated core materials in a controlled manner can pave the way for fabricating novel types of multifunctional coatings that can repair themselves. The increasing use of multifunctional coatings across various industries is largely inspired by their intellectual features. In this chapter, the recent research and development of a variety of stimuli-responsive, self-healing, barrier, significant impermeability, in addition to anti-corrosive properties, with the incorporation of micro or nano container techniques for impregnating or embedding corrosion inhibitors that consider innovative as well as greener approaches have been highlighted explicitly briefing, the anti-corrosive mechanisms.

Keywords: Microcapsules, Thin Film, Coating, Corrosion Inhibition, Metal Surface

12.1 Introduction

Every year, metal corrosion causes enormous economic losses in several industrial sectors across the globe as well as poses negative impact on everyday lives [1]. One of the simplest as well as most effective methods of safeguarding metal from deterioration is to apply anti-corrosive coating. The main purpose of conventional corrosion protection coatings is to generate a physical barrier that hinders the intrusion of water and other corrosive ionic species. Passive corrosion prevention is another name for this particular

Acknowledgements: PB is thankful to the Department of Higher Education, Science and Technology and Biotechnology, Government of West Bengal, India for providing the financial assistance (vide sanction order number 78(sanc.)/ST/P/S&T/6 G-1/2018 dated 31.01.2019 and project number GAP-225612). SM acknowledges University Grants Commission (UGC), New Delhi, Government of India, for his fellowship (212/CSIR-UGC NET DEC.2017). SZ acknowledges Department of Science and Technology (DST), New Delhi, Government of India, for her fellowship (IF 200407). MM is very thankful to Ministry of Tribal Affairs, New Delhi, India for his National Fellowship for Higher Education of Scheduled Tribes candidates, NFST (vide award letter no. F1-17.1/2014-15/RGNF-2014-15-ST-JHA-71,559).

https://doi.org/10.1515/9783111016160-012

physical barrier effect. Despite the physical barrier effect, the coated architectures may still sustain damage due to corrosive substances. It may deteriorate faster due to alterations in the external environment during extended exposure to harsh conditions. All of these factors facilitate the creation of microscopic pores and fractures on the coating surface. The defects created by the microscopic pores or fractures allow aggressive ionic species to diffuse and pass through the coating, ultimately reaching the metal surface. This process can lead to the degradation of the coating substrate and cause corrosion of the underlying metals. The micro-corrosion sites become larger as the service duration increases, resulting in the breakdown of the coating materials and corrosion of the metallic surface. Thus, coatings that solely possess physical barrier properties are frequently unable to provide long-term protection from corrosion, especially in presence of harsh environment, which is accompanied by mechanical stresses. Various techniques have been improvised in order to enhance the effectiveness of coating in preventing corrosion and extending its service life. The idea of "smart, next-generation self-healing coatings" has just been brought out and is currently the subject of extensive research. These coatings can be developed using intrinsic or extrinsic self-healing techniques. The intrinsic self-healing property of a material relies on the breaking and rearrangement of chains at the molecular level, allowing for autonomous self-healing effects to occur. In contrast, the extrinsic self-healing approach uses active substances (such as corrosion inhibitors and healing agents) to prevent corrosion or restore coating structures. The inclusion of active species can help minimize the occurrence of corrosion processes at the metal-electrolyte interface when the physical barrier effect of a coating is compromised. These coatings can create a physical barrier to prevent the intrusion of water, oxygen, and aggressive corrosive ions [2]. The "passive corrosion protection" method is often transformed into "active corrosion protection" with the addition of active healing agents to the coating, where corrosion inhibitory activities begin after the coating gets damaged [3]. The primary disadvantage of using this type of coating is that it has a temporal self-healing capability that ends after the active species are entirely consumed. Furthermore, the effectiveness of the coating is further hampered because of the incompatibility of the integrated active species with the polymeric matrix as well as potential interactions between active species and the matrix.

In recent times, active corrosion prevention coatings loaded with micro/nanocontainers have been developed. In this type of coating, the active ingredients are initially loaded into the micro/nanocontainers before being distributed throughout the coating matrix. Those containers can respond to various environmental changes and provide prompt feedback. Consequently, the encapsulated active substances can be subjected to release and repair the micro/nanocracks in the coatings, thus preventing further degradation. This type of "active corrosion protection" technology for metal substrates has also steadily gained popularity in research, and a wide variety of micro- and nanocontainers are developed to carry out their various activities. In this chapter, a summary of the latest developments in the field of microcapsules-embedded corrosion inhibitors in functionalized thin film coatings has been discussed. In-depth analytical explanations of the different stimulus-responsive coatings of encapsulated pay-loads and release models have also been highlighted. Finally, a thorough discussion of the use of micro containers in functionalized thin–film coatings and future possibilities have been presented.

12.2 Corrosion inhibitor and their functions

The use of corrosion inhibitors is likely the simplest, most practical, and most efficient corrosion control strategy among several methods now in use in several industries [4–7]. Corrosion inhibitors are often categorized based on their ability and mechanism of suppressing corrosion. Accordingly, there are adsorption inhibitors that undergo chemisorption onto the metallic surface, film-forming inhibitors, which are classified into passivation inhibitors (oxidizing or non-oxidizing), and precipitation inhibitors (deposition of the three-dimensional film). However, the distinction between anodic, cathodic, and mixed-type inhibitors is most often made based on which half-reaction they obstruct during corrosion reactions. Most of the organic inhibitors get chemisorbed on the metallic surface and act as mixed-type inhibitors. Inorganic inhibitors like phosphate slow down the reduction reaction during corrosion, and therefore, act as cathodic inhibitors, while benzoate, azelates inhibit corrosion of metal by film-formation [8]. The addition of corrosion inhibitors within the coating can suppress the electrochemical reactions due to corrosion. On the contrary, if the corrosion inhibitors are introduced directly into the coating matrix without encapsulation, those inhibitors may react directly with the metal substrate or the coating composition prematurely. This can lead to the coating losing its self-healing protective capability and potentially having negative impacts [9, 10]. Encapsulating inhibitors within micro/nanocontainers is the method used to get rid of these problems [11]. Inspired by the drug delivery systems in the medicinal field, researchers first introduced self-healing material loaded micro/nanocontainer (capsule) for the fabrication of self-repairable coating. Whenever the coating is damaged externally, the healing agents get released from the ruptured shells. Cross-linking reactions facilitate the repairing of the as-developed cracks within the coating matrix. However, this causes the corrosion inhibitors to be discharged from the containers due to mechanical damage, which will cause an uncontrollable one-time release of corrosion inhibitors. The self-healing protective capability of coatings will soon cease to exist and cannot ensure long-term service of the coating. Furthermore, microcapsules are not appropriate for thin coatings as they typically have dimensions between a few and hundreds of microns. Additionally, the performance of final composites is destroyed by the poor mechanical qualities and incompatibility of the microcapsules with the coating matrix.

12.3 Fabrication methods of microcapsules

A process known as microencapsulation enables the formulation of customized micro-particles, each of which is composed of an active core substance and an outer shell material. Microencapsulation might be employed to ensure that the encapsulated core substance is well-protected, to regulate its flowing out, or even to make the use of liquid items easier. There are numerous types of encapsulation methods. Each method has its corresponding encapsulation rate, capsule morphology, and range of particle sizes. The encapsulation methods might be categorized depending on various methodologies and criteria, as follows: (i) process involved in microencapsulation, (ii) nature of dispersing medium, (iii) materials used, and (iv) organic solvent usages [12]. According to the first criterion, the microencapsulation process is further classified into three main classes. These three classes are chemical processes, physical processes, and physicochemical processes. The classifications have been schematically presented in Figure 12.1.

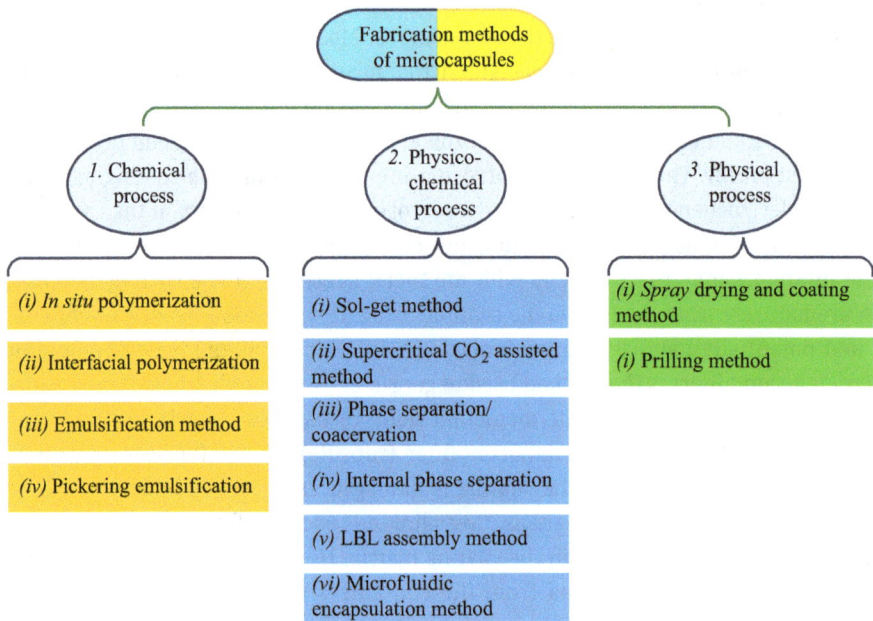

Figure 12.1: Schematic representation of the fabrication methods of microcapsules.

12.3.1 Chemical processes of microencapsulation

12.3.1.1 In situ polymerization method

The widely manifested method to fabricate microcapsules is in situ polymerization. Due to its simple procedure, the in situ polymerization method is used in the industrial production of microcapsules. This method aids in the formation of a shell wall by adding a reactive polymerizable species with or outside of the core material [13, 14].

The core material is generally non-reactive, which may be considered the idiosyncratic feature of this process. In this process, the monomeric forms of the shell material are allowed to get dissolved in the continuous phase. Polymerization starts accordingly, followed by the deposition or precipitation of monomer in the interface of the core material in a controlled manner with the aid of the addition of precipitant at a specified temperature and pH [15]. Finally, the microcapsule with a polymeric shell is fabricated. Depending on how soluble the monomers are, in situ polymerization can take place in three different forms, as reported by Arshady and George [16]. In the first case, when the monomer is not soluble in the continuous phase, it is suspended as droplets within the continuous phase. Consequently, microcapsules are developed via suspension polymerization. In the second case, when the polymer is soluble in the dispersion medium rather than the monomer, precipitation and polycondensation take place. In this process, the polymer with a low molecular weight might result in particles with enormous sizes and irregular shapes due to flocculation and aggregation. In the third case, when the monomer is well-soluble in the continuous phase instead of the polymer, microencapsulation with a narrow size distribution occurs through dispersion polycondensation. Till date, in situ or interfacial polymerization in an oil-in-water (O/W) emulsion has been used to produce microcapsules for self-healing material fabrication. Microcapsules are frequently synthesized via in situ polymerization using formaldehyde components with urea, or melamine. Veatch and Burhans reported hollow microcapsules based on phenol, formaldehyde and melamine [17]. This research work has become one of the pioneering works in the fabrication of microcapsule-based self-healing material. Later, spherical poly(urea-formaldehyde) (PUF) microcapsules with dicyclopentadiene (DCPD) core for self-healing epoxy were reported by Brown et al. [18]. Microcapsules were obtained with an average diameter between 10 and 1,000 mm and shell thickness of 160 and 200 nm for agitation rate of 200 and 2000 rpm. This process has been accepted as one of the most used procedures for its easy pathway [19].

12.3.1.2 Interfacial polymerization method

This process is just similar to that of the previously mentioned in situ polymerization method. Reaction-diffusion rather than the thermodynamic equilibrium process serves as the basis for interfacial polymerization. This method primarily relies on the Schotten-Baumann reaction, which explains the irreversible polymerization of two multifunctional monomers at the boundary between two incommensurable phases in a heterogeneous system [20]. Interfacial polymerization is a type of poly-condensation process where highly reactive monomers are dispersed in two liquids that do not mix or combine. In this process the shell-forming materials would come from both the continuous phase along with the dispersed phase. Interfacial polymerization is a well-known expedient approach for encasing a liquid core material since it does not need pre- and post-treatment of the template [21]. Oil-in-water (O/W) or water-in-oil (W/O) systems are the main parts of this process [22–24]. Interfacial polymerization is a two-step process. In the first step, two monomer components are primarily dissolved in two immiscible phases, viz., continuous phase and dispersed phase. In the second step, two monomeric forms react with each other in the interface of the two liquid phases via condensation polymerization. Here, condensation polymerization is mainly involved, which includes the reactions of amine with aldehydes, acid chlorides, or isocyanates, and so on. The main drawbacks of this process are the thin shell wall and poor mechanical strength of the as-fabricated microcapsules. Another limitation is that the core material must be soluble either in aqueous or in organic phases as such types of bi-phase systems are involved. Thus, the solubility profile of the core agent restricts the formation of microcapsules via interfacial polymerization. Nevertheless, thin shell-wall-based microcapsules are getting more attention nowadays because of their high core-to-shell ratio. Thus, achieving high loading capacity as well as the desirable release of the core material requires an efficient and stimuli-responsive release process. In order to accomplish these goals, pH has frequently been employed as an appropriate stimulus. Currently, pH-based systems significantly depend on specified pH values to initiate the release of the core material. This method makes it very simple to create the polymer shell, which makes it ideal for encapsulating a variety of compounds including medicines, natural oil, pesticides, adhesives, agrochemicals, sweeteners, perfumes, and inks [25, 26].

12.3.1.3 Emulsification method

A dispersion consisting of two immiscible liquids is known as an emulsion that is stabilized by an emulsifier or surfactant. Emulsification results in the creation of several systems. When oil droplets are allowed to be dispersed in an aqueous phase, then O/W emulsion is created; similarly, when water droplets are allowed to be dispersed in oil phase, then W/O emulsion is created [27, 28]. Water-in-oil-in-water (W/O/W) or oil-in-water-in-oil (O/W/O) systems are referred to as double emulsion systems, as shown in Figure 12.2.

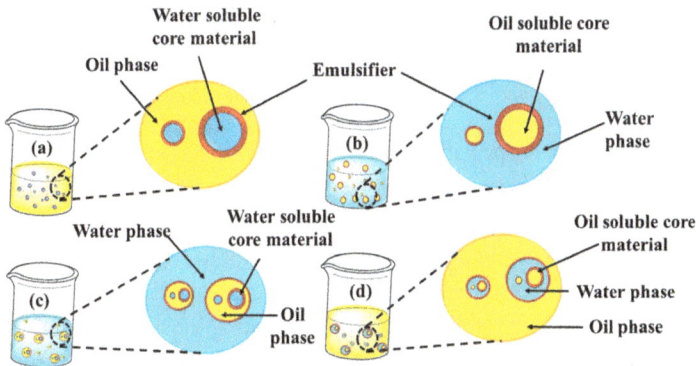

Figure 12.2: (a) W/O emulsion, (b) O/W emulsion, (c) W/O/W emulsion, and (d) O/W/O emulsion.

The idea of emulsification may be used in conjunction with other processes like coacervation and extrusion. Emulsification plays a pivotal role in the microencapsulation process. One of the issues to regulate in this procedure is the stability of the emulsions. Because of this, emulsifiers are often introduced to the emulsion system and the mixture is homogenized using mechanical homogenizers to create a stable emulsion. This process gains superior attention to generate spherical micro/nano capsules because of its cost-effectiveness and simple and easy procedure [29], though there are some lacunas regarding the controlled release of core material along with the stability of the microcapsules [30].

12.3.1.4 Pickering emulsification method

When immiscible liquids get dispersed and the system gets stabilized by solid particles instead of conventional surfactant molecules that system is referred to as Pickering emulsion. This is a two-step process. In the first step, colloidal particles are allowed to come together at the interface by applying mechanical forces. Then in the second step, the colloidal particles are immobilized by cross-linking (physical/ chemical) to allow the generation of a stable and compact shell structure. Two phenomena are capable of ensuring the stability of this kind of emulsion. The formation of solid shells around the droplets results in adherence of the particles at the liquid–liquid interface by the formation of a three-dimensional network. Henceforth, the use of solid particles leads to the formation of a very stable interface. This process has become highly demanding because of less toxicity, easy recovery, easy encapsulation, and controlled release of the active core material.

12.3.2 Physicochemical processes of microencapsulation

12.3.2.1 Sol–gel method

The sol–gel technology for microencapsulation is a promising technique for stabilizing natural scents and odors over a longer period [31–33]. Inorganic micro/nano capsules can be prepared using the sol–gel process at mild or relatively low temperatures. The sol–gel method mostly involves assembling reactions, herein, polycondensation reactions that incorporate several types of alkoxide precursor to create a variety of products. Sol–gel fabrication is renowned as a flexible method for creating inorganic materials with well-crafted microstructures at low temperatures. It is worthwhile to mention that this procedure is ideal for the production of organic–inorganic hybrid materials as it produces homogeneous molecules at the molecular level at low temperature. The initial step of the process involves making the shell material soluble in aqueous phase to form a solution with low viscosity. This enables thorough mixing of the shell forming inorganic material at a molecular level before the gelling stage, allowing uniform modification of the shell with additional functional groups or elements. The main disadvantages of this process are that this method is time-consuming, and if the process is not carefully regulated, pores or cracks in micro level may develop on the finished capsule shell after drying.

12.3.2.2 Supercritical CO_2-assisted method

Recently, supercritical fluids have received huge interest and have been used in a variety of industries. One of the advantageous usages of this method is microencapsulation. Polymer encapsulation has been the major subject of research and improvement in microencapsulation aided by supercritical fluids up to this point [34, 35]. The use of supercritical CO_2 is preferred in these methods due to its low critical point, and non-toxic and non–flammable nature. Consequently, it permits a tidy and ecologically friendly approach. Though this method has several advantages, the poor solubility of polymers and active substances in supercritical CO_2 limits the wider utilization of this process [36–38].

12.3.2.3 Phase separation/ coacervation method

Coacervation is a technique that causes phase separation in a solution by dissolving macromolecules. Accordingly, the coacervate and the supernatant are generated. The basic principle of simple coacervation is the desolvation of a single polymer brought on by alteration in temperature, the addition of an electrolyte, or even a second incompatible polymer. As a result, there are more interactions between polymers than the interactions between the polymers and solvents, which reduce the amount of solvation of the polymer. Both aqueous and organic media can be used to conduct it. The dispersion phase is where the active agent is added. The ability to encapsulate both water-soluble and insoluble materials is thus made feasible [39, 40]. The complex coacervation process involves the incorporation of two different polyelectrolytes with opposite charges, resulting in the formation of a coacervate phase that is rich in polymer and a supernatant phase that is rich in solvent. This process is initiated by adjusting the pH to neutralize the charges of the two polymers, causing the active ingredient to adsorb onto the surface of the droplets. Gelatin, which is cationic, and gum arabic, alginates, or carrageenans, which are anionic, are the polyelectrolytes that are most often used. With this method, large encapsulation rates are possible while using little to no organic solvent. The two main limitations of this process are as follows: only liposoluble active compounds may be encapsulated, and the material options are limited to pairs of polyelectrolytes having opposing charges [41].

12.3.2.4 Internal phase separation method

An effective method for creating containers with a liquid core and polymeric shells is internal phase separation. But the use of this method is still relatively limited to hydrophobic core materials [42]. For the manufacture of large quantities of particles, internal phase separation is applicable to the majority of polymers. This method is based on the dissolution of a polymer in a combination of volatile and non-volatile solvents. To ensure that the polymer dissolves, a sufficient quantity of a reliable solvent is required. An oil-in-water emulsion will then be created by dispersing this solution into a solution of water and emulsifiers. The volatile-solvent evaporation will alter the composition of the droplets, and within the emulsion droplets, the polymer-rich phase will get separated into smaller droplets. The polymer-rich droplets migrate to the oil–water interface where they combine to form polymer particles. It has been reported that thermodynamics controls the shape of the microparticles to achieve the lowest possible total interfacial free energy [43].

12.3.2.5 Layer-by-layer assembly method

Microcapsules have been the subject of several research investigations in recent years because of their intriguing and significant applications in several biomedical, cosmetic, and anti-corrosive self-healing coating areas, including their controlled-release, targeted distribution, biocompatibility, and biodegradability [44, 45]. Layer-by-layer (LBL) assembly is a method of creating membranes via the self-assembly of polymers having opposing charges. It entails alternating adherence of negatively and positively charged polyelectrolytes to the organic or mineral core component. Calcite, latex, silica, and other materials can be used as the core materials. The core size can vary from nanometers to micrometers, depending on its composition. Hollow nano/micro capsules might be produced once the core material is destroyed in an appropriate solution. There may be more than one layer of each charged polymer deposited when a polyelectrolyte layer is formed. It is possible to build up layers successively, which inevitably increases the final size of the microcapsules. The majority of research on LBL assembly-based anti-corrosion materials focused on the impact of pH variations brought on by redox interactions between corrosive chemicals and metallic substrates [46, 47].

12.3.2.6 Microfluidic encapsulation method

Fluid flow through channels with a micrometric size is referred to as a microfluidic process. The basic components of the system are a junction and a microfluidic droplet generator. Till date, corrosion inhibitor-containing capsules are not yet produced using microfluidic encapsulation techniques for use as an end product in anti-corrosion self–healing coatings.

12.3.3 Physical processes of microencapsulation

12.3.3.1 Spray drying and coating method

Using spray drying method, a liquid ingredient containing the active species and the shell forming agent is prepared into dry particles. To get the required microparticles, a number of actions are to be taken. The process begins by atomizing a liquid solution composed of a solvent, a coating, and an active ingredient using either a rotatory atomizer or a nozzle. Once atomized, a heated gas is applied to the droplets through the atomized feed, which is connected to a gas sparger, causing the solvent to evaporate. A quick evaporation of the liquid causes the formation of microparticles, which fall to the bottom of the chamber. The collection of the microcapsule (with particle size range in size from 1–100 mm) powder from the exhaust gases will finally be made possible by a cyclone or bag filter. This process is mainly used in food, anti-corrosive applications in self–healing material formation, cosmetic, and pharmaceutical industries [48–51].

12.3.3.2 Prilling method

A solution, dispersion, or emulsion of active substances in aqueous phase of polymers is used to gelate droplets. These polymers must be able to form gels when temperature or other chemical or physical variables (such as ionic concentration, pH, or the presence of chemicals that cause precipitation) are altered. Solubilization or dispersion of the active ingredient within a coating substance while it is still molten results in the freezing of droplets. For these two procedures, the droplets are created using a vibrating nozzle and collected in a liquid that allows the coating material to gel or solidify. Alginate, chitosan, or agarose are polymers often used for gelation [52, 53].

12.4 Microcapsules-embedded corrosion inhibitor in functionalized thin film coatings

Self-healing coatings have received a lot of interest recently because of their potential to repair damage by themselves. Microcapsules embedded in a matrix or the reconstruction of inherent dynamic chemical linkages in the molecular structures may both be employed for developing coatings with self–healing properties [54–59]. Corrosion-resistant self-healing coatings based on microcapsules have been extensively researched. Self-healing corrosion coatings have been developed by microencapsulating a number of corrosion-protective agents and adding them to a matrix. A few examples of reactive materials that have been liberated from microcapsules as a consequence of damage and then polymerized under the influence of a catalyst include dicyclopentadiene (DCPD), epoxy, isophoronediisocyanate (IPDI), hexamethylenediisocyanate (HDI), cinnamide moiety-containing polydimethylsiloxane (CA–PDMS), and hydrolysable organic silane [60–64].

Li et al. have developed polysulfone-based microcapsules containing tung oil for the application of self-lubricating coating with self-healing characteristics [65]. In this work, tung oil-filled polysulfone (PSF) microcapsules were used to develop a dual-purpose coating. Linseed oil and tung oil are both drying oils with certain characteristics in common. An elaeostearic acid conjugated triene glyceride is the major component. This conjugated, highly unsaturated system is the main cause of the rapid polymerization. The cross-linking of tung oil molecules makes the surface impermeable and chemically resistant. The solvent evaporation process was employed in order to fabricate PSF microcapsules that were filled with tung oil. The self-healing performances of the developed material have been assessed by corrosion test. The developed epoxy-based coatings were manually scratched using a razor blade, and after that the coatings were left to heal for 24 h . The scratched coatings were submerged in a 10 wt% NaCl solution for 1, 4, and 7 days in order to study the corrosion process. The authors noted that the surfaces of steel plates coated with the self-healing material showed almost no signs of corrosion after 1 and 4 days, even in regions that had been scratched. However, there was a small amount of rust observed after 7 days. The idea put forth was that when the microcapsules ruptured and the core components, including tung oil, were released, and a secondary layer of coating was formed in the areas of the coating that had been scratched. The structure of the scratched region is presented in Figure 12.3 for both the control specimen and the specimen that was coated with a self-healing coating.

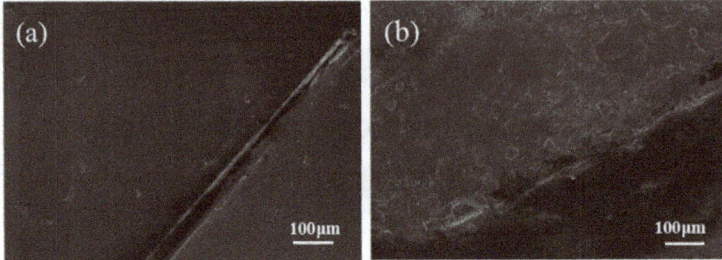

Figure 12.3: FE-SEM images of the damaged areas of (a) a control specimen and (b) a self-healing coating coated with 10 wt% microcapsules (reproduced with permission from ref. [65]).

A new self-healing layer was noticed on the self-healing coated sample, as depicted in Figure 12.3. (b). The self-healing process is schematically displayed in Figure 12.4. As the microcapsules burst, tung oil is released, which reacts with the oxygen in the air to automatically seal and repair the fracture.

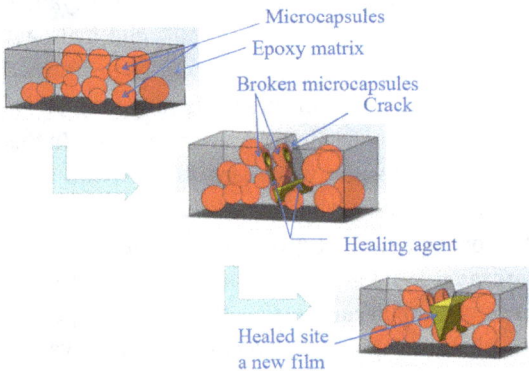

Figure 12.4: Schematic mechanism of self-healing coating (reproduced with permission from ref. [65]).

In another work reported by Li et al. polyurea/polyaniline (PU/PANI)-based microcapsules loaded with IPDI as the core substance by the combination of Pickering emulsion formation and interfacial polymerization in O/W emulsion [66] have been developed. The self-repairing property and corrosion inhibition behaviour of three distinct epoxy resin coatings (first type: pure epoxy coating without microcapsules, second type: a coating containing 10% IPDI-loaded PU microcapsules, and third type: a coating containing 10% IPDI-loaded PU/PANI microcapsules) were evaluated on mild steel surfaces. To assess the corrosion process at an accelerated rate, the samples were intentionally scratched with a razor blade and then submerged in a 10 wt.% NaCl aqueous solution for a period of 30 days without interruption. Figure 12.5 demonstrates that the first type of coating displays some rust after being immersed in a salt solution for one day. In contrast, the third type of coatings were free of rust after 10 days of salt solution immersion. The second type of coating exhibited some corrosion after 20 days of being immersed in a salt solution, while this phenomenon was not observed for the third type of coating samples even after 30 days of immersion. According to their findings, the third type of self-healing coatings exhibited very good corrosion inhibition efficiency.

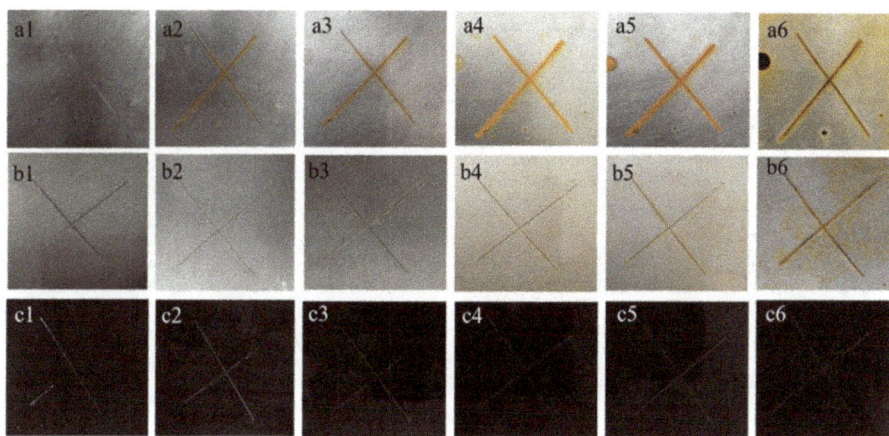

Figure 12.5: The results of the corrosion tests for scratched coatings are presented in the following figures: (a) only epoxy coating, (b) self-healing coating containing 10 wt% IPDI-incorporated PU microcapsules, and (c) self-healing coating having 10 wt% IPDI-incorporated PU/PANI microcapsules after immersion in a 10 wt% NaCl solution for (1) 0 day, (2) 1 day, (3) 5 days, (4) 10 days, (5) 20 days, and (6) 30 days (reproduced with permission from ref. [66]).

EIS testing was done in order to assess anti-corrosion capabilities more thoroughly. In this study, the scratched coatings were submerged in a 3.5 wt% NaCl aqueous solution for 15 days at room temperature. The initial impedance moduli of the second and third type of self-repairable coatings at 0.01 Hz were all approximately 1,011 $\Omega \cdot cm^2$ in this study. This value was considerably greater than the initial impedance moduli of pure epoxy coatings, which were only 10^5 $\Omega \cdot cm^2$ (vide Figure 12.6) and which shows that the first type of coatings cannot maintain anti-corrosive properties. On the other hand, the liberated IPDI and the PANI successfully repaired the micro cracks in the self-healing coating shell. The steady penetration of the electrolyte into the scratches of the coatings throughout the immersion duration was responsible for the gradual drop in impedance modulus at low frequency for all coatings. The third type of coating exhibited an impedance modulus 10^6 $\Omega \cdot cm^2$ at 0.01 Hz. The recorded impedance value was greater than the impedance modulus of the first type of coating (10^4 $\Omega \cdot cm^2$) and second type of coating (10^5 $\Omega \cdot cm^2$). According to the investigation's findings,

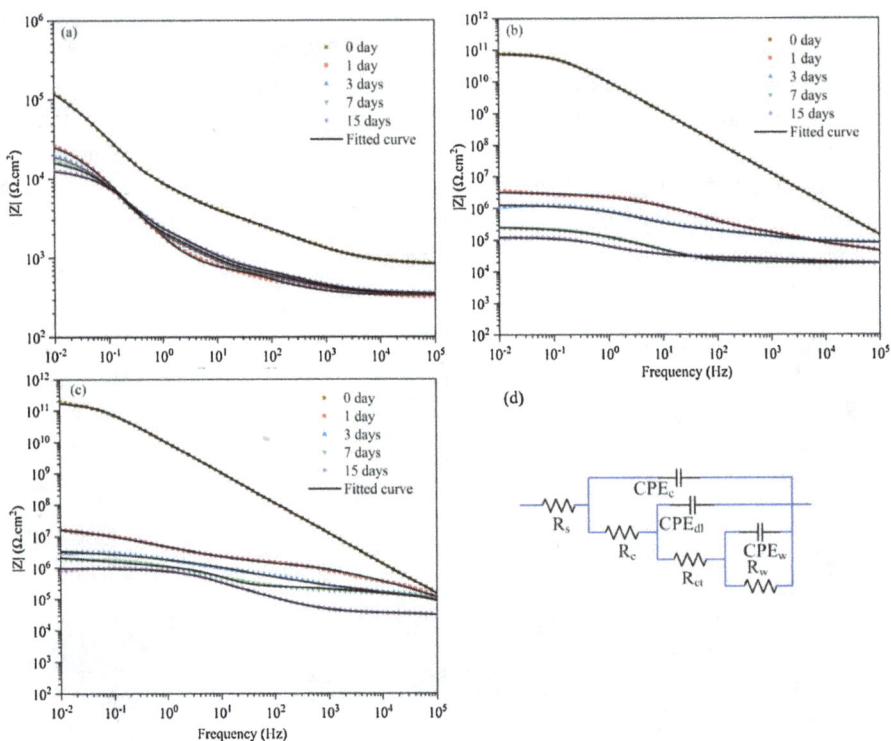

Figure 12.6: The Bode plots for the scratched coatings are shown in the figure, which includes: (a) only epoxy coating, (b) self-healing coating with PU microcapsules incorporated with IPDI (10 wt%), and (c) self-healing coating with PU/PANI microcapsules incorporated with IPDI (10 wt%) after 15 days immersion in 3.5 wt% NaCl solution. Additionally, the figure also includes (d) an electrical equivalent circuit employed to fit the impedance result (reproduced with permission from ref. [66]).

using PANI that has been loaded with IPDI significantly increased the ability of the self-healing coatings to impede corrosion.

The anti-corrosion efficacy of the microcapsule-embedded coatings corresponds with their capacity for self-healing. It has been demonstrated that the interaction between IPDI and PANI enhances the self-repairing and anti-corrosive effectiveness of polymer coatings. The healing ingredient IPDI was initially released into the fracture zone by the bursting of microcapsules caused by the cracks in polymer coatings. The fractures were then fixed by a new polymer film developed by IPDI's subsequent interaction with atmospheric H_2O. The newly produced polymer layer functioned as a barrier that safeguarded the underlying metal by obstructing the penetration of the electrolyte and antagonistic ions. Second, PANI was added to the wall of the microcapsules, which helped prevent corrosion. Previous studies have shown that the cyclic, reversible changes in PANI's chemical structure from emeraldine base to leucoemerald base by interaction with iron ions were responsible for the formation of a passivation layer on the metal surface [67–69].

In order to use linseed oil in epoxy based self-healing coatings, Navarchian et al. have fabricated microcapsules made of poly(methyl methacrylate) with modified surfaces [70]. The existence of an amine functional group on the outermost portion of the modified polymethyl methacrylate (PMMA) microcapsules implies that it can function as an epoxy matrix curing agent. As a result, at the interface of the epoxy matrix and the microcapsule shell, additional chemical bonds are formed. This would considerably increase the efficiency of the interaction between the matrix and the microcapsule. Figure 12.7 shows the chemical reaction between hexamethylene diamine (HDMA) and epoxy matrix.

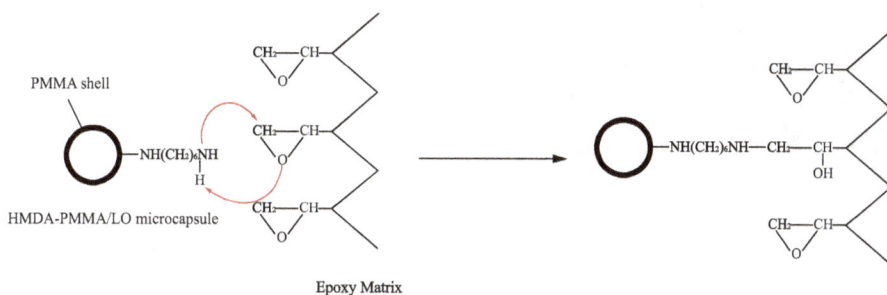

Figure 12.7: The way that HMDA and epoxy matrix interacts chemically (adapted with permission from ref. [70]).

Optical microscopy was used to assess the self-healing capabilities of an epoxy coating incorporating PMMA/LO microcapsules. The healing process is triggered by the release of Lipid Oxidation (LO) from ruptured microcapsules onto the damaged region. After 48 h, the borderline had nearly completely healed due to the PMMA microcapsules in the epoxy matrix, which could efficiently repair the fracture. Nevertheless, HMDA-PMMA/LO microcapsules have excellent self-healing capabilities. The borderline is practically gone and has completely integrated with the epoxy matrix, suggesting that the modified microcapsules performed better throughout the healing process. The improved interfacial adhesion that exists between the modified microcapsule shell and epoxy resin, leading to more effective microcapsule rupturing, is the most plausible explanation for this.

Li et al. have developed urea-formaldehyde (UF) based microcapsule for the application of self-healing anti-corrosion coating [71]. They have used tung oil as corrosion-protective as well as self-healing agent. The FE-SEM pictures of the developed microcapsules have been presented in Fig. 12.8 (a). The size distribution was ascertained using a collection of more than 500 microcapsules, and the outcome is depicted in Fig. 12.8 (b). The smaller mean diameter (8.5 µm) and limited dispersion of the microcapsules confirm their suitability for use in a thin coating. More significantly, if the coating were placed with the same weight per cent of microcapsules, the amount of smaller microcapsules would be higher than that of bigger microcapsules. If the damage is done, there will be greater chance of microcapsules bursting inside the coatings because of the enormous dispersion of microcapsules that they will generate. The microcapsules exhibit a distinctive core-shell structure as depicted in Figure 12.8 (c). Figure 12.8 (d) demonstrates that the shell's thickness was approximately 1 µm. This made it possible for microcapsules to survive during both the preparation and processing pathways, which was critical for the self-healing effect to function.

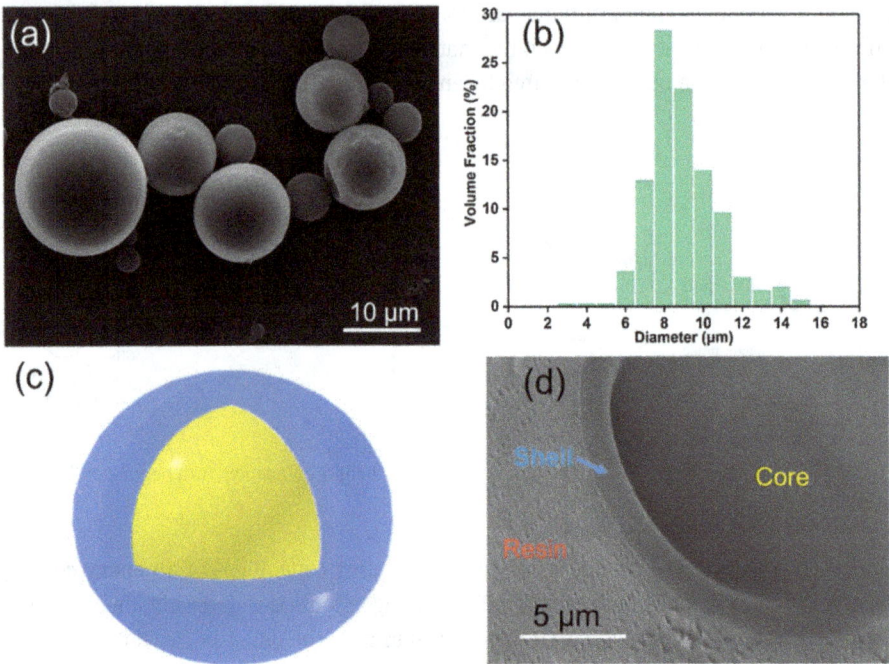

Figure 12.8: (a) FE-SEM images of UF microcapsules; (b) size distribution of the developed microcapsules; (c) schematic representations of the core-shell architecture; and (d) SEM image of shell thickness of microcapsules (reproduced with permission from ref. [71]).

Adhesion and EIS experiments were performed to establish how the microcapsules influenced self-healing protection. Long-term anti-corrosion ability was assessed using salt spray experiments. As indicated in the schematic design in Figure 12.9 (a) the self-healing coating operates to restore itself after being scratched. When a coating is damaged, the microcapsules within the coating rupture, enabling the encapsulated tung oil to flow out and create a new layer at the scratched region by capillary action. For the coating without microcapsules, things are very different since there is a gap that cannot be filled because the plain epoxy coating lacks a polymerizable ingredient (vide Figure 12.9 (b)). The incorporated microcapsules that store the tung oil rupture when a scratch occurs on a self-healing coating, releasing it automatically to the damaged area where it integrates with oxygen to produce a new layer that repairs the wounded region. The barrier function of the self-healing coating can be recovered by the production of a new layer (vide Figure 12.9 (c)).

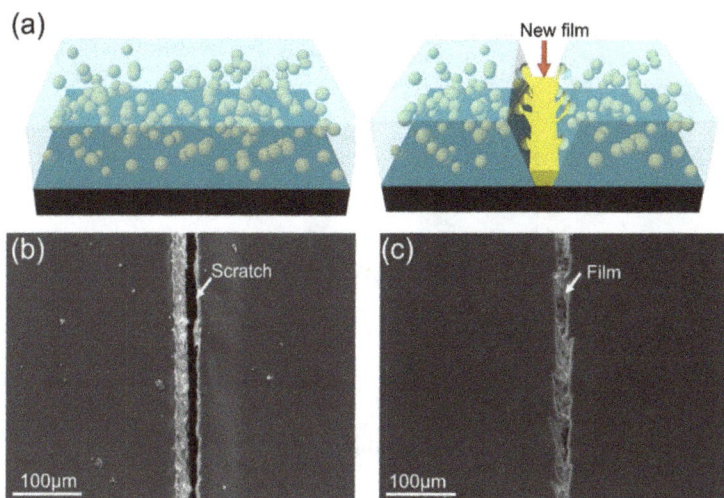

Figure 12.9: (a) Diagrammatic representation of the self-healing mechanism after scratching, (b) SEM images of the pure epoxy coating that has been scratched, and (c) FE-SEM images of the repaired coating (reproduced with permission from ref. [71]).

Dong et al. developed a special form of PANI microcapsules that are durable, two-shell constructed, packed with inhibitors, and have dual anti-corrosion properties to successfully prevent corrosion of carbon steel [72]. By using modified SiO_2 nanoparticles as Pickering emulsifiers to stabilize the oil-in-water (O/W) emulsion, the PANI microcapsules were developed. The oil phase contained aniline, a photo cross-linkable monomer (glycidyl methacrylate, GMA), and a cross-linker (1,6-hexanediol diacrylate, HDDA). As a result of photo polymerizing GMA and HDDA in the oil phase in the presence of UV light, a strongly cross-linked poly (glycidyl methacrylate) (PGMA) shell was created in the first stage. To build the PANI wall, the second step included oxidation of aniline at the oil-water contact. The organic–inorganic hybrid structure of the PGMA/PANI shell, which contains SiO_2 nanoparticles at the oil–water interface, provides the microcapsules with the necessary stability and toughness. Mechanical durability and adaptable functioning are advantages of this distinctive two-shell construction. The 2-mercaptobenzothiazole (MBT) was then incorporated into the PGMA@PANI microcapsules via the impregnation technique, functioning as a corrosion inhibitor. Figure 12.10 provided a schematic representation of the creation of MBT-encapsulated PGMA@PANI microcapsules. Following their mixing with epoxy resin, they were applied to carbon steel surfaces.

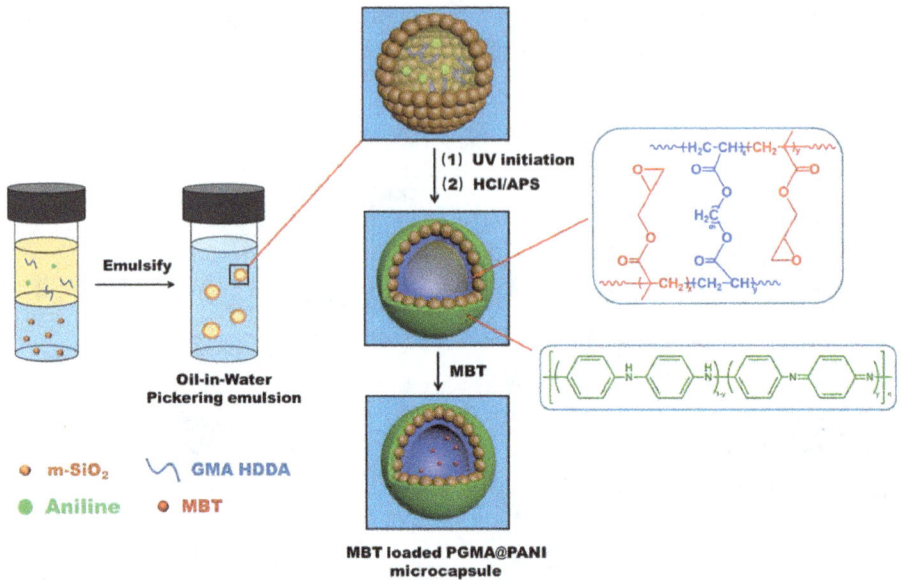

Figure 12.10: Schematic representation of the fabrication process of MBT-encapsulated PGMA@PANI microcapsules with double anti-corrosion capabilities (reproduced with permission from ref. [72]).

The effectiveness of the anti-corrosion properties of coatings was analyzed using EIS and salt spray test. A thick coating of Fe_2O_3 may form as a result of oxidizing the metal surface with PANI. The passivation coating serves as a barrier, obstructing moisture, O_2, ions, and other corrosive elements from reaching the surface of the substrate. Furthermore, PANI has the ability to shift between the oxidation and reduction states by consuming the oxygen that exists in the coating that comes before the metal. As a consequence, it is possible to considerably limit the anodic reaction, which contributed to the development of corrosion products, as well as coating breakdown from the defect area. MBT will discharge and migrate to steel surfaces when the environment's pH level decreases or the coating experiences damage like microcracks. MBT has the potential to create a thick adsorption coating on metal surfaces, raising the activation energy of corrosion. However, the adsorbed coating can also function as a barrier to stop the transfer of corrosive environment [72]. Huang et al. have disclosed a self-healing, corrosion-resistant polymer covering consisting of microencapsulated novel organic silane that is perfluorooctyltriethoxysilane (1 H, 1 H, 2 H, and 2 H-perfluorooctyl triethoxysilane, or POTS) [64]. A self-healing coating's healing behaviour may be quickly demonstrated using electrochemical testing. A coating that is intact performs well as a capacitor, but a coating that has been scribed will deliver high current because the metal substrate comes in direct contact with the electrolyte solution to create a circuit. As a result of self-healing, if the scribes need to be resealed, the newly formed materials will obstruct the circuit's ability to conduct current, reducing the amount of current that can be measured. The test was performed using a standard three-electrode setup with the coated steel substrate operating as the working electrode, vide Figure 12.11. The major objective of this experiment was to assess the capacitive characteristics of the scribed coatings following various degrees of corrosion in the electrolyte solution. This was done by comparing the current flows through the control sample with the self-healing specimen at the same potential.

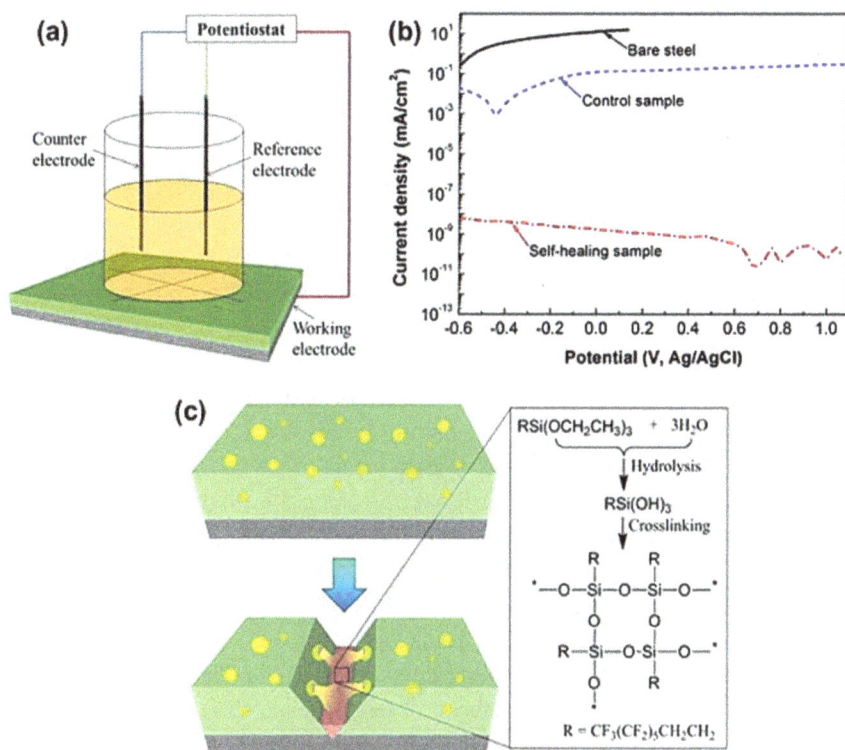

Figure 12.11: Self-healing anti-corrosion mechanisms and electrochemical testing (a) a diagram of the electrochemical test. (b) Findings for the bare steel panel, the scribed control, and the self-healed specimens, and (c) schematic of anti-corrosion mechanism of self-healing coatings (reproduced with permission from ref. [64]).

This led to the adoption of a high 20 mV/s potential sweep rate .It is has been observed that following the healing process during immersion, the current flowing through the scribed self-healing coating was nearly zero amperes (0.2 pA/cm^2). In contrast, just 0.3 mA/cm^2 of current flowed through the scribed control coating, but a large amount of current flowed through the naked steel panel. During the immersion process, the scribes in the self-healing coating spontaneously re-sealed, but the scribes in the control specimen were still open, as evidenced by the significantly different electrochemical behaviour of the two samples. It should be highlighted that in addition, whereas the healed coating performed effectively as a capacitor despite being an inappropriate specimen for the electrochemical test, the precise current density value of the self-healing coating as determined by the test may not be accurate. However, the recovered capacitance property clearly demonstrated that the scribes experienced self-healing during the immersion. The embedded microcapsules broke as the coating was scribed, releasing the POTS liquid that had been kept within. When POTS came into contact with water or moisture in the air, processes called polycondensation and hydrolysis were triggered, resulting in a silane-based substance that deposited in the scribed regions. Corrosion was stopped by the repaired damage acting as a barrier between the corrosive environment outside and the steel substrate below.

In order to enhance the anti-corrosion properties of waterborne epoxy coating, Wu et al. designed a brand-new class of robust PANI microcapsules [73]. In a single pot, interfacial aniline polymerization and photo polymerization were used to produce the PANI microcapsules. The synthetic strategy has been demonstrated in Figure 12.12 The photo polymerization monomer (GMA) and cross-linker (HDDA) are all present in the oil phase together with the healing agent, tung oil. GMA and HDDA were polymerized inside the oil droplets after being subjected to UV light, producing a cross-linked PGMA shell. The addition of aniline to the water phase and aniline's subsequent adsorption on the oil–water interface were both caused by the electrostatic interaction between lignosulfonate and aniline. Aniline was then chemically oxidized and polymerized to create PANI shell. Thus, PANI and PGMA made up both layers of the wall of the as-prepared PANI microcapsules. While the PANI shell has anti-corrosion properties, the strongly cross-linked PGMA shell has superior mechanical properties. This particular hybrid shell included the advantages of solvent resistance, toughness, and anti-corrosion functionality. The primary component (tung oil), which forms a quick film when exposed to oxygen, has the ability to cure itself. After being fabricated, the waterborne epoxy resin was combined with the tung oil-containing microcapsules, and the mixture was then coated on a carbon steel plate. Electrochemical test, and salt spray test were used to assess the self-healing and anti-corrosion characteristics of the smart coatings.

PANI functioned as a corrosion inhibitor in the microcapsule shell, oxidizing the metal surface and encouraging the development of a thick Fe_2O_3 layer. Oxygen, water, and ions may not be able to penetrate the thick Fe_2O_3 layer and reach the metal. The microcapsules were ruptured by an external force, and the released tung oil filled the scratched regions with an entirely new layer that was formed by oxidative polymerization with oxygen in the air, vide Figure 12.13.

In order to strengthen corrosion resistance of cement-based materials, patch up their micro cracks, and absorb chloride ions already existing in the matrix, Wang et al. developed Ca-Al-based LDH hybrid self-healing microcapsules [74]. The layered Ca-Al LDH was applied to the surface of hybrid microcapsules, and the physical technique of microcapsule production stabilized the layered Ca-Al LDH, preventing the chemical structure from being broken down. In addition to achieving chloride ion adsorption, the hybrid microcapsule prevented Ca-Al LDH from aggregating together in

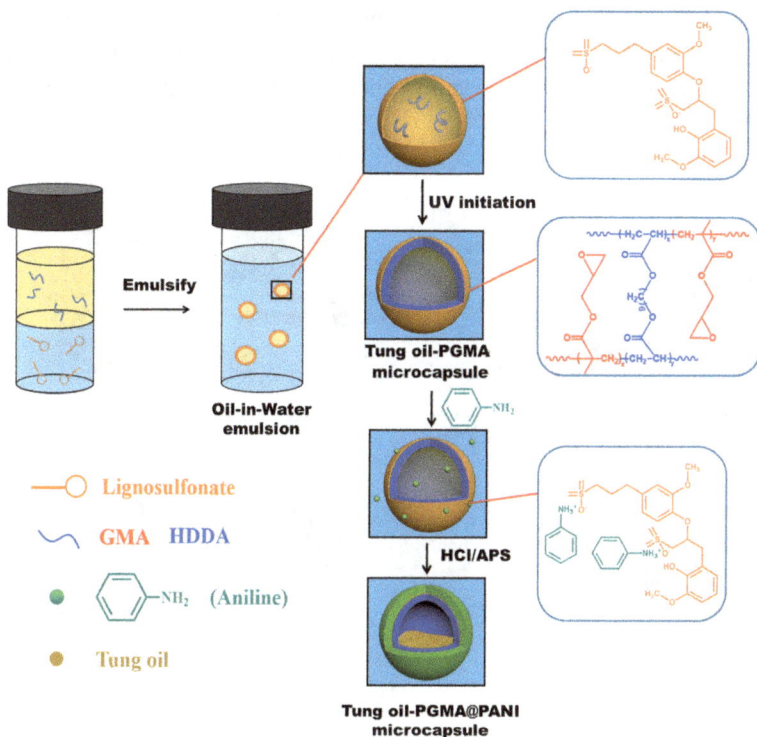

Figure 12.12: The fabrication of self-healing, anti-corrosion tung oil-PGMA@PANI microcapsules (reproduced with permission from ref. [73]).

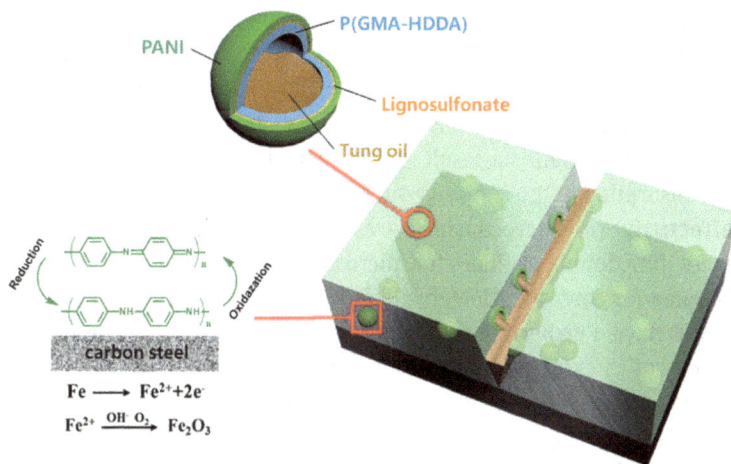

Figure 12.13: Mechanism of self-healing and corrosion prevention in smart coatings employing microcapsules filled with tung oil (reproduced with permission from ref. [73]).

cement-based products. The method for creating microcapsules involves several processes, including the formulation of the wall mixture, the core mixture, encapsulation, and solidification. The preparation method and microcapsule production mechanism are shown in Figure 12.14.

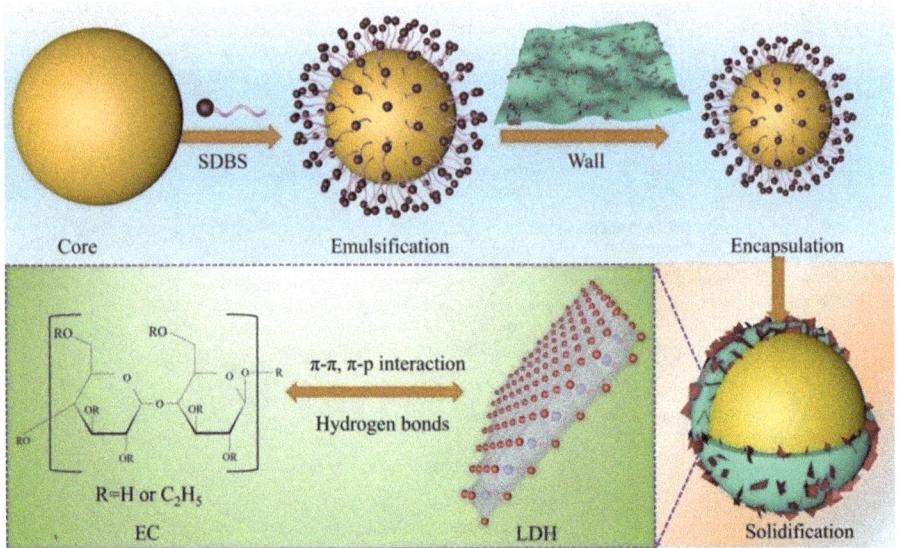

Figure 12.14: Microcapsule preparation and formation mechanisms (reproduced with permission from ref. [74]).

By combining the advantages of the wall and core materials, hybrid microcapsules improve the susceptibility of cement-based materials to chloride ion penetration and may be used with cement-based materials in chloride-rich conditions. Figure 12.15 depicts the wall and core's cooperative anti-corrosion mechanism. The resistance to corrosion provided by the hybrid wall is a result of the labyrinth effect, physical barrier effect, and chlorine adsorption to the corrosive substance.

(a) **Labyrinth effect:** Hybrid microcapsules are distributed throughout cement-based substances, functioning as fillers for microscopic pores and cracks to make it more complicated for corrosive elements like O_2, H_2O, and chlorine to move towards the steel surface. Furthermore, the nano-layered Ca-Al LDH of the hybrid microcapsules can offer extra nucleation sites for cement hydration, enhancing bonding effectiveness and minimizing micropores between microcapsules and hydration products. The migration of aggressive species to the steel surface is significantly more difficult because of the close proximity to the wall and hydration products, which complicates their transport path. Overall, the hybrid microcapsules behave like a labyrinth when it comes to protecting steel components against corrosion in chloride environment.

(b) **Physical barrier:** A hybrid wall's Ca-Al LDH film is made up of an inner dense film and an exterior porous film. Physical defense against corrosive elements like H_2O, O_2, Cl, etc. is provided by the inner thick membrane. Tung oil, ethyl cellulose, as well as other microcapsule components also possess highly hydrophobic characteristics. The Ca-Al LDH's physical barrier effect, along with ethyl cellulose and tung oil's super hydrophobic properties, gives the microcapsules outstanding impermeability. By preventing corrosive elements from penetrating, the microcapsules demonstrate exceptional corrosion resistance.

(c) **Chlorine adsorption effect:** Chloride ions can first be absorbed into the Ca-Al LDH corridor via an anion exchange mechanism. Second, because of its positive charge, Ca-Al LDH may absorb chloride ions on its surface. The corrosiveness of cement-based compounds to steel bars can be lessened, and the service life of steel rods can be increased, by reducing the concentration of free chloride ions.

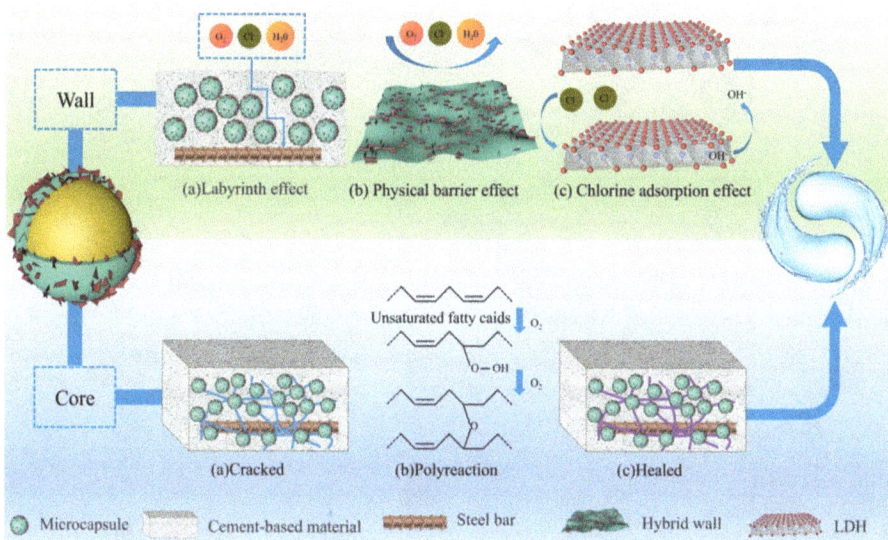

Figure 12.15: Wall and core have complementary anti-corrosion mechanisms (reproduced with permission from ref. [74]).

The multi-chain combined triene structure of tung oil molecules gives them significant chemical activity and enables them to cure at room temperature. In the presence of oxygen, tung oil can produce free radicals that might lead to the polymerization of its own double bonds.

For the application of smart coatings, Ghahremani et al. developed cerium-based pH-responsive microcapsules [75]. Cerium nitrate is a low-toxic cathodic inhibitor, but whether it possesses substantial inhibition capabilities, particularly for copper-rich aluminium alloys, has been strongly contested. Additionally, in order to enhance anti-corrosion effectiveness of coatings, cerium nitrate-containers have been added. Cerium nitrate, a well-known corrosion inhibitor with microscopic self-healing properties, is included in ethylene glycol- and melamine-formaldehyde-based microcapsules. The chemical interaction between melamine and formaldehyde that produces the melamine-formaldehyde compound is shown in Figure 12.16 (a). The composition and structure of the microcapsules shell are shown in Figure 12.16 (b).

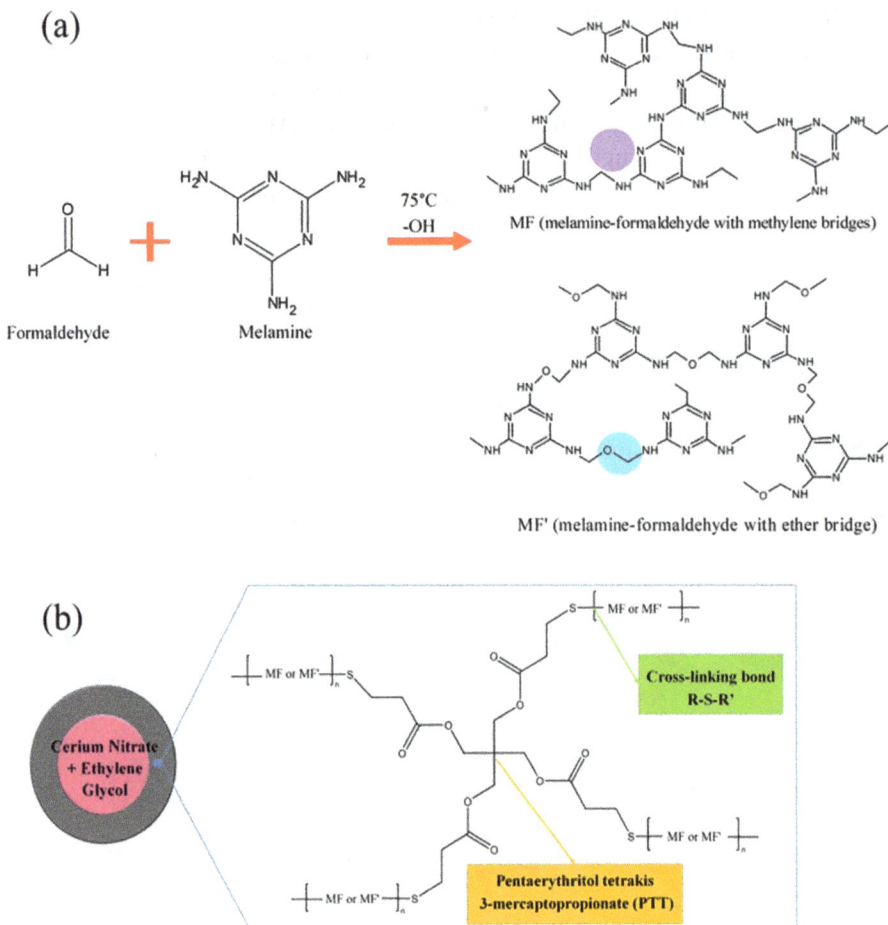

Figure 12.16: (a) Depiction of melamine reacting with formaldehyde producing major product melamine-formaldehyde (MF) and the minor product MF'. (b) Melamine-formaldehyde microcapsule structure and chemical formula (reproduced with permission from ref. [75]).

Incorporating 10 wt% microcapsules greatly improved the anti-corrosion capability of the epoxy coatings, according to the EIS data. This is explained by the fact that cerium ions generated by melamine-formaldehyde microcapsules prevent corrosion. The pH-sensitive pentaerythritol-tetrakis 3-mercaptopropionate (PTT) and melamine-formaldehyde matrix, which make up the microcapsule shell can diminish the cross-linking connections and change them into a thiol structure in an alkaline environment. In response to the stimulation, cross-linked structures in the microcapsule shell were chemically cleaved, allowing the core material to be released. The release mechanism of the pH-sensitive melamine-formaldehyde microcapsules and their operation inside epoxy coatings are schematically shown in Figure 12.17.

Cerium ions discharged from pH-responsive microcapsules work as a self-healing material as the development of insoluble cerium oxide/hydroxide restricts the propagation of corrosion reaction on the Al surfaces.

Taheri et al. propose that the incorporation of dual-functional materials into microcapsules as a self-healing agent permits the early identification of corrosion before the biggest degradation of the metal surface takes place [76]. They used the direct emulsion method to oxidatively polymerize aniline to produce PANI microcapsules in their research. The 8-hydroxyquinoline (8-HQ), a corrosion inhibitor that may also behave as a fluorescent indicator of Fe ions was present in the core of the microcapsules. As a luminous indicator 8-HQ increases the coating's corrosion resistance by producing a fluorescent complex and is particularly sensitive to the Fe ions produced during anodic activities. By using EIS and a salt spray test, the epoxy coating's anti-corrosion performance was assessed. Furthermore, the effectiveness of 8-HQ's detection and inhibition was assessed using an ICP Spectrometer and a fluorescence microscope. An illustration of the passivation mechanism of iron surfaces by PANI coating is shown in Figure 12.18. When the corrosion process starts and the coating is damaged (vide Figure 12.18 (a)), Fe atoms gets oxidized to Fe^{2+} ions and electrons are liberated. At the interface of the substrate and PANI coating, the conductive PANI (ES) is then transformed into the nonconductive PANI (EB). The recurrent reduction–oxidation cycle therefore results in the formation of a protective and thick covering formed of metal oxide [77]. Figure 12.18 (b) attributes the formation of passive film to reduction-oxidation processes. Additionally, PANI coatings may serve as a barrier against corrosive substances. The formation of a passive layer, the large impedance of EB, and the division of the cathodic and anodic operations by conductive PANI are all possibilities for the barrier protection process of PANI. PANI microcapsules additionally discharge the corrosion inhibitor by rupturing the coating. As shown in Figure 12.18 (c), 8-HQ, a smart corrosion indicator, may interact with Fe ions produced during anodic processes to create a fluorescent complex within the damaged areas. The development of a fluorescent complex repairs the fracture and shields the substrate. As seen in Figure 12.18 (d), both PANI microcapsules and encapsulated 8-HQ promote the ability of the injured region to heal on its own and protect the substrate from corrosive substances.

By incorporating fluorescence-labelled self-healing microcapsules, Wang et al. have observed the self-healing behaviors of microcracks in cement-based materials [78]. The self-healing microcapsules with fluorescent labels were incorporated into the resin matrix using the intercalation technique. Microcracks may be effectively re-

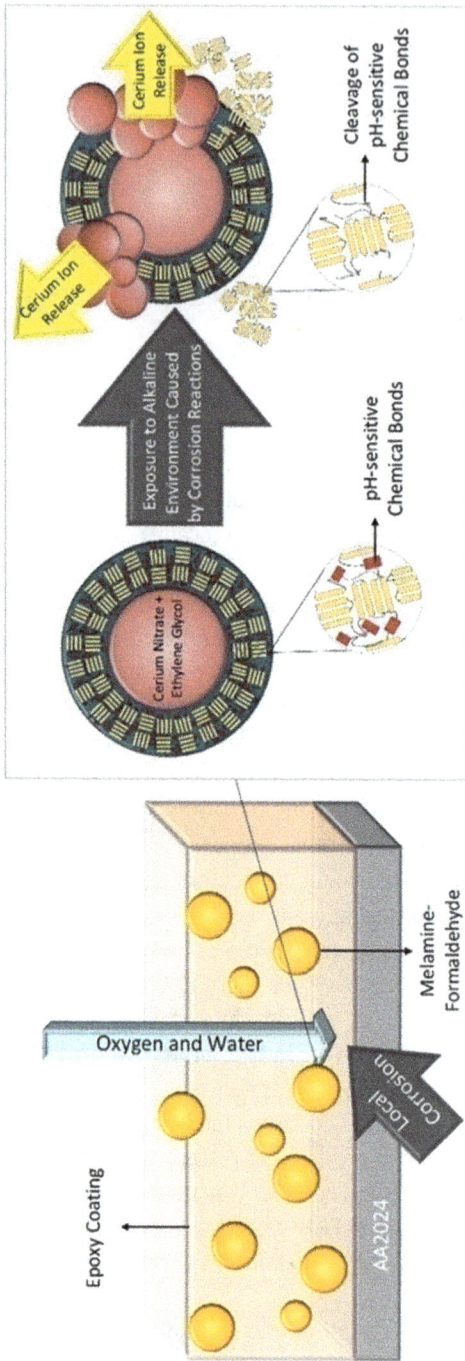

Figure 12.17: Schematic representation of the pH-responsive melamine-formaldehyde release behaviour (reproduced with permission from ref. [75]).

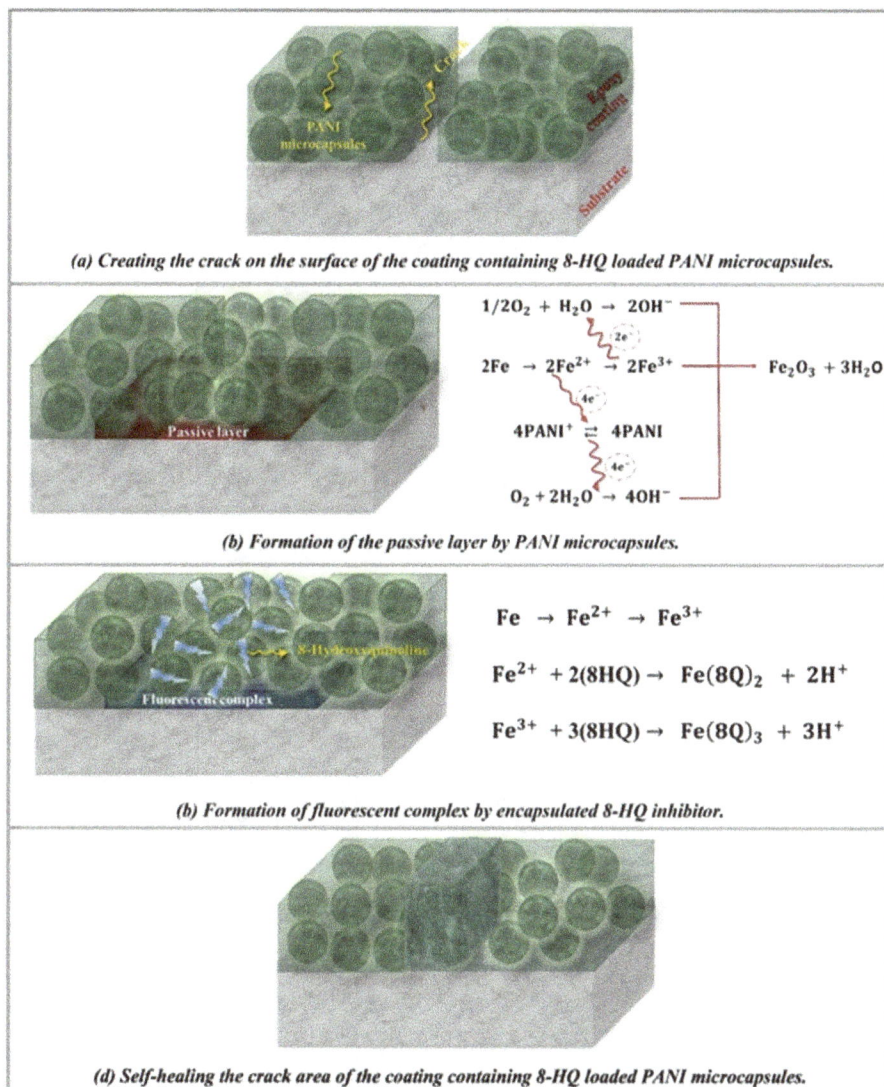

(a) *Creating the crack on the surface of the coating containing 8-HQ loaded PANI microcapsules.*

$$1/2O_2 + H_2O \rightarrow 2OH^-$$

$$2Fe \rightarrow 2Fe^{2+} \rightarrow 2Fe^{3+}$$

$$Fe_2O_3 + 3H_2O$$

$$4PANI^+ \rightleftharpoons 4PANI$$

$$O_2 + 2H_2O \rightarrow 4OH^-$$

(b) *Formation of the passive layer by PANI microcapsules.*

$$Fe \rightarrow Fe^{2+} \rightarrow Fe^{3+}$$

$$Fe^{2+} + 2(8HQ) \rightarrow Fe(8Q)_2 + 2H^+$$

$$Fe^{3+} + 3(8HQ) \rightarrow Fe(8Q)_3 + 3H^+$$

(b) *Formation of fluorescent complex by encapsulated 8-HQ inhibitor.*

(d) *Self-healing the crack area of the coating containing 8-HQ loaded PANI microcapsules.*

Figure 12.18: Diagram illustrating how 8-HQ-loaded PANI microcapsules operated at the crack site after their dissolution in 3.5 wt% NaCl solution (reproduced with permission from ref. [76]).

paired using fluorescence-labelled self-healing microcapsules, which also allow for the visualization of the repair procedure. Using a laser scanning confocal microscope (LSCM) and step profiler, the healing of interior and exterior microcracks were examined and quantified. Once the fluorescence-labelled microcapsules rupture, there is a precise relationship between the different microcapsules, and the core substance flows out gradually until it is used up. According to Figure 12.19, when the light illuminates into a fluorescent probe, the fluorescent probe absorbs the light energy, which causes some of the electrons around its nucleus to change from their original orbit to a higher-energy orbit and enter the excited as well as unstable state. The energy will then be released as light when the fluorescent probe de-excites and returns to the ground state by emitting light that has a larger wavelength than the incoming light.

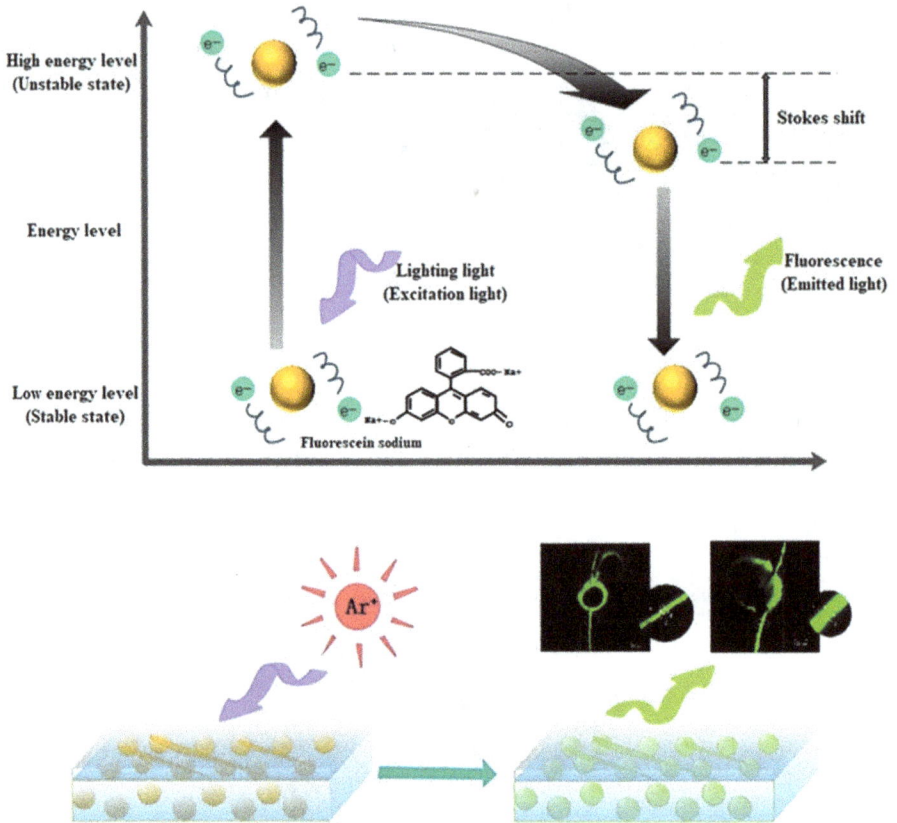

Figure 12.19: Principle diagram of fluorescence (reproduced with permission from ref. [78]).

In the resin matrix intercalated with fluorescence-labelled self-healing microcapsules, the microcracks in the microcapsules allow the core component to flow into the microcracks. The fluorescently connected probes in the microcapsules generate fluorescence, and the laser-emitting LSCM collects this fluorescence and shows fluorescence images of the small cracks throughout the matrix at various intervals. It has been demonstrated that the fluorescein sodium-labelled core material slowly diffuses out to seal up microcracks. The core material that has been associated with fluorescein sodium still has high matrix flexibility and can fill and repair micro cracks. According to Figure 12.20, when cement-based substances are coupled with fluorescently labelled self-healing microcapsules, the fluid inside the epoxy resin core component will spread out along the cracks and ultimately fill them. The microcapsules will break as a result of fractures in the cement materials. The surface height of fractures in cement-based substances that have healed to varying degrees is measured using a step profiler to determine the amount of microcracks at different stages.

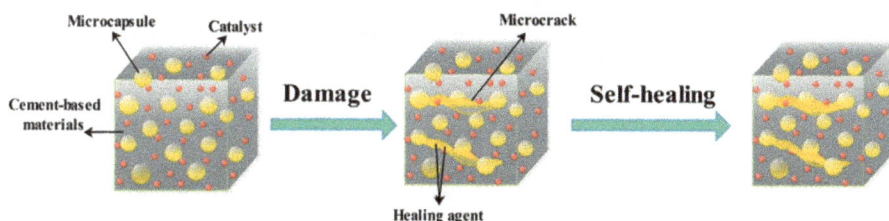

Figure 12.20: Diagram demonstrating the self-healing mechanism for cement-based substances (reproduced with permission from ref. [78]).

The epoxy resins cross-linking process may cause the microcapsules to harden, generating strength, filling in microcracks and allowing for the preliminary assessment of the depth of the microcracks. Cheng et al. have showcased an improved epoxy coating delivery method with self-healing capabilities inspired by mussels [79]. In that work, self-healing and anti-corrosion performances for carbon steel were achieved by putting the corrosion inhibitor BTA in polydopamine (PDA) microcapsules' shell structure and then embedding it in epoxy matrices. With the help of PDA-Fe^{3+}'s chelating activity and the fast dissipation of BTA in an acidic medium brought on by local corrosion, protective layers can form on bare steel surfaces, providing coatings the potential to cure themselves. Herein, Fig. 12.21 illustrates an easy and reliable method of producing PDA-based microcapsules by using dopamine's oxidative self-polymerization on the SiO$_2$ surface and then removing the template cores. Then, using the adhesive affinity between BTA and catechol groups in PDA, BTA was applied to the shells of microcapsules.

Furthermore, a detailed schematic illustration of the self-healing protective mechanism of PMB/EP coatings has been depicted in Figure 12.22.

When preparing a coating, certain microcracks and micropores are unavoidably created. Through these flaws, corrosive media progressively enter the coating matrix and start to cause metal corrosion. The developed microcapsules have good epoxy-coating dispersion, which seals the pores and significantly reduces the entry route for the corrosive electrolyte medium. An acidic environment promotes BTA release from PMB, according to UV-vis studies. As a result of the collaborative inhibitory action between BTA and PDA, the loading system immediately produces a twofold protective effect in response to the pH trigger brought on by localized corrosion. In order to

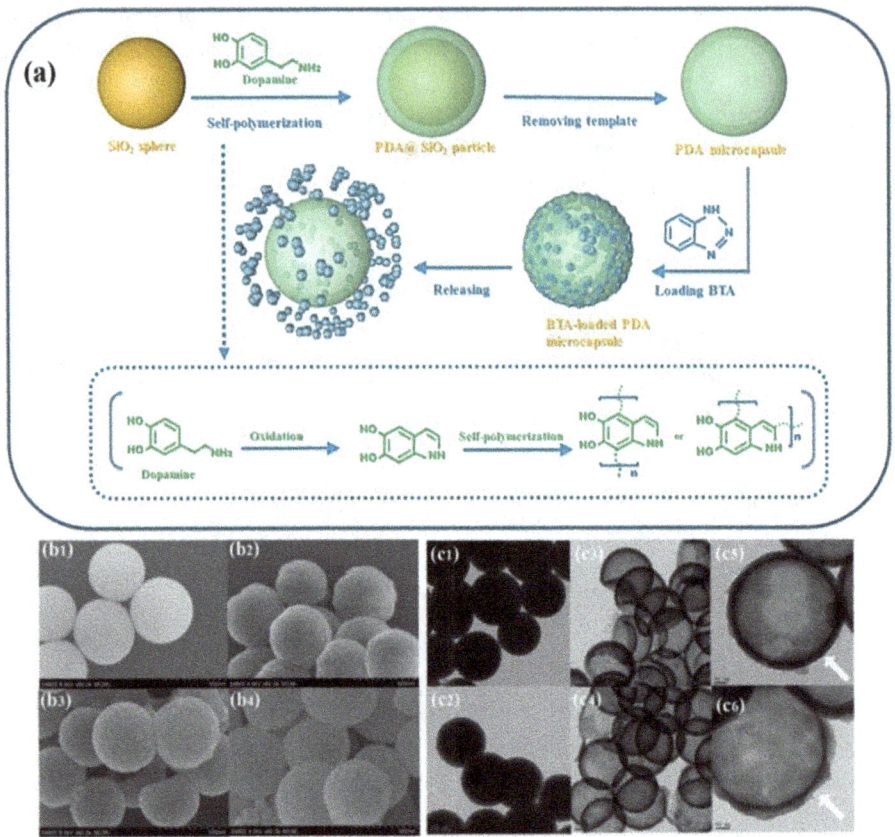

Figure 12.21: (a) The PDA container production process and the BTA loading technique; (b, c) Images of SiO$_2$ nanoparticle from SEM and TEM (b1, c1), PDA@SiO$_2$ particles (b2, c2), PMs (b3, c3) and PMBs (b4, c4); (c5, c6) are the enlargements of (c3, c4), respectively (reproduced with permission from ref. [79]).

block the diffusion of corrosion-causing substances, it can mix with Fe^{2+} and settle on a corrosion pit as shown in eq. (12.1).

$$Fe^{2+} + 2BTA \rightarrow BTAFeBTA + 2H^+ \qquad (12.1)$$

Additionally, by reacting directly with Fe, BTA could form an effective barrier at the non-corroded region as shown in eq. (12.2).

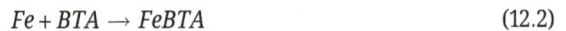

$$Fe + BTA \rightarrow FeBTA \qquad (12.2)$$

It is appropriate to note that, in comparison to pure epoxy coating, epoxy coating boosted by PDA-based microcapsules demonstrates strong corrosion protection and good self-healing ability.

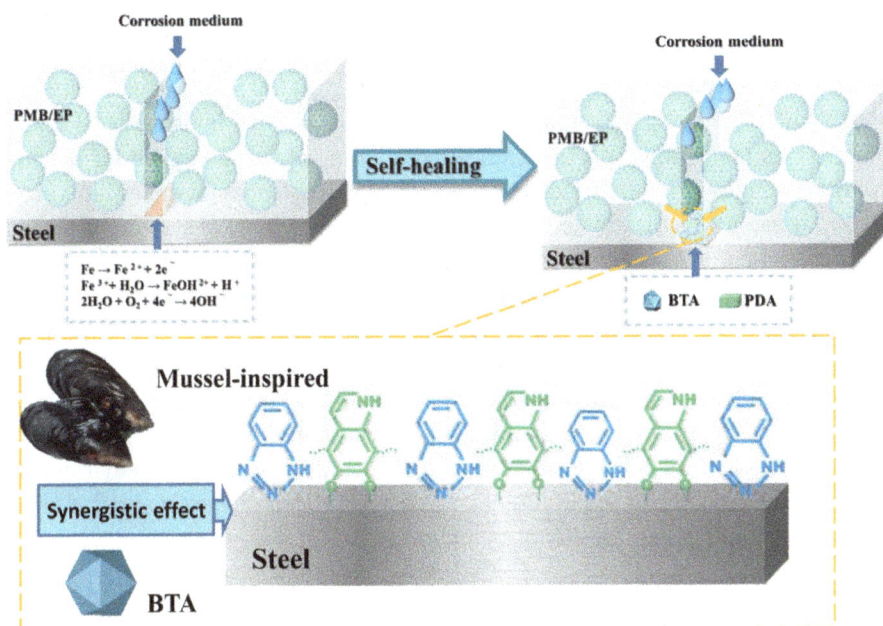

Figure 12.22: Molecular mechanisms of self-healing in epoxy coatings containing PMB (reproduced with permission from ref. [79]).

12.5 Conclusion and future perspectives

Self-healing coatings based on micro/nanocontainers are research fields with significant theoretical and practical application potential, from the production of carriers and active ingredients to the development of sustainable organic coating systems. Corrosion or other degradation processes cause self-healing coatings to release the corresponding active agents in response to environmental or mechanical attacks/stress. The active chemicals in self-healing coatings are required to be loaded appropriately in order to enhance their shelf-life and sensitivity. The concept and production of multifunctional organic, inorganic, or organic–inorganic hybrid containers demonstrated possibilities for the future of smart coatings. Under various stimuli, they execute self-healing functions. This kind of self-healing coating has even more potential, which fosters rapid advancement of "green and smart" corrosion inhibitors.

This chapter outlined the recent development of intelligent anti-corrosion coatings over the past few years. These anti-corrosive coatings are embedded with active substances like corrosion inhibitors in a variety of containers with varying architectures and forms. Additionally, a thorough summary of literature revealed the stimulus-responsive self-healing processes for prolonging the anti-corrosion life of coatings.

In addition to self-healing coating, there are lot of interests in exploring self-reporting corrosion-protective coatings owing to autonomous response to recognize impending structural problems instantly. The corrosion-sensing coatings connected to various metal substrates have been revealed through corrosion-related fluorescence or color indicators as the sensing species. It suggested that an alternate method for demonstrating coating deterioration and promptly alerting the potential for under film corrosion is to use mechanically activated indicators. These coating systems provided effective and non-destructive ways to identify coating degradation as well as corrosion initiation using visualization indications. It is useful to avert future structural collapse and increase the safety as well as the longevity of protective coatings. Despite the enormous advancements in the field of research, there are still many issues that need to be resolved at this time. The corrosion-detecting species are first and foremost material-specific. It may not function well on all metal surfaces. Therefore, it is very important to use novel indicators that react to diverse corrosion-induced environmental changes. Extending the application domains of corrosion-sensing coatings also depends on improving color sensitivity and intensity.

References

[1]   Chen Z, Scharnagl N, Zheludkevich ML, Ying H, Yang W. Micro/nanocontainer-based intelligent coatings: Synthesis, performance and applications – A review. Chem Eng J [Internet] 2023;451:138582. Available from: https://linkinghub.elsevier.com/retrieve/pii/S1385894722040633.

[2]   Carneiro J, Tedim J, Fernandes SCM, Freire CSR, Gandini A, Ferreira MGS, et al. Functionalized chitosan-based coatings for active corrosion protection. Surf Coat Technol [Internet] 2013;226:51–9. Available from: https://linkinghub.elsevier.com/retrieve/pii/S0257897213002983.

[3]   Zheludkevich ML, Tedim J, Ferreira MGS. "Smart" coatings for active corrosion protection based on multi-functional micro and nanocontainers. Electrochim Acta [Internet] 2012;82:314–23. Available from: https://linkinghub.elsevier.com/retrieve/pii/S0013468612006652.

[4]   Bockris JO, Conway BE. Hydrogen overpotential and the partial inhibition of the corrosion of iron. J Phys Colloid Chem [Internet] 1949;53(4):527–39. Available from. https://pubs.acs.org/doi/abs/10.1021/j150469a009.

[5]   Reeves NJ, Mann S. Influence of inorganic and organic additives on the tailored synthesis of iron oxides. J Chem Soc Faraday Trans [Internet] 1991;87(24):3875. Available from: http://xlink.rsc.org/?DOI=ft9918703875.

[6]   El Ibrahimi B, Jmiai A, Bazzi L, El Issami S. Amino acids and their derivatives as corrosion inhibitors for metals and alloys. Arab J Chem [Internet] 2020;13(1):740–71. Available from: https://linkinghub.elsevier.com/retrieve/pii/S1878535217301430.

[7]   Singh A, Ansari KR, Bedi P, Pramanik T, Ali IH, Lin Y, et al. Understanding xanthone derivatives as novel and efficient corrosion inhibitors for P110 steel in acidizing fluid: Experimental and theoretical studies. J Phys Chem Solids [Internet] 2023;172:111064. Available from: https://linkinghub.elsevier.com/retrieve/pii/S0022369722004814.

[8]   McCafferty E. Introduction to Corrosion Science [Internet], New York, NY: Springer New York, 2010, Available from: http://link.springer.com/10.1007/978-1-4419-0455-3.

[9]   Zhang F, Ju P, Pan M, Zhang D, Huang Y, Li G, et al. Self-healing mechanisms in smart protective coatings: A review. Corros Sci [Internet] 2018;144:74–88. Available from: https://linkinghub.elsevier.com/retrieve/pii/S0010938X17322308.

[10] Saji VS. Supramolecular concepts and approaches in corrosion and biofouling prevention. Corros Rev [Internet] 2019;37(3):187–230. Available from: https://www.degruyter.com/document/doi/10.1515/corrrev-2018-0105/html.

[11] White SR, Sottos NR, Geubelle PH, Moore JS, Kessler MR, Sriram SR, et al. Autonomic healing of polymer composites. Nature [Internet] 2001;409(6822):794–7. Available from: https://www.nature.com/articles/35057232.

[12] Ouarga A, Lebaz N, Tarhini M, Noukrati H, Barroug A, Elaissari A, et al. Towards smart self-healing coatings: Advances in micro/nano-encapsulation processes as carriers for anti-corrosion coatings development. J Mol Liq [Internet] 2022;354:118862. Available from: https://linkinghub.elsevier.com/retrieve/pii/S0167732222004007.

[13] Hwang J-S, Kim J-N, Wee Y-J, Yun J-S, Jang H-G, Kim S-H, et al. Preparation and characterization of melamine-formaldehyde resin microcapsules containing fragrant oil. Biotechnol Bioprocess Eng [Internet] 2006;11(4):332–6. Available from: http://link.springer.com/10.1007/BF03026249.

[14] Kage H, Kawahara H, Hamada N, Kotake T, Ogura H. Operating conditions and microcapsules generated by in situ polymerization. Adv Powder Technol [Internet] 2002;13(3):265–85. Available from: https://linkinghub.elsevier.com/retrieve/pii/S0921883108600679.

[15] Nguon O, Lagugné-Labarthet F, Brandys FA, Li J, Gillies ER. Microencapsulation by in situ Polymerization of Amino Resins. Polym Rev [Internet] 2018;58(2):326–75. Available from: https://www.tandfonline.com/doi/full/10.1080/15583724.2017.1364765.

[16] Arshady R. Suspension, emulsion, and dispersion polymerization: A methodological survey. Colloid Polym Sci Internet 1992;270(8):717–32. Available from: http://link.springer.com/10.1007/BF00776142.

[17] Burhans FV, RW. Process of producing hollow particle and resulting product. US; US Patent 2797201, 1957.

[18] Brown EN, Kessler MR, Sottos NR, White SR. In situ poly(urea-formaldehyde) microencapsulation of dicyclopentadiene. J Microencapsul [Internet] 2003;20(6):719–30. Available from: http://www.tandfonline.com/doi/full/10.3109/02652040309178083.

[19] Jialan Y, Chenpeng Y, Chengfei Z, Baoqing H. Preparation process of epoxy resin microcapsules for self-healing coatings. POC [Internet] 2019;132:440–4. Available from: https://linkinghub.elsevier.com/retrieve/pii/S0300944018311895.

[20] Tan Z, Chen S, Peng X, Zhang L, Gao C. Polyamide membranes with nanoscale Turing structures for water purification. Science (80-) [Internet] 2018;360(6388):518–21. Available from: https://www.science.org/doi/10.1126/science.aar6308.

[21] Chen W, Liu X, Lee DW. Fabrication and characterization of microcapsules with polyamide–polyurea as hybrid shell. J Mater Sci [Internet] 2012;47(4):2040–4. Available from: http://link.springer.com/10.1007/s10853-011-6004-8.

[22] Ichiura H, Morikawa M, Ninomiya J. Preparation of smart paper part I – formation of nylon microcapsules on paper surface using interfacial polymerization. J Mater Sci [Internet] 2006;41 (21):7019–24. Available from. http://link.springer.com/10.1007/s10853-006-0789-x.

[23] Zhang L, Liu P, Ju L, Wang L, Zhao S. Polypyrrole nanocapsules via interfacial polymerization. Macromol Res [Internet] 2010;18(7):648–52. Available from. http://link.springer.com/10.1007/s13233-010-0713-8.

[24] Raaijmakers MJT, Benes NE. Current trends in interfacial polymerization chemistry. Prog Polym Sci [Internet] 2016;63:86–142. Available from: https://linkinghub.elsevier.com/retrieve/pii/S0079670016300363.

[25] Zhang Y, Rochefort D. Characterisation and applications of microcapsules obtained by interfacial polycondensation. J Microencapsul [Internet] 2012;29(7):636–49. Available from: http://www.tandfon line.com/doi/full/10.3109/02652048.2012.676092.

[26] Kobašlija M, McQuade DT. Polyurea microcapsules from oil-in-oil emulsions via interfacial polymerization. Macromolecules [Internet] 2006;39(19):6371–5. Available from: https://pubs.acs.org/doi/10.1021/ma061455x.

[27] Bakry AM, Abbas S, Ali B, Majeed H, Abouelwafa MY, Mousa A, et al. Microencapsulation of oils: A comprehensive review of benefits, techniques, and applications. Compr Rev Food Sci Food Saf [Internet] 2016;15(1):143–82. Available from: https://onlinelibrary.wiley.com/doi/10.1111/1541-4337.12179.

[28] Low LE, Siva SP, Ho YK, Chan ES, Tey BT. Recent advances of characterization techniques for the formation, physical properties and stability of Pickering emulsion. Adv Colloid Interface Sci [Internet] 2020;277:102117. Available from: https://linkinghub.elsevier.com/retrieve/pii/S0001868619304336.

[29] Grigoriev DO, Haase MF, Fandrich N, Latnikova A, Shchukin DG. Emulsion route in fabrication of micro and nanocontainers for biomimetic self-healing and self-protecting functional coatings. Bioinspired Biomim Nanobiomater [Internet] 2012;1(2):101–16. Available from: https://www.icevirtual library.com/doi/10.1680/bbn.11.00017.

[30] McClements DJ, Decker EA, Park Y, Weiss J. Structural Design Principles for Delivery of Bioactive Components in Nutraceuticals and Functional Foods. Crit Rev Food Sci Nutr [Internet] 2009;49 (6):577–606. Available from. http://www.tandfonline.com/doi/abs/10.1080/10408390902841529.

[31] Ashraf MA, Khan AM, Ahmad M, Sarfraz M. Effectiveness of silica based sol-gel microencapsulation method for odorants and flavors leading to sustainable environment. Front Chem [Internet] 2015;3: Available from: http://journal.frontiersin.org/Article/10.3389/fchem.2015.00042/abstract.

[32] Allain LR, Sorasaenee K, Xue Z. Doped thin-film sensors via a sol–gel process for high-acidity determination. Anal Chem [Internet] 1997;69(15):3076–80. Available from: https://pubs.acs.org/doi/10.1021/ac970258g.

[33] Lee H-Y, Kim J, Hwang C-H, Kim H-E, Jeong S-H. Strategy for preparing mechanically strong hyaluronic acid-silica nanohybrid hydrogels via in situ sol-gel process. Macromol Mater Eng [Internet] 2018;303(9):1800213. Available from: https://onlinelibrary.wiley.com/doi/10.1002/mame.201800213.

[34] Zacchi P, Pietsch A, Voges S, Ambrogi A, Eggers R, Jaeger P. Concepts of phase separation in supercritical processing. Chem Eng Process Process Intensif [Internet] 2006;45(9):728–33. Available from. https://linkinghub.elsevier.com/retrieve/pii/S0255270106000626.

[35] Jung J, Perrut M. Particle design using supercritical fluids: Literature and patent survey. J Supercrit Fluids [Internet] 2001;20(3):179–219. Available from: https://linkinghub.elsevier.com/retrieve/pii/S089684460100064X.

[36] Chattopadhyay P, Huff R, Shekunov BY. Drug Encapsulation Using Supercritical Fluid Extraction of Emulsions. J Pharm Sci 2006;95(3):667–79. Available from: https://linkinghub.elsevier.com/retrieve/pii/S0022354916319669.

[37] Meziani MJ, Pathak P, Sun Y-P. Supercritical fluid technology for nanotechnology in drug delivery. In: Nanotechnology in Drug Delivery [Internet] (pp. 69–104). New York, NY: Springer New York; 2009. Available from: http://link.springer.com/10.1007/978-0-387-77668-2_3.

[38] Gao H, Hu G, Liu K, Wu L. Preparation of waterborne dispersions of epoxy resin by ultrasonic-assisted supercritical CO2 nanoemulsification technique. Ultrason Sonochem [Internet] 2017;39:520–7. Available from: https://linkinghub.elsevier.com/retrieve/pii/S1350417717302547.

[39] Chemtob C, Chaumeil JC, N'Dongo M. Microencapsulation by ethylcellulose phase separation : Microcapsule characteristics. Int J Pharm [Internet] 1986;29(1):1–7. Available from: https://linkinghub.elsevier.com/retrieve/pii/0378517386901936.

[40] Lazko J, Popineau Y, Legrand J. Soy glycinin microcapsules by simple coacervation method. Coll Surf B: Biointerfac [Internet] 2004;37(1-2):1–8. Available from: https://linkinghub.elsevier.com/retrieve/pii/S0927776504001596.

[41] Thies C. Microcapsules. In: Encyclopedia of Food Sciences and Nutrition [Internet] (pp. 3892–903), Elsevier; 2003. Available from: https://linkinghub.elsevier.com/retrieve/pii/B012227055X013699.

[42] Andersson Trojer M, Ananievskaia A, Gabul-Zada AA, Nordstierna L, Blanck H. Polymer core-polymer shell particle formation enabled by ultralow interfacial tension via internal phase separation: morphology prediction using the van Oss formalism. Colloid Interface Sci Commun [Internet] 2018;25:36–40. Available from: https://linkinghub.elsevier.com/retrieve/pii/S221503821830044X.

[43] Wang Y, Guo B-H, Wan X, Xu J, Wang X, Zhang Y-P. Janus-like polymer particles prepared via internal phase separation from emulsified polymer/oil droplets. Polymer (Guildf) [Internet] 2009;50 (14):3361–9. Available from: https://linkinghub.elsevier.com/retrieve/pii/S0032386109002407.

[44] Wang Y, Zhou J, Guo X, Hu Q, Qin C, Liu H, et al. Layer-by-layer assembled biopolymer microcapsule with separate layer cavities generated by gas-liquid microfluidic approach. Mater Sci Eng C [Internet] 2017;81:13–9. Available from: https://linkinghub.elsevier.com/retrieve/pii/S092849311731216X.

[45] Akyuz L, Sargin I, Kaya M, Ceter T, Akata I. A new pollen-derived microcarrier for pantoprazole delivery. Mater Sci Eng C [Internet] 2017;71:937–42. Available from: https://linkinghub.elsevier.com/retrieve/pii/S0928493116309845.

[46] An S, Lee MW, Yarin AL, Yoon SS. A review on corrosion-protective extrinsic self-healing: Comparison of microcapsule-based systems and those based on core-shell vascular networks. Chem Eng J [Internet] 2018;344:206–20. Available from: https://linkinghub.elsevier.com/retrieve/pii/S1385894718303942.

[47] Bah MG, Bilal HM, Wang J. Fabrication and application of complex microcapsules: a review. Soft Matter [Internet] 2020;16(3):570–90. Available from. http://xlink.rsc.org/?DOI=C9SM01634A.

[48] Munin A, Edwards-Lévy F. Encapsulation of natural polyphenolic compounds; a review. Pharmaceutics [Internet] 2011;3(4):793–29. Available from: http://www.mdpi.com/1999-4923/3/4/793.

[49] Shirokawa K, Taguchi Y, Yokoyama H, Ono F, Tanaka M. Preparation of Temperature and Water Responsive Microcapsules Containing Hydroquinone with Spray Drying Method. J Cosmet Dermatological Sci Appl [Internet] 2013;03(03):49–54. Available from: http://www.scirp.org/journal/doi.aspx?DOI=10.4236/jcdsa.2013.33A2012.

[50] Kosaraju SL, D'ath L, Lawrence A. Preparation and characterisation of chitosan microspheres for antioxidant delivery. Carbohydr PolymInternet 2006;64(2):163–7. Available from: https://linkinghub.elsevier.com/retrieve/pii/S0144861705005473.

[51] Soottitantawat A, Takayama K, Okamura K, Muranaka D, Yoshii H, Furuta T, et al. Microencapsulation of l-menthol by spray drying and its release characteristics. Innovative Food Sci Emerg Technol [Internet] 2005;6(2):163–70. Available from: https://linkinghub.elsevier.com/retrieve/pii/S1466856404001134.

[52] Poncelet D. Microencapsulation: fundamentals, methods and applications. In: Surface Chemistry in Biomedical and Environmental Science [Internet] (pp. 23–34). Springer Netherlands. Available from: http://link.springer.com/10.1007/1-4020-4741-X_3.

[53] Green MD, D'Souza MJ, Holbrook JM, Wirtz RA. In vitro and in vivo evaluation of albumin-encapsulated primaquine diphosphate prepared by nebulization into heated oil. J Microencapsul [Internet] 2004;21(4):433–44. Available from. http://www.tandfonline.com/doi/full/10.1080/02652040410001729232.

[54] Shchukin DG, Zheludkevich M, Yasakau K, Lamaka S, Ferreira MGS, Möhwald H. Layer-by-layer assembled nanocontainers for self-healing corrosion protection. Adv Mater [Internet] 2006;18 (13):1672–8. Available from: https://onlinelibrary.wiley.com/doi/10.1002/adma.200502053.

[55]   Chen T, Chen R, Jin Z, Liu J. Engineering hollow mesoporous silica nanocontainers with molecular switches for continuous self-healing anticorrosion coating. J Mater Chem A [Internet] 2015;3 (18):9510–6. Available from: http://xlink.rsc.org/?DOI=C5TA01188D.

[56]   Borisova D, Akçakayıran D, Schenderlein M, Möhwald H, Shchukin DG. Nanocontainer-based anticorrosive coatings: effect of the container size on the self-healing performance. Adv Funct Mater [Internet] 2013;23(30):3799–812. Available from: https://onlinelibrary.wiley.com/doi/10.1002/adfm.201203715.

[57]   Coulibaly S, Roulin A, Balog S, V. BM, Foster EJ, Rowan SJ, et al. Reinforcement of optically healable supramolecular polymers with cellulose nanocrystals. Macromolecules [Internet] 2014;47(1):152–60. Available from: https://pubs.acs.org/doi/10.1021/ma402143c.

[58]   Postiglione G, Turri S, Levi M. Effect of the plasticizer on the self-healing properties of a polymer coating based on the thermoreversible Diels–Alder reaction. POC [Internet] 2015;78:526–31. Available from: https://linkinghub.elsevier.com/retrieve/pii/S0300944014001945.

[59]   Wang Z, Yang Y, Burtovyy R, Luzinov I, Urban MW. UV-induced self-repairing polydimethylsiloxane–polyurethane (PDMS–PUR) and polyethylene glycol–polyurethane (PEG–PUR) Cu-catalyzed networks. J Mater Chem A [Internet] 2014;2(37):15527. Available from: http://xlink.rsc.org/?DOI=C4TA02417F.

[60]   Liu X, Zhang H, Wang J, Wang Z, Wang S. Preparation of epoxy microcapsule based self-healing coatings and their behavior. Surf Coat Technol [Internet] 2012;206(23):4976–80. Available from: https://linkinghub.elsevier.com/retrieve/pii/S0257897212005488.

[61]   Liao LP, Zhang W, Zhao Y. Preparation and healing property evaluation of self-repairing polymer coating. Surf Eng [Internet] 2014;30(2):138–41. Available from: http://www.tandfonline.com/doi/full/10.1179/1743294413Y.0000000216.

[62]   Wang W, Xu L, Li X, Yang Y, An E. Self-healing properties of protective coatings containing isophorone diisocyanate microcapsules on carbon steel surfaces. Corros Sci [Internet] 2014;80:528–35. Available from: https://linkinghub.elsevier.com/retrieve/pii/S0010938X13005325.

[63]   Song Y-K, Chung C-M. Repeatable self-healing of a microcapsule-type protective coating. Polym Chem [Internet] 2013;4(18):4940. Available from: http://xlink.rsc.org/?DOI=c3py00102d.

[64]   Huang M, Zhang H, Yang J. Synthesis of organic silane microcapsules for self-healing corrosion resistant polymer coatings. Corros Sci [Internet] 2012;65:561–6. Available from: https://linkinghub.elsevier.com/retrieve/pii/S0010938X12003757.

[65]   Li H, Cui Y, Wang H, Zhu Y, Wang B. Preparation and application of polysulfone microcapsules containing tung oil in self-healing and self-lubricating epoxy coating. Colloids Surfaces A Physicochem Eng Asp [Internet]. 2017;518:181–7. Available from: https://linkinghub.elsevier.com/retrieve/pii/S0927775717300687.

[66]   Li H, Feng Y, Cui Y, Ma Y, Zheng Z, Qian B, et al. Polyurea/polyaniline hybrid shell microcapsules loaded with isophorone diisocyanate for synergetic self-healing coatings. Prog Org Coatings [Internet] 2020;145:105684. Available from: https://linkinghub.elsevier.com/retrieve/pii/S0300944020301867.

[67]   Gupta G, Birbilis N, Cook AB, Khanna AS. Polyaniline-lignosulfonate/epoxy coating for corrosion protection of AA2024-T3. Corros Sci [Internet] 2013;67:256–67. Available from: https://linkinghub.elsevier.com/retrieve/pii/S0010938X12005069.

[68]   Grgur BN, Elkais AR, Gvozdenović MM, Sž D, Trišović TL, Jugović BZ. Corrosion of mild steel with composite polyaniline coatings using different formulations. Prog Org Coatings [Internet] 2015;79:17–24. Available from: https://linkinghub.elsevier.com/retrieve/pii/S0300944014003427.

[69]   Shao Y, Huang H, Zhang T, Meng G, Wang F. Corrosion protection of Mg-5Li alloy with epoxy coatings containing polyaniline. Corros Sci [Internet] 2009;51(12):2906–15. Available from: https://linkinghub.elsevier.com/retrieve/pii/S0010938X09003783.

[70] Navarchian AH, Najafipoor N, Ahangaran F. Surface-modified poly(methyl methacrylate) microcapsules containing linseed oil for application in self-healing epoxy-based coatings. Prog Org Coatings [Internet] 2019;132:288–97. Available from: https://linkinghub.elsevier.com/retrieve/pii/S0300944018310798.

[71] Li J, Shi H, Liu F, Han E-H. Self-healing epoxy coating based on tung oil-containing microcapsules for corrosion protection. Prog Org Coatings [Internet] 2021;156:106236. Available from: https://linkinghub.elsevier.com/retrieve/pii/S0300944021001077.

[72] Dong J, Pan W, Luo J, Liu R. Synthesis of inhibitor-loaded polyaniline microcapsules with dual anti-corrosion functions for protection of carbon steel. Electrochim Acta [Internet] 2020;364:137299. Available from: https://linkinghub.elsevier.com/retrieve/pii/S0013468620316923.

[73] Wu K, Gui T, Dong J, Luo J, Liu R. Synthesis of robust polyaniline microcapsules via UV-initiated emulsion polymerization for self-healing and anti-corrosion coating. Prog Org Coatings [Internet] 2022;162:106592. Available from: https://linkinghub.elsevier.com/retrieve/pii/S030094402100463X.

[74] Wang X, Zhu J, Zou F, Zhou N, Li Y, Lei W. Ca-Al LDH hybrid self-healing microcapsules for corrosion protection. Chem Eng J [Internet] 2022;447:137125. Available from: https://linkinghub.elsevier.com/retrieve/pii/S138589472202616X.

[75] Ghahremani P, Sarabi AA, Roshan S. Cerium containing pH-responsive microcapsule for smart coating application: Characterization and corrosion study. Surf Coat Technol [Internet] 2021;427:127820. Available from: https://linkinghub.elsevier.com/retrieve/pii/S0257897221009944.

[76] Taheri N, Sarabi AA, Roshan S. Investigation of intelligent protection and corrosion detection of epoxy-coated St-12 by redox-responsive microcapsules containing dual-functional 8-hydroxyquinoline. Prog Org Coat [Internet]. 2022;172:107073. Available from: https://linkinghub.elsevier.com/retrieve/pii/S0300944022003708.

[77] Zhang Y, Shao Y, Zhang T, Meng G, Wang F. High corrosion protection of a polyaniline/organophilic montmorillonite coating for magnesium alloys. Prog Org Coat [Internet] 2013;76(5):804–11. Available from: https://linkinghub.elsevier.com/retrieve/pii/S0300944013000106.

[78] Wang X, Li Y, Zhang C, Zhang X. Visualization and quantification of self-healing behaviors of microcracks in cement-based materials incorporating fluorescence-labeled self-healing microcapsules. Constr Build Mater [Internet] 2022;315:125668. Available from: https://linkinghub.elsevier.com/retrieve/pii/S0950061821034036.

[79] Cheng L, Liu C, Wu H, Zhao H, Mao F, Wang L. A mussel-inspired delivery system for enhancing self-healing property of epoxy coatings. J Mater Sci Technol [Internet] 2021;80:36–49. Available from: https://linkinghub.elsevier.com/retrieve/pii/S1005030220310161.

Richika Ganjoo, Shveta Sharma, Praveen K. Sharma, Nancy George,
Abhinay Thakur, Ashish Kumar

13 Surface functionalized bio ceramic coatings for anti-corrosion performance

Abstract: Due to sports injuries, aging-related problems, accidents, and the need for revision surgery, the use of trauma fixation devices has substantially expanded in contemporary times. The human body's missing or damaged components have been replaced with a variety of materials, including ceramics, polymers, titanium, stainless steel, and Co-Cr alloy. After implantation, bodily fluids (Na^+, K^+, and Cl^-), proteins, and blood cells interact with the surface of metallic implants, favoring the release of ions from the metallic surface to nearby human tissues and causing a hypersensitive response. Body pH, temperature, and immune cell contact all contribute to metal ion leaching as well as loss of host cell interaction and efficient mineralization for greater durability. Additionally, microbial invasion is a significant problem that results in extracellular chemicals on the surface of the biomaterial, which the microbes use to evade the anti-microbial agents. Surface modification, which typically involves the use of chemical vapor deposition (CVD), physical vapor deposition (PVD), the sol–gel method, and electrochemical deposition, is a prerequisite for improving the mechanical, corrosion resistance, anti-microbial, and biocompatibility properties of materials. The surface functionalization of metallic implants is best suited to the characteristics of bio ceramics, such as chemical inertness, bioactivity, biocompatibility, and corrosion protection. As far as we are aware, there is not much literature on the subject of how pH, temperature, and body proteins interact with bio ceramic coatings. This chapter emphasizes is on the corrosion behaviour of several ceramic composite coating materials under various environmental circumstances. Potential developments for coatings in biomedical applications are described.

Keywords: Corrosion, Bio ceramic Coatings, Alloys, Electrochemical Impedance Spectroscopy, Hydroxyapatite (HAp).

13.1 Introduction

Metallic implantable materials have been employed for surgical and implant fixation in the implant industries for a few decades, including stainless steel(SS), titanium(Ti),Co-Cr, Ni-Cr, and Ti alloy [1, 2]. Triphasic Ti-6Al-4 V alloy is less ductile than commercially pure titanium when used in prosthodontics implants for teeth or crowns (bridges or dentures). Ti and its alloys are widely used in dental medicine [3]. Comparable to that, 316 L stainless steel, which is suitable for use in medicine, is a popular material for car-

https://doi.org/10.1515/9783111016160-013

diovascular stents, femoral implants, ankle fractures, etc. because of its cheap cost, simplicity of production, strong corrosion resistance, and mechanical qualities [4]. Despite being used in biomedical field, 316 L stainless steel and Ti-based alloys have minimal surface roughness, low abrasion resistance, and a high friction coefficient. Furthermore, implant endurance is negatively impacted by their propensity for localized corrosion inside the human body [5]. When discharged into the circulation or the tissues around an implant, nickel exhibits more toxicity than chromium or iron in 316 L stainless steel. Sachiko et al. investigated the carcinogenic and allergenic effects of nickel ions in rats in biological settings. The importance of biomedical materials has grown significantly as a result of more operations, implantations as people age, accidents, and health issues [6].

This chapter provides a comprehensive review of the state-of-the-art research on the anti-corrosion behaviour of several bio ceramic coatings in an in vitro physiological system for use in medical application. Past research on PVD-applied bio ceramic coatings is also examined, and its effects on biomechanical integrity, biocompatibility, and corrosion resistance performance are explored. Corrosion caused by microbes on bio ceramic surfaces and the significance of ceramic surface modification for efficient bone restoration are also discussed.

13.2 Importance of ceramic coatings

Surface tailoring is a necessary step in the process of simulating foreign substance as the host system. The modern implant industry has discovered a means to increase corrosion protection, biocompatibility, structural behaviour, and antibacterial behavior by modifying the surface of metallic implants with the appropriate materials. In order to replicate the in vivo environment and enhance optimum bio implants, surface-tailored implant materials are necessary for in vitro testing. The surface of desirable metallic implants has been customized using a variety of techniques, including electrodeposition, dip coating, sol–gel coating, physical vapor deposition (PVD), and chemical vapor deposition (CVD). The PVD process is one of the many ways to deposit the material and is a useful tool for achieving precise stoichiometry, coating density, strong adhesion ability, and homogeneity.

Ceramics including the following five important ceramic substances are categorized on the basis of macroscopic surface characteristics and chemical stability: zirconia, alumina, bioactive glass, glass- and calcium-based ceramics. The majority of ceramic materials are often employed for different body sections. Among the different applications, the following are some of the particular applications: (a) the chemically inert bio ceramic's surface on the rough surface has a strong bond with the human cells; (b) body cells will firmly adhere to ceramic surfaces after which a chemical bond will develop between the surface of the cell and the substance; and (c) the host bone will eventually replace the bio absorbable ceramic components as they combine with it over time. Ceramic materials are only useful in bulk for applications that solely apply compressive stresses due to their poor tensile strength and fracture toughness.

As per the ex vivo push-out test results, before coming into touch with the tissue/ ceramic bond, the metal/ceramic bond would collapse and this is acknowledged as vulnerability in the physical environment. Therefore, there is a good case for metal/ ceramic bonding using bulk ceramic materials given their dependability at the inter- face during the course of functional loading. Numerous surface-modifying materials, including polymers, ceramics, as well as metals and their alloys, have been employed for shock and replacement surgical applications in contemporary industries through- out the last few decades. Among the many surface-tailored materials, ceramics pro- vide a number of benefits, including being non-allergenic, chemically stable, resistant to corrosion, having a high mechanical strength, and having improved biocompatibil- ity. Due to its very excellent aesthetics, wear resistance, and improved biocompatibil- ity, ceramic implants have grown in favor among physicians and patients.

Dental implant applications employ Yttria-stabilized zirconia (YSZ), which offers exceptional toughness and mechanical strength. Yttria (Y_2O_3) was partly stabilized with zirconia as YSZ to stabilize the phase. Our earlier research included coating tita- nium samples with YSZ cubic phase and observing the biological reactions. El-Ghany and Huseinsherief exhaustively examined and proclaimed the applications of t-YSZ pertaining to dental implant and hip reconstructive surgery. In light of this, there is debate about whether ZrO_2 coatings – monoclinic, t-YSZ, and cubic – perform best [7]. These materials are not only functionally adequate but also aesthetically beautiful. Furthermore, transition metal nitrides, such as ZrN and TiN, have been widely em- ployed in implant, ornamental, and defensive coatings such as gold coatings for their attractive appearance [8].

13.2.1 Ceramic coatings: electrochemical behaviour

The performance of SS, and Ti and its alloys for usage in implants has been improved by considering various surface modification approaches to increase the corrosion resistance characteristic of implant substrates. PVD approaches have drawn a lot of attention among the different techniques owing to their strong adhesion capability, enhancement of defect-free films and strong packing density. Typically, PVD technologies were used to deposit ceramic materials such as ZrO_2, TiO_2, ZnO, CeO_2, etc. Using the electron beam evaporation (EBPVD) technique, they previously investigated the corrosion behaviour of H_2O_2 and lactic acid on a hydrophilic ZrO_2 ceramic-covered 316 L SS material. The experi- ment revealed that H_2O_2 and artificial blood plasma including lipoic acid (LA) provided good corrosion protection in ZrO_2 coating. As ceramic materials are chemically inert, they demonstrate superior corrosion resistance than uncoated or bare surfaces [9]. Jansen in- vestigated the use of magnetron flashing to produce ceramic coverings for implants that are biodegradable. In order to increase corrosion protection towards microbial infections, the anatase phase of TiO_2 covering on Ti material was synthesized via hydrothermal treatment [10]. In comparison to Ti surfaces, *E. coli*'s expression of the green fluorescent protein was reduced by over 50%. At pH 10 and in presence of tetramethylammonium hydroxide, Ti nanotubes significantly reduced bacterial adsorption in specimens that had been exposed to UV radiation [11].

The surface engineering method was used by Gurappa et al. to create a cheap and effective biomaterial that improved the corrosion resistance of 316 L stainless steel [12]. A magnetron sputtering technique was used by Rajan and colleagues to deposit thin film metallic glasses (TFMG) with various compositions [13]. They experimented with electrochemical corrosion in solutions that were made to mimic bodily fluids (SBF). In comparison to bare crystalline titanium (Ti) substrate, they discovered that TFMG coating composed of Zr maintains corrosion protection while having a lower penetration level. Using techniques for atmospheric plasma spraying, Wang et al. developed ceramic glass with coatings consisting of the elements sphene (SP) and hardystonite (HT). The coating had greater hardness and tensile strength [14].

Similar to this, Subramanian used a magnetron to sputter $Ti_{40}Cu_{36}Pd_{14}Zr_{10}$ TFMG while researching the material's corrosion, biocompatibility, and mechanical properties [15]. For electrochemical investigations, they compared materials with and without coatings in SBF solution and discovered that Ti-based TFMG coatings had greater corrosion resistance. Additionally, bio ceramics like HAp are used to prevent surface coatings from corroding metallic substrates. Two distinct kinds of ceramic coatings with a varied Ca/P ratio were created by Mohedano et al. for dental implant applications on titanium surfaces. By using plasma electrolyte oxidation, the bioactive coatings were applied on Ti (PEO) [16]. Before and after the samples were immersed in SBF at 37 °C for four weeks, and their shape and composition were analyzed. Similar ceramic coatings of bioactive silicate glass for use in orthopaedic and dentistry have been proven by Brunello et al. [16]. Additionally, coatings applied to samples of magnesium alloy were examined for their potential to serve as specialized protective barriers for magnesium alloy and retain mechanical integrity over an extended length of time [17]. YSZ-reinforced hydroxyapatite (HAp) was created by Yugeswaran et al. using a plasma spray process [18]. Meanwhile HAp has poor mechanical qualities; YSZ was mixed with HAp in various amounts (10, 20, and 30 wt%) before being placed. The generated layers of the HAp-YSZ complex were studied to comprehend the electrochemical properties in SBF medium. According to the corrosion study, the admixed HAp with a greater YSZ content demonstrated improved corrosion resistance to bare surfaces. Novel composites of zirconia and bioactive glass with exceptional mechanical durability and bioactivity were developed by Zhang et al. to create the perfect dental implant [19]. Bian et al. examined the impact of Gd and Zn on Mg alloy elements. They discovered that adding varied amounts of Gd to Mg alloys might increase their corrosion-resistant capabilities compared to pure Mg samples in biological fluid [20]. Particularly, an alloy with 2.0% Gd and Mg outperformed pure Mg in terms of performance. Numerous microbial floras have been found to cling to the implant surfaces in the oral environment. The contact between the bacterial cell and the implant causes the bacteria to proceed through many phases of attachment, resulting in the formation of a biofilm that causes implant infection. Mouthwashes and other prophylactic antimicrobial medications are used to avoid microbial-focused illnesses for lowering bacterial contamination. According to this theory, Gobi Saravanan et al. studied how ZrO_2 coating on 316 L stainless steel samples behaved when exposed to artificial saliva that had been acidified without and with the addition of sodium fluoride (NaF) solutions [21]. Owing to the chemical stability and oxide nature of the ZrO_2 coated surface, the findings indicated that 316 L SS with ZrO_2 placed on it demonstrated superior anti-corrosion properties than bare materials. The extensive study of surface-modified ceramic coatings and their characteristics demonstrates their value in the biological environment. Particularly, it is anticipated that the biocompatibility and chemical neutrality of ceramic coatings may lessen immunogenic responses in nearby tissues.

In addition, the coatings significantly impede the progression of corrosion brought on by the lactic acid, enzymes, and proteins present in human plasma as well as hydrogen peroxide and lactic acid. Due to their smooth surfaces, these ceramic coatings also prevent human pathogenic germs including Porphyromonas gingivalis, Pseudomonas aeruginosa, and bacteria that precipitate Ca from adhering early [22]. Antimicrobial and bone regeneration treatments use nanostructured grains with nanolayer-thick ceramic coatings extensively [23]. Owing to the nanolayer coating's microporous structure, they were unable to withstand the destructive ions existing in the bodily fluid, resulting in pitting corrosion. To develop denser ceramic coatings, thermal as well as plasma spray coatings are often used, resulting in a rougher surface. Even while the surface coarseness increases bone cell contact, there is risk of harmful bacteria adherence and ensuing biofilm development [24]. In addition, a bio ceramic covering with greater thickness provides stronger corrosion protection; nevertheless, it suffers from significant adhesion issues and loses its nanostructured features since it acts as a massive substance.

13.2.2 Hydroxyapatite (HAp)

Calcium phosphate (Ca_3PO_4), is frequently used in implant substrates due to its high degree of resemblance with mineral components of human bone. Additionally, owing to its better biological reaction, calcium phosphate in the form of HAp (($Ca_{10}(PO_4)_6$ $(OH)_2$)) is extensively employed in a physiological state of the human body. Generally speaking, bone is composed of collagen, HAp, non-collagenous proteins, and water. HAp has a Ca/P ratio of 1.67, which is very similar to hard tissues like bone and teeth. HAp is employed as the principal implant substrate because of its high biocompatibility [25]. Lacefield investigated osteoconductive bioactive ceramic coatings using permeable $Ca_3(PO_4)_2$ to enhance bone development [26]. Roy et al. coated HAp on Cp-Ti substrates employing RF magnetron sputtering and examined their physical and biocompatibility properties [27]. The HAp coating exhibited improved biocompatibility under both in vivo and in vitro settings.

Ding et al. produced ceramic coatings incorporating hydroxyapatite (HAp) by introducing HAp particles to the electrolytes during micro-arc oxidation (MAO) on a sustainable $Mg_{66}Zn_{29}Ca_5$ magnesium alloy [28]. In Hank's solution at 37 °C, electrochemical studies and immersion tests were conducted to determine the corrosion protection of the coatings. Figure 13.1 illustrates the potentiodynamic polarization curves of HA-containing coating, MAO coating, and bare $Mg_{66}Zn_{29}Ca_5$ alloy in Hank's solution at 37 °C degrees Celsius. Both of Hank's solution's coatings have anti-corrosion properties. The E_{corr} of the coatings, such as HAp-containing coating (−1.215 V) and MAO coating (−1.263 V), increased substantially compared to that of exposed $Mg_{66}Zn_{29}Ca_5$ alloy (−1.422 V); the I_{corr} of the coatings decreased significantly with values of 6.138×10^{-8} $A \cdot cm^{-2}$ and 8.831×10^{-7} $A \cdot cm^{-2}$, demonstrating better resistance to corrosion than $Mg_{66}Zn_{29}Ca_5$.

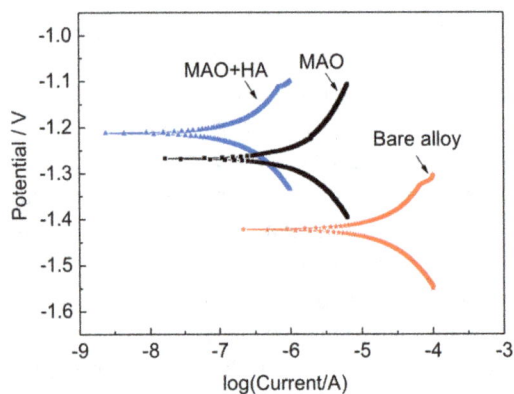

Figure 13.1: Hank's solution potentiodynamic polarization curves obtained at 37 °C [28]. Source attained from Open Access (CCBY).

Further, SEM was also performed to get an idea of how HAp has an effect on the platelets attached to the coating surface. It is clear that a lower blood platelet count was detected on the coating that contained HAp when compared to the MAO coating. In addition to this, the platelets on the MAO coating were distributed in an uneven manner, with certain places displaying a higher platelet density than those on the coating containing HAp. Figure 13.2 illustrates the adherence of the platelets and results even disclosing that the HAp-containing coating was better as literature also suggested that material with lower density of platelets has the better compatibility with the blood.

In order to study the hierarchical architectures of HAp, Cheng et al. used hydrothermal processes to create nanoscale hydroxyapatite (HAp) MAO coatings. Utilizing preosteoblast cell lines, the cell activity of these structures was assessed [29]. Wan et al.

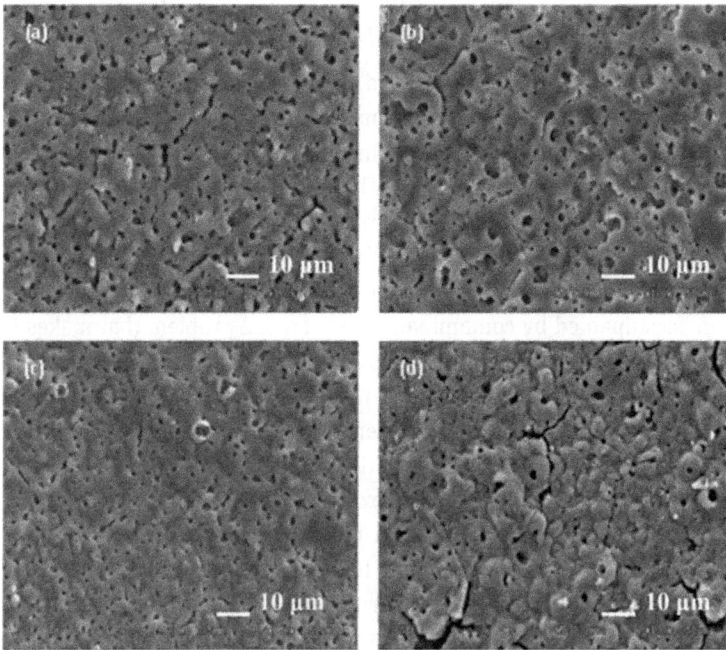

Figure 13.2: SEM analysis of coatings at HAp with varied concentration in g/l. [28]. Source attained from Open Access (CCBY).

used a variety of discharge powers, argon partial pressures, and targets to synthesize HAp coatings. Following the sputtering procedure, the coatings were hydrothermally heated to produce a homogenous, even, and thick morphology of HAp film [30] . Using the RF-magnetron sputtering approach, Dinu et al. coated a variety of Mg substrates with HAp and assessed the coating's tribological behavior with physiological saline [31]. In comparison to uncoated substrates, they noticed coated films had a decreased rate of wear. For the use of implants, bioactive ceramic layer comprised of calcium phosphate was created by Furko et al. Sankar and colleagues investigated the electrochemical behavior of HAp coating produced using pulsed laser deposition (PLD) and electrophoretic techniques [32]. They investigated the HAp coatings corrosion by employing the Hank's Balanced Salt Solution (HBSS) at 7.4 pH . According to the corrosion data, electrophoretically coated HAp coatings outperformed HAp coatings applied using the PLD approach in terms of corrosion protection.

The HAp-titania nanostructured composite coatings were created by Yugeswaran et al. using a gas tunnel plasma spray mechanism. To assess the corrosion pattern of nanocomposites and HAp coatings, the researchers investigated electrochemical behaviour in SBF medium. According to the findings, HAp-TiO$_2$ nanocomposite coatings provide more corrosion protection than HAp coating alone because of their very compact covering [33].

13.2.2.1 Metal-doped HAp

Of late, a variety of metallic implants is employed for internal fixation devices, and complete joint arthroplasty procedures are becoming more popular all around the globe. There is considerable need for orthopaedic implants with good biocompatibility and corrosion resistance for a variety of reasons. The frequent problems that patients experience, however, result in revisionary procedures that lead to aseptic loosening, infection, erosion, periprosthetic breakage, and growth retardation. The formation of biofilms on the surface of implant materials that create secondary metabolites, namely the microbial incursion accompanied by contamination, is a critical problem that makes it possible for bacteria to resist the effects of antimicrobial drugs. Numerous researchers have thoroughly investigated the performance of metallic nanoparticle-based bio ceramics that have antibacterial action against bacteria that cause implant-allied infections for use in dental and orthopaedic implants in this respect.

By co-sputtering Ag-doped HAp with a magnetron, Chen et al. were able to deposit the substance and test its bactericidal effects on the RP 12 and Cowan I strains of *S. aureus* [34]. The coatings that had been applied had hydrophilic surfaces and effectively combatted *S. aureus* and *S. epidermidis* germs. The growing integration layer (GIL) technique of the reactive integration layer was shown by Huang et al. to be capable of causing deposition of a-Ti alloy with Sr-doped hydroxyapatite (Sr-HAp) specimen surfaces at room temperature [35]. Tests using EIS and potentiodynamic polarization (PDP) produced positive corrosion potential findings. The GIL procedure uses less electricity and does not need vacuum or other expensive machinery. Larger scientific and technological fields like biological engineering and biomedical research might benefit from the adoption of GIL technology. Additionally, they noticed that adding silver would cause the HAp coating to delaminate.

To customize structural stability and enhance biocompatibility, several ions have been co-doped into HAp. In 2019, Wang and his colleagues created Sr and F-doped HAp. According to the study, the addition of Sr strengthened the HA's lattice, increasing structural stability and promoting osteogenic cell development. Similar to this, F ions doping into HAp greatly stopped *S. aureus* from growing. Similar improvements were made to the antibacterial and osteointegration characteristics of HAp doped with ZnO comprising polypyrrole (PPy) coatings by in situ oxidative polymerization. Ca^{2+} and Zn^{2+} ion release was greatly reduced in multipurpose composite coatings formed using PPy encapsulation and enhanced antibacterial capabilities against mesenchymal stem cell adhesion and growth, as well as *E. coli* and *S. aureus*.

Similar to this, additional microbicidal substances including copper and zinc have been added to HAp for applications in biomedicine. By using chemical synthesis, Zn-loaded amorphous calcium phosphate nanorods were created, and they demonstrated antibacterial activity against the pathogens of the oral cavity Aggregatibacter actinomycetemcomitans, Fusobacterium nucleatum, and Streptococcus mutans. The mechanical strength of HAp is decreased; thus ZrO_2 nanoparticles are introduced to the Hap particles to increase the structural rigidity. By lysing bacteria against *E. coli*, copper substitution HAp nanoparticles have also shown good antibacterial activity. The antibacterial efficacy of the HAp-ZrO$_2$ composite materials was also improved by the addition of sodium fluoride.

Overall, HAp is a material that cannot be avoided, particularly in bone ceramics because of its structural and chemical resemblance to actual bone. The osteoblast cell's development, adhesion, proliferation, and differentiation for higher healing rates are all considerably facilitated by the correct structure, phase, constitution, in addition to the topology of the HAp film.

Particularly, HAp with admixed Ag, Cu, Zn, and Mg enhances wide range antibacterial action as well as little cytotoxicity with mammalian cells. For example, Hap doped with Sr increased bone resorption and decreased bone formation. In a similar vein, Mg-doped HAp demonstrated stronger osteoblast cell attachment than pure HAp.

The greater stiffness of HAp makes it more brittle even if it resembles real bone and encourages osteoconductivity. As it was less compact throughout in vitro testing, the HAp coating doped with metal was unable to discharge the metal ions over time. An immunogenic reaction is then brought on by the abrupt metallic ion discharge that has gathered in the nearby tissues and cells of the body. In order to increase the coating's compactness, scientists have been creating metal-doped HAp coatings using polymers including collagen, alginate, agarose, and chitosan. But because of the relatively small pore size of the polymeric matrix that encases the metal-doped HAp, it is less efficient for effective bone regeneration and has other features.

2.2.2 HAp and nanocomposite coatings: electrochemical behavior

By using a blast-coating approach at room temperature, Dunne et al. coated crystalline HAp on Mg alloy. In the 48 MPa range, the adhesive strength of the deposited films was improved. In an SBF, the crystalline HAp layer showed a reduced current density of corrosion than the Mg alloy. To compare the enhanced qualities of pulsed electrodeposition and electrodeposition by electrodeposition, Meng et al. conducted comparative investigations. Due to the negative impacts of electrodeposited HAp, they discovered that pulse electrodeposition performed well in terms of biological behavior and anti-corrosive qualities [36].

The same goes for HAp coating, which was created via plasma electrolytic oxidation in order to investigate their mechanical, film thickness, and corrosive behavior [37]. According to the research, corrosion resistance of titanium-modified HAp coatings was further increased by using direct current mode and a 10 Hz unipolar pulse. In a similar vein, Sankar et al. investigated a comparison of HAp coatings made by electrodeposition and PLD [38]. In HAp coating created using the PLD process, slower degradation was attained. Similar to how electrodeposited HAp coating enhanced corrosion resistance property by 51.5%, PLD-applied Hap coating greatly increased anti-corrosion property by 96.97%.

Based on their observations, they came to the conclusion that the PLD's dense and porous-free coatings successfully increased corrosion resistance performance. The fact is that HAp coverings are employed as synthetic bones because they are strong, chemically stable, biologically active, as well as able to withstand high stress and unnatural pH levels. Studies using bone morphogenetic proteins (BMP) as well as additional gene expression have shown that these coatings improve the connection between the film and the host system to speed up the process of healing. Additionally, improved mineralization and hemocompatibility were seen. Additionally, a nature that was resistant to corrosion and could endure bodily fluids was produced. The HAp coatings, on the other hand, have poor tensile strength and low fracture toughness, which limits their use in high-load applications. Additionally, its broad uses are limited by coating thickness, debris release, and ceramic coating degrading behavior. Therefore, superior mechanical qualities are elevated by the creation of composite coatings for metallic implants, such as polymer/ceramic or ceramic/ceramic layers.

Composites are made by combining more than two substances or phases to get the desired metal attributes in individual composites or metal alloys and to use the key properties of every component. The primary need for compounded metals is that each component is biocompatible and resists degradation at the component's contacts [39]. Since it is exceedingly difficult to create composites in the correct form, they have not yet reached technological progress for these applications [40]. Various implant applications have used composite materials in clinical practice. Since a few decades ago, dental fillings have been replaced by polymethyl methacrylate (PMMA) polymer in dental implant applications. The better mechanical and compressive qualities of dental amalgam composite materials, however, still make them a popular choice for dental fillers. The monomer to polymerization process has an impact on the mechanical stability and biocompatibility of implant materials. In the tested conditions during the polymerization reaction, microcavities, high stiffness, polymerization contraction, and wear resistance of the materials all play a significant impact. The cytotoxic impact on biological systems is caused by a low degree of polymerization conversion.

Due to their greater benefits over metals and alloys, composite materials like SiC, carbon or carbon-reinforced fiber are primarily employed for endosseous implants in dental implant applications. In maxillofacial surgery, PTFE, carbon, and alumina composites are employed. These materials cause noticeable inflammatory reactions. The majority of uses for nanocomposite-based scaffolds involve bone grafts. Despite being employed in many implant applications, HAp has a limited range of uses due to its mechanical properties and has subpar antibacterial and anti-cancer activity capabilities.

Htun et al. investigated how to increase the toughness and endurance of HAp by using calcium oxide and zirconia (ZrO_2)-augmented HAp that was heat-treated with a modest quantity of CaF to enhance the interability and phase steadiness of the composites [41]. According to the research, adding ZrO_2 strengthened the HAp and increased its resilience from 35.70 MPa to 52.88 MPa. In order to study the corrosion behavior of magnesium-fluorapatite composites, Razavi et al. produced them using the compression technique. According to the findings, magnesium-fluorapatite composites had greater anti-corrosion ability as compared to Mg solely [42]. Fares et al. examined the possibility of covering Ti grafts with SiC to see if the coating would hold up after the simulated implantation of the implant. Similar research was conducted by Khalajabadi et al. on corrosion and mechanical behavior of cold annealed Mg/HAp/MgO composite materials. [43] Mg and HAp powders in varying weight percentages were ball-milled, fused, and heated throughout the annealing process. We looked into the sintered specimen's mechanical and corrosion properties. Different weight percentages of MgO were added to Mg and HAp particles throughout the corrosion studies.

Finally, they disclosed that the inclusion of MgO particles caused corrosion resistance. For the first time, an electrodeposition process was used by Huang et al. to develop a hydroxyapatite/magnesium phosphate/zinc phosphate (HMZ) coating. They conducted cytocompatibility and hemolysis experiments in addition to the electrochemical and deterioration testing. They discovered that the HMZ coating had better corrosion resistance qualities than coatings made of zinc phosphate/hydroxyapatite and magnesium phosphate/hydroxyapatite owing to its dense structure and reduced porosity [44].

Using an electrophoretic technique, Kwok et al. investigated the corrosion behavior of HAp reinforced with carbon nanotubes (CNT) deposited on Ti_6Al_4V alloy. Three different sizes of HAp (in the form of flakes, needles, and spheres) were applied with CNT during the deposition. In comparison to electrodeposited HAp films, the developed HAp-CNT composite layers demonstrated greater adhesion capability, smoother film, anti-corrosion properties [45].

The mechanical properties of the materials are examined together with Lacefield's investigation on the biocompatibility of dental implant substances. A metal implant has protective oxide coatings, yet ions nevertheless come out more often due to the constitution, passivation state, and corrosion propensity of the materials [46]. These titanium dental alloys, which also include strong ceramics such as zirconia and niobium, may have certain unique benefits over standard materials (reduced modulus of elasticity). When Shojaee and Afshar evaluated the impact of ZrO_2 on HAp coating for bonding strength, coating, and corrosion behaviour, they got a flake-like structure. They looked at whether a greater ZrO_2 concentration increased adhesion strength compared to a HAp coating alone. Further, it was shown that HAp-ZrO_2 coating may be used as a suitable substrate for biomedical implants since greater concentrations of ZrO_2 including HAp coating greatly reduced corrosion rate [47].

With the electrophoretic deposition approach, Chakraborty et al. introduced various CaP ratios to multi-walled CNT (MWCNT) and obtained a 6–10 GPa Young's modulus value with excellent compatibility, which is comparable to genuine bone. They also looked at how HAp-MWCNT composite coatings behaved biologically and in terms of corrosion, and they found that the composite coatings had superior biologi-cal and corrosion-resistant qualities [48]. Alumina (Al_2O_3) coating has also been em-ployed extensively in biological applications.

To improve the material's resistance on corrosion, Wang et al. applied an Al_2O_3 coating to NiTi alloy substrates. In order to improve the bioactive characteristics, a thin HAp film was further deposited onto Al_2O_3 using polydopamine (PDA) as an in-ducer. According to research, the coating enhanced osteogenic cell differentiation [49]. The specimens were then submerged in SBF medium for around 56 days; electro-chemical characteristics revealed that the nickel component discharge was lower for HAp-Al_2O_3 coating (0.6 mg/mL) than Al_2O_3 covering alone (1.9 mg/mL). Similar to this, the same research team used the cathodic plasma electrolytic deposition (CPED) tech-nology to create HAp-Al_2O_3 composite coatings with two distinct phases of Al_2O_3, and the composite layers showed greater anti-corrosive behavior than the sole HAp coat-ing. Electrochemical corrosion tests showed that CPED-HT coating significantly re-duced corrosion current (0.7 mg/mL) compared to untreated NiTi alloy (18.0 mg/mL).

Numerous investigations have demonstrated that the use of biomedical implants like Ti and its alloys, SS, and Ni-Cr alloys may result in health problems owing to the discharge of metal ions (Ni, V, Al, Cr) to adjacent tissues as a result of repeated con-nections with corrosive body fluids (Na^+, Cl^-). Surface changes by hydroxyapatite cre-ate a favored option to improve implant performance because they defend the metal ion discharge and enhance osteointegration. The use of numerous metal oxides (TiO_2, ZnO, ZrO_2, Al_2O_3) and carbon-based materials as possible bio implants has successfully improved the anti-corrosion resistance of HAp.

There are numerous coating processes, including electrodeposition, dip coating, sol–gel coating, PVD coatings, and plasma spray coating. Pure HAp coating experiences significant dissolving in the bodily environment, which reduces performance and dura-bility and results in implant failure [50]. To improve the performance of ceramic coat-ings, many metal and metal oxide replacements to HAp are proposed [51]. Although the HAp coating showed improved bioactivity, it was unable to demonstrate its anti-corrosion abilities. The anti-corrosive behavior of bio ceramic layer is mostly influenced by the deposition quality, including adhesion capability, film density, coating width, etc. The HAp and its composite films made using the electrodeposition, dip-coating, and sol–gel methods have weak adhesion capability and coating non-uniformity, which allow corrosive ions to readily interact with the surface and cause corrosion to advance. On the other hand, because of enhanced adhesive strength, HAp coating applied by PVD and plasma spray might last over time. Despite having certain benefits when creat-ing a HAp coating using the PVD process, it has a highly poor rate of deposition, and a very permeable structure that is prone to corrosion [52]. Similar to the lack of homoge-neity in the thermal spray approach, the HAp coating's poor adherence to the substrate material and crack development restricts the method's ability to be corrosion-resistant.

13.3 Conclusion

Surface tailoring is a method that shows promise for increasing longevity of implant material. Numerous coating processes, including the sol–gel technique, spin covering, pulsed electrodeposition, electrochemical deposition, etc., are often used in surface modification. By using magnetron sputtering, oxides, ternary, multicomponent, and metal nitrides films exhibit outstanding wear resistance, corrosion resistance, and biocompatible qualities. Ceramic coatings with improved corrosion resistance, mechanical toughness, and biocompatibility include those made of HAp, ZrO_2, YSZ, and MgSZ using the plasma spray and EBPVD methods.

Additionally, ZrO_2 coatings and its allotropes (m-ZrO_2, t-ZrO_2, and c-ZrO_2) demonstrate superior corrosion resistance qualities in biological fluid comprising H_2O_2, LA, and BSA proteins. Polymeric coverings with antibiotics embedded in them exhibit sustained drug release that significantly slows the development of harmful microorganisms. Moreover, microorganisms that have stuck to surfaces are successfully eliminated by photocatalytic coatings made of TiO_2 and ZnO. From the viewpoints of antibacterial property, corrosion resistance, mechanical properties, anti-cancer properties, and biocompatibility, HAp that has been metal-doped and its nanocomposite films perform better than solely HAp coating. Furthermore, bioactive glass films have advanced with better degradation tolerance and biocompatibility. The antibacterial and wound-healing rates and capabilities of coatings and hybrid scaffolds made of polymers with growth stimulant and drug-hydrolyzing capabilities are greatly improved. Overall, it can be said that surface modification is advantageous for medicinal implants since it allows for the surface combination of numerous qualities via surface coatings for increased endurance.

However, certain ceramics, such as hydroxyapatite, tricalcium phosphate, and Al_2O_3, have poor tribological and mechanical characteristics. Bioactive coatings are vigorously engaged in improving biological attachment between medicinal implants and bone. Ceramic nanocomposite coatings like zirconia allotrope, hydroxyapatite-metal oxide, magnesia-stabilized zirconia (MgSZ), and zinc oxide-stabilized zirconia (ZnSZ) films are used to improve tribological, structural, antimicrobial, and bioactivity characteristics, that are crucial factors for implant uses in medicine. Ceramic coatings with a certain mix of characteristics must be created in order to meet these specifications. Additionally, the use of additive printing enables the creation of intricate, 3D ceramic structures made specifically for each patient that come in a variety of forms, sizes, and styles and are useful for orthopaedic and dental implants. This sector is currently developing optimal implants using a variety of production techniques in an effort to discover greater performance and encourage the spread of these coatings on a worldwide scale

References

[1]   Alzubaydi TL, AlAmeer SS, Ismaeel T, AlHijazi AY, Geetha M. In vivo studies of the ceramic coated titanium alloy for enhanced osseointegration in dental applications. J Mater Sci Mater Med 2009;20 (1):35–42.

[2]   Kumar P, Dehiya BS, Sindhu A. Bioceramics for hard tissue engineering applications: A review. Int J Appl Eng Res 2018;13(5):2744–52.

[3]   Kaliaraj GS, Ramadoss A, Sundaram M, Balasubramanian S, Muthirulandi J. Studies of calcium-precipitating oral bacterial adhesion on TiN, TiO2 single layer, and TiN/TiO2 multilayer-coated 316L SS. J Mater Sci 2014;49(20):7172–80.

[4]   So S, Harris IA, Naylor JM, Adie S, Mittal R. Correlation between metal allergy and treatment outcomes after ankle fracture fixation. J Orthop Surg 2011;19(3):309–13.

[5]   Nagarajan S, Rajendran N. Surface characterisation and electrochemical behaviour of porous titanium dioxide coated 316L stainless steel for orthopaedic applications. Appl Surf Sci 2009;255 (7):3927–32.

[6]   Hiromoto S, Onodera E, Chiba A, Asami K, Hanawa T. Microstructure and corrosion behaviour in biological environments of the new forged low–Ni Co–Cr–Mo alloys. Biomaterials 2005;26 (24):4912–23.

[7]   El–Ghany OS A, Sherief AH. Zirconia based ceramics, some clinical and biological aspects: Review. Fut Dent J 2016;2(2):55–64.

[8]   Chen C-Y, Chiang C-L. Preparation of cotton fibers with antibacterial silver nanoparticles. Mater Lett 2008;62(21):3607–9.

[9]   Madaoui N, Saoula N, Kheyar K, Nezar S, Tadjine R, Hammouche A, et al. The effect of substrate bias voltage on the electrochemical corrosion behaviors of thin film deposited on stainless steel by r. f magnetron sputtering. Prot Met Phys Chem Surf 2017;53(3):527–33.

[10]   Jansen J, Wolke J, Swann S, Van Der Waerden J, De Groof K. Application of magnetron sputtering for producing ceramic coatings on implant materials. Clin Oral Implants Res 1993;4(1):28–34.

[11]   Lorenzetti M, Dogša I, Stošicki T, Stopar D, Kalin M, Kobe S, et al. The influence of surface modification on bacterial adhesion to titanium–based substrates. ACS Appl Mater Interfaces 2015; 7(3):1644–51.

[12]   Gurappa I. Development of appropriate thickness ceramic coatings on 316 L stainless steel for biomedical applications. Surf Coat Technol 2002;161(1):70–8.

[13]   Rajan ST, Karthika M, Bendavid A, Subramanian B. Apatite layer growth on glassy Zr48Cu36Al8Ag8 sputtered titanium for potential biomedical applications. Appl Surf Sci 2016;369:501–9.

[14]   Wang G, Lu Z, Liu X, Zhou X, Ding C, Zreiqat H. Nanostructured glass–ceramic coatings for orthopaedic applications. J R Soc Interface 2011;8(61):1192–203.

[15]   Subramanian B. In vitro corrosion and biocompatibility screening of sputtered Ti40Cu36Pd14Zr10 thin film metallic glasses on steels. Mater Sci Eng C 2015;47:48–56.

[16]   Mohedano M, Matykina E, Arrabal R, Pardo A, Merino MC. Metal release from ceramic coatings for dental implants. Dent Mater 2014;30(3):e28-40.

[17]   Brunello G, Elsayed H, Biasetto L. Bioactive glass and silicate-based ceramic coatings on metallic implants: Open challenge or outdated topic?. Materials 2019;12(18):2929.

[18]   Yugeswaran S, Yoganand C, Kobayashi A, Paraskevopoulos K, Subramanian B. Mechanical properties, electrochemical corrosion and in-vitro bioactivity of yttria stabilized zirconia reinforced hydroxyapatite coatings prepared by gas tunnel type plasma spraying. J Mech Behav Biomed Mater 2012;9:22–33.

[19]   Zhang K, Van Le Q. Bioactive glass coated zirconia for dental implants: a review. J Compos Comp 2020;2(2):10–7.

[20] Bian D, Deng J, Li N, Chu X, Liu Y, Li W, et al. In vitro and in vivo studies on biomedical magnesium low–alloying with elements gadolinium and zinc for orthopedic implant applications. ACS Appl Mater Interfaces 2018;10(5):4394–408.

[21] Kaliaraj GS, Vishwakarma V, Kirubaharan K, Dharini T, Ramachandran D, Muthaiah B. Corrosion and biocompatibility behaviour of zirconia coating by EBPVD for biomedical applications. Surf Coat Technol 2018;334:336–43.

[22] Kaliaraj GS, Kumar N. Oxynitrides decorated 316L SS for potential bioimplant application. Mater Res Express 2018;5(3):036403.

[23] Kaliaraj GS, Muthaiah B, Alagarsamy K, Vishwakarma V, Kirubaharan AK. Role of bovine serum albumin in the degradation of zirconia and its allotropes coated 316L SS for potential bioimplants. Mater Chem Phys 2021;258:123859.

[24] Del Fabbro M, Taschieri S, Canciani E, Addis A, Musto F, Weinstein R, et al. Osseointegration of titanium implants with different rough surfaces: a histologic and histomorphometric study in an adult minipig model. Implant Dent 2017;26(3):357–66.

[25] Vallet-Regi M, González-Calbet JM. Calcium phosphates as substitution of bone tissues. Prog Solid State Chem 2004;32(1-2):1–31.

[26] Lacefield WR. Current status of ceramic coatings for dental implants. Implant Dent 1998;7(4):315–22.

[27] Roy M, Bandyopadhyay A, Bose S. Induction plasma sprayed nano hydroxyapatite coatings on titanium for orthopaedic and dental implants. Surf Coat Technol 2011;205(8–9):2785–92.

[28] Ding H-Y, Li H, Wang G-Q, Liu T, Zhou G-H. Bio-Corrosion Behavior of Ceramic Coatings Containing Hydroxyapatite on Mg-Zn-Ca Magnesium Alloy. Appl Sci 2018;8(4):569.

[29] Chen K-C, Lee T-M, Kuo N-W, Liu C, Huang C-L. Nano/micro hierarchical bioceramic coatings for bone implant surface treatments. Materials 2020;13(7):1548.

[30] Wan T, Aoki H, Hikawa J, Lee JH. RF–magnetron sputtering technique for producing hydroxyapatite coating film on various substrates. Biomed Mater Eng 2007;17(5):291–7.

[31] Dinu M, Ivanova AA, Surmeneva MA, Braic M, Tyurin AI, Braic V, et al. Tribological behaviour of RF–magnetron sputter deposited hydroxyapatite coatings in physiological solution. Ceram Int 2017;43(9):6858–67.

[32] Furko M, Balázsi C. Calcium phosphate based bioactive ceramic layers on implant materials preparation, properties, and biological performance. Coatings 2020;10(9):823.

[33] Yugeswaran S, Kobayashi A, Ucisik AH, Subramanian B. Characterization of gas tunnel type plasma sprayed hydroxyapatite–nanostructure titania composite coatings. Appl Surf Sci 2015;347:48–56.

[34] Chen W, Liu Y, Courtney H, Bettenga M, Agrawal C, Bumgardner J, et al. In vitro anti-bacterial and biological properties of magnetron co-sputtered silver–containing hydroxyapatite coating. Biomaterials 2006;27(32):5512–7.

[35] Huang CH, Yoshimura M. Direct ceramic coating of calcium phosphate doped with strontium via reactive growing integration layer method on α–Ti alloy. Sci Rep 2020;10(1):1–12.

[36] Dunne CF, Levy GK, Hakimi O, Aghion E, Twomey B, Stanton KT. Corrosion behaviour of biodegradable magnesium alloys with hydroxyapatite coatings. Surf Coat Technol 2016;289:37–44.

[37] Lederer S, Sankaran S, Smith T, Fürbeth W. Formation of bioactive hydroxyapatite-containing titania coatings on CP-Ti 4+ alloy generated by plasma electrolytic oxidation. Surf Coat Technol 2019;363:66–74.

[38] Sankar M, Suwas S, Balasubramanian S, Manivasagam G. Comparison of electrochemical behavior of hydroxyapatite coated onto WE43 Mg alloy by electrophoretic and pulsed laser deposition. Surf Coat Technol 2017;309:840–8.

[39] Fernando RH, Sung L–P. Nanotechnology applications in coatings: American Chemical Society. Washington, DC; 2009.

[40] Gotman I. Characteristics of metals used in implants. J Endourol 1997;11(6):383–9.

[41] Htun ZL, Ahmad N, Thant AA, Noor A-FM. Characterization of CaO–ZrO2 reinforced HAP biocomposite for strength and toughness improvement. Procedia Chem 2016;19:510–6.

[42] Razavi M, Fathi M, Meratian M. Bio-corrosion behavior of magnesium–fluorapatite nanocomposite for biomedical applications. Mater Lett 2010;64(22):2487–90.

[43] Fares C, Hsu S–M, Xian M, Xia X, Ren F, Mecholsky JJ, Jr, et al. Demonstration of a SiC protective coating for titanium implants. Materials 2020;13(15):3321.

[44] Huang W, Xu B, Yang W, Zhang K, Chen Y, Yin X, et al. Corrosion behavior and biocompatibility of hydroxyapatite/magnesium phosphate/zinc phosphate composite coating deposited on AZ31 alloy. Surf Coat Technol 2017;326:270–80.

[45] Kwok CT, Wong P, Cheng F, Man HC. Characterization and corrosion behavior of hydroxyapatite coatings on Ti6Al4V fabricated by electrophoretic deposition. Appl Surf Sci 2009;255(13–14):6736–44.

[46] Lacefield WR. Materials characteristics of uncoated/ceramic-coated implant materials. Adv Dental Res 1999;13(1):21–6.

[47] Shojaee P, Afshar A. Effects of zirconia content on characteristics and corrosion behavior of hydroxyapatite/ZrO2 biocomposite coatings codeposited by electrodeposition. Surf Coat Technol 2015;262:166–72.

[48] Saha P, Datta S, Raza MS, Pratihar DK. Effects of heat input on weld-bead geometry, surface chemical composition, corrosion behavior and thermal properties of fiber laser–welded nitinol shape memory alloy. J Mater Eng Perform 2019;28(5):2754–63.

[49] Wang X, Liu F, Song Y, Sun Q. Enhanced corrosion resistance and bio-performance of Al2O3 coated NiTi alloy improved by polydopamine–induced hydroxyapatite mineralization. Surf Coat Technol 2019;364:81–8.

[50] Huang Y, Ding Q, Pang X, Han S, Yan Y. Corrosion behavior and biocompatibility of strontium and fluorine co–doped electrodeposited hydroxyapatite coatings. Appl Surf Sci 2013;282:456–62.

[51] Mucalo M. Animal–bone derived hydroxyapatite in biomedical applications. Hydroxyapatite (hap) for biomedical applications (pp. 307–42). Elsevier; 2015.

[52] Maleki-Ghaleh H, Khalil–Allafi J. Characterization, mechanical and in vitro biological behavior of hydroxyapatite-titanium-carbon nanotube composite coatings deposited on NiTi alloy by electrophoretic deposition. Surf Coat Technol 2019;363:179–90.

Humira Assad, Ishrat Fatma, Abhinay Thakur, Ashish Kumar

14 Functionalized thin film coatings for reinforced concrete engineering

Abstract: The possibility of concrete's steel reinforcement (SR) corroding has long been a source of concern. Corrosion can still only be mitigated to a limited extent, despite having a significant negative influence on the infrastructural strength and service life. Numerous techniques have been developed with the aim of reducing the decomposition of metals and extending their serviceability. The techniques comprise coating the concrete substrate, coating the reinforcement, using alternative reinforcement, using corrosion inhibitors, etc. Moreover, surface functionalization is a technique that can be used to cut the connection between the destructive substances and the metal substances, outspreading the useful lifespan of the constituents. Nevertheless, in some difficult service conditions, a physical barrier may not be enough to guarantee the protracted performance of reinforced concrete. Generally, the deterioration of materials can be effectively prevented by applying a robust and resilient coating as a physical barrier. In this situation, it is crucial to give coating the appropriate functional qualities such as superhydrophobicity and self-healing qualities. In this chapter, we outline novel methods for studying corrosion processes from a fundamental perspective as well as the preparation techniques and the protective methods of functional coatings (FCs). Additionally, a prospective exploration route is briefly suggested to help steer practical methods and stimulate additional discoveries.

Keywords: Corrosion, Concrete, Functional Coatings, Inhibition, Physical Barrier

14.1 Introduction

A popular steel and concrete amalgamation that has been expended around for more than a century is reinforced concrete [1]. Steel gives the set its tensile strength, and concrete, acting as a physical barrier and a substance with a high alkalinity level, guards the steel against corrosion. A passive coating that safeguards the metal reinforcement from deterioration can form because of the concrete's alkalinity, which Abdurrahman and his team [2] claim has a pH range of roughly 12–14 [3]. However, the security provided by the concrete is insufficient due to the substance's porosity and cracks, which allow the passage of aggressive chemicals such as chloride ions, which cause the reinforcement to corrode. Another important factor to consider is the carbonation reaction that occurs when carbon dioxide (CO_2) interacts with cement hydration products. This reaction produces calcium carbonate, which lowers the pH and weakens the reinforcing passivating coating, making it more vulnerable to corrosion [4–7]. Due to its fre-

https://doi.org/10.1515/9783111016160-014

quent occurrence in specific sorts of structures and the high expense in restoring them, the corrosion of metallic substances, particularly reinforcing steel, in concrete (Figure 14.1) has drawn more attention in recent years.

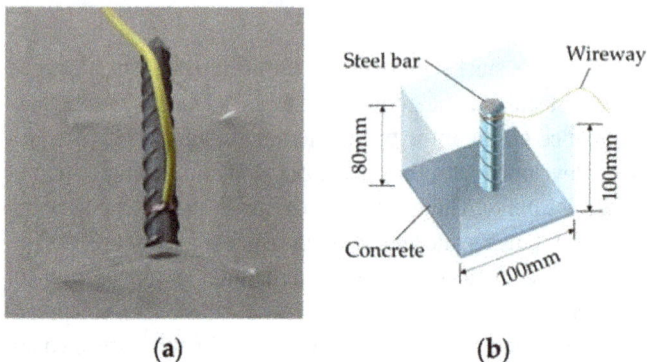

(a)   (b)

Figure 14.1: Schematic representation of reinforcement in concrete. Figure adapted from ref. [8]. Source attained from Open Access (CCBY).

Corrosion of SR was primarily noted in chemical manufacturing facilities and marine constructions [9]. The issue has freshly gained a degree of consideration due to the multiple instances of its prevalence in viaduct decks, parking garages, and in buildings susceptible to Cl⁻ [10–14]. However, there are some situations where corrosion can still happen without chloride ions. For instance, carbonation of concrete lowers the concrete's alkalinity, allowing embedded steel to corrode [15]. In concretes with a small H_2O-to-cementitious constituent's ratio (w/cm), carbonation is often a gradual process. Corrosion brought on by carbonation does not occur as frequently as corrosion brought on by chloride ions. The use of the research's outcomes is projected to lead to a decrease in corrosion in modern reinforced concrete buildings and an improvement in procedures for restoring corrosion-related deterioration to structural elements. Since the compact (solid) stuffs of rust (corrosion) contain a higher volume than the primary steel and place significant expanding strains on the concrete substrate, rusting of the reinforcing steel causes concrete to deteriorate [16, 17]. Concrete discoloration, splitting, and raveling are some of the visible signs of rusting. The cross-sectional area (CSA) of the RS (reinforcing steel) is decreased synchronously. When reinforcing steel and concrete lose their bond over time, owing to splitting and spalling or when the CSA of the steel is diminished, structural distress may develop. In constructions made of high-strength pre-stressing steel, where even a minute amount of metal damage could instigate failure, the latter effect can be very problematic. In order for these advancements to take place, practice and scientific information should be made available to those in charge of concrete structural systems – its production and management. Nevertheless, the issues brought on by the oxidation of the entrenched RS and other metals, remains.

Although some complexes have been constructed utilizing stainless steel, careful consideration ought to be given to its use as reinforcement for structures subjected to chlorides [18]. The usage of corrosion inhibitors (CIs), thin film Coatings (Cs) on the reinforcing steel, and cathodic fortification are strategies that are currently being employed and are being progressively researched [19–21]. Each of these approaches has generally proven effective and, over many years, has been used widely. Coated SR is typically utilized in mild -to-severe disclosure situations, such as in the building of bridges and marine and coastal structures, industrial buildings, H_2O purification and chemical controlling plants, and power plants. In less demanding applications, coated reinforcement is frequently used in structures and edifice for both cast-in-place and molded material (concrete) parts. Yet, in some challenging service conditions, physical barrier coating alone might not be sufficient to ensure coated reinforcements' long-term performance. In this situation, it is crucial to give the coating the appropriate functional qualities, like superhydrophobicity and self-healing capabilities [22]. Blowing ZnO nanoparticles (NPs), enhanced by SA onto the AZ31 magnesium alloy platform, was the straightforward approach Li et al. [23] used for creating superhydrophobic Cs with enhanced resilience on Mg alloys. In numerous physical and chemical degradation testing scenarios, the coating demonstrated good dimensional integrity and exhibited strong anti-corrosion properties [23]. Therefore, it is critically necessary to develop novel, functionally efficient corrosion prevention measures for reinforced concrete construction. This chapter provides an overview of the recent advancements in functionalized Cs for metals to increase resistance to oxidation.

14.2 Corrosion mechanism in concrete

Steel in concrete produces a passive protective layer as soon as the cement begins to hydrate; it consists of γ-Fe_2O_3 that adheres firmly to the steel and has a thickness between 103 and 101 μm [24]. This coating slows down corrosion by preventing ions from moving between the steel and the nearby concrete [25]. This oxide coating shields steel from deterioration. Only at high pH levels, namely 12–14, is it stable [17]. This layer needs to be damaged in order for corrosion to occur. This happens in the existence of carbonation, Cl^- ions, or subpar concrete, and corrosion happens in the presence of H_2O and O_2 [26]. Eqs. 14.1–14.5 can help us understand the rusting process.

– **Anodic Reaction**

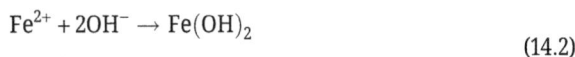

$$Fe \rightarrow Fe^{2+} + 2e^-$$

(14.1)

$$Fe^{2+} + 2OH^- \rightarrow Fe(OH)_2$$

(Ferroushydroxide)

(14.2)

$$4Fe(OH)_2 + 2H_2O + O_2 \rightarrow 4Fe(OH)_3 \quad (14.3)$$
$$\textbf{(Ferric hydroxide)}$$

$$2Fe(OH)_3 \rightarrow Fe_2O_3 \cdot H_2O + 2H_2O \quad (14.4)$$
$$\textbf{(Hydrated Ferric oxide)}$$

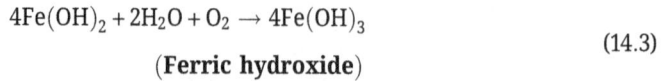

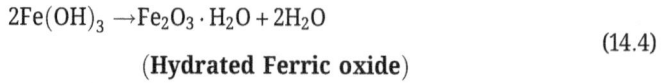

– **Cathodic Reaction**

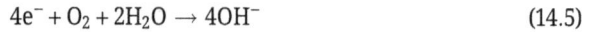

$$4e^- + O_2 + 2H_2O \rightarrow 4OH^- \quad (14.5)$$

These interactions culminate in the formation of hydrated ferric oxide or rust, which is extremely spongy and has a volume 6–10 times that of steel, leading to fracturing and raveling. Both iron reduction and O_2 removal befalls both the terminals of electrochemical cell (cathode and anode), via one of following mechanisms.

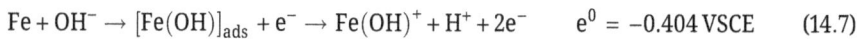

$$Fe \rightarrow Fe^{2+} + 2e^- \qquad\qquad e^0 = -0.688 \text{ VSCE} \quad (14.6)$$

$$Fe + OH^- \rightarrow [Fe(OH)]_{ads} + e^- \rightarrow Fe(OH)^+ + H^+ + 2e^- \quad e^0 = -0.404 \text{ VSCE} \quad (14.7)$$

$Fe(OH)^+$ generation is dependent on the presence of OH^- ions. As Fe^{2+} ions are generated, OH^- ions move from the mass toward the interface in order to preserve the electro-neutrality. Eq. 14.7 is preferable to eq. 14.6 on the surfaces of iron at high pH. The strength of $Fe(OH)^+$ ions at the RS rises as a consequence of the electrode potential shifting in a more anodic way, while keeping $Fe(OH)^+$ ions. A barrier oxide layer is created when ferric oxide and the $Fe(OH)^+$ oxidize (eq. 14.8). This is the steel's passive film of defense. This passive layer needs to be broken through by aggressive substances like Cl^- ions or a drop in pH for corrosion to start.

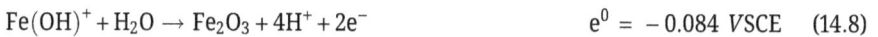

$$Fe(OH)^+ + H_2O \rightarrow Fe_2O_3 + 4H^+ + 2e^- \qquad\qquad e^0 = -0.084 \text{ } VSCE \quad (14.8)$$

14.2.1 Corrosion induced by chloride

When steel is immersed in a layer of stable concrete, it persists in a passive state, which is immune from corrosion, but when the material around it disintegrates, it changes to an active state (where corrosion begins). Chloride ions may enter the reinforcement from the outside or may be combined within. According to eq. 14.5, as chloride infiltrates concrete, the alkalinity close to the reinforcement rises. Chloride and hydroxide ions diffuse to the contact to preserve electro-neutrality. Since they move more, the Cl^- ion strength will increase from the local to the exterior, steeping the (Fe^{2+}) and (Cl^-) contact. As a result, less $Fe(OH)^+$ will develop, changing the potential in a cathodic way [16]. Critical Cl^- content refers to the quantity of chloride desirable for steel de-passivation and corrosion onset. The protective film is locally damaged, and localized pitting corrosion results if the Cl^- ion intensity exceeds this threshold value as shown in Figure 14.2.

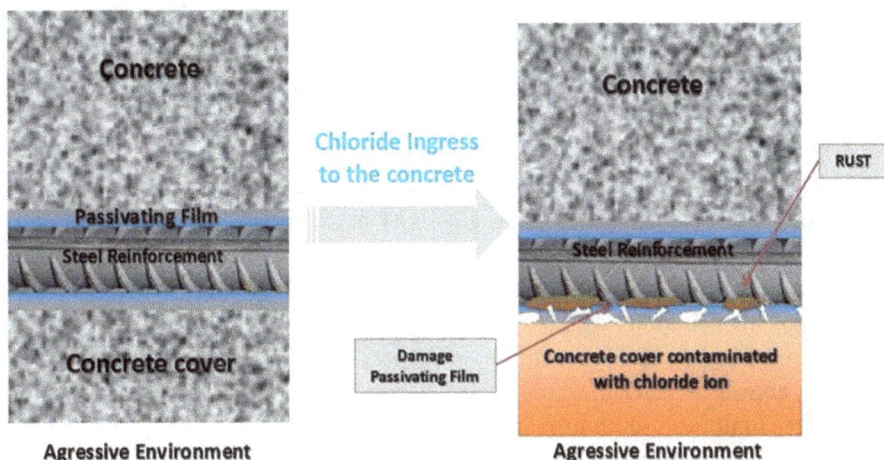

Figure 14.2: Mechanism of the corrosion process initiated by chlorine. Figure adapted from ref. [27]. Source attained from Open Access (CCBY).

When chloride ions attack steel, the surface turns into an anodic terminal while the passivated surface turns into a cathodic terminal. The reactions implicated in the processes are illustrated in eqs. 14.9 and 14.10.

$$Fe_2{}^+ + 2Cl^- \rightarrow FeCl_2 \tag{14.9}$$

$$FeCl_2 + 2H_2O \rightarrow Fe(OH)_2 + 2HCl \tag{14.10}$$

Both the Fe_3O_4 and magnetite stratums on the steel are broken by the processes in eqs. 14.9 and 14.10. Total chlorides, also known as bonded chlorides, are those chlorides that are solvable in nitric acid and are chemically adjoined to cement hydration outcomes like C3A or C4AF, or weakly attached to C-S-H. Steel only reacts with residual chlorides, specifically free or water-soluble chlorides, and this causes corrosion [28]. The amount of chloride has an effect on the passivity of steel. As the chloride concentration rises, the pitting potential of steel decreases. If a framework becomes [Cl⁻] poisoned, the pitting potential (E_{pit}) decreases from 500 to − 500 mV in a chloride-free material. The necessary overall Cl⁻ level ranges between 0.4–1% by the cement's weight for a normal oxidation potential [29]. The maximum Cl⁻ concentration for pre-stressed and reinforced cementitious materials is, respectively, 0.4 and 0.1%, according to BS 1504-9:2008. However, several researches revealed that the ratio [Cl⁻]/[OH⁻] is more crucial for corrosion and an improved depiction of the chloride limit in concrete [30]. A greater chloride threshold ratio will result from a greater chloride affinity for a given total chloride level. There have been reports of threshold ratios ranging from 0.3 to 40.0

[31]. Corrosion can occur in the presence of chlorides even at very basic pH levels, or approximately 12. However, more recent studies revealed that, depending on the treatment and environmental factors, this maximum border might be as low as 0.2% or less or even more than 1%. As a result, there is no consensus among scientists on the corrosion threshold level.

Therefore, corrosion vulnerability should be assessed for each construction based on the actual location situations rather than presuming any secure limitations. According to the researchers' professional exposure, the hazard of rusting should not only be calculated based on half-cell or chloride readings but also on the prevailing situation of the constructions. When evaluating corrosion risk, it is important to take into account the impact of temperature and moisture on Cl^- ion transit, concrete resistivity, and corrosion pace. Varying areas or segments of a structure may have different chloride ion pollution threshold levels. In actuality, the following strategy is used:

- Safeguarding of reinforcement is not necessary when Cl^- ionic strength by cement mass is less than 0.4%.
- The top limit over which prompt intervention is necessary is 1% Cl^- ion and there are signs of reinforcement deterioration and concrete delamination.
- Remediation may be postponed between 0.4% and under 1.0% if the threat and impact of corrosion are deemed to be low.
- It is crucial to observe corrosion and flaws.

The following choices should be taken into account and evaluated based on whole the life cost analysis for chloride ion concentrations of more than 0.4%:

- Replacement of tainted masonry and its evacuation.
- Galvanic anodes placed around the edges of concrete repair patches serve as patch repairs.
- No treatment is administered to chloride-contaminated masonry, but prospective degeneration is monitored.
- Utilization of impressed current cathodic protection (ICCP).
- Minor Cl^- concentrations – as little as 0.2% – can cause stress corrosion splintering in post-tensioned or pre-stressed constructions, particularly if there are cavities in the tendon duct systems.

14.3 Carbonization

Concrete can have pores as small as a micron or as large as a nanometer [32]. In addition to liquid water, the pores of concrete also include adsorbed water and structural water, which influence the various mechanical and structural qualities of the concrete. Concrete is vulnerable to carbonation, a type of natural degradation caused by the porous structure, and inherent reactivity of concrete. In other words, carbonation is the term used to describe the process by which the alkaline cement paste constituents react with the atmospheric carbon dioxide (CO_2) to neutralize concrete. Concrete's pH is decreased to around nine where the passive strata is unstable and deterioration is possible [33]. Along with $Ca(OH)_2$, C-S-H gel and the un-hydrated cement elements C_3S and C_2S are likewise attacked by CO_2. Owing to the widespread disintegration of the passive strata around the metal, corrosion caused by carbonation can befall across the entire outer layer of the steel bars. Eqs. 14.11 and 14.12's mechanism governs the carbonation process.

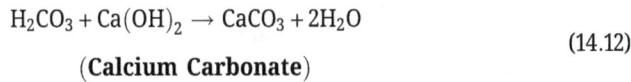

$$CO_2 + H_2O \rightarrow H_2CO_3$$

$$(\textbf{CarbonicAcid})$$

$$(14.11)$$

$$H_2CO_3 + Ca(OH)_2 \rightarrow CaCO_3 + 2H_2O$$

$$(\textbf{Calcium Carbonate})$$

$$(14.12)$$

Since the reaction takes place in a mixture, the primary sign of carbonation is a drop in pore solution's pH to 8.5 [34], a condition at which the protective layer on RS is unstable. Concrete commonly experiences carbonation as a frontage, behind which the material is unaffected and the pH is unaffected. When the reinforcing steel is exposed to the carbonation frontage, usual de-passivation can take place across sizable sections of the steel surface or perhaps the entire steel surface, which can lead to general corrosion. The good news is that sound concrete often has modest carbonation rates. However, because of the higher CO_2 concentration found in industrial areas, concrete, located in or close to an industrial region, may carbonate at a faster rate. The atmospheric intensity of CO_2 in the air is 0.03% under natural settings; in urban areas, this number is often raised to ten times that amount, and in industrial locations, it can reach 100 times the levels found in nature. Although the reaction process between the gaseous moieties and the reinforced substance occurs in solution and is more intense at extreme relative dampness, the intrusion of gases is larger at low relative moistness. Therefore, alternating wet and dry cycles and high temperatures will create the most hostile environment for concrete neutralization. Ambient relative moisture of 60% has consistently been the utmost beneficial for carbonation [35]. Initiation durations for carbonation-induced corrosion are also greatly influenced by three additional significant factors: thin concrete cover, the existence of fractures, [36] and extreme sponginess, correlated with a petite cement factor and high w/cm.

14.4 Reinforcing bar

Reinforcing bars are typically manufactured of billet steel that adheres to ASTM A 615/A 615 M or ASTM A 706/A 706 M. When uncovered steel comes into contact with steel that is encased in concrete, it can be problematic to employ uncoated bars. The uncovered steel becomes anodic and the buried steel serves as the cathode in this arrangement, which function as a galvanic pair. The ratio of the cathodic terminal area to the anodic terminal area generally determines the rate of corrosion. The rate of deterioration of the accessible steel can be very severe because the quantity of embedded steel is frequently far more than the amount of exposed steel. Epoxy-coated steel or galvanized steel are the current substitutes for uncoated bars. Although pricey and uncommon, stainless steel and nonmetallic steel substitutes are being considered. Since around 1973, epoxy-coated reinforcing bars have been used extensively in harsh situations to prevent corrosion caused by chloride infiltration. Coating application and testing were detailed in the standard ASTM A 775 and AASHTO40. Epoxy-coated bars have been the subject of numerous field and laboratory research [37]. Epoxy-coated steel reinforcement needs to have few coating cracks and defects, strong electrical resistance, limited corrosion to bare areas, resistance to undercutting, and resistance to the passage of ions, oxygen, and water in order to offer long-term corrosion resistance.

ASTM A 775 addresses these problems. The coated bar must not split when bent around a standard mandrel, and the thickness of the coating must be between 130 and 300 microns. Additionally, there must be no more than six pinhole faults per meter and no more than 2% of the bar's surface area should be damaged. Perhaps the most well-known indication of poor workability of epoxy-layered slabs was in a number of redeveloped structures in the Florida Keys [38]. Florida researchers found that neglecting to properly prepare the slabs afore covering and de-bonding the covering prior to setup in the frameworks were the leading reasons of deterioration. Since 1991, there has been a considerable enhancement in the caliber of epoxy-glazed slabs and knowledge of coating grip to RS, largely as a consequence of the extensive study and plant certification programs (CP). The "Concrete Reinforcing Steel Institute (CRSI)" launched a volunteer CP for facilities that coat reinforcement with epoxy in 1992. Over the past five years, a lot of study has been done on epoxy-coated reinforcing bars, and numerous state agencies have also undertaken field investigations. According to these tests, systems with epoxy-layered slabs are more resilient than those with uncovered slabs. New coating substances and test procedures may increase the protracted longevity of concrete structures, according to laboratory studies [39]. These novel test methods ought to be converted into consensus standards in order to assess the protracted durability of epoxy coating materials [40]. Since it maintains passivation at far lower pH levels than black steel does, galvanized RS has been utilized as building material for the past five decades. This makes it especially suitable for shielding concrete against carbonation. Unfortunately, hydrogen (H_2) is produced as a cathodic reaction when zinc dissolves in a solution with a high pH.

If precautions are not taken to counteract it, a porous surface of concrete may develop around the reinforcing bar when zinc-coated (galvanized) steel is employed in concrete. If the concrete enclosing the galvanized bars has carbonation, the effectiveness of such bars is severely reduced.

Fresh concrete may have a tiny quantity of CrO_4^{2-} supplemented to it to stop H_2 evaporation [41], and $Ca(NO_2)_2$ has been utilized to stop H_2 evaporation of galvanized precast concrete forms. The use of stainless steel as a reinforcement material for constructions in extremely hostile settings is being researched. Although ASTM A 304 stainless steel may withstand more chlorides, it is necessary to utilize the costlier ASTM A 316 L grade to obtain remarkably improved qualities, especially in bar mats of welded reinforcing steel [42].

14.5 Functionalities of coatings

Among the most intriguing methods for creating highly efficient anti-corrosion technologies for diverse industries is the deployment of functional and intelligent coatings. Encapsulation/loading of functional chemical species in host haulers or modification of the coating structural properties for the addition of substantial and/or superficial functional groups represent the primary strategies investigated to incorporate the necessary functionalities into coverings. Following is an overview of the most emerging advancements of the two techniques' (see Figure 14.3) for corrosion resistant coatings.

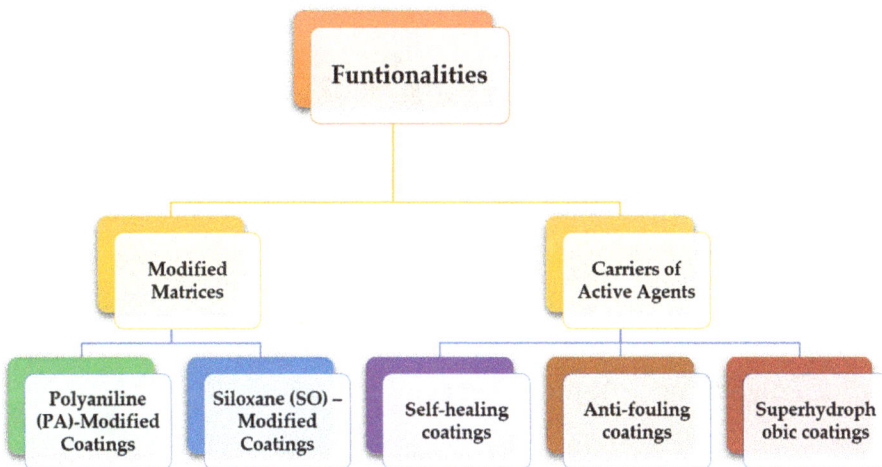

Figure 14.3: Functionalization approaches for creating proficient anti-corrosion coatings.

14.5.1 Based on modified matrices

To effectively suppress corrosion mechanisms, pigments must be present in Cs for corrosion protection. These pigments must also provide strong adherence to the substrate and compatibility with other layers, like top coats. Some of these criteria, as was previously said, can be developed by using carriers that have functional agents added to them and then applied to the coating. Functionalization of the coated outer-layer or bulk matrix through manipulation of its molecular framework and/or compound make-up is another approach that has been investigated [43]. In recent years, attention has been focused on novel functional covering formulations with reduced volatile organic substances and iso-cyanate-free compounds. Cyclic carbonates, amines, and siloxanes-modified chemistries are gaining a great deal of attention for producing functional barrier Cs for multi-substrate implementation since they are simple to scale up and are thought to be "greener". By adjusting the cross-linking and molecular framework of polymeric matrices like polyur-ethanes, acrylics, or epoxies with the addition of various additives (silanes, amines, pol-yaniline), it is possible to create functional Cs that are thicker and slimmer but have significantly improved barrier properties, increasing corrosion protection, economic vi-ability, and longevity as shown in Figure 14.4. In order to emphasize the most recent developments in the adoption of diverse capabilities, this section will change the com-position of traditional anti-corrosion Cs.

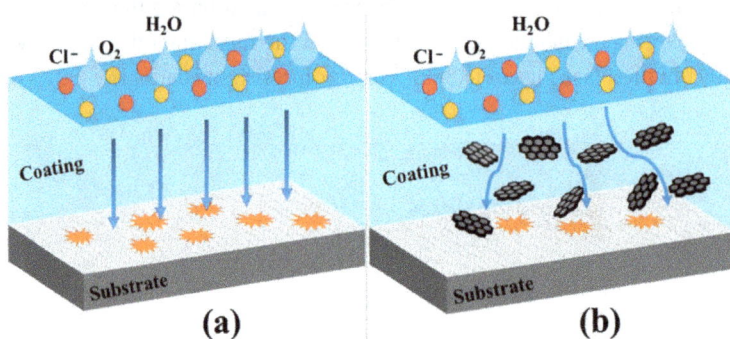

Figure 14.4: Mechanism of corrosion protection for (a) pure epoxy coating and (b) functionalized epoxy coating. Figure adapted from ref. [44]. Source attained from Open Access (CCBY).

14.5.2 Polyaniline (PA)-modified coatings

Modifying corrosion-resistant coverings with filler materials like graphite, fullerenes, car-bon black and nanotubes, metallic particles etc. is the most common method of function-alizing Cs in aspects of electroactivity. The conductivity of corrosion defensive Cs can be changed by adding conductive polymers, including polyaniline (PANI), polypyrrol, and thiophenes. PANI (and its different variants) has been one of these polymers that has had the most success [45]. Williams et al. investigated the incorporation of polyaniline salts to organic Cs [46]. In addition to the composition of the bare metal, the involvement of polyaniline salt and interfacial pH, which may have an important buffering impact and prevent corrosion process, the work demonstrates pertinent discoveries. Galvanized steel is less likely to have cathodic delamination when polyaniline is added to polyvinyl-butyral Cs. The process is connected to a postponement of the ZnO barrier strata's alkaline disintegration. To determine if coated AA2024 was cathodically delaminating, PA emeraldine salts (ES) of $C_7H_8O_3S$ (Para toluene sulphonic acid) were investigated [47]. It is demonstrated that those Cs successfully prevent filiform deterioration of coated AA2024 coupons using the scanning Kelvin probe technique. More durable and protective oxide layers that form at the metal-coating contact are linked to the protective mechanisms. Identical coatings were utilized to assess the resistance of coated cold-rolled steel surfaces to cathodic delamination, and it was discovered that the anti-oxidant property is determined by the type of acid employed in the manufacturing of the PANI-modified coating [47].

A sol–gel matrix was altered by Wang et al. [48] using emeraldine salt powders. Salt spray experiments and electrochemical impedance spectroscopy (EIS) were utilized to analyze the coating's effectiveness. The findings demonstrate that the sol–gel milieu to the conductive polymer ratio affects the anti-corrosion effect. After 500 h of salt spray, neither filiform decomposition nor creep near the scraped boundaries was seen. The authors also suggest a defense mechanism that links the creation of a preventive film at the interface, to a mechanism of "self-repair.". Epoxy and polyimide coatings' functionalization was also reported by Huang et al. [49]. In order to add electro-activity, an amine-capped $C_6H_5NH_2$ trimmer was added to the organic matrices. A $C_3H_7NO_2$ composite covering encompassing the dispersed SiO_2 NPs can be created by adding siloxanes. The silica nanoparticles improve the barrier qualities while the redox characteristics of the $C_6H_5NH_2$ moieties encourage the production of steady and shielding interfacial oxides. According to reports [50], polyaniline- improved coverings are efficient anti-corrosion structures for a variety of metallic units, including Zn, steel, Al, and Mg. PANI, an electro-active chemical, changes the coating's conductivity and encourages the development of more stable and noble interfacial barrier layers that halt the spread of corrosion and delamination [50].

14.5.3 Siloxane (SO)-modified coatings

Since SO complexes are primarily inorganic, they are very resilient to the majority of conventional ageing practices, including oxidation, deterioration, UV dilapidation, heat inclines, and motorized stressors. Si-O bonds are stronger than C-C linkages, and the impediment fortification impact of mediums, amended with Si-O, is significantly increased. Siloxanes also contribute to lower VOC levels, have low viscosities, and effectively improve adhesion, which makes Cs less disposed to delamination. SO-founded preparations do not require isocyanates, and improved veneers can be preserved at minor temperatures, comprising room temperature, even when there is moisture present. This makes it easier to paint or repaint structural components in-place. The siloxane concentration in epoxy and acrylic paints, the most widely used SO-centered paints, varies widely, contingent on the kind of siloxane, the polymeric milieu, and the desired covering qualities [52]. When the mass lessening of the tinted elements is taken into account, siloxane-based coatings have a competitive advantage because they are highly resistant to the effects of weathering even at reduced coating thickness. Additionally, the effectiveness of these coatings provides the opportunity to cut down the number of corrosion protection system coats. This guarantees more cost-effective solutions that are also more competitive.

As per Diaz et al. [53], hybrid epoxy-siloxane Cs outperform traditional denser epoxy-polyurethane Cs in augmented decomposition assessments by having diminished H_2O and O_2 penetrability, lesser capacitances, and augmented ionic forbearance. These Cs can deliver excellent decomposition safeguarding for a variety of metallic materials, including steel substrates. According to Brusciotti et al. [54], the constitutional makeup of the functional silane had an influence on the protective effectiveness of epoxy coverings treated with silane. By combining silanes with the amine functional group, the best corrosion protection was attained. Owing to the bonding of the $-NH_2$ moieties in the DETA (diethylenetriamine) with the glycidyl unit in the epoxy fragment, the presence of this group aids in the formation of a thicker and more cross-linked hybrid medium.

In accelerated corrosion testing, hybrid epoxy-siloxane coatings outperform conventional, thicker epoxy-polyurethane coatings because they had lesser O_2 and H_2O perviousness, lower capacitances, and higher ionic forbearances, according to Diaz et al. [53]. Steel substrates can benefit from the improved corrosion protection that these coatings can offer for a range of metallic substrates. The chemical composition of the functional silane has an effect on the protective performance of epoxy Cs treated with silane, according to Brusciotti et al. [54]. For an optimum corrosion protection, silanes and the amine functional group were combined. This group's existence contributes to the creation of a hybrid matrix that is thicker and even more cross-linked, because the amine moieties in the DETA bind with the $C_3H_6O_2$ unit in the epoxy fragment [55].

According to the literature [56, 57], siloxanes are a useful tool for modifying corrosion-resistant organic chemistries because they have high barrier properties, which improve corrosion resistance, low/room temperature, defiance to weathering, protracted withholding of superficial characteristics, compound steadiness under a broad pH array, and enhanced automated impedance. A variety of polymeric matrices, including epoxy, $C_3H_7NO_2$, fluoro-polymers, $-C_2H_4$ and phenolic resins, can be mixed with SO chemistries, which are very supple and can be used to create Cs, adhesives, and composites with specialized properties for corrosion fortification. SO-altered frameworks allow for the reduction of layer thickness or number, as well as a reduction in application time. As a result of improved long-term stability, repainting is required less frequently, which saves money in large-scale applications. Additionally, because of their adaptable chemistry, Cs with effective anti-graffiti, self-cleansing, hydrophobic/superhydrophobic, anti-wiping, anti-botching, and ice-nauseating qualities that are harmonic with a variety of materials can be designed and tailored.

14.6 Based on carriers of active agents

The covering composition can include functional entities carried on carriers of active agents that will scatter in the coating matrix after being introduced to the covering composition. When not in use, these carriers serve as repositories where the functional species are kept. Clays, nanotubes, porous and hollow particles, polymeric casings, a wide variety of highly porous inorganic materials, and polymeric sachets are all acceptable carriers. A high interoperability between the guest hauler and the host covering substrate is crucial when incorporated to corrosion-shielding veneers. This is essential for sustaining the preventive coating's anticipated barrier performance. Additionally, the guest transporters must be reliable and able to keep the functional component in storage for a lengthy duration, discharging it only when necessary. This method enables a number of characteristics to be incorporated into coating materials. Encapsulating functional compounds for the creation of self-healing coverings is a well-known illustration. In the past 20 years, there has been a lot of research done on this category of functional coverings [58–60], but very little of it has scale up capability, as listed below;
-   Self-healing coatings,
-   Anti-fouling coatings, and
-   Superhydrophobic coatings.

For a complete or partial restoration of covered areas harmed by ageing or unanticipated hostile occurrences, self-healing is necessary. One of the most promising methods for inserting functionalities that repair the polymeric matrix is the encapsulation of polymerizable entities into shells or nanoparticles, responsive to external stimulation. By combining multiple electrochemical and physico-chemical characterization techniques, the basis for the theory has been shown. Silyl ester was encased in urea-formaldehyde microspheres by Garcia et al. [61]. Using qualities that change over time, such as coating hydrophilicity, hydrophobicity, stickiness, water reactivity, and densification, the clever approach is able to achieve its goals. EIS and regional electrochemistry were used in a short-term test to validate the self-healing mechanism. The moisture that exists in this system causes the polymerizing ingredient to discharge. As corrosion-induced defects always contain humidity, this is particularly useful. Encapsulated silyl ester also eliminates the need for catalysts or cross-linkers in the covering, giving it a strategic advantage in the protective layer formation, since electrochemical and corrosion experiments, along with physico-chemical characterization, have been used to show the inhibitory effectiveness of different healing methods. Only a few researches have been published on this subject, and it is hard to figure out how well smart carriers buried in coating materials can repair and how quickly they will do so. Modern numerical simulations that relate the mending kinetics to the covering makeup were presented by Javierre et al. [62]. According to the authors, release kinetics may be anticipated and described by concurrently taking into account the healing agent, its molecular makeup and dispersion within the covering, the carrier qualities, and the molecular makeup and architecture of the protective layer. It is fascinating to consider that expansible clays found in the interior coating strata can repair flaws; however, the procedure can lead the upper layers to shift unintentionally [63].

Moreover, corrosion-resistant films were given the addition of microspheres or polymer particles loaded with silver, which serve as anti-fouling elements. The active substance alters the cellular membranes of living things and prevents the cell activity-regulating enzymes from working. The modified coatings reveal a gradual release profile of the active species while inhibiting fouling activities on the steel surfaces. Relative to a condition where the active ingredient is evenly spread throughout the coating, this offers longer-lasting anti-fouling ability [64]. Szabó, et al. [65] have documented a one-pot manufacturing of gelatin-based, slow-release polymer microscopic particles, incorporating silver nanoparticles and their employment in solvent-based acrylic paints, which yielded efficient outcomes in a seven-month exposure test in seas. The catalytic oxidative responsiveness of silver, obstruction of transfer of electrons, and detrimental consequences on biological organisms' DNA are all credited with its anti-fouling properties. Ag/SiO_2 core-shell nanoparticles were evaluated for their ability to minimize fouling since they exhibit significantly less corrosion than coatings with the same percentage of Cu_2O particles (40 wt%). Huang et al. [66] recently revealed the creation of a superhydrophobic composite coating comprised of styrene copolymers, methyl methacrylate, and silica nanoparticles; the exterior of which had earlier been changed with hexamethyldisilazane. Research shows that adding silica particles to a composite coating enables the creation of superhydrophobic surfaces with contact angles above 160° and minimal hysteresis. Superhydrophobic films for metallic items may be produced on a wide scale using the technique, which is seen to be very promising. Besides, increased super hydrophobicity can also improve the anti-fouling defense. These combined qualities can also be delivered by surface modification and active agent entrapment. The simplest and most efficient ways to accomplish these goals are through the sol–gel method and the use of silica particles. The variety of organic matrices that can be altered using haulers laden with functional reagents is increased by the ability to host various functions.

14.7 Functional coatings for reinforced concrete infrastructure

Coated reinforcement is understood to be standard reinforcing steel that has had a surface coating that may be metallic, non-metallic, or a combined effect of these implemented to it before being integrated in concrete. This reinforcement may take the form of straight bar lengths, constituted parts, or prefabricated segments. It should be mentioned that the best quality concrete that is readily accessible and suited for the intended use should not be sacrificed in order to use coated reinforcing for concrete's mitigation of corrosion. As a result, it is plausible to believe that the covering of strengthening provides a minimum of the foregoing benefits above the regular traditional SR, provided that the concrete components and mix design are specified with care, and professional craftsmanship and concrete practice supervision are also taken into account [67]:

- Long before corrosion of the reinforcement begins, there is a noticeably lower danger of concrete cracking, rust discoloration, and spalling throughout this time.
- A little bit more tolerance for subpar concrete that has already been put, bad concreting techniques, and subpar craftsmanship, such as misplaced reinforcement in the formwork, inappropriate compression, and/or insufficient concrete curing.
- Increased tolerance for negative consequences brought on by the reinforcement's lowered concrete cover is neither advisable nor desired, even when done so on purpose, and
- Protection against corrosion of the reinforcement before it is entrenched in concrete, or where it swells up from the concrete mass when it is in use.

All Cs on steel offer barrier-type safeguarding because they are designed to keep the metal separate from the surrounding environment, and keep the species that cause corrosion from getting close to the metal [67]. Thus, the accomplishment of coatings typically depends on the integrity of the covering because, should the coating be destroyed and the steel be revealed, the primary mechanism of corrosion prevention will be gone unless some type of supplemental protection is present, typically a coating that is sacrificed to safeguard the substance.

14.8 Metallic coatings

Metallic Cs (stainless steel, Pb, tin, Ni, and Cu) for steel are either more noble or cathodic than steel, or more active (i.e., anodic) than steel, like Cd, Al, Zn, and Mg [68–71]. The coating must be totally intact for noble metals to shield the steel material. If it is destroyed, the electrolytic activity between the uncovered steel and the covering will speed up the disintegration of the steel, since the coating functions cathodically and protects, while the exposed steel is anodic and so oxidizes and corrodes. However, the intrinsic sacrificial character of the coating, besides shield effect, also provides additional corrosion prevention at the steel-coating surface when active metallic substances are plastered onto steel. In this situation, the coating acts anodically while the steel is cathodic, allowing for the tolerability of local coating damage and the preservation of the complex so long as the uncovered steel maintains an altogether cathodic relationship with the coating metal. Numerous coating techniques, such as zinc, cadmium, nickel, and stainless steel, have been researched for metal-coated SR and, in some instances, employed in reinforced concrete construction. The most popular coating metal for this purpose is zinc, which is typically applied as a hot-dip galvanized coating. Galvanized reinforcement has been extensively studied and tested [72, 73].

14.9 Non-metallic coatings

Non-metallic Cs, whether they take the shape of organic or inorganic films, often rely almost exclusively on their less reactivity or inertness and the shield they provide between the metal and its surroundings. Non-metallic Cs are typically thought to solely offer barrier-type protection; as a result, damage to such Cs almost invariably causes the protection mechanism to fail. The technique of electrostatic powder coating has advanced significantly over the past 30 to 40 years and, today, a significant amount of steel pipes, reinforcing bars, and other consumer goods are coated in this way. Thermoplastic, thermosetting, and vitreous enamel powders are the three main types of powders used on steel. Typically, vitreous enamels are associated with specialized goods and applicators, while thermoplastic powders are associated with plastics commerce. Because raw ingredients, particularly, pigments could be found in large quantities and prevailing paint stowing apparatus could be utilized for curing powders; thermosetting powder coating has mostly grown inside the paint business [67]. Epoxy resin was used to create the first thermoset powder Cs but since then, various resin systems have been created; today, the only options available are $(C_{10}H_8O_4)n$, epoxy, and hybrid epoxy-polyester varieties. $(C_{10}H_8O_4)n$ powder Cs provide outstanding external sturdiness, particularly when it comes to color and gloss retention. They also have good UV protection and chalking resistance. Epoxies have a benefit over powder coatings made of polyester, in that they cure more quickly and have improved chemical and solvent resistance. When polyester's excellent visual appearance, including movement and luster, is needed but UV resistance is not, hybrid epoxy-polyester powders are utilized. Other organic substances such as epoxy and alkyl resins, $C_3H_7NO_2$, phenolic $-C_2H_4$, $(C_3H_6)n$, and $(C_{12}H_{22}N_2O_2)n$, as well as bitumens, petroleum, and several mineral components have been researched and employed for coating SR. Epoxy resins have distinguished themselves among these coatings as being the most adaptable and congruent, to the point where they are currently essentially the only non-metallic coatings in widespread usage for SR. Numerous notable symposiums on the development and application of epoxy coatings for strengthening have taken over the past few years [74, 75]. Thus, a passive coating that is generated in the alkaline pore mixture of concrete (pH > 12.5) protects reinforcing steel that is implanted in concrete, from excessive corrosion [28]. However, the pore solution's mobile Cl^- ions (and pH drops such as those brought on by carbonation) [76] can damage the passive layer, with the pace and degree of disintegration being proportional to the $[Cl^-]$. Hence, the risk of steel corrosion is frequently expressed as the molar ratio of ($[Cl^-]/[OH^-]$), which takes into account the passivating and de-passivating influences of OH^- ions and chloride units, respectively. Additionally, it has been demonstrated that NO_3^- and NO_2 moieties in pore solution inhibit Cl^-'s corrosive effects. By reacting Fe^{2+} units with Fe^{3+} ions, which sediment down and recreate passivating Cs, these species prevent anodic corrosion activities [77].

Furthermore, it is known that the cementitious phases in concrete can bind Cl^- species to slow metal corrosion. Falzone and coworkers [78] have recently shown how functional coatings (Cs) made of NO_3^- AFm-enriched formulations may bind a lot of Cl^- groups while instantaneously liberating NO_3 units that prevent corrosion. Here, thermodynamic selectivity controls the ion exchange process that releases the anion that was previously present in the AFm interlayer. When functional, ion-exchange Cs are used and corrosion in reinforced structural systems is significantly slowed down and delayed. When used in applications like bridge decks where they can significantly reduce the costs of protracted bridge operation, maintenance, and overhaul, the placement of functional top sheets is predominantly alluring. Since such FCs only have a slim covering, the reasonably elevated price of $CaAl_2O_4$ cement (CAC), which costs around five times as much as OPC, should not be anticipated to be a major bottleneck. This is certainly relevant when the potential savings linked to deferred corrosion commencement are taken into consideration. Through Cl^- binding isotherms, the advantages of utilizing CN-dosed CAC Cs are underlined. Indeed, at similar temperatures and water: cement (w/c) ratios, CN-dosed CAC preparations connect large amounts of Cl^-, which costs around four times as much as OPC. The following Langmuir expression (eq. 14.13) can be used to characterize the binding isotherms [79]:

$$C_{cl,b} = \frac{a.C_{cl,f}}{1 + \beta.C_{cl,f}} \tag{14.13}$$

Where,

$C_{cl,b}$ represents the bound $[Cl^-]$ (mol Cl^-/kg paste),

$C_{cl,f}$ denotes the free $[Cl^-]$ (mol Cl-/L solution) in the pore solution, and

a and β both are coefficients of Langmuir isotherm.

The hydrated phase constitutions of the CAC formulations account for their increased binding capacity. Multispecies ionic transference modeling of basic geometry, simulating a bridge-deck bare to sea H_2O, clearly illustrates the advantages of CAC + CN coatings. This demonstration makes two presumptions:

- The sphere is homogenous and fissure -free; and
- Ion transportation is controlled by dispersion and electrical immigration (i.e., advection and elemental movement influences are neglected).

Using these presumptions, the Nernst-Planck (NP) equation, which is resolved using the finite element process, may be used to characterize the intensities of various ionic moieties in construction aperture medium as a function of space (x) and time (t). The simulations use Langmuir isotherms to simulate Cl^- binding. By multiplying the dispersion coefficient of an charged units in H_2O by the materialization parameter, which takes into consideration the impacts of permeability, constrictively, and surface roughness, one can obtain the diffusion coefficients of the pertinent ionic groups (such as Cl^-, NO_3^-, and OH^-) in a restricted cementitious crystalline structure [77]. Simulations for OPC concrete configurations subjected to sea H_2O with and devoid of CAC uppermost film of 25 mm (xc) were done to statistically illustrate the corrosion slowing capacity of FCs. It is evident that a functional covering significantly lowers the permeability of Cl⁻moieties; for instance, when a film made of 30% CN + CAC is employed, Cl⁻ units only breach 9 mm in the initial 1.5 decades of contact. For the CAC-only covering, the main reasons for these consequences include:

– Reduced ionic diffusivity and ion-exchange; and
– Replacement of NO_3^- by Cl⁻ ions from the AFm's interlayer sites, (Ionic exchange when a film made of 30% CN + CAC).

These findings are important because they show that chemical seizure of invasive Cl⁻ moieties, which can be easily accomplished by FCs, is a potent strategy to defer corrosion and lengthen the serviceability of concrete architecture. This strategy goes beyond a simple reduction in wettability (or diffusivity). The quick dynamics of negative ion capturing and replacement by the AFm substances will guarantee that these FCs will execute positive influences even when the covering is fractured [80].

14.10 Conclusion

Effective corrosion engineering principles are used to isolate reinforced concrete from the aggressive atmosphere, modify the environment, or regulate the flow of electrical current within the surrounding as corrosion control measures for existing reinforced-concrete structures. Measures to isolate construction material reduce the C_R and the entry of new caustic agents, but they also capture the current levels of these caustic chemicals. It has been demonstrated that a variety of corrective actions can control corrosion on existing structures. For the purpose of protecting the reinforcing steel in concrete frameworks that are visible to the elements, benchmarks and recommendations have been founded for the assortment, creativity, edifice, and maneuver of these arrangements. The necessity of coatings with water repellent properties, such as hydrophobicity, superhydrophobicity, and ice-repellence, is paramount. New goods are rapidly expanding, and numerous approaches and solutions are being put out. Finding novel autonomous solutions that combine many capabilities such as superhydrophobicity, bioentangling, reduced wear, corrosion prevention, etc. is a problem in this field. Polysiloxane chemistry's adaptability and flexibility enable the creation of slightly elevated corrosion-protective coverings that meet the most recent environmental friendly standards. The benefits include the ability to apply fewer coats and thinner layers than with traditional coating techniques. Siloxanes' adaptable chemistry enables the development of surface qualities with extra functionality, including superhydrophobicity. Other intriguing interface capabilities are also being studied like aqueous soft touch Cs, which are highly desirable for the market for transportable devices as well as for the cores of vehicles and aircraft. Additionally, these coatings must offer improved impedance to corrosion and scratching. The creation of these devices necessitates molecular tweaking of

H_2O-based coating formulations (polyurethanes being the most prevalent) while incorporating the necessary functional ingredient.

References

[1]   Shan H, Xu J, Wang Z, Jiang L, Xu N. Electrochemical chloride removal in reinforced concrete structures: Improvement of effectiveness by simultaneous migration of silicate ion. Const Build Mater 2016;127:344–52.

[2]   Abdulrahman AS, Ismail M, Hussain MS. Corrosion inhibitors for steel reinforcement in concrete: A review. Sci Res Essays 2011;6(20):4152–62.

[3]   Elfmarkova V, Spiesz P, Brouwers HJ. Determination of the chloride diffusion coefficient in blended cement mortars. Cement Concr Res 2015;78:190–9.

[4]   Zhang D, Ghouleh Z, Shao Y. Review on carbonation curing of cement-based materials. J CO2 Util 2017;21:119–31.

[5]   Thakur A, Kumar A, Sharma S, Ganjoo R, Assad H. Computational and experimental studies on the efficiency of Sonchus arvensis as green corrosion inhibitor for mild steel in 0.5 M HCl solution. Mat Today: Proc 2022;66:609–21.

[6]   Assad H, Kumar A. Understanding functional group effect on corrosion inhibition efficiency of selected organic compounds. J Mol Liq 2021;344:117755.

[7]   Sharma S, Ganjoo R, Saha SK, Kang N, Thakur A, Assad H, Kumar A. Investigation of inhibitive performance of Betahistine dihydrochloride on mild steel in 1 M HCl solution. J Mol Liq 2022;347:118383.

[8]   Yang W, Ye X, Li R, Yang J. Effect of Stray Current on Corrosion and Calcium Ion Corrosion of Concrete Reinforcement. Materials 2022;15(20):7287.

[9]   Tremper B, Beaton JL, Stratfull RF. Causes and repair of deterioration to a California bridge due to corrosion of reinforcing steel in a marine environment. Part ii: fundamental factors causing corrosion. Highway Res Board Bull 1958(182).

[10]   Tuutti K. Corrosion of steel in concrete. Cement–och betonginst.; 1982.

[11]   Angst U, Elsener B, Larsen CK, Vennesland Ø. Critical chloride content in reinforced concrete – A review. Cement Concr Res 2009;39(12):1122–38.

[12]   Assad H, Ganjoo R, Sharma S. A theoretical insight to understand the structures and dynamics of thiazole derivatives. In: Journal of Physics: Conference Series (Vol. 2267, No. 1, p. 012063). IOP Publishing, 2022.

[13]   Thakur A, Sharma S, Ganjoo R, Assad H, Kumar A. Anti-corrosive potential of the sustainable corrosion inhibitors based on biomass waste: A review on preceding and perspective research. In Journal of Physics: Conference Series 2022 May 1 (Vol. 2267, No. 1, p. 012079). IOP Publishing.

[14]   Sharma S, Saha SK, Kang N, Ganjoo R, Thakur A, Assad H, Kumar A. Multidimensional analysis for corrosion inhibition by Isoxsuprine on mild steel in acidic environment: Experimental and computational approach. J Mol Liq 2022;357:119129.

[15]   González F, Fajardo G, Arliguie G, Juárez CA, Escadeillas G. Electrochemical realkalisation of carbonated concrete: An alternative approach to prevention of reinforcing steel corrosion. Int J Electrochem Sci 2011;6(12):6332–49.

[16]   Popov BN. Corrosion engineering: principles and solved problems. Elsevier; 2015.

[17]   Michel A, Otieno M, Stang H, Geiker MR. Propagation of steel corrosion in concrete: Experimental and numerical investigations. Cem Conc Compos 2016;70:171–82.

[18] McDonald DB, Sherman MR, Pfeifer DW, Virmani YP. Stainless steel reinforcing as corrosion protection. Concr Int 1995;17(5):65–70.

[19] Söylev TA, Richardson MG. Corrosion inhibitors for steel in concrete: State-of-the-art report. Constr Build Mater 2008;22(4):609–22.

[20] Razaqpur AG, Isgor OB. Prediction of reinforcement corrosion in concrete structures. Front Technol Infrastruct Eng Struct Infrastruct 2009;4:45–69.

[21] Cicek V. Corrosion engineering and cathodic protection handbook: With extensive question and answer section. John Wiley & Sons; 2017.

[22] Li B, Zhang Z, Liu T, Qiu Z, Su Y, Zhang J, Lin C, Wang L. Recent progress in functionalized coatings for corrosion protection of magnesium alloys – a review. Materials 2022;15(11):3912.

[23] Li L, Li X, Chen J, Liu L, Lei J, Li N, Liu G, Pan F. One-step spraying method to construct superhydrophobic magnesium surface with extraordinary robustness and multi-functions. J Magnesium Alloys 2021;9(2):668–75.

[24] Kurdowski W. Chloride corrosion in cementitious system. w Structure and performance of cements (pp. 295–309), 2nd ed., London & NY.: Spon Press; 2002.

[25] Berrocal CG, Lundgren K, Löfgren I. Corrosion of steel bars embedded in fibre reinforced concrete under chloride attack: State of the art. Cement Concr Res 2016;80:69–85.

[26] Ebell G, Burkert A, Fischer J, Lehmann J, Müller T, Meinel D, Paetsch O. Investigation of chloride-induced pitting corrosion of steel in concrete with innovative methods. Mater Corros 2016;67 (6):583–90.

[27] Sutrisno W, Hartana IK, Suprobo P, Wahyuni E. Cracking process of reinforced concrete induced by non-uniform reinforcement corrosion. J Teknologi 2017;79(3).

[28] Ahmad S. Reinforcement corrosion in concrete structures, its monitoring and service life prediction––a review. Cem Concr Compos 2003;25(4–5):459–71.

[29] Wesselsky A, Jensen OM. Synthesis of pure Portland cement phases. Cem Concr Res 2009;39(11):973–80.

[30] Figueira RB, Sadovski A, Melo AP, Pereira EV. Chloride threshold value to initiate reinforcement corrosion in simulated concrete pore solutions: The influence of surface finishing and pH. Construct Build Mater 2017;141:183–200.

[31] Ann KY, Song HW. Chloride threshold level for corrosion of steel in concrete. Corros Sci 2007;49(11):4113–33.

[32] Nóvoa XR. Electrochemical aspects of the steel-concrete system. A review. J Solid State Electrochem 2016;20:2113–25.

[33] Aperador W, Bautista-Ruiz J, Chunga K. Determination of the efficiency of cathodic protection applied to alternative concrete subjected to carbonation and chloride attack. Int J Electrochem Sci 2015;10(9):7073–82.

[34] Roberts MH. Carbonation of concrete made with dense natural aggregates. Building Research Establishment; 1981.

[35] Tuutti K. Corrosion of steel in concrete. Cement-och betonginst.; 1982.

[36] Beeby AW. Cracking, cover, and corrosion of reinforcement. Concr Int 1983;5(2):35–40.

[37] Bentur A. Steel corrosion in concrete: fundamentals and civil engineering practice. CRC press; 1997.

[38] Manning DG. Corrosion performance of epoxy–coated reinforcing steel: North American experience. Construct Build Mater 1996;10(5):349–65.

[39] McDonald DB, Pfeifer DW, Sherman MR. Corrosion evaluation of epoxy-coated, metallic-clad and solid metallic reinforcing bars in concrete. 1998.

[40] McDonald DB, Pfeifer DW. Epoxy-coated bars-state-of-art. In: Proceedings of the Second Regional Conference and Exhibition, American Society of Civil Engineers, Saudi Arabia Section (Vol. 2). 1995.

[41] Boyd WK, Tripler AB. Corrosion of reinforcing steel bars in concrete. Mate Prot 1968.

[42] Sykes JM. Electrochemical studies on steel in concrete. In: Materials Science Forum (Vol. 192, pp. 833–842). Trans Tech Publications Ltd; 1995.

[43] Kunst SR, Cardoso HR, Oliveira CT, Santana JA, Sarmento VH, Muller IL, Malfatti CF. Corrosion resistance of siloxane–poly (methyl methacrylate) hybrid films modified with acetic acid on tin plate substrates: Influence of tetraethoxysilane addition. Appl Surf Sci 2014;298:1–1.

[44] Dou B, Xiao H, Lin X, Zhang Y, Zhao S, Duan S, Gao X, Fang Z. Investigation of the anti-corrosion properties of fluorinated graphene-modified waterborne epoxy coatings for carbon steel. Coatings 2021;11(2):254.

[45] Pour-Ali S, Dehghanian C, Kosari A. In situ synthesis of polyaniline–camphorsulfonate particles in an epoxy matrix for corrosion protection of mild steel in NaCl solution. Corros Sci 2014;85:204–14.

[46] Williams G, Holness RJ, Worsley DA, McMurray HN. Inhibition of corrosion–driven organic coating delamination on zinc by polyaniline. Electrochem Commun 2004;6(6):549–55.

[47] Williams G, Gabriel A, Cook A, McMurray HN. Dopant effects in polyaniline inhibition of corrosion-driven organic coating cathodic delamination on iron. J Electrochem Soc 2006;153(10):B425.

[48] Wang H, Akid R, Gobara M. Scratch–resistant anticorrosion sol–gel coating for the protection of AZ31 magnesium alloy via a low temperature sol–gel route. Corros Sci 2010;52(8):2565–70.

[49] Huang KY, Shiu CL, Wu PS, Wei Y, Yeh JM, Li WT. Effect of amino–capped aniline trimer on corrosion protection and physical properties for electroactive epoxy thermosets. Electrochimica Acta 2009;54(23):5400–7.

[50] Huang KY, Shiu CL, Wu PS, Wei Y, Yeh JM, Li WT. Effect of amino–capped aniline trimer on corrosion protection and physical properties for electroactive epoxy thermosets. Electrochimica Acta 2009;54(23):5400–7.

[51] Chen F, Liu P. Conducting polyaniline nanoparticles and their dispersion for waterborne corrosion protection coatings. ACS Appl Mater Interfaces 2011;3(7):2694–702.

[52] Kunst SR, Cardoso HR, Oliveira CT, Santana JA, Sarmento VH, Muller IL, Malfatti CF. Corrosion resistance of siloxane–poly (methyl methacrylate) hybrid films modified with acetic acid on tin plate substrates: Influence of tetraethoxysilane addition. Appl Surface Sc 2014;298:1–1.

[53] Díaz I, Chico B, De La Fuente D, Simancas J, Vega JM, Morcillo M. Corrosion resistance of new epoxy–siloxane hybrid coatings. A laboratory study. Prog Org Coat 2010;69(3):278–86.

[54] Brusciotti F, Snihirova DV, Xue H, Montemor MF, Lamaka SV, Ferreira MG. Hybrid epoxy–silane coatings for improved corrosion protection of Mg alloy. Corros Sci 2013;67:82–90.

[55] Qian M, Soutar AM, Tan XH, Zeng XT, Wijesinghe SL. Two-part epoxy–siloxane hybrid corrosion protection coatings for carbon steel. Thin Solid Films 2009;517(17):5237–42.

[56] Seo JY, Han M. Multi-functional hybrid coatings containing silica nanoparticles and anti-corrosive acrylate monomer for scratch and corrosion resistance. Nanotechnology 2010;22(2):025601.

[57] Ahmad S, Gupta AP, Sharmin E, Alam M, Pandey SK. Synthesis, characterization and development of high performance siloxane–modified epoxy paints. Prog Org Coat 2005;54(3):248–55.

[58] Plawecka M, Snihirova D, Martins B, Szczepanowicz K, Warszynski P, Montemor MF. Self healing ability of inhibitor–containing nanocapsules loaded in epoxy coatings applied on aluminium 5083 and galvanneal substrates. Electrochimica Acta 2014;140:282–93.

[59] Li GL, Zheng Z, Möhwald H, Shchukin DG. Silica/polymer double-walled hybrid nanotubes: synthesis and application as stimuli-responsive nanocontainers in self-healing coatings. ACS Nano 2013;7(3):2470–8.

[60] Snihirova D, Lamaka SV, Montemor MF. "SMART" protective ability of water based epoxy coatings loaded with CaCO3 microbeads impregnated with corrosion inhibitors applied on AA2024 substrates. Electrochimica Acta 2012;83:439–47.

[61] García SJ, Fischer HR, White PA, Mardel J, González-García Y, Mol JM, Hughes AE. Self-healing anticorrosive organic coating based on an encapsulated water reactive silyl ester: Synthesis and proof of concept. Prog Org Coat 2011;70(2–3):142–9.

[62] Javierre E, García SJ, Mol JM, Vermolen FJ, Vuik C, Van Der Zwaag S. Tailoring the release of encapsulated corrosion inhibitors from damaged coatings: controlled release kinetics by overlapping diffusion fronts. Prog Org Coat 2012;75(1–2):20–7.

[63] Rey R, Javierre E, Garcia SJ, Van der Zwaag S, García-Aznar JM. Numerical study of the scratch-closing behavior of coatings containing an expansive layer. Surf Coat Technol 2012;206(8–9):2220–5.

[64] Szabó T, Molnár-Nagy L, Bognár J, Nyikos L, Telegdi J. Self-healing microcapsules and slow release microspheres in paints. Prog Org Coat 2011;72(1–2):52–7.

[65] Calabrese L, Bonaccorsi L, Caprì A, Proverbio E. Adhesion aspects of hydrophobic silane zeolite coatings for corrosion protection of aluminium substrate. Prog Org Coat 2014;77(9):1341–50.

[66] Huang YF, Huang C, Zhong YL, Yi SP. Preparing superhydrophobic surfaces with very low contact angle hysteresis. Surf Eng 2013;29(8):633–6.

[67] Yeomans SR. Coated steel reinforcement for corrosion protection in concrete. HKIE Trans 1995;2(2):17–28.

[68] Ganjoo R, Bharmal A, Sharma S, Thakur A, Assad H, Kumar A. Imidazolium based ionic liquids as green corrosion inhibitors against corrosion of mild steel in acidic media. In: Journal of Physics: Conference Series (Vol. 2267, No. 1, p. 012023). IOP Publishing; 2022.

[69] Ganjoo R, Sharma S, Thakur A, Assad H, Sharma PK, Dagdag O, Berisha A, Seydou M, Ebenso EE, Kumar A. Experimental and theoretical study of Sodium Cocoyl Glycinate as corrosion inhibitor for mild steel in hydrochloric acid medium. J Mol Liq 2022;364:119988.

[70] Bertolini L, Elsener B, Pedeferri P, Redaelli E, Polder RB. Corrosion of steel in concrete: prevention, diagnosis, repair. John Wiley & Sons; 2013.

[71] Deflorian F, Fedel M. UV-curable organic polymer coatings for corrosion protection of steel. In: Handbook of Smart Coatings for Materials Protection (pp. 530–559). Woodhead Publishing; 2014.

[72] Yeomans SR. Considerations of the characteristics and use of coated steel reinforcement in concrete. US Department of Commerce, National Institute of Standards and Technology, Building and Fire Research Laboratory; 1993.

[73] Swamy RN. In-situ behaviour of galvanized reinforcement. Durab Build Matl Comp 2006:325–38.

[74] Safier AS. Development and use of electrostatic, epoxy-powder coated reinforcement. Struct Eng 1989;67:95–8.

[75] Gustafson DP. Steel reinforcement: purpose, types, and uses. ASTM(American Society for Testing Materials) Stand News 1990;18(12):38–42.

[76] Poursaee A. Corrosion measurement and evaluation techniques of steel in concrete structures. In: Corrosion of Steel in Concrete Structures (pp. 219–244). Woodhead Publishing; 2023.

[77] Gaidis JM. Chemistry of corrosion inhibitors. Cem Concr Compos 2004;26(3):181–9.

[78] Falzone G, Balonis M, Bentz D, Jones S, Sant G. Anion capture and exchange by functional coatings: New routes to mitigate steel corrosion in concrete infrastructure. Cem Concr Res 2017;101:82–92.

[79] Engel T, Reid PJ. Thermodynamics, statistical thermodynamics, and kinetics. Upper saddle River: Prentice Hall; 2010.

[80] Caré S. Influence of aggregates on chloride diffusion coefficient into mortar. Cem Concr Res 2003;33(7):1021–8.

Sonam Singh, Sayantan Guha

15 Modeling of SH waves in a functionally graded piezo-poroelastic structure with sensitive coating in presence of point source of disturbance

Abstract: We aim to model SH wave propagation in a stratum-substrate structure comprising a piezo-poroelastic material (PPM) layer overlying a functionally graded PPM substrate. In this model, a source of disturbance of point size influences wave propagation. The model's free surface is assumed to be coated with an infinitesimally thin layer for analyzing the mass loading sensitivity (MLS). Admissible boundary conditions are used to determine the proper Green's functions for the layered structure components. The SH wave's dispersion relation is obtained, which is in well-agreement with the classical result of Love wave when it is reduced to an isotropic elastic material case. For a $BaTiO_3$-crystal layer and PZT-5 H substrate, the effects of various prevalent parameters, like functional gradient parameters and piezo-porous coupling parameter, on SH wave dispersion are illustrated by means of graphs and analyzed. Different materials such as ZnO, SiO_2, Si, and PZT-2 are considered to be coated on the free surface to analyze and compare the model's MLS.

Keywords: SH Wave, Piezo-poroelastic Material, Mechanical Stiffness, Mass Loading Sensitivity, Functionally Graded Material

15.1 Introduction

Monolithic piezoelectric materials are continuously used in a variety of fields, necessitating several advancements in their material properties, including high magnitudes of piezoelectric coefficients, coupling parameters, stiffness, etc. Due to their brittle nature and inability to typically withstand high deformations needed in applications of naval and aerospace engineering, these materials need to have some of their material properties improved. This can be done by customizing them with other materials exhibiting piezoelectric/non-piezoelectric properties. Thus, the concept of piezoelectric-composite material was introduced. In order to study three-dimensional harmonic plane wave propagation in a piezoelectric-poroelastic medium, [1] developed a mathematical model for its electrical and mechanical dynamics. In their work [2] examined the effect of piezoelectric interaction and porosity on the energy ratios of the reflected/transmitted waves. In recent times, some studies have been conducted dealing with wave propagation/reflection/refraction phenomenon in different types of smart

https://doi.org/10.1515/9783111016160-015

piezo-porous and other types of composite materials as reported by [3–14]. Some other works focusing on isotropic and piezo-composite materials considering wave propagation in them, as well as their vibration analysis, may be referred to in the works of [15–19].

The compelling need for better materials that can fulfill the expanding performance demands imposed by developing technologies like aeronautical/ceramic engineering, and nuclear fusion is what is driving the rebirth of interest in gradient materials. Other examples of graded material applications include electronic devices and materials, biomaterials, thermal barrier coatings for gas turbines, coatings to prevent corrosion and wear, and case-hardened steel. The necessity of functionally graded materials and their production procedure are described in [20]. Refs. [21–24] studied transference of Love wave in a FGM piezoelectric system. Some recent articles of [25, 26] have also considered functional gradient material in their model. In such models, when a wave is propagating, the existence of any point-size source having impulsive nature disturbs the natural motion of the wave with high impact. This impact is better analyzed with the help of Green's function technique as it is a significant tool to deal with such force concentrated at a point source. This similar technique is applied by [27] in their work.

Equipment for measuring Love waves is made to be highly sensitive. Acoustic sensors or SAW devices' MLS is primarily what their principal sensing mechanism is focused on. The mass sensitivity and velocity expressions in a complex layered system were reported in the study done by [28]. Ref. [29] examined SAW device's MLS for a piezoelectric layered structure having interfacial imperfections. A work of [30] analyzed the propagation characteristics of shear waves in a piezoelectric fiber-reinforced poroelastic composite structure with a sandwiched functionally graded buffer layer using an advanced power series technique. In this work, MLS is also analyzed and shown graphically, considering a fine SiO_2 coating on the free surface.

The present work aims on analyzing the MLS of a FGPPM structure due to SH wave propagation influenced by a point-sized source of impulsive nature which creates a disturbance to the propagating wave. The structure comprises a guiding layer of thickness (h) of PPM overlying a FGPPM substrate in which the source is placed at their interface. A layer of infinitesimally small thickness is coated at the free surface. This study's primary objectives are as follows:

– To derive SH wave dispersion relation, including the properties of coated layer and gradient parameter.
– To investigate the effect of functionally gradient parameters on the dispersion relation.
– To study the dispersion relation influenced by the piezo-porous coupling parameter.
– To analyze the model's MLS due to a coating of a layer with infinitesimally small thickness.

The SH wave dispersion relation is obtained using Fourier and Inverse Fourier Transform, appropriate Green's functions, and relevant conditions at the boundary of the undertaken model. The obtained results are graphically analyzed for the numerical data of $BaTiO_3$ as layer and PZT-5 H as half-space. The MLS is analyzed and compared for deposition of various materials like ZnO, SiO_2 and Si, on the free surface.

15.2 Governing equations

Let us consider a PP material which is transversely isotropic and homogeneous. The constitutive stress-strain relations in the solid skeleton and interstitial fluid following [31] are given by

$$\sigma_{ij}^{(n)} = c_{ijkl}^{(n)} \varepsilon_{kl}^{(n)} + m_{ij}^{(n)} \varepsilon^{*(n)} - e_{kij}^{(n)} E_k^{(n)} - \zeta_{kij}^{(n)} E_k^{*(n)},$$

$$\sigma^{*(n)} = m_{ij}^{(n)} \varepsilon_{ij}^{(n)} + R^{(n)} \varepsilon^{*(n)} - \zeta_k^{(n)} E_k^{(n)} - e_k^{*(n)} E_k^{*(n)},$$

(15.1)

and the corresponding electric field displacements of solid and fluid phases are

$$D_i^{(n)} = e_{ikl}^{(n)} \varepsilon_{kl}^{(n)} + \zeta_i^{(n)} \varepsilon^{*(n)} + \xi_{il}^{(n)} E_l^{(n)} + A_{il}^{(n)} E_l^{*(n)},$$

$$D_i^{*(n)} = \zeta_{ikl}^{(n)} \varepsilon_{kl}^{(n)} + e_i^{*(n)} \varepsilon^{*(n)} + A_{il}^{(n)} E_l^{(n)} + \xi_{il}^{*(n)} E_l^{*(n)}.$$

(15.2)

Here, $\sigma_{ij}^{(n)}, \sigma^{(n)}$ denote the stress tensors; $\varepsilon_{ij}^{(n)}, \varepsilon^{*(n)}$ denote the strain tensors; $D_i^{(n)}$, $D_i^{*(n)}$ denote the electric displacements, and $E_i^{(n)}, E_i^{*(n)}$ are electric field vectors. The terms $C_{ijkl}^{(n)}, m_{ij}^{(n)}, R^{(n)}$ represent material constants; $e_{ijkl}^{(n)}, e^{*(n)}, \varsigma_{ijkl}^{(n)}, \varsigma_i^{*(n)}$ are piezoelectric constants, $\xi_i^{(n)}, \xi_{ij}^{*(n)}, A_{ij}^{(n)}$ are dielectric constants, and the superscript $n = 1, 2$ stands to denote the quantities associated with the layer and substrate, respectively.

Now the equations of motion for an inviscid and non-dissipative fluid saturated porous media without body force are given by [1] as

$$\sigma_{ij,j}^{(n)} = (\rho_{11})_{ij}^{(n)} \ddot{u}_i^{(n)} + (\rho_{12})_{ij}^{(n)} \ddot{U}_i^{(n)},$$

$$\sigma_{,i}^{*(n)} = (\rho_{12})_{ij}^{(n)} \ddot{u}_i^{(n)} + (\rho_{22})_{ij}^{(n)} \ddot{U}_i^{(n)},$$

(15.3)

where $(\rho_{11})_{ij}^{(n)}, (\rho_{12})_{ij}^{(n)}, (\rho_{22})_{ij}^{(n)}$ are the medium's dynamical mass coefficients; $u_i^{(n)}, U_i^{(n)}$ are the mechanical displacement in the solid and fluid frames, with $i, j = 1, 2, 3$.

The electric displacements in the solid and fluid phases are governed by the Maxwell's equations as

$$D_{i,i}^{(n)} = 0, \quad D_{,i}^{*(n)} = 0.$$

(15.4)

Further, the strain-displacement relation and electric field-electric displacement relation are

$$e_{ij}^{(n)} = \frac{1}{2}(u_{i,j}^{(n)} + u_{j,i}^{(n)}),\, e^{*(n)} = U_{i,i}^{*(n)},$$

$$E_i^{(n)} = -\phi_{,i}^{(n)},\, E_i^{*(n)} = -\phi_{,i}^{*(n)},$$ (15.5)

where ϕ and ϕ^* denote the solid and fluid phases' electric potential functions.

For SH wave propagation along the y-direction, which causes displacement along the z-direction, the mechanical displacements and electric potentials associated with the layer and the substrate are

$$(u_1^{(1)}, u_2^{(1)}, u_3^{(1)}, \phi^{(1)}, U_1^{(1)}, U_2^{(1)}, U_3^{(1)}, \phi^{(1)*})$$
$$= (0, 0, w_1(x,y,t), \phi_1(x,y,t), 0, 0, W_1^*(x,y,t), \phi_1^*(x,y,t)),$$
$$(u_1^{(2)}, u_2^{(2)}, u_3^{(2)}, \phi^{(2)}, U_1^{(2)}, U_2^{(2)}, U_3^{(2)}, \phi^{(2)*})$$
$$= (0, 0, w_2(x,y,t), \phi_2(x,y,t), 0, 0, W_2^*(x,y,t), \phi_2^*(x,y,t)).$$ (15.6)

15.3 Equation of motion and charge

15.3.1 Equations of layer

The motion and charge equations (non-vanishing) for the layer are given by

$$C_{44}^{(1)}\nabla^2 w_1 + e_{15}^{(1)}\nabla^2\phi_1 = \left((\rho_{11})_{33}^{(1)} - \frac{((\rho_{12})_{33}^{(1)})^2}{(\rho_{22})_{33}^{(1)}}\right)\frac{\partial^2 w_1}{\partial t^2},$$

$$e_{15}^{(1)}\nabla^2 w_1 - \left(\xi_{11}^{(1)} - \frac{(A_{11}^{(1)})^2}{\xi_{11}^{*(1)}}\right)\nabla^2\phi_1 = 0.$$ (15.7)

Assume that, due to the point source, the source distribution function be $4\pi\sigma_1(r,t)$.
Then eq. (15.7) reduces to

$$\mu_1\nabla^2 w_1 - \bar{\rho}_1\frac{\partial^2 w_1}{\partial t^2} = 4\pi\sigma_1(r,t),$$ (15.8)

where $\mu_1 = C_{44}^{(1)} + \frac{(e_{15}^{(1)})^2}{\left(\xi_{11}^{(1)} - \frac{(A_{11}^{(1)})^2}{\xi_{11}^{*(1)}}\right)}$ and $\bar{\rho}_1 = \left((\rho_{11})_{33}^{(1)} - \frac{((\rho_{12})_{33}^{(1)})^2}{(\rho_{22})_{33}^{(1)}}\right).$

The displacement and source distribution functions are assumed as

$$w_1(x,y) = w_1(x,y)e^{i\omega t},\quad \sigma_1(r,t) = \sigma_1(r)e^{i\omega t}.$$ (15.9)

Using eq. (15.9) in eq. (15.8), we have

$$\nabla^2 w_1 + \frac{\overline{\rho_1}\omega^2}{\mu_1} w_1 = \frac{4\pi\sigma_1(r)}{\mu_1} e^{-a_1 x}, \tag{15.10}$$

where ω ($= kc$) denotes angular frequency.

Using Dirac-delta function, the source distribution function is defined as

$$\sigma_1(r) = \delta(y)\delta(x - H).$$

Fourier Transform of $w(x,y)$ is given by

$$W(f,x) = \frac{1}{2\pi} \int_{-\infty}^{\infty} w(x,y) e^{ify} dy.$$

On the other hand, Inverse Fourier Transform is given by

$$w(x,y) = \int_{-\infty}^{\infty} W(f,x) e^{-ify} df.$$

Now, taking Fourier Transform of eq. (15.10) w.r.t. y, we get

$$\frac{d^2 W_1}{dx^2} - a_1^2 W_1 = \frac{2\delta(x - H)}{\pi\mu_1}, \tag{15.11}$$

where $a_1^2 = (f^2 - \frac{\omega^2}{\beta_1^2})$ and $\beta_1^2 = \frac{\mu_1}{\rho_1}$.

15.3.2 Equations of the substrate

The rigidity and mass density in the FGPPM substrate's solid phase is considered to be a function of depth in terms of gradient parameters, say a_1 and a_2, respectively, as

$$C_{44}^{(2)} = C_{44}^{(02)} + a_1(x - H)^2, (\rho_{11})_{33}^{(2)} = (\rho_{11})_{33}^{(02)} + a_2(x - H)^2. \tag{15.12}$$

Thus, the non-vanishing motion equation and charge equation are given by

$$C_{44}^{(02)} \nabla^2 w_2 + e_{15}^{(2)} \nabla^2 \phi_2 - \overline{\rho_2} \frac{\partial^2 w_1}{\partial t^2} = -(x-H)^2 \left(a_1 \nabla^2 w_2 - a_2 \frac{\partial^2 w_1}{\partial t^2} \right) - 2a_1(x-H) \frac{\partial w_2}{\partial x}, \tag{15.13}$$

and

$$e_{15}^{(2)} \nabla^2 w_2 - \left(\xi_{11}^{(2)} - \frac{\left(A_{11}^{(2)}\right)^2}{\xi_{11}^{*(2)}} \right) \nabla^2 \phi_2 = 0. \tag{15.14}$$

Using eq. (15.14) in eq. (15.13) we have

$$\mu_2 \nabla^2 w_2 - \overline{\rho_2}\frac{\partial^2 w_2}{\partial t^2} = \left(-\alpha_1 \nabla_2 w_2 - \frac{2\alpha_2}{x-H}\frac{\partial w_2}{\partial x} + \alpha_2 \frac{\partial^2 w_2}{\partial t^2}\right)(x-H)^2, \tag{15.15}$$

where

$$\mu_2 = \left(C_{44}^{(02)} + \frac{\left(e_{15}^{(2)}\right)^2}{\left(\xi_{11}^{(2)} - \frac{\left(A_{11}^{(2)}\right)^2}{\xi_{11}^{*(02)}}\right)}\right) \text{ and } \overline{\rho_2} = \left((\rho_{11})_{33}^{(02)} - \frac{\left((\rho_{12})_{33}^{(2)}\right)^2}{(\rho_{22})_{33}^{(2)}}\right).$$

On substituting $w_2 = w_2(x,y)e^{i\omega t}$, we have

$$\mu_2 \nabla^2 w_2 + \overline{\rho_2}\omega^2 w_2 = \left(-\alpha_1 \nabla_2 w_2 - \frac{2\alpha_1}{x-H}\frac{\partial w_2}{\partial x} - \alpha_2\omega^2\right)(x-H)^2. \tag{15.16}$$

Taking Fourier Transform of eq. (15.16) w.r.t. y, we have

$$\frac{d^2 W_2}{dx^2} - a_2^2 W_2 = -\frac{\alpha_1(x-H)^2}{\mu_2}\frac{d^2 W_2}{dx^2} - 2\frac{\alpha_1(x-H)}{\mu_2}\frac{dW_2}{dx} + \frac{(x-H)^2}{\mu_2}(\alpha_1 f^2 - \alpha_2\omega^2)W_2$$

$$= 4\pi\sigma_2(x),$$

$$\tag{15.17}$$

where $a_2^2 = \left(f^2 - \frac{\omega^2}{\beta_2^2}\right)$, $\beta_2^2 = \frac{\mu_2}{\overline{\rho_2}}$, and $4\pi\sigma_2(x) = -\frac{d^2 W_2}{dx^2}\frac{\alpha_1(x-H)^2}{\mu_2} - 2\frac{dW_2}{dx}\frac{\alpha_1(x-H)}{\mu_2} + W_2\frac{(x-H)^2}{\mu_2}$ $(\alpha_1 f^2 - \alpha_2\omega^2)$.

15.4 Boundary conditions

15.4.1 At the free surface ($x = 0$)

The added layer's elastic properties have no significant effect. Thus, the stress-free mechanical boundary condition, which is only affected by mass loading, is as follows:

$$\tau_{13}^{(1)} = -\rho' H' \ddot{W}_1(0), \tag{15.18}$$

No electric potential exists at $x = 0$, i.e.,

$$\phi_1^{(1)} = 0, \tag{15.19}$$

15.4.2 At the interface (x = H)

The shearing stress, the electric displacements, and the mechanical displacement are all continuous at the interface ([2]). These conditions are mathematically given by

$$\tau_{13}^{(1)} = \tau_{13}^{(2)},$$ (15.20)

$$D_1^{(1)} = D_1^{(2)},$$ (15.21)

$$D_1^{*(1)} = D_1^{*(2)},$$ (15.22)

$$W_1 = W_2$$ (15.23)

As mentioned in the work of [32], the pores of layer and substrate are completely disconnected at $x = H$. Thus, we have

$$\phi_1^* = 0,$$ (15.24)

$$\phi_2^* = 0,$$ (15.25)

15.5 Green's function for the solution of the problem

Let $G_1(\frac{x}{x_0})$ denote the Green's function corresponding to PPM layer.

Then, $G_1(\frac{x}{x_0})$ must satisfy the boundary conditions eqs. (15.18) and eq. (15.19) at $x = 0$ and $x = H$, which results in the following condition

$$\frac{dG_1^M}{dx} = -\rho H \ddot{W}_1(0),$$ (15.26)

Also, the equation satisfied by $G_1(\frac{x}{x_0})$ is

$$\frac{d^2 G_1}{dx^2} - a_1^2 G_1 = \delta(x - x_0),$$ (15.27)

where the point x_0 can be chosen independently in the layer.

Now, we multiply eq. (15.11) by $G_1(\frac{x}{x_0})$ and eq. (15.26) by W_1. Then we subtract the latter equation from the former equation, and integrate the result w.r.t. x from 0 to H to have

$$\int_0^H \left(\frac{d^2 W_1}{dx^2} G_1\left(\frac{x}{x_0}\right) - \frac{d^2 G_1}{dx^2} W_1 \right) dx = \int_0^H \left(\frac{2}{\mu_1} \delta(x - H) G_1\left(\frac{x}{x_0}\right) - \delta(x - x_0) W_1(x) \right) dx.$$

(15.28)

In view of eq. (15.26), eq. (15.28) gives

$$W_1(x_0) = \frac{2}{\mu_1} G_1\left(\frac{H}{x_0}\right) - G_1\left(\frac{H}{x_0}\right)\left[\frac{dW_1}{dx}\right]_{x=H}. \tag{15.29}$$

Using the Green's function symmetric property, i.e., $G_1\left(\frac{x}{x_0}\right) = G_1\left(\frac{x_0}{x}\right)$, and exchanging x with x_0 in eq. (15.29) we obtain

$$W_1(x) = \frac{2}{\mu_1} G_1\left(\frac{x}{H}\right) - \left[\frac{dW_1}{dx}\right]_{x=H} G_1\left(\frac{x}{H}\right). \tag{15.30}$$

Similarly, let $G_2\left(\frac{x}{x_0}\right)$ be the Green's function associated with the FGPPM substrate holding the condition that $\frac{dG_2}{dx} = 0$ at $x = H$ and tends to 0 as $x \to \infty$, where the point x_0 can be taken independently from the substrate. Then, we have

$$\frac{d^2 G_2}{dx^2} - a_2^2 G_2 = \delta(x - x_0). \tag{15.31}$$

At this stage, we multiply eq. (15.17) by $G_2\left(\frac{x}{x_0}\right)$ and eq. (15.31) by W_2. Then we subtract the latter equation from the former equation, and integrate the result w.r.t. x from H to$\infty$to have

$$\int_H^\infty \left(\frac{d^2 W_2}{dx^2} G_2\left(\frac{x}{x_0}\right) - \frac{d^2 G_2}{dx^2} W_2\right) dx = \int_H^\infty \left(4\pi\sigma_2(x)G_2\left(\frac{x}{x_0}\right) - \delta(x - x_0)W_2(x)\right) dx. \tag{15.32}$$

On simplifying the above equation further, we get

$$W_2(x) = G_2\left(\frac{x}{H}\right)\left[\frac{dW_2}{dx}\right]_{x=H} + \int_H^\infty 4\pi\sigma_2(x_0)G_2\left(\frac{x}{x_0}\right) dx_0. \tag{15.33}$$

15.6 Evaluation of green's function

The two independent solutions of

$$\frac{d^2 U}{dx^2} - a_1^2 U = 0, \tag{15.34}$$

owing the condition that the solutions must tens to zero as $x \to -\infty$ and $x \to \infty$ are $U_1(x) = e^{a_1 x}$ and $U_2(x) = e^{-a_1 x}$, respectively.

Thus, considering an infinite medium, the solution (using the property of Wronskian) of eq. (15.34) can be written as

$$\frac{U_1(x)U_2(x_0)}{W} \text{ considering } x < x_0, \quad \frac{U_1(x_0)U_2(x)}{W} \text{ considering } x > x_0.$$

Here, Wronskian W is defined as $W = U_1 U_2' - U_1' U_2 = -2a_1$.

Thus, for an infinite medium, a solution of eq. (15.34) is $-\frac{e^{-a_1|x-x_0|}}{2a_1}$.

Therefore, we assume the solution of eq. (15.27) as

$$G_1\left(\frac{x}{x_0}\right) = -\frac{e^{-a_1|x-x_0|}}{2a_1} + Ae^{a_1 x} + Be^{-a_1 x}, \tag{15.35}$$

In view of eq. (15.27) and eq. (15.35), we have

$$G_1\left(\frac{x}{x_0}\right) = \frac{-T_0}{N_1},$$

Moreover, $G_1\left(\frac{x}{H}\right) = \frac{-T_1}{N_1}$, and $G_1\left(\frac{H}{H}\right) = \frac{-T_2}{N_1}$.

The unknown terms T_0, T_1, T_2, N_1 are given in the Appendix.

In a similar manner, we have $G_2\left(\frac{x}{x_0}\right) = \frac{-1}{2a_2}\left[e^{-a_2|x-x_0|} + e^{-a_2(x+x_0-2H)}\right]$ and $G_2\left(\frac{H}{H}\right) = -\frac{1}{a_2}$.

15.7 Dispersion relation

The SH wave dispersion relation is obtained using the boundary conditions given by eqs. (15.20)–(15.25), Green's function expressions for layer and substrate (i.e., the expressions of $G_1\left(\frac{x}{H}\right)$, $G_1\left(\frac{H}{H}\right)$, $G_2\left(\frac{x}{H}\right)$, and $G_2\left(\frac{H}{H}\right)$), the expression of $W_1(x)$, and the Inverse Fourier Transformation, as

$$N_2{}^* N_1\left(1 - \frac{\sigma^2 a_1 MI}{N_2 \mu_2 (C_5 + \sigma C_6^M)^2}\left(\frac{2G_1^M\left(\frac{H}{H}\right)}{\mu_1} + \frac{\rho' H' \omega^2 W_1(0)G_1^M\left(\frac{0}{H}\right)}{C_{44}^{(01)}}\right)\right) = 0, \tag{15.36}$$

where N_2 may be referred from Appendix.

15.8 Special cases

15.8.1 Case 1

In the absence of piezoelectricity and poroelasticity in both the layer and substrate, $a_1 = a_2 = \varepsilon$, and without sensitive coating, eq. (15.36) reduces to

$$\tan\left(kH\sqrt{\frac{c^2}{\beta_1^2} - 1}\right) = \frac{\mu_2\sqrt{1 - \frac{c^2}{\beta_2^2}}}{\mu_1\sqrt{\frac{c^2}{\beta_1^2} - 1}} + \frac{\varepsilon}{4k^2\mu_1\sqrt{1 - \frac{c^2}{\beta_2^2}}\sqrt{\frac{c^2}{\beta_1^2} - 1}}, \tag{15.37}$$

which is similar to the dispersion relation derived by [33].

15.8.2 Case 2

in the absence of piezoelectricity, poroelasticity, and functional gradedness in the layered structure without sensitive layer and having a perfectly bonded interface, eq. (15.36) reduces to

$$
\tan\left[kH\sqrt{\frac{c^2}{(\beta_1^0)^2}-1}\right]=\frac{c_{44}^{(02)}\sqrt{1-\frac{c^2}{(\beta_2^0)^2}}}{c_{44}^{(01)}\sqrt{\frac{c^2}{(\beta_1^0)^2}-1}}, \tag{15.38}
$$

where $\beta_1^0=\sqrt{\frac{c_{44}^{(1)}}{\rho_{11}^{(1)}}}$ and $\beta_2^0=\sqrt{\frac{c_{44}^{(02)}}{\rho_{11}^{(02)}}}$ denote the layer and substrate's shear wave velocities. Here eq. (15.38) represents the classical Love wave equation given by [34].

15.9 Numerical discussions

We have obtained the dispersion relation eq. (15.36) in the earlier sections for the SH wave's propagation in the PPM/FGPPM layer-half-space structure impacted by an interface-located disturbance. The result that is attained has been verified in special instances by comparison with classical Love wave dispersion equation.

In addition to displaying a relationship among wave number and phase velocity, the dispersion equations also show how the piezo-porous coupling parameter as well as different functional gradient parameters affect the dispersion curves. The data shown in Table 15.1 are taken into consideration for the numerical simulations.

Table 15.1: Material properties.

Material	Elastic constants $(10^{10}\ N/m^2)$	Piezoelectric constants (C/m^2)	Dielectric constants $(10^{-10}F/m)$	Mass density (Kg/m^3)
BaTiO$_3$-Layer	$c_{44}^{(1)}=4.386$	$e_{15}^{(1)}=11.4$	$\xi_{11}^{(1)}=108$	$\rho_{11}^{(1)}=3876$
			$\xi_{11}^{*(1)}=118$	$\rho_{12}^{(1)}=-741$
			$A_{11}^{(1)}=128$	$\rho_{22}^{(1)}=3762$
PZT-5H substrate	$c_{44}^{(02)}=2.3$	$e_{15}^{(2)}=17$	$\xi_{11}^{(2)}=277$	$\rho_{11}^{(02)}=4950$
			$\xi_{11}^{*(2)}=299$	$\rho_{12}^{(2)}=-1125$
			$A_{11}^{(2)}=112$	$\rho_{22}^{(2)}=4800$

15.9.1 Analysis of function gradient parameter

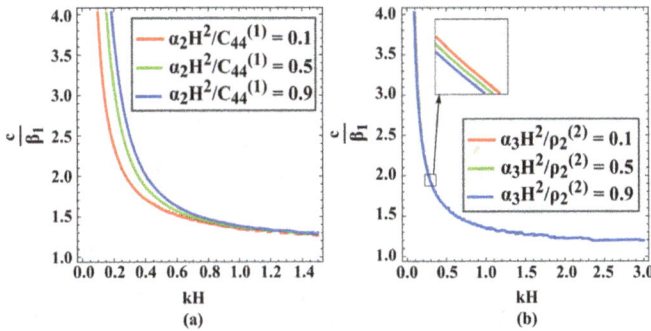

Figure 15.1: Variation of gradient parameter (a) $\left(a_1 H^2/c_{44}^{(1)}\right)$ and (b) $\left(a_2 H^2/\rho_2^{(2)}\right)$, on the dispersion curve.

The variation of gradient parameters $a_1 H^2/c_{44}^{(1)}$ and $a_2 H^2/\rho_2^{(2)}$ is discussed in Figure 15.1 (a) and 1(b) respectively. These figures portray the impact of said gradient parameters on dispersion of the surface wave plotted in terms of phase velocity against wave number. All the terms are taken in dimensionless form. The influence of $a_1 H^2/c_{44}^{(1)}$ on dispersion is increasing which means a low value of $a_1 H^2/c_{44}^{(1)}$ is preferable of low phase velocity. Whereas the influence of $a_2 H^2/\rho_2^{(2)}$ is decreasing implying that a high value of $a_2 H^2/\rho_2^{(2)}$ is preferable of low velocity wave. The impact of $a_2 H^2/\rho_2^{(2)}$ is less significant than that of $a_1 H^2/c_{44}^{(1)}$ as in case of $a_2 H^2/\rho_2^{(2)}$ all the three curves are overlapped and only a magnified subfigure can show the variation clearly.

15.9.2 Analysis of piezo-porous coupling parameter

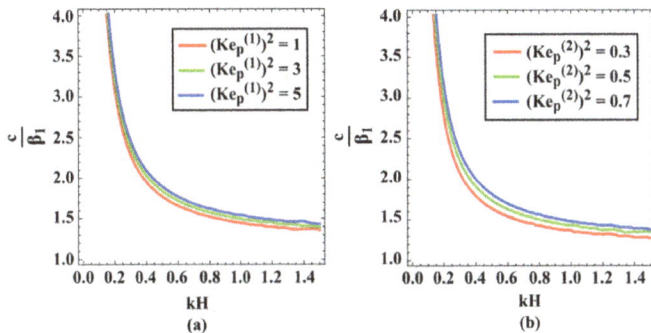

Figure 15.2: Effect of $(Ke_p)^2$ for (a) the PPM layer and (b) the FGPPM substrate, on the dispersion curve.

The piezo-porous coupling parameters are defined as

$$(Ke_p^{(1)})^2 = \frac{(e_{15}^{(1)})^2}{\mu_1(\xi_{11}^P)^{(1)}} \text{ and } (Ke_p^{(2)})^2 = \frac{(e_{15}^{(2)})^2}{\mu_2(\xi_{11}^P)^{(2)}},$$

where $\mu_1 = C_{44}^{(1)} + \frac{(e_{15}^{(1)})^2}{(\xi_{11}^P)^{(1)}}$, $\mu_2 = C_{44}^{(02)} + \frac{(e_{15}^{(2)})^2}{(\xi_{11}^P)^{(2)}}$, $(\xi_{11}^P)^{(1)} = \xi_{11}^{(1)} - \frac{(A_{11}^{(1)})^2}{(\xi_{11}^*)^{(1)}}$, $(\xi_{11}^P)^{(2)} = \xi_{11}^{(2)} - \frac{(A_{11}^{(2)})^2}{(\xi_{11}^*)^{(2)}}$.

Figure 15.2(a–b) represents the variation of c/β_1 against kH for different values of the piezo-porous coupling parameters associated with the layer and substrate.

Increasing values of $(Ke_p^{(1)})^2$, as observed from Figure 15.2(a), enhances the dispersion curve of SH wave.

Figure 15.2(b) reveals that increasing values of $(Ke_p^{(2)})^2$ also enhance the dispersion curve of SH wave.

Comparative analysis of both the figures reveals that the phase velocity is affected more by $(Ke_p^{(2)})^2$ than by $(Ke_p^{(1)})^2$.

15.9.3 Analysis of MLS ($|S_c|$)

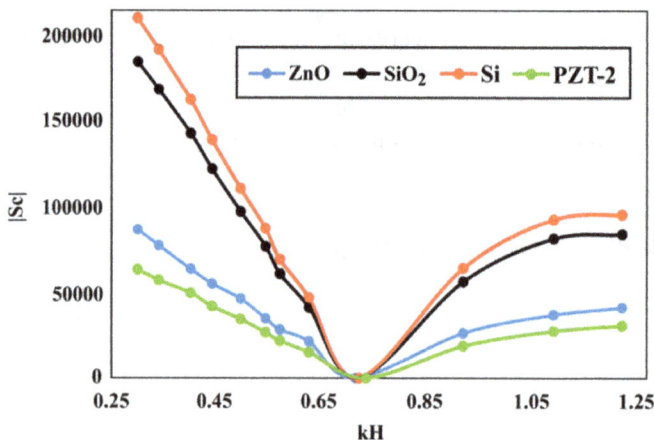

Figure 15.3: $|S_c|$ owing to the coating of a thin layer of various materials (such as $ZnO, SiO_2, Si, PZT-2$).

Here, we have analyzed $|S_c|$ of the undertaken model owing to a layer (infinitesimally thin) coating on the free surface. The sensitivity of the model is analyzed and compared for coating by various materials on the free surface such as ZnO, SiO_2, Si, and $PZT-2$. Following [29], S_c is defined as

$$S_c = \frac{1}{c_0} \lim_{\Delta m \to 0} \frac{c - c_0}{\Delta m},$$

where c and c_0 represent the phase velocities prior to and post the coating, respectively. The term m represents the free surface's mass per unit area.

Before the coating is applied, the top of the PPM layer is supposed to be independent of any mass. Thus, the change in mass after coating is denoted by $m = \Delta m (= H'\rho')$. The non-vanishing thickness and the mass of the coated layer is termed as H' and ρ' respectively. Hence, the above equation reduces to

$$S_c = \frac{c - c_0}{c_0 \rho' H'}.$$

Figure 15.3 shows the comparison of variation of $|Sc|$ of the model due to coating of a thin film of ZnO, SiO_2, Si, and $PZT - 2$, respectively, at the free surface. $|Sc|$ is observed to start with high values for lower range of kH. As kH increases, $|Sc|$ starts decreasing and assumes a minimum value close to zero, past which, it starts increasing again. The graph suggests that sensitivity of the model is highest for Si and lowest for $PZT - 2$.

15.10 Conclusion of the present work

In this work, we have extensively analyzed SH wave propagation behavior in a layered structure consisting of a PPM ($BaTiO_3$) layer resting atop a FGPPM ($PZT - 5H$) substrate being influenced by an impulsive point source and sensitive coating. The study is carried out using Green's Function Technique, Fourier Transformation, and Dirac-delta function. Using suitable boundary conditions, the SH wave dispersion relation is obtained. Numerical simulations are performed to analyze the various affecting parameters such as piezo-porous coupling parameter of the layer $\left(Ke_p^{(1)}\right)$, piezo-porous coupling parameter of the substrate $\left(Ke_p^{(2)}\right)$, functional gradient parameter related to the rigidity of the substrate's solid phase $\left(a_1 H^2 / c_{44}^{(1)}\right)$, functional gradient parameter related to the mass density of the substrate's solid phase $\left(a_2 H^2 / \rho_2^{(2)}\right)$, and MLS ($|S_c|$) of the model, and their impacts are graphically illustrated. The obtained outcomes are encapsulated as follows:

- Increasing values of $a_1 H^2 / c_{44}^{(1)}$ increase the phase velocity.
- Increasing values of $a_2 H^2 / \rho_2^{(2)}$ show a discouraging effect on the phase velocity.
- The influence of $a_1 H^2 / c_{44}^{(1)}$ on the phase velocity is comparatively greater than $a_2 H^2 / \rho_2^{(2)}$.
- Increasing values of $Ke_p^{(1)}$ increase the phase velocity.
- Increasing values of $Ke_p^{(2)}$ increase the phase velocity.
- $|S_c|$ is maximum in an initial range of wave number and sharply approaches zero as kH increases. After attaining almost zero sensitivity, it increases again.
- The model's $|S_c|$ is the highest for Si and lowest for $PZT - 2$.

Appendix

$$C_1 = \frac{e^{-a_1 H} e_{15}^{(2)}}{\xi_{11}^{(2)} \left(\frac{A_{11}^{(01)}}{A_{11}^{(2)}} - \frac{\xi_{11}^{(01)}}{\xi_{11}^{(2)}} \right)}, \quad C_2 = \frac{e_{15}^{(01)}}{\xi_{11}^{(2)} \left(\frac{A_{11}^{(01)}}{A_{11}^{(2)}} - \frac{\xi_{11}^{(01)}}{\xi_{11}^{(2)}} \right)}, \quad C_3 = \left(C_{44}^{(01)} - C_2 \left(e_{15}^{(01)} - e_{15}^{(2)} \frac{A_{11}^{(01)}}{A_{11}^{(2)}} \right) \right),$$

$$C_4 = C_{44}^{(02)} + C_1 \left(-e_{15}^{(01)} + e_{15}^{(2)} \frac{A_{11}^{(01)}}{A_{11}^{(2)}} \right), \quad C_5 = \left(C_{44}^{(01)} - C_2 e_{15}^{(01)} + M C_1 e_{15}^{(01)} \right),$$

$$M = \frac{C_3}{C_4}, \quad C_6 = \sigma \left(G_1 \left(\frac{H}{H} \right) + M G_2 \left(\frac{H}{H} \right) \right),$$

$$T_0 = (a_1^2)(e^{a_1 H} - e^{a_1 H}) e^{a_1 |x - x_0|} + (a_1) e^{a_1 x} \left[(a_1) e^{-a_1 (x+H)} + (a_1) e^{-a_1 (H-x)} \right]$$

$$+ (a_1) e^{-a_1 x} \left[(a_1) e^{a_1 (H-x)} + (a_1) e^{-a_1 (H-x)} \right],$$

$$T_1 = 2(1 - e^{2a_1 H})(a_1^2 - \frac{a_1^2}{4}) e^{-a_1 |x - x_0|} + \frac{4 a_1 \rho' H' \omega^2 W_1(0)}{C_{44}^{01}} \left[(a_1)(e^{a_1 H} e^{a_1 H} + e^{2a_1 H}) e^{-a_1 x} \right.$$

$$+ (a_1)(1 - e^{\frac{a_1 H}{2}} e^{a_1 H})] - 2 a_1 e^{a_1 x_0} [(a_1) e^{a_1 x} + (a_1) e^{-a_1 x}]$$

$$- 2 a_1 e^{-a_1 x_0} \left[(a_1) e^{a_1 x} + (a_1) e^{-a_1 x} e^{2a_1 H} \right],$$

$$N_1 = 4 a_1 (1 - e^{2a_1 H})(a_1^2 - \frac{a_1^2}{4}), \quad I = \left(\frac{(a_1)^2 + \frac{a_2 \omega^2}{a_1} - f^2}{4(a_2)^3} + \frac{1}{2 a_2} \right),$$

$$N_2 = \left(\frac{2 G_1^M \left(\frac{H}{H} \right)}{\mu_1} + \frac{\rho' H' \omega^2 W_1(0) G_1^M \left(\frac{0}{H} \right)}{C_{44}^{(01)}} \right),$$

$$T_2 = 2(1 - e^{2a_1 H})(a_1^2) + \frac{4 a_1 \rho' H' \omega^2 W_1(0)}{C_{44}^{01}} \left[(a_1)(e^{a_1 H} + e^{2a_1 H}) e^{-a_1 H} + (a_1) \right.$$

$$(1 - e^{a_1 H})] - 2 a_1 e^{a_1 H} \left[(a_1) e^{a_1 H} + (a_1) e^{-a_1 H} \right] - 2 a_1 e^{-a_1 H} \left[(a_1) e^{a_1 H} + (a_1) e^{-a_1 H} e^{2a_1 H} \right],$$

$$T_3^M = 2(1 - e^{2a_1 H})(a_1^2) e^{-a_1 H} + \frac{4 a_1 \rho' H' \omega^2 W_1(0)}{C_{44}^{01}} \left[(a_1)(e^{a_1 H} + e^{2a_1 H}) e^{-a_1 H} + (a_1) \right.$$

$$(1 - e^{a_1 H})] - 2 a_1 \left[(a_1) e^{a_1 H} + (a_1) e^{-a_1 H} \right] - 2 a_1 \left[(a_1) e^{a_1 H} + (a_1) e^{-a_1 H} e^{2a_1 H} \right].$$

References

[1] Sharma M. Piezoelectric effect on the velocities of waves in an anisotropic piezo–poroelastic medium. Proc R Soc A: Math Phys Eng Sci 2010;466(2119):1977–92.

[2] Vashishth AK, Gupta V. Reflection and transmission of plane waves from a fluid–porous piezoelectric solid interface. J Acoust Soc Am 2011;129(6):3690–701.

[3] Kar-Gupta R, Venkatesh TA. Electromechanical response of porous piezoelectric materials: effects of porosity distribution. Appl Phys Lett 2007;91(6):062904.

[4] Guha S, Singh AK. Effects of initial stresses on reflection phenomenon of plane waves at the free surface of a rotating piezothermoelastic fiber-reinforced composite half-space. Int J Mech Sci 2020;181:105766.

[5] Singh S, Singh AK, Guha S. Impact of interfacial imperfections on the reflection and transmission phenomenon of plane waves in a porous–piezoelectric model. Appl Math Model 2021;100:656–75.

[6] Singh AK, Rajput P, Chaki MS. Analytical study of love wave propagation in functionally graded piezo–poroelastic media with electroded boundary and abruptly thickened imperfect interface. Waves Random Complex Media 2022;32(1):463–87.

[7] Guha S, Singh AK. Influence of varying fiber volume fractions on plane waves reflecting from the stress-free/rigid surface of a piezoelectric fiber–reinforced composite half–space. Mech Adv Mater Struct 2021;1–5.

[8] Singh AK, Rajput P, Guha S, Singh S. Propagation characteristics of love-type wave at the electro-mechanical imperfect interface of a piezoelectric fiber-reinforced composite layer overlying a piezoelectric half-space. Eur J Mech A Solids 2022;93:104527.

[9] Singh S, Singh AK, Guha S. Reflection of plane waves at the stress-free/rigid surface of a micro-mechanically modeled piezo-electro-magnetic fiber–reinforced half-space. Waves Random Complex Media 2022:1–30.

[10] Guha S, Singh AK. Plane wave reflection/transmission in imperfectly bonded initially stressed rotating piezothermoelastic fiber-reinforced composite half-spaces. Eur J Mech A Solids 2021;88:104242.

[11] Singh AK, Guha S. Mathematical study of reflection and transmission phenomenon of plane waves at the interface of two dissimilar initially stressed rotating micro-mechanically modeled piezoelectric fiber–reinforced composite half–spaces. Wave Dyn 2022;131–62.

[12] Singh AK, Mahto S, Guha S. Analysis of plane wave reflection phenomenon from the surface of a micro-mechanically modeled piezomagnetic fiber–reinforced composite half-space. Waves Random Complex Media 2021:1–22.

[13] Singh AK, Mahto S, Guha S. Analysis of plane wave reflection and transmission phenomenon at the interface of two distinct micro-mechanically modeled rotating initially stressed piezomagnetic fiber–reinforced half-spaces. Mech Adv Mater Struct 2022;29(28):7623–39.

[14] Singh P, Singh AK, Chattopadhyay A, Guha S. Mathematical study on the reflection and refraction phenomena of three-dimensional plane waves in a structure with floating frozen layer. Appl Math Comput 2020;386:125488.

[15] Guha S, Singh AK. Transference of SH waves in a piezoelectric fiber-reinforced composite layered structure employing perfectly matched layer and infinite element techniques coupled with finite elements. Finite Elem Anal Des 2022;209:103814.

[16] Singh S, Singh AK. Anti-plane surface and interfacial waves influenced by layer reinforcement in piezo–electro–magnetic structures with surface energy. EPJ Plus 2021;136(3):1–20.

[17] Singh AK, Singh S. Application of polynomial functions in analyzing anti-plane wave profiles in a functionally graded piezoelectric–viscoelastic–poroelastic structure with buffer layer. In Polynomial paradigms: trends and applications in science and engineering 2022. IOP Publishing.

[18] Guha S, Singh AK. Frequency shifts and thermoelastic damping in distinct micro-/nano-scale piezothermoelastic fiber-reinforced composite beams under three heat conduction models. Joe 2022.

[19] Guha S, Singh AK. Frequency shifts and thermoelastic damping in different types of nano-/micro-scale beams with sandiness and voids under three thermoelasticity theories. J Sound Vib;2021;510:116301.

[20] Rabin BH, Shiota I. Functionally gradient materials. MRS Bulletin 1995;20(1):14–8.

[21] Du J, Jin X, Wang J, Xian K. Love wave propagation in functionally graded piezoelectric material layer. Ultrasonics 2007;46(1):13–22.

[22] Liu J, Wang ZK. The propagation behavior of love waves in a functionally graded layered piezoelectric structure. Ures SMS 2004;14(1):137.

[23] Qian ZH, Jin F, Lu T, Kishimoto K, Hirose S. Effect of initial stress on love waves in a piezoelectric structure carrying a functionally graded material layer. Ultrasonics 2010;50(1):84–90.

[24] Cao X, Shi J, Jin F. Effect of gradient dielectric coefficient in a functionally graded material (FGM) substrate on the propagation behavior of love waves in an fgm–piezoelectric layered structure. IEEE Trans Ultrason Ferroelectr Freq Control 2012;59(6):1253–7.

[25] Biswas M, Sahu SA. Analysis of love–type acoustic wave in a functionally graded piezomagnetic plate sandwiched between elastic layers. Acta Mech 2022;1–6.

[26] Biswas M, Sahu SA. Analysis of love–type acoustic wave in a functionally graded piezomagnetic plate sandwiched between elastic layers. Acta Mech 2022;1–6.

[27] Singh AK, Singh S, Kumari R, Ray A. Impact of point source and mass loading sensitivity on the propagation of an SH wave in an imperfectly bonded FGPPM layered structure. Acta Mech 2020;231(6):2603–27.

[28] Wang Z, Jen CK, Cheeke JD. Mass sensitivity of two–layer shear horizontal plate wave sensors. Ultrasonics 1994;32(3):209–15.

[29] Li P, Jin F. Excitation and propagation of shear horizontal waves in a piezoelectric layer imperfectly bonded to a metal or elastic substrate. Acta Mech 2015;226(2):267–84.

[30] Singh S, Singh AK, Guha S. Shear waves in a piezo-fiber–reinforced–poroelastic composite structure with sandwiched functionally graded buffer layer: power series approach. Eur J Mech A Solids 2022;92:104470.

[31] Vashishth AK, Dahiya A. Shear waves in a piezoceramic layered structure. Acta Mech 2013;224(4):727–44.

[32] Sharma MD. Wave propagation across the boundary between two dissimilar poroelastic solids. J Sound Vib 2008;314(3–5):657–71.

[33] Chattopadhyay A, Pal AK, Chakraborty M. SH waves due to a point source in an inhomogeneous medium. Int J Nonlinear Mech 1984;19(1):53–60.

[34] Ewing WM, Jardetzky WS, Press F, Beiser A. Elastic waves in layered media. Phys Today 1957;10(12):27.

Amita Somya, Amit Prakash Varshney

16 Functionalized thin film coatings for automotive coatings

Abstract: Automobile coatings are an excellent example of cutting-edge technology that can provide long-lasting surfaces, satisfy customers' expectations for design, maximize efficiency, and adhere to environmental requirements. These successes are the result of a century's worth of experience, investigation methods, technological and methodological developments, and theoretical evaluations. Thin coatings are used in the automotive sector to save both resources and the environment. This is accomplished by lightening the weight of construction materials now in use, extending their useful lives, and raising the manufacturing standards for the materials as a result. In this chapter, automotive thin-layer coating has been described with the focus on specific functions of each layer. State-of-the-art automotive coating procedures have also been explained herewith.

Keywords: Automobile, Coatings, Functionalization, Hydrophobic, Self-Life, Durability

16.1 Introduction

Automotive sector is often split into tier-one and tier-two suppliers [1]. Sub-assemblies are supplied to automakers by the first-tier vendors. Their activity is connected to product innovation, from product design to research and development. First-tier suppliers are frequently involved in the early stages of automobile evolution/development and design by automakers. The second-tier unit manufacturers provide vehicle manufacturers with unassembled components and the first-tier companies with parts for their sub-assemblies. By steadily improving the structural, thermal, electrical, magnetic, optical and, catalytic capabilities of materials as well as their biocompatibility, new performance features are considerably enhanced. The product line in the automobile industry includes technology for the full range of ornamental and functional surface treatments for a variety of materials, inclusive of steel, aluminum, plating on plastics and, many more potential applications – starting from power trains and engines to various electrical gadgets. Products must adhere to the local regulations and

Acknowledgments: The authors are grateful and convey thanks to the Chancellor, Vice-Chancellor, Pro-Vice Chancellor, Dean, School of Engineering, Presidency University, Bengaluru, Karnataka, India for the above project work. Thanks are also due to Miss. Anushi Varshney and Mr. Anirudh for supporting to compile this work.

https://doi.org/10.1515/9783111016160-016

laws with regard to their characteristics and the environment in order to be successful on a global scale. Because of this, consumers in the automotive sector are among
the pickiest in the world.

Almost 100 years ago, at the dawn of the automobile industry, automotive parts
were coated using varnish-like substance, which was sheathed over automotive surfaces. And, the coating was then sanded and buffed before the varnish was repeatedly
coated and polished to create multiple films. Followed by the application of several
coats of varnish, automobiles were then polished to create gleaming surfaces. To
completely cover and preserve different areas of a car, few manufacturers like Ford,
with the Model T line, used a combo of sheathing, dipping, and evenly pouring [2].
The entire coating process was carried out manually and it was not unusual for it to
take up to 40 days from the start to the point at which coating was dried and the vehicle was got ready for sale. Between the 1920s and the 1940s, alkyd resin-based "stoving
enamels" and spray apparatus were introduced in automotive coating technologies,
which cut usages and drying periods – up to seven to eight days or lesser than that.
The surface finishes were more uniform and required less sanding, as a result of the
recently developed spray coating technology. Nitrocellulose lacquer systems with several color options and better employability for the usage of spray paint guns were
created by E. I. DuPont De Nemours in 1923 [3]. These lacquer systems had a relatively
low resistance to chemical solvents like hydrochloric acid due to their chemical makeup and required an application of three to four coats to attain the necessary surface
features. This drawback made it harder for coatings to withstand acidic environments
and other chemicals. The advent of "alkyd" enamel paint on a few car models in the
early 1930s was another notable advancement in paint technology [4]. The molecular
bonding events that took place meant paint was sprayed onto the automobiles and
was then dried or desiccated in an oven, which caused these enamels to form an extremely durable coating. With the advent of acrylic stoving enamels in the 1960s, the
longevity of enamel coatings was significantly increased [5]. They were sprayed on
with a paint sprayer, cooked in the oven, and left with a tough, glossy finish. Safety,
environmental, and processing problems hindered the use of the dip-coating technique in the 1950s. Explosion and fire risks plagued the usage of solvent- or water-
based dip tanks [6]. A novel wet-on-wet finish with a thin basecoat and a thicker clear
coat was created and released in the late 1970s to further enhance the appearance
and toughness of the coatings [7]. In the late 1980s, basecoat and/or clear coat processing had gained popularity [3] and was used on the majority of vehicles produced of
the day. Costs were brought down by improvements in the materials and in processing methods. Moreover, the very first water-based basecoats and primer surfacers
were established by Opel in Germany in the 1980s and 1990s, respectively [8]. Nowadays, a two-component (2 K) formulation is the foundation for the majority of clear
coatings in Europe. OH-functional acrylic resin and a reactive polyurethane cross-
linker are both included in this composition. All around the globe, coating primarily
employs a one-component composition grounded on acrylic resins and melamine

cross-linkers [9]. Improvements in processing and paint chemistry have been made with novel advances in paint pigments. For instance, interference pigments that change color, contingent on the angle from which they are observed (often referred to as the "flip flopping" effect), and flake-based aluminum pigments have improved the brilliance, color, look, and consumer satisfaction of vehicle coatings [10]. The novel pigments were first difficult to employ with spray gun technology, but to address these difficulties, latest spray guns along with spray gun configurations have been established. Automotive coatings are still evolving in order to either fulfil or anticipate meeting customer demands, environmental restrictions, and reducing the cost of ownership.

Automotive coatings have traditionally had two purposes: adornment and preservation. The former entails not only giving a colored and smooth surface but also emphasize the contour of the automotive body using a brightness and/or color that is dependent on the angle of sight. These optical sequels are primarily based on microscopic mirrors made of coated mica platelets or metal flakes-like aluminum that are more or less uniformly spread inside one coating layer, with their surface normal being primarily orthogonal to the substrate's surface. Typically, the particle's lateral diameters are in the range of multiple hundred micron. The primary function of coatings in maintaining a metallic automobile body is to protect the substrate from electrochemical deterioration (corrosion), either directly or by acting as a barrier layer. They also serve to limit the degree to which coating defects are brought about by significant mechanical perturbations such as gravel stones slamming into the automobile. On the other hand, mounted plastic parts or body panels typically need to be safeguarded against chemical deterioration brought on by UV-vis light exposure, physical erosion brought by solvents such as fuel or aqua, determined by the characteristics of polymers and the morphological structure of the bulk, as well as catastrophic collapse of parts upon mechanical impacts (called as brittle crash behavior). The techniques used to accomplish the latter are analogous to those utilized in the development of stone-chip-resistant coatings for automobile bodies.

16.2 State-of-the-art automotive coating procedures

Since the industrialization of the automobile production process a few decades ago, recent automotive coatings have indeed been enhanced through an evolutionary development, bringing in a conventional stack of extremely specialized specific coating films (depicted in Figure 16.1) consisting of five major steps as follows:
1) Pretreatment process
2) The electrodeposition (ED) of a corrosion preventive layer. This consists of the cathodic electrodeposited material that first acts as an adhesive and active corrosion inhibitor (ED coat),

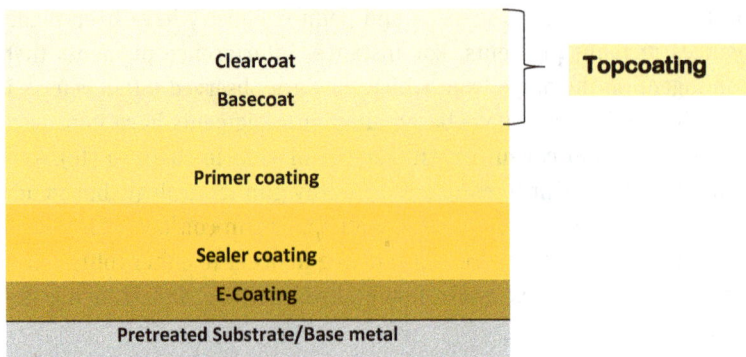

Figure 16.1: Multifunctional coating layers in automotive coatings.

3) Application of sealer such as Poly Vinyl Chloride (PVC) to mitigate corrosion, control of aqua leakages and, minimize chipping and vibrating sounds, which is then covered by a spray-coated layer (primer) to smoothen it out and shield it from UV rays.
4) Next, a primer is utilized to help the surface and the basecoat adhere to one another. It, moreover, provides a glossy finish for the subsequent layers and also has anti-chipping capabilities.
5) The final step is applying the topcoats, which consist of a basecoat and clear coat. These coats offer the desired surface qualities, such as colors, look, brightness, texture, and weather resistance.

Two layers are successively sprayed onto the primer surface: first is a basecoat that imparts color; next is a transparent topcoat (also known as a "clear coat") that makes the entire system bright, smooth, and resistant to chemicals (tree resin, bird droppings, tar, acid rain, etc.), mechanical smack associated with the surface, such as scratching during car washing, and surface-related chemicals. The automobile body must undergo three baking cycles at 170 °C (elctrodeposition) or at 140 °C because all layers other than the basecoat are chemically processed, following physical drying and film creation (combined base and clear layer).

Other unique coating systems that consist of stacks with fewer layers such as those without the primer [11] have recently been created in an effort to make this coating process more environmentally friendly [11]. These lean coatings demonstrate the inadequate capacity to retain a high degree of performance of their sophisticated technology with more advanced counterparts under challenging situations. This is the environment in which new material concepts are being developed in an effort to find ways to get over the limitations imposed by the most newest coating systems, which often still rely on conventional technology. In relation to the conservation functionality of automotive coatings, it must be noted that in contrast to what is occasionally implied by books and inventions, some traits have not been solely referred to as one single layer such as the relationship between the properties of the primer layer and the coating's resistance to stone chips. The prospective uses of the recent materials theories presented here should not be seen as being limited to certain prototypes, substrates, or applications, even though they are normally evaluated and introduced within one coating layer.

16.2.1 Pretreatment process

Prior to real electrodeposition, the base metal requires to be prepared in order to be cleansed, degreased, and activated as the electrodic surface; only after that can galvanic deposition be done. The proper substrate surface preparation is essential to all metal finishing applications [12, 13]. Well-established plating processes on aluminum and/or magnesium pretreatment offer the potential to handle/establish these light metals with high ornamental and functional finishing, with weight reduction being a top automotive concern. Cleaning is a crucial and basic first step to succeed with any of the metal finishing procedures, regardless of the ultimate finish. For surface preparation and appropriate cleaning, we can utilize the following technologies that are both affordable and environmentally friendly:
- Metal pretreatment systems
- Acidic and/or alkaline soak cleaners
- Spray cleaners
- Phosphating
- Electro cleaners
- Fluid cleaners
- Long life spray cleaners
- Frosty cleaners

Rarely, but increasingly frequently, multilayer depositions are used to coat the substrate with just one metal. As peeling occurs as a result of internal tensions, this method is recommended rather than straightaway depositing the final alloy over the substrate, which frequently causes adhesion issues. Although there are a number of galvanic methods used in the metal finishing industry, traditionally (for ZAMA, brass, and bronze base metals) a thick layer (of 5 to 20 micron) of Ni or Cu is initially deposited and this interlayer offers a surface that may be easily levelled. These metals typically are not the last layer to be applied; instead, more layers are typically added until the desired hue is achieved. The inter-diffusion of these metals between several layers, which results in a gradual switch in the product's characteristics, is another crucial consideration. The phenomenon of diffusion is significantly present in the matter of Au coating over Au [14], with the resultant reddening and loss in grade of the invaluable finish and, because of this intention, one or more barrier layers are often placed between these two depositing layers.

Originally, nickel deposits, between 1 and 5 micron, were the most typical barrier layer; however, due to environmental and allergic issues, bronze deposits between 2 and 5 micron increasingly took its position, followed by palladium deposits (between 0.5 and 1 micron). Similar steps are applied to aluminium substrates as is done for brass, but in order to prevent adhesion issues caused by the passivation layers that develop on this metal, a Zn electro-less coating must be applied before the real galvanic procedure. A different procedure is used for silver substrates; first, a layer of galvanic Ag metal is applied to remove the exemplary porosity that results from lost-wax casting stocks. Next, a layer of Ro (rhodium) is applied, interspersed with a layer of Pd (palladium), to keep the metal from oxidizing and turning black, while maintaining its silvery appearance. Sometimes, a flash deposition (usually 0.05–0.02 micron) is applied after the precious layer (generally 0.5–1 micron) to give the surface a unique hue. As a result, we use terms like yellow gold, light gold, rose gold, black gold, etc. This flash deposit may also be partly mechanically eliminated to achieve a particular effect, exposing the underlying layer.

16.3 Electrodeposition

Automobiles have coatings applied to the metal frames and underbody to stop corrosion, but not to other surfaces like the roof. The base metals or any other materials are primed before further coatings are applied to avoid corrosion. The electrodeposition (ED) coat, often known as the E-coat, was initially introduced in the 1960s, and protects against rust and corrosion. Subsequently, the usage of E-coat has rapidly increased, and 10% of all automobiles had electrocoating by 1970, and 90% had it by 1990. It is currently the most widely utilized coating method in the automobile industry. Since initially launched, the E-coat process has expanded quickly and has seen significant development. The anodic E-coat technique was utilized from 1964 to 1972, and the cathodic E-coat process has been used since 1976. A combination of resin, binder, a paste containing the colors, and a solvent make up the E-coat composition. Metal components are positively charged during anodic electrodeposition, whereas the paint is negatively charged; this affects the performance attributes of the coating, owing to migration of little quantities of metal ions into the paint layer. As a result, anodic coatings that provide excellent color and gloss control are typically employed on interior surfaces. The metal component is negatively charged during cathodic electrodeposition, whereas the paint is positively charged. The cathodic process, in contrast to the anodic process, limits the amount of iron that can get into the paint coating. As a result, cathodic coatings in North America have largely reinstated anodic coatings because of their extraordinary performance and outstanding corrosion resistance. Because it only utilizes 0.5%

solvent, the E-coat method is thought to be reasonably eco-friendly [15]. The contamination of coated surfaces by water spots is a prevalent issue in the present E-coating technique. If water smears on the air-dried surface from the conveyor or from any other source, these spots may appear, especially if aqua in the tanks has not been appropriately deionized and has a high level of conductivity. The aqua spots are typically a by-product of undesirable ions in the aqua and will interfere with the adhesion and appearance of subsequent coating applications. Therefore, it is essential to utilize adequately deionized water and to regularly check its conductivity. To manage or remove water spots, operators may additionally add surface-active agents or ultra-filtrates to the water cleansing zone.

16.4 Sealer/PVC

The third phase involves applying an underbody coating (UBC) and sealing seams with urethane and PVC (polyvinyl chloride), which are useful for a variety of purposes in automobiles, such as, sealing, noise proofing, interior and exterior sealing, and vibration deadening. The sealant is placed all around and within the doors, trunk, hood and, front dash, as well as to the outside and interior of metal joints and the outer area of the back wheel as well. It is applied manually or by robots, and it stops the entry of air and water and inhibits the growth of rust. In order to add noise-proofing and vibration-dampening, PVC and urethane/acryl sealants have also recently been employed to the underside zones in a procedure known as a Dampening Coat (DC). The underside sealants lessen sound transfer into the car's traveler compartment. Sound and vibrations are communicated from the suspension system, engine, drive train, road sounds from tires, and circulating air [16]. A robot fitted with an airless sprayer is commonly used to apply the underside coating, which offers anti-corrosion and chipping safeguard too. In the third phase, a soft-tip priming coat that increases resistance to chip is often applied as well (resistance toward chipping done by flying stones or trash). This coat is employed to the anterior side of the hood that is susceptible to chipping shock, employing a higher elastic resin that is sandwiched between the ED and the primer coating. A rather dull black pigment, known as a "blackout coating", is also applied to the body's underside, radiator supports, wheel housings, and rear sections during the third step.

16.5 Primer

Applying a primer surfacer, also known as just primer, is the fourth step in the coating process. It may be powder- or solvent-borne, or it may be waterborne. Up to 1990, increasing weather protection, aesthetics and chip resistance were the key reasons for applying primer. Since 1990, primers in solvent-borne, waterborne, and powdered forms have been created (explicitly to reduce the quantity of volatile organic compounds (VOC) released to the environment or contaminant seize equipment). The present day's primers need to defend against chipping, offer better paint look, and be environmentally friendly in order to comply with emission standards. They also need to increase adhesion between the E-coat and the topcoat. The important function of the primer is filling and smoothening minor scratches and imperfections that might be created during prior steps. Moreover, pressing, stamping, and welding processes sometimes result in surface flaws in automotive bodies, and the ED process tends to highlight these flaws. Hence, even though the primer's primary function is to operate as a leveler to create a smoother final surface, it also offers supplementary barriers against rust and maximizes adhesion between the surface, the E-coat, and the basecoat, increasing paint endurance. The smoothing properties of primers, however, lose significance as pressing and stamping processes get more refined and produce fewer flaws. Primers are often examined for surface elasticity and hardness during the development process as well as for their ability to adhere to base coating and ED coating materials [17]. Testing for surface hardness is essential because it has a big impact on how well defect-correction procedures, like sanding work etc. Elasticity gives a clue as to how well the priming surfacer coat could be able to protect against stone chips. For corrosion to be reduced or prevented and to provide the best possible surface texture, adhesion to ED coatings and basecoats is too crucial. The chemical reactivity of primers to ultraviolet (UV) radiation and utmost climatic conditions, like excessive humidity and high/low temperatures, must also be evaluated because automobiles are continuously prone to such conditions throughout their lifetime [17]. This is in addition to their mechanical and adhesive properties. At an automobile assembling facility, primer coating procedures are identically performed in the three following steps: interior coating, external coating, and microwave curing. The interior of the doors, engine compartment, and baggage or boot room is painted manually using spray paint. Several of these places do not get basecoat spraying. For being proficient and to present uniformity in coating layers between all the interior and exterior layers, it is critical to match the primer color with the basecoat.

16.6 Topcoat

Applying the topcoat, which is composed of the basecoat and clear coat in two coats, is the last stage in the automotive coating procedure. The basecoat consists of the main coloring pigment, and the clear coat offers protection from the elements, corrosion, and UV light deterioration; it promotes unparalleled color retention and offers an even, glossy unblemished, and uniform texture [18]. A basecoat, which is either water or solvent-borne, is primarily applied to the body. When the producer employs a wet-on-wet procedure, the clear coat is then applied over the basecoat after a brief flash-off and is then baked to cure. The basecoat is restored before the clear coat is applied to the surface in the absentia of a wet-on-wet technique. Irrespective of the method opted, identical thermal cure periods and temperatures range from 30 to 40 min at 125 °C are selected [19].

16.6.1 Basecoat

The basecoat, which is a component of the topcoat, is the third layer to be applied to the automobile, following the primer and before the clear coat. It gives the automobile its color; there are currently about 40,000 known basecoat colors, and each year, approximately 1,000 new colors are added to this list [20]. Automobiles typically have three different basecoat types: waterborne, solvent-borne high solids (HS), and solvent-borne medium solids (MS). The spray viscosity of the MS basecoats at 1,000 rpm is 100 mPa/s, and they have solid content levels of 15%–20%. The HS basecoats were created from MS basecoats in response to the U.S. Environmental Protection Agency's (EPA) request for solvent-borne spray paints to have higher solid levels and lower amounts of organic solvents in order to reduce VOC emissions. Due to their favorable environmental effects, waterborne basecoats have replaced conventional basecoats since the year 2000 in the entire U.S. automotive production sector.

16.6.2 Clear coat

The clear coat, a final coating to be applied to an automobile's bodywork, provides durability, resistance to environmental etching, and scratch resistance to the coating's overall finish. In addition to providing protection against damage, like fading brought on by ultraviolet radiations from the Sun, it also adds gloss and depth to objects that might not otherwise, be as vibrant. Repairs and upkeep are further simplified by the clear coat. Environmental etch is a visual problem caused by the appearance of stains caused by substances, bird droppings, and tree resin, or other substances that come in contact with the surface of a car and leave behind indelible water spots or stains that cannot be removed. Etching causes the clear coat surface to distort or develop pits, which contributes to the physical harm that follows. Automotive clear coats are created with a combination of UVA (ultraviolet light absorbers) and HALS (hindered amine light stabilizer) compounds to provide them UV endurance. These substances' main purposes are to absorb UV light with a wavelength between 290 and 400 nm and to inhibit UV-induced breakdown of the polymeric backbone (via HALS). After four years of Sun exposure, it has been observed that the concentration of UVAs in clear coats drops by roughly 50%; this depletion is due to photochemical breakdown and the subsequent diffusion from the clear coat. The long lasting finish of an automobile's surface is also influenced by the clear coats' scratch resistance, which has been linked to the density and flexibility of the polymer chain's cross-linking [21]. Because of their exceptional performance for automotive coating and well-understood and perfected application processes and chemistry, liquid clear coats are the most widely used type in the world. However, due to their advantages for the environment, such as the lack of VOC emissions during application, powder clear coats are becoming more and more common in the automotive sector [22].

16.7 Types/usage/applications of coatings

16.7.1 Decorative coatings

The complete gamut of ornamental/embellishment and functional surface treatments for a number of materials such as, steel, aluminum, and even, plastics are enfolded by the product line for decorative coatings. Starting with jewellery to different parts for the automobile [23, 24], sanitary, or electronics divisions, grade and dependability of dazzling finishes, rust-free safeguard, and durability are essential. Currently, plastics and metals are used in applications ranging from radiator grills to door handles and bumpers to automotive wheels. Variety of decorative coatings have been created using: coating on plastics [25], nickel [26], precious metals [26], copper [26–28], zinc [29–31], chromium [32, 33], bronze [34, 35], palladium [36–38], gold [39–45], silver [46], etc.

Plastic as well as plastic composite materials are still in demand because they incorporate best aspects of both worlds. Plastics are lightweight and corrosion-free, and they can be transformed into almost any shape. Even sophisticated components could be mass produced inexpensively. When plastics are applied on some kind of metal surface, they appear elegant and high-end. Because of this, a growing number of sectors are starting to grasp the advantages of plating plastics [47]. Non-chrome substitutes for such decorative top-coat applications include electrolytic nickel-tin, cobalt-tin, white bronze, and bright, and high-speed electro-less nickel. The printing technique known as rotogravure [48] (sometimes known as engraving) is well-known for producing excellent colors and gloss when reproducing vast numbers of copies repeatedly. Rotogravure is the process of transferring a blueprint on a printing image carrier, which is a steel or aluminum base cylinder with copper plating and a chrome finish. To produce the required hardness and wear resistance for printing, a thin layer of chromium is required. The surface is either chemically etched or, more frequently, electronically engraved to create different-sized cells of varying depth and shape. Silver electrolytes have a wide range of uses, including post-treatment procedures for outstanding anti-tarnish on Ag and Ag-plated products, rack and barrel plating, and other plating methods. Au (gold) is the perfect material for a broad range of uses because of its distinct visual appeal and durability. Gold provides a dependable plating for the technical and industrial reasons like those in the electrical and electronics industries because it is also a superior conductor and almost completely resistant to corrosion and wear. Pd and Pd-Ni coatings are excellent choices for both decorative and practical purposes. Many of the characteristics of gold coatings are mirrored by palladium coatings, which are also far less expensive. Diffusion barriers and conditions requiring a nickel-free coating both call for pure Pd coatings. The amazing visual effect is produced by ruthenium's anthracite-colored, highly transparent visual approach. The decorative effects that may be accomplished with Ru are incomparable and give great wear resistance, particularly when combined with gold or palladium coating. Acid Copper is the most often used method of decorative copper plating. The high ductility of acid coppers reduces issues with flexing and thermal expansion. Applications on Zn, Al, brass, bronze, steel, different other metal alloys, and ABS, ABS/PC, or the majority of other regularly plated substrates are suitable for acid Cu. Due to its ductility, acid copper is perfect for plating plastics to create a layer that will absorb the difference in thermal expansion between the layers of plastic and nickel. The Cu layer is simple to polish, which is frequently needed for complicated cast Al components like Al wheels. Acid Copper is adaptable and simple to use. Even in places with low current density, it is quite bright to make troubleshooting simpler. The technology is particularly resistant to burning in locations with high current densities.

16.8 Functional coatings: corrosion resistant

Corrosion is the destructive attack on the metal and its environment caused by a chemical or an electrochemical reaction. Corrosion is a natural process in which any unstable metal reacts with the surroundings (oxygen, moisture, or impurities) to form a stable product on the exterior. This stable byproduct results from a chemical process that results in metal loss, and is used to estimate corrosion. Due to fluctuations in temperature and relative humidity, the environment predominantly damages metals during transit and in storage. The metals used in the automotive industry to make the body, body panels, and frameworks exhibit a wide range of approaches in fabrication, safety improvement, and enhanced level of flexibility. As a result, a wide range of various metals and intermetallic alloys have been created using steel that is zinc, aluminum, or magnesium coated. The majority of the naturally occurring oxide or oxide-hydroxide passivation layers on these metal or alloy surfaces do not show enough resistance to chemical deterioration in the severe environments where they frequently form during the course of the corrosion processes. One more thing to consider is that unexpectedly, a metal substrate made of just one metal does not have a chemically uniform surface. Another factor that affects corrosion protection is the possibility of micrometer-scale imperfections like porosity, point defects, and splits in practically every thick organic coating system, as well as in regionally variable cross-linked densities in reactive systems. Furthermore, capillary forces and diffusion channels for aqueous solutions may be present in heavily packed pigmented coatings such as dipole surfaces or on electrical dual layers at the surfaces of seeping pigments or filler granules. Consequently, in order to create a fine, stable layer on the nano- to micrometer scale, the conversion coating, the metal surface, or even the passive film layer should be modified or regenerated, respectively, prior to adding such coatings. For a worldwide business with significant environmental stipulations, corrosion-resistant coatings are utilized as functional coatings. Corrosion-control coatings are often about Zn (Zinc) and Zn alloy, conversion coatings, and various sealers. The global automobile sector is compelled to prolong vehicle warranties due to the rising demands of consumers. So, it is becoming more and more crucial to increase corrosion protection from brilliant acid Zn baths to cyanide-free alkaline Zn electrolytes that are suitable for the environment to high performance electrolytes for Zn alloys and mechanical plating solutions. For instance, through a consistent thickness distribution, the Zn-Fe and Zn-Ni procedures offer outstanding corrosion resistance even for intricately formed objects, and these very malleable layers also permit machining. By better corrosion resistance, zinc coatings improve the performance and prolong the service life of components. They make it possible to build materials like steel to be integrated with their useful bulk properties due to their advantageous surface qualities. Applications of zinc must be done in ways that are both economically and environmentally sound. In the present scenario, industrial applications frequently preclude the utilization of conventional Zn coatings due to their performance limitations; therefore, advancements in corrosion protection are a crucial factor to take into account. Zinc alloys offer more corrosion resistance than pure zinc. The necessity to increase the component's service life through improved coating qualities led to the development of alloy usage. The coatings are chromated or passivated as part of a subsequent process to shield the zinc from "white rust" deterioration. An additional treatment that stabilizes the conversion coating is the application of a sealer. Surfaces are coated with conversion coatings through immersion. There are several colors of traditional chromates made with hexavalent chromium. Varying chromate colors signify variations in the film's thickness and makeup, which have an impact on the level of corrosion protection that can be expected. Where a chromium-like finish is sought, clear blue films are typically utilized. Films that are yellow and olive green are chosen because they offer the best corrosion protection. Black films are frequently employed to produce a decorative surface that is both aesthetically pleasing and effective against corrosion. As an alternative to chromates, passivates based on trivalent chromium are utilized. A pleasant chromium-like color and high levels of corrosion protection are provided by thin film blue passivates. Iridescent passivates with a thicker coating produce a thickness akin to yellow chromates and provide greater corrosion protection, especially following heat treatment. Further, there are many new efficient, low-cost, environmental friendly corrosion inhibitors [49–51] that have been introduced and they have proven to be excellent corrosion mitigators on mild-steel surfaces in acidic media. And, they can be explored for the same purposes in automotive sector.

16.9 Functional coatings: wear resistant

The next functional surface use involves applications for increased wear resistance and reduced friction. Optimal wear protection is required for a long life because vehicle parts often encounter harsh environments. Delivered for automotive applications are hard Cr and electro-less Ni coatings with high wear resistance. Functional coatings that are wear-resistant can be created by Hard Chrome and Electro-less Nickel. Because of its higher hardness characteristics and lower friction coefficient, the high-efficiency hard chromium method offers exceptional wear performance in piston rings, engine valves, and suspension parts. Maximum wear resistance, apex levels of corrosion prevention, and the smallest coefficient of friction are all features of the hard chrome process. The intended surface characteristics can be varied to a great extent with carefully controlled modifications to the electrolytic and deposition parameters. The superior performance of this method for applications that involve dry friction, lubricating sliding wear, as well as corrosion mitigation in suspension systems has indeed been proven by scientific investigations and practical experience. The hard-chrome layers typically display high micro-hardness. Micro-hardness values for chrome coatings with a matte or satin finish, like hot chrome, are often at the lower end of the spectrum and exhibit excellent resistance to a wide range of acidic and alkaline environments. The layers and channel systems' believable corrosion resistance is based on this aspect. A high level of corrosion resistance is offered by hard-chrome coatings that contain microcracks. Besides these beneficial surface qualities, the network of microcracks also offers increased oil adherence on hydraulic components, for instance. Brakes, fuel systems, planetary gears, valve stems, and steering components are just a few examples of the many applications where electro-less nickel plating methods offer the best wear performance. These procedures provide a great degree of pliability to precisely supply the required hardness and affinity to lubricants to lessen wear on highly mobile elements and to enable them function at their best as a component of the entire system. Electro-less nickel coatings are non-magnetic and can sometimes be manufactured so that they can be soldered. They are also resistant to oxidation and chemical sensitivity. In addition to creating wear-resistant surfaces for a range of automotive components, these solutions are also utilized to extend the appliance's long-lasting life and improve its all-inclusive grade for the automobile manufacturing industry, particularly in processing and stamping plants. Electro-less Ni coatings are extremely resistant to chemicals; they are hard, non-magnetic, resistant to wear, corrosion resistant, resistant to oxidation, and can be soldered.

16.10 Functional electronics' coatings

The special demands of automotive electronics have a consequence on the production techniques and procedures of coatings for functional electronics . Electronic components are crucial in today's cars, trucks, and motorbikes because they regulate all of the mechanical, electrical, and chemical processes involved in movement and operation. These electronic devices are pivotal for the onboard amusement as well as for communication and safety. Automotive applications, including occupant protection, driver assistance, and transmission control systems, are major development sectors in this industry. The contribution of electrical components to the entire cost of manufacturing is expected to increase dramatically. Functional coatings provide coating to the high-tech electronics that the automotive sector specifically demands. The complete range consists of both materials and methods for pre- and post-treatment as well as plating with precious and non-precious metals. Not only these coatings need to meet the highest quality standards, but they must also abide with the current environmental laws. Vehicles are becoming much more sophisticated, owing to the increase in their technologically enabled features. The main issue for automobile makers has always been ensuring the operational integrity of electronic components. Quality and yield are the challenges from the manufacturing perspective. High-tech electronic components are necessary to meet the demanding requirements of automotive applications in terms of temperature, humidity, endurance (life-cycle), and mechanical shocks. Ceramic substrates are perfect for a variety of automotive applications due to their thermal and dimensional stability. Functional electronics' coatings are concerned with high-speed Cu/Ni plating, and coating with Ag and Pd/Ni alloys, Sn and Sn/Pb alloys, and Au and Au alloys.

16.11 Conclusion

By enabling researchers to modify the properties of the current materials in an extremely and granular manner, thin film technology is playing an important part in cutting-edge research. Several harmful coating techniques have already been replaced by thin film technology in the automobile sector, and improvements in speed and uniformity will encourage more manufacturers to switch to environmentally friendly thin film procedures. Manufacturers may shift away from wasteful procedures toward more sustainable, environmentally friendly products that will last longer and be more readily produced by enhancing the durability of consumer electronics, optical displays, and mechanical components. Based on the contemporary procedures of automotive coating, an important aspect that has to be considered is the development and further research along the lines of cost effectiveness, sustainability, and environmental friendly materials meeting the requirements of customers world wide.

References

[1] Louda P. Applications of thin coatings in automotive industry. J Achiev Mater Manuf Eng 2007;24 (1):51–56.

[2] Kensium. History Timeline and Types of Automotive Paint. Available online: http://www.eastwood. com/blog/eastwood-chatter/history-timeline-and-types-of-automotive-paint

[3] Khanna AS. High-performance organic coatings. Cambridge, UK: Woodhead Publishing Limited; 2008.

[4] Standeven H. The development of decorative gloss paints in Britain and the United States C. 1910–1960. J Am Inst Conserv 2006;45:51–65.

[5] Learner T. A review of synthetic binding media in twentieth-century paints. Conservator 2000;24:96–103.

[6] Streitberger H-J, Dossel K-F. Automotive paints and coatings. Weinheim, Germany: Wiley-VCH Verlag GmbH & Co. KGaA; 2008.

[7] Fettis G. Automotive Paints and Coatings. New York, NY, USA: Wiley-VCH; 2008.

[8] Jürgens U. Implanting Change: The Role of 'Indigenous Transplants' in Transforming the German Productive Model. In: Boyer, R, Charron, E, Jurgens, U, Tolliday, S (Eds) The transfer and hybridization of productive models in the international automobile industry. Oxford, UK: Oxford University Press. pp. 319–41; 1998.

[9] Melchiors M, Sonntag M, Kobusch C, Jürgens E. Recent developments in aqueous two-component polyurethane (2K-PUR) coatings. Prog Org Coat 2000;40:99–109.

[10] Maile FJ, Pfaff G, Reynders P. Effect pigments – past, present and future. Prog Org Coat 2005;54:150–63.

[11] Kreis W. 26th Eurpean Car Body Conference, Bad Nauheim. 27 Febr. 2008.

[12] Larson C. Global comparisons of metal finishing sectors: Part 2, some technology and operational variations. Trans IMF 2012;90:232–36.

[13] Veinthal R, Kulu P, Žikin A, Sarjas H, Antonov M, Podgurski V, Adoberg E. Coatings and surface engineering. Industry oriented research. Est J Eng. 2012;18:176–84.

[14] Siu C, Man H, Yeung C. Electrodeposition of Co-Mo-P barrier coatings for Cu/Au coated systems. Surf Coat Tech 2005;200:2223–27.

[15] Loop FM. Cathodic automotive electrodeposition; No. 780189. SAE Technical Paper, SAE International: Warrendale, PA, USA. 1978.

[16] Tomalino M, Bianchini G. Heat-expandable microspheres for car protection production. Prog Org Coat 1997;32:17–24.

[17] Misev TA, Van der Linde R. Powder coatings technology: New developments at the turn of the century. Prog Org Coat 1998;34:160–68.

[18] Akafuah NK Automotive paint spray characterization and visualization. Automotive Painting Technology. Dordrecht, The Netherlands: Springer. pp. 121–65; 2013.

[19] Wu YH, Surapaneni S, Srinivasan K, Stibich P. Automotive vehicle body temperature prediction in a paint oven. No. 2014-01-0644; SAE Technical Paper, SAE International: Warrendale, PA, USA. 2014.

[20] Wu YH, Surapaneni S, Srinivasan K, Stibich P. Automotive vehicle body temperature prediction in a paint oven. No. 2014-01-0644; SAE Technical Paper, SAE International: Warrendale, PA, USA. 2014.

[21] Wu YH, Surapaneni S, Srinivasan K, Stibich P. Automotive vehicle body temperature prediction in a paint oven. No. 2014-01-0644; SAE Technical Paper, SAE International: Warrendale, PA, USA. 2014.

[22] Wu YH, Surapaneni S, Srinivasan K, Stibich P. Automotive vehicle body temperature prediction in a paint oven. No. 2014-01-0644; SAE Technical Paper, SAE International: Warrendale, PA, USA. 2014.

[23] Presuel-Moreno F, Jakab MA, Tailleart N, Goldman M, Scully JR. Corrosion–resistant metallic coatings. Mater Today 2008;11:14–23.

[24] Nuss P, Eckelman MJ. Life cycle assessment of metals: A scientific synthesis. PLoS One 2014;9: e101298.

[25] Krug T, Tietema R. Decorative PVD coatings on automotive plastic. Coat Tech Veh Appl 2015;215–30.

[26] Giurlani W, Zangari G, Gambinossi F, Passaponti M, Salvietti E, Di Benedetto F, Caporali S, Innocenti M. Electroplating for decorative applications: recent trends in research and development. Coatings 2018;8(8):260.

[27] Benner HL, Wernlund CJ. The high efficiency cyanide copper bath. J Electrochem Soc 1941;80:355–65.

[28] Willis WJ. Electrodeposition of copper, acidic copper electroplating baths and additives therefor. U.S. Patent 4,347,108, 31 August 1982.

[29] Grossblatt GH. Zinc plating baths. U.S. Patent 3,883,405, 13 May 1975.

[30] Grill CD, Kollender JP, Hassel AW. Preparation and investigation of combinatorially electrodeposited zinc-nickel, zinc-cobalt, and zinc-nickel-cobalt material libraries. Phys Status Solidi Appl Mater Sci 2017;214:1600706.

[31] Hegde AC, Venkatakrishna K, Eliaz N. Electrodeposition of Zn-Ni, Zn-Fe and Zn-Ni-Fe alloys. Surf Coat Tech 2010;205:2031–41.

[32] Carneiro E, Parreira NMG, Vuchkov T, Cavaleiro A, Ferreira J, Andritschky M, Carvalho S. Cr-based sputtered decorative coatings for automotive industry. Materials 2021;14(19):5527.

[33] Ponte F, Sharma P, Figueiredo NM, Ferreira J, Carvalho S. Decorative chromium coatings on polycarbonate substrate for the automotive industry. Materials 2023;16(6):2315.

[34] Hovestad A, Tacken RA, Mannetje HH. Electrodeposited nanocrystalline bronze alloys as replacement for Ni. Phys Status Solidi C 2008;5:3506–09.

[35] Dos Santos WAT, Dos Santos WIA, De Assis SL, Terada M, Costa I. Bronze as alternative for replacement of nickel in intermediate layers underneath gold coatings. Electrochim Acta 2013;114:799–804.

[36] Atkinson RH, Raper AR. The Electrodeposition of Palladium. Trans IMF 1932;8:10–11. 10–24

[37] Pushpavanam M, Natarajan SR, Balakrishnan K, Sharma LR. Electrodeposition of palladium-nickel alloy. Bull Electrochem 1990;6:761–64.

[38] Baumgärtner ME, Gabe DR. Palladium-iron alloy electrodeposition. Part I single metal systems. Trans IMF 2000;78:11–16.

[39] Renner H, Schlamp G, Hollmann D, Lüschow HM, Tews P, Rothaut J, Dermann K, Knödler A, Hecht C, Schlott M, et al. Gold, Gold Alloys, and Gold Compounds. Available online: https://onlinelibrary. wiley.com/doi/abs/10.1002/14356007.a12_499

[40] Drost E, Haußelt J. Uses of gold in jewellery. Interdiscip Sci Rev 1992;17:271–80.

[41] Klotz UE, Tiberto D, Held F. Optimization of 18-karat yellow gold alloys for the additive manufacturing of jewelry and watch parts. Gold Bull 2017;50:111–21.

[42] MacCormack IB, Bowers JE. New white gold alloys – Their development on the basis of quantitative colour assessment. Gold Bull 1981;14:19–24.

[43] Cretu C, Van Der Lingen E. Coloured gold alloys. Gold Bull 1999;32:115–26.

[44] Fier-Bühner J, Basso A, Poliero M. Metallurgy and processing of coloured gold intermetallics – Part II: Investment casting and related alloy design. Gold Bull 2010;43:11–20.

[45] Corti CW. Blue, black and purple! the special colours of gold. In: Proceedings of the International Jewellery Symposium, St Petersburg, Russia. 2006.

[46] Missey RJ. Gold and silver plating basics. Prod Finish 2010;75:142–47.

[47] Olivera S, Muralidhara HB, Venkatesh K, Gopalakrishna K, Vivek CS. Plating on acrylonitrile-butadiene-styrene (ABS) plastic: a review. J Mater Sci 2016;51:3657–74.

[48] Bohan M, Claypole T, Gethin DT. The effect of process parameters on product quality of rotogravure printing. Proc Ins Mech Eng Part B J Eng Manuf 2000;214:205–19.

[49] Sowmyashree AS, Somya A, Rao S, Pradeep Kumar CB, Al-Romaizan AN, Hussein MA, Khan A, Marwani HM, Asiri AM. Potential sustainable electrochemical corrosion inhibition study of Citrus limetta on mild steel surface in aggressive acidic media. J Mater Res Technol 2023;24:984–94.

[50] Sowmyashree AS, Somya A, Kumar S, Pradeep Kumar CB, Rao S. Discotic anthraquinones as novel corrosion inhibitor for mild steel surface. J Mol Liq 2022;347:118194.

[51] Sowmyashree AS, Somya A, Pradeep Kumar CB, Rao S. Novel nano corrosion inhibitor, integrated zinc titanate nano particles: Synthesis, characterization, thermodynamic and electrochemical studies. Surfaces and Interfaces 2021;22:100812.

Elyor Berdimurodov, Khasan Berdimuradov, Ilyos Eliboev,
Anvar Khamidov, Oybek Mikhliev, Abduvali Kholikov,
Khamdam Akbarov

17 Influence of inorganic/organic additives on mechanical and electrochemical properties of functionalized thin film coatings

Abstract: The development of functionalized thin film coatings containing organic and inorganic additives is crucial for enhancing corrosion protection and performance in various industries. This book chapter explores the influence of organic and inorganic additives on the mechanical and electrochemical properties of coatings, as well as the synergistic effects achieved by combining these additives. We discuss the importance of understanding the influence of additives to tailor coatings for specific applications and optimize performance. The chapter also highlights the potential for further advancements in functionalized thin film coatings for corrosion protection by incorporating novel materials, advanced manufacturing techniques, and emerging technologies. Practical applications of these coatings in the aerospace industry, offshore structures, and electronics and sensors are also examined. This comprehensive review offers valuable insights into the development and application of high-performance coatings containing organic and inorganic additives for effective and sustainable corrosion protection.

Keywords: Thin Film Coatings, Organic Additives, Inorganic Additives, Corrosion Protection, Mechanical Properties, Electrochemical Properties, Synergistic Effects

17.1 Introduction

Functionalized thin film coatings have grown in popularity recently because of their numerous advantages and prospective uses. The introduction of chemicals that affect these coatings' qualities, such as mechanical strength, adhesion, and electrochemical activity, is crucial to their effectiveness. Thin film coatings may be improved by adding both inorganic and organic compounds, and their combination can have synergistic effects that raise the effectiveness of corrosion protection even higher.

This chapter seeks to offer an overview of the effect of inorganic and organic additives on the mechanical and electrochemical characteristics of these coatings given

Elyor Berdimurodov, Faculty of Chemistry, National University of Uzbekistan, Tashkent, 100034, Uzbekistan; Medical School, Central Asian University, Tashkent, 111221, Uzbekistan

https://doi.org/10.1515/9783111016160-017

the significance of functionalized thin film coatings in corrosion protection and the crucial role that additives play in their performance. The types of additives, how they affect the characteristics of coatings, and prospective uses in various sectors will all be covered in this chapter. We will also talk about the difficulties and prospects for developing improved functionalized thin film coatings for corrosion prevention. The issues created by corrosion in terms of the economy, structures, and environment can be addressed by functionalized thin film coatings. These coatings can demonstrate outstanding mechanical and electrochemical capabilities by combining inorganic and organic additives, making them perfect for a variety of applications. To create formulas that are optimal and fit the unique needs of various sectors, it is crucial to comprehend how additives affect the performance of these coatings [1–3].

Economic Effects. Corrosion is a widespread and expensive issue that impacts many different sectors and infrastructure throughout the world. Corrosion has a huge financial impact, with yearly expenses that top trillions of dollars worldwide. These costs range from direct expenditures like material replacement, upkeep, and repairs to indirect costs like lost productivity, downtime for operations, and environmental cleanups. Numerous sectors are significantly impacted financially by corrosion [4, 5]. The replacement of materials as well as maintenance and repairs, which can be expensive, are direct expenditures related to corrosion. For instance, it is estimated that the yearly direct cost of corrosion in the United States is around $276 billion, or 1.6% of the nation's GDP. According to estimates, the yearly cost of corrosion-related damage is in the trillions of dollars worldwide [6, 7].

Corrosion can entail indirect costs linked to lost productivity, downtime, and environmental cleanups in addition to direct expenses. For instance, corrosion in oil and gas pipelines may result in oil leaks, which may have serious negative effects on the environment and the economy. A blowout preventer's deterioration led to the *Deepwater Horizon* oil disaster in 2010, which cost more than $60 billion to clean up.

Corrosion has an economic impact that is not exclusive to a particular sector or area. Infrastructure, transportation, industry, and energy are just a few of the industries it has a large impact on, with serious ramifications for both established and emerging countries. By prolonging the lifespan of metallic materials and lowering maintenance and repair costs, employing efficient corrosion protection measures, such as the use of functionalized thin film coatings, can help lessen the economic impact of corrosion [8, 9].

Structural Integrity. Metallic materials' structural integrity can be compromised by corrosion, which increases the risk of fatal accidents, catastrophic breakdowns, and other negative outcomes. It is a crucial issue in sectors like aircraft, building, and transportation where the security and dependability of components and structures are crucial. Functionalized thin film coatings improve the overall security and toughness of these crucial applications by avoiding or reducing corrosion. The structural integrity of metallic materials can be negatively impacted by corrosion, which might result in catastrophic failures and possible fatalities. Corrosion-induced metal deterioration can result in cracking, pitting, and thinning, which can reduce a material's overall toughness and endurance. The loss of structural integrity can have serious repercussions in crucial applications including aircraft, building, and transportation [10].

For instance corrosion in airplanes can result in the weakening of vital parts, reducing their capacity to carry loads and making them more susceptible to fatigue cracking. This may jeopardize the aircraft's dependability and safety, endangering the lives of both passengers and crew. Similar to how it may weaken structural elements, corrosion in bridges and other infrastructure can increase the risk of collapse and jeopardize public safety. The issue of corrosion-induced metal deterioration can be solved by functionalized thin film coatings. Superior corrosion protection may be provided by these coatings, reducing the possibility of structural failure. Additionally by making metallic materials more durable, these coatings can increase the lifespan of buildings and components, lowering the frequency of maintenance and repair. The introduction of functionalized thin film coatings can improve the overall safety and dependability of the systems in demanding industries like aerospace and transportation.

Environment-Related Effects. Corrosion may have serious negative effects on the environment because it can cause toxins and harmful elements to be released into the air and water. The quality of the land, water, and air may be negatively impacted by this pollution, endangering both human health and ecosystems. Effective corrosion prevention measures assist to reduce the effects of corrosion on the environment while also preserving the durability of metallic materials.

Benefits over Conventional Techniques. When compared to conventional corrosion protection techniques like painting or galvanizing, functionalized thin film coatings provide a number of benefits. These coatings can provide the substrate improved durability, resistance to corrosion, and adhesion. Their efficacy can also be increased by engineering them to have self-healing capabilities or to be adapted to certain environmental circumstances. Additionally thin film coatings have a lesser environmental effect than conventional techniques and typically need less maintenance [11].

Versatile in Their Use. Functionalized thin film coatings are appropriate for a variety of applications and sectors due to their adaptability. These coatings may be tailored to fulfill certain performance needs and can be used on a variety of metallic substrates, including steel, aluminum, and copper. They are a desirable option for a variety of sectors, including automotive, aerospace, and marine as well as oil and gas, electronics, and renewable energy because of their versatility, which enables them to meet various corrosion concerns. Thin film functionalized coatings are useful in a variety of sectors and applications due to their adaptability in usage. Numerous metallic substrates, such as steel, aluminum, copper, and titanium, among others, can get these coatings. Additionally they can be modified to satisfy certain application needs, such as those for corrosion resistance, wear resistance, and thermal stability [12].

The capacity of functionalized thin film coatings to offer specialized corrosion protection is one of its main advantages. These coatings can be created to safeguard particular metallic component locations where corrosion is more likely to happen, such as joints and edges. This focused defense can lessen the frequency of repairs and replacements while extending the life of metallic components. Aerospace, automotive, construction, and the energy sectors are just a few of the industries that can benefit from functionalized thin film coatings. These coatings can be utilized in the aircraft sector to shield crucial components from corrosion and wear, such as engine parts and landing gear. They can be utilized in the automobile sector to enhance the performance and longevity of braking systems, transmission components, and engine parts. These coatings can be utilized in the building sector to increase the corrosion resistance of steel components and structures like pipelines and bridges. They may be applied to the energy sector to enhance the efficiency and dependability of pipelines, oil rigs, and other infrastructure [3, 8, 13].

17.2 Inorganic additives

17.2.1 Types of inorganic additives

Metal Oxides. A typical kind of inorganic additive utilized in coatings for a variety of purposes, including corrosion prevention, is metal oxide. Metal oxides are oxygen and metal compounds that have special qualities that make them appropriate for use in coatings. The capacity of metal oxide compounds to prevent corrosion is one of their main benefits. Zinc oxide and titanium dioxide are two examples of metal oxides that are well-known for their capacity to coat metallic surfaces with a protective coating that stops the underlying metal from corroding. The durability and longevity of metallic components can be increased with the aid of this protective coating. Metal oxide additions can offer advantages including UV protection, antibacterial capabilities, and enhanced adhesion in addition to their abilities to prevent corrosion. Since titanium dioxide can deflect UV radiation, it is frequently employed in coatings to lower the risk of fading and damage brought on by exposure to sunshine. Contrarily, zinc oxide possesses antibacterial qualities that make it ideal as a coating for medical equipment and supplies. Coatings can be made using metal oxide additions by a number of techniques, including as physical mixing, chemical synthesis, and sol–gel processing. The procedure used will depend on the precise application needs as well as the coating's intended qualities [14–18].

Metallic Salts. Metal salts are another type of inorganic additive often used in coatings for a variety of reasons, including corrosion protection. Ionic compounds called metal salts are made up of metal cations and non-metal anions. They are perfect for use in coatings because of their unique characteristics. One of the key advantages of metal salt additions is their ability to stop corrosion. Metal salts, such as zinc phosphate and iron phosphate, are known for their ability to alter the surface of metallic substrates into a more stable condition and prevent the underlying metal from corroding. The process of "phosphating" leaves a layer of metal phosphate on the surface of the substrate, acting as a barrier against corrosion. In addition to preventing corrosion, metal salt additives have the power to encourage adhesion and surface preparation [19]. For example zinc phosphate is widely used to clean the surface of metallic substrates before painting them. By improving the coating's adhesion to the substrate, this procedure can help the coating become more durable and long-lasting. Metal salt additives can be added to coatings via a variety of methods, including chemical synthesis, physical mixing, immersion coating, and others. The method used will rely on the particular application requirements as well as the desired attributes of the coating [20–24].

Metal-Organic Frameworks (MOFs). Inorganic additives known as metal-organic frameworks (MOFs) have attracted interest because of their distinctive characteristics and possible use in a wide range of industries. MOFs are porous materials made of organic ligands that connect metal ions or clusters to build a three-dimensional network of cages and channels. The large surface area and porosity of MOFs, which make them suited for use in processes like gas storage, separation, and catalysis, are among its main features. Additionally MOFs may be created to have certain characteristics like selectivity for particular gases or catalytic activity, which makes them valuable in a range of sectors. MOFs can be employed as coating additives for a variety of applications in addition to their ability to store gas and act as catalysts. For instance MOFs can be added to coatings for metallic substrates to act as corrosion inhibitors. Since MOFs are porous they may absorb and release corrosion inhibitors, preventing corrosion over an extended period of time. Different techniques, including as physical mixing, sol–gel processing, and spray coating, can be used to integrate MOFs into coatings [25–28]. The procedure used will depend on the precise application needs as well as the coating's intended qualities [29, 30]. For a variety of uses, such as gas storage, catalysis, and corrosion prevention, MOFs provide a flexible and efficient solution. They may be used in a variety of industries, including the automotive, aerospace, and energy sectors, because to their special qualities. To fully realize MOFs' potential and enhance their qualities for particular applications, more study is necessary [31–33]. Figure 17.1 shows: The Cu-TCPP MOFs provide a useful alternative for applications due to their lower toxicity and thinner thickness. In the experiment Cu-TCPP MOF ultrathin nanosheets and bulk material

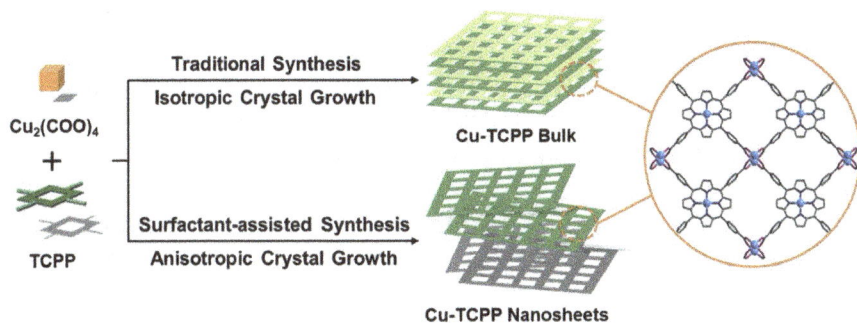

Figure 17.1: Schematic illustration of conventional and surfactant-assisted techniques for Cu-TCPP MOF production [34].

are prepared utilizing conventional and improved surfactant-assisted techniques. The improved approach, which uses polyvinylpyrrolidone (PVP) as a surfactant, helps limit the development of Cu-TCPP MOFs in the z-axis and provides structural support to prevent the framework from collapsing. The original method only produces isotropic growth and bulk crystals [34].

17.2.2 Effects on mechanical properties

Adhesion Strength. The capacity of a coating to stick to the surface of a substrate is determined by its adhesion strength, which is an essential mechanical feature of coatings. Depending on the kind and quantity of inorganic additives utilized, the adhesion strength of the coating might be impacted in a variety of ways. For instance metal oxide nanoparticles like zinc oxide and titanium dioxide are frequently utilized as inorganic coating additives. By making the substrate's surface rougher and creating more surface area for the coating to cling to, these nanoparticles can strengthen the coating's adherence. Additionally the substrate and nanoparticles can create chemical connections, which strengthens the coating's ability to adhere. On the other side overusing inorganic compounds might weaken the coating's ability to adhere. This could happen if the coating additives are not evenly distributed, which might lead to agglomeration, void formation, and other coating flaws. These flaws may degrade the coating and lessen its ability to adhere [35]. The technique of preparation can also have an impact on the adhesion strength of coatings incorporating inorganic additives. For instance the development of chemical bonds between the coating and the substrate during sol–gel processing can increase the adhesion strength of coatings. In contrast because there is no chemical connection between the coating and the substrate when the additives are physically mixed, the adhesion strength may be weak. Depending on a number of variables, including the kind and concentration of the additive, the technique of manufacture, and the dispersion of the addition in the coating, the use of inorganic additives can have both favorable and unfavorable impacts on the adhesion strength of coatings. Inorganic additives can be used effectively to increase coating durability and adhesion strength [18, 25, 33].

Hardness and Wear Resistance. The hardness and wear resistance of coatings are essential mechanical characteristics as well. These qualities can be greatly enhanced by inorganic additions by strengthening the coating matrix and raising its resistance to wear and abrasion. Inorganic coating additives like aluminum oxide, silicon dioxide, and titanium dioxide are frequently employed to increase the hardness and wear resistance of coatings. Due to their great hardness these nanoparticles can serve as reinforcements, boosting the coating matrix's strength and avoiding chipping and breaking. Other inorganic additions, such as carbon nanotubes and graphene, can also increase the hardness and wear resistance of coatings in addition to metal oxide nanoparticles. While graphene has a large surface area and can operate as a barrier to prevent wear and abrasion, carbon nanotubes have a high tensile strength and can act as reinforcing agents in the coating matrix. The coating's hardness and wear resistance are significantly influenced by the concentration and dispersion of inorganic additions. Overusing additives can cause agglomeration and the production of coating flaws, which can reduce the coating's mechanical strength [36]. On the other hand appropriate additive dispersion can result in more uniform reinforcement distribution throughout the coating, enhancing its hardness and wear resistance. The technique used to incorporate inorganic compounds into the coating can also have an impact on the coating's hardness and wear resistance. For instance, sol–gel processing can provide a coating matrix that is more homogenous and dense and has better mechanical qualities. On the other side physical mixing might lead to a weaker coating matrix and worse dispersion of the chemicals [20, 23, 27].

Elastic Modulus. The elastic modulus, commonly referred to as Young's modulus, is a mechanical characteristic that characterizes a material's stiffness or rigidity. The elastic modulus of coatings governs their capacity to withstand stretching or deformation under stress. Depending on the kind and quantity of inorganic additives employed, the elastic modulus of the coating might be affected in a variety of ways. For instance by strengthening and stiffening the coating matrix, metal oxide nanoparticles like alumina and silica can raise the coating's elastic modulus. On the other side overusing inorganic chemicals can lower the coating's elastic modulus. This can happen if the coating's ingredients are not evenly distributed, which can result in voids and other coating flaws. The coating may become less elastic as a result of these flaws [37]. The process used to include inorganic compounds into the coating can also have an impact on its elastic modulus. For instance sol–gel processing can provide a coating matrix that is more homogeneous and dense and has a greater elastic modulus. On the other side physical mixing could lead to a coating matrix that is less elastic and weaker. Environmental elements like temperature and humidity can also have an impact on the elastic modulus of coatings. By enhancing the coating's stability and longevity in various environmental settings, inorganic additions can produce a more constant and predictable elastic modulus [25, 32, 33].

17.2.3 Effects on electrochemical properties

Polarization Resistance. The degree to which a coating can withstand corrosion in an electrochemical environment is measured by its polarization resistance. By altering a coating's surface characteristics and boosting its corrosion resistance, inorganic additions can make it more resistant to polarization. Metal oxide nanoparticles are a typical inorganic addition used in coatings to increase polarization resistance. These nanoparticles can create a shield on the coating's surface, blocking the entry of corrosive chemicals and slowing the pace of corrosion. The polarization resistance of coatings can also be increased by adding additional inorganic ingredients like carbon nanotubes and graphene. These additives have the ability to create a conductive network on the coating's surface, improving ion and electron transport while slowing corrosion. The coating's polarization resistance may vary depending on the concentration and dispersion of inorganic additions. The coating may aggregate and develop flaws as a result of excessive additive usage, which may lower the coating's resistance to polarization [38]. A more uniform distribution of reinforcement throughout the coating due to proper additive dispersion might increase polarization resistance. The coating's polarization resistance may also be impacted by the process used to include inorganic chemicals. For instance, sol–gel processing can result in the creation of a coating matrix that is more homogeneous and dense, improving corrosion resistance. On the other side physical mixing might lead to a weaker coating matrix and worse dispersion of the chemicals [14, 16, 26].

Electrochemical Impedance Spectroscopy (EIS). The electrochemical characteristics of coatings can be investigated using the potent electrochemical method known as electrochemical impedance spectroscopy (EIS). EIS is applying a low-amplitude AC signal to a coating submerged in a solution and evaluating the current response that results as a function of frequency. By altering the surface characteristics of the coating and the behavior of the coating/solution interface, inorganic additives can influence the EIS response of coatings. For instance, metal oxide nanoparticles can raise the coating's impedance, which lowers the rate of charge transfer and improves the coating's resistance to corrosion. By creating a conductive network on the surface of the coating, which can facilitate the movement of electrons and ions and lower the coating's resistance, carbon nanotubes and graphene can also boost the EIS responsiveness of coatings. The coating's EIS response may vary depending on the inorganic additives' content and dispersion. High additive concentrations can cause coating flaws and agglomeration, which can have a negative impact on impedance response. A more uniform distribution of reinforcement throughout the coating and better EIS response may come from proper additive dispersion. The coating's EIS reaction may also depend on how inorganic chemicals are added. An better impedance response can be attained by using sol–gel processing to create a coating matrix that is more homogeneous and dense. A weaker coating matrix and worse additive dispersion, on the other hand, might lead to physical mixing and a subpar EIS response [20, 21]. The Bode graphs of the blank and Cu-MOF coated steel specimens submerged in a 3.5% NaCl solution are shown in Figure 17.2. Figure 17.2a shows a single time constant in the Bode phase diagram after a one-day immersion, proving that the coating's ionic resistance or barrier effect is what essentially controls the electrochemical activities. Notably the Cu-MOFNS0.25% and Cu-MOFNS0.5% specimens' Bode modulus graphs exhibit a straight-line behavior with a slope of -1 across a wide frequency range and a high impedance modulus ($|Z|0.1$ Hz) surpassing 1,010 cm^2 at low frequencies. This implies that the coatings have outstanding barrier qualities and continue to adhere well to the steel substrate.

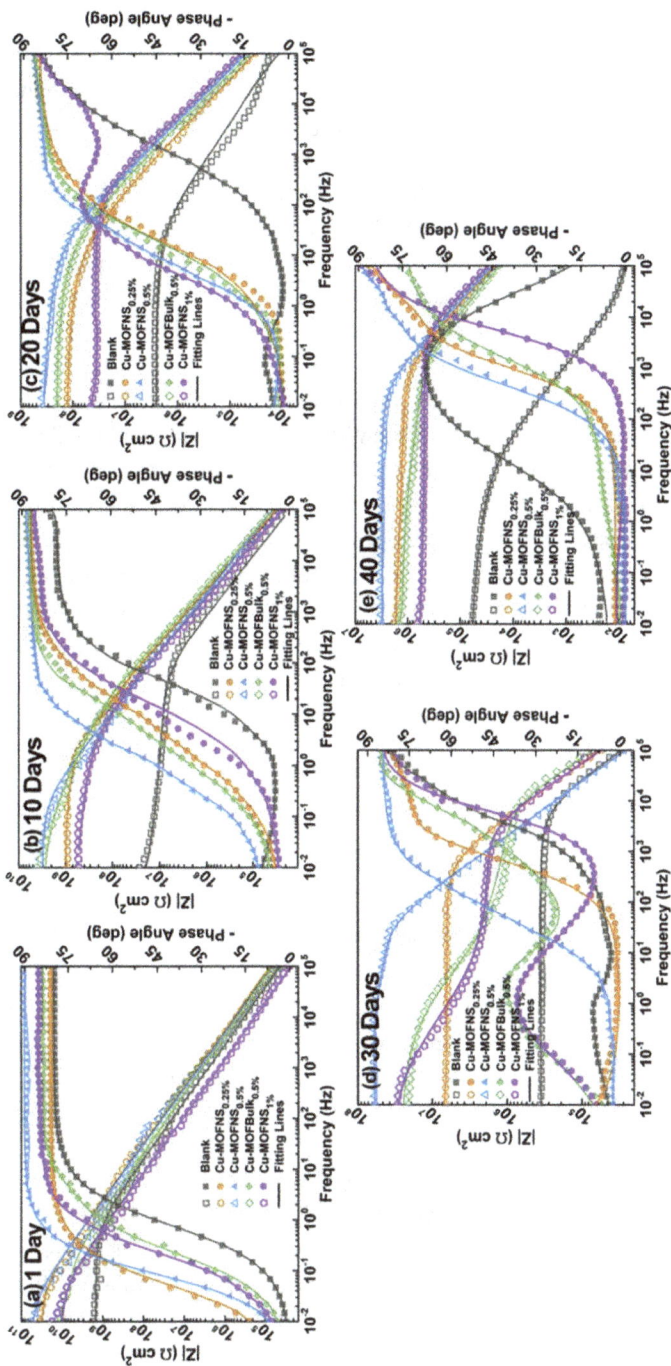

Figure 17.2: Bode plots of Blank, Cu-MOFNS$_{0.25}$%, Cu-MOFNS$_{0.5}$%, Cu-MOFBulk$_{0.5}$% and Cu-MOFNS$_{1}$% after (a) 1 day, (b) 10 days, (c) 20 days, (d) 30 days, and (e) 40 days immersion [34].

Potentiodynamic Polarization Curves. Potentiodynamic polarization curves are an electrochemical test that may be used to gauge how well coatings resist corrosion. When applying a linear increase in potential (or voltage) to a coating in an aqueous solution, the procedure entails measuring the current response of the coating. By altering coatings' electrochemical behavior and corrosion resistance, inorganic additions can change the potentiodynamic polarization curves of such coatings. For instance metal oxide nanoparticles can improve the coating's polarization resistance, which lowers the corrosion rate and strengthens the coating's resistance to corrosion. By creating a conductive network on the coating's surface that can facilitate the flow of electrons and ions and lower the rate of corrosion, carbon nanotubes and graphene can also increase the corrosion resistance of coatings. The coating's potentiodynamic polarization curves may be impacted by the concentration and dispersion of inorganic additions. High additive concentrations can cause the coating to aggregate and produce flaws, which can reduce its ability to resist corrosion. A more uniform distribution of reinforcement throughout the coating due to proper additive dispersion might increase corrosion resistance. The coating's potentiodynamic polarization curves may be impacted by how inorganic additives are included. A more homogeneous and thick coating matrix may be formed during the sol–gel manufacturing process, increasing corrosion resistance. A weaker coating matrix and worse additive dispersion, on the other hand, might result from physical mixing and lower corrosion resistance [24, 25].

17.3 Organic additives

17.3.1 Types of organic additives

Organic additives can be used to improve the adhesion, corrosion resistance, and surface characteristics of coatings to improve performance. These three classes of organic coating additives – polymers, graphene and other carbon-based compounds, and organic inhibitors – will be covered in this section [2, 8, 39, 40].

Polymers. Due to their capacity to alter the mechanical, thermal, and chemical characteristics of the coating, polymers are a family of organic additives that are frequently utilized in coatings. Polymers can boost a coating's resistance to abrasion and impact while also enhancing its adhesion, flexibility, and durability. Epoxies, acrylics, polyurethanes, and polyester resins are among the polymers that are frequently employed in coatings [41–43]. Polyurethanes provide good abrasion resistance and chemical resistance, whereas acrylics are particularly valuable for their outstanding weatherability and UV resistance. While polyester resins are prized for their high gloss, hardness, and scratch resistance, epoxies are renowned for their superior adherence and durability [1, 44, 45].

Graphene and Carbon-Based Materials. Due to its distinctive qualities, such as high strength, electrical conductivity, and thermal stability, graphene and other carbon-based compounds, such as carbon nanotubes, are increasingly employed as organic additions in coatings. Graphene can enhance the coatings' barrier qualities, preventing water and other corrosive elements from penetrating the coating and so enhancing the coating's corrosion resistance [46–48]. Additionally the mechanical characteristics of coatings, such as their flexibility and abrasion resistance, as well as their electrical conductivity, can be enhanced by the use of carbon nanotubes and graphene. The durability and endurance of coatings can be enhanced by using these ingredients, which can help improve coating adherence to the substrate [3, 49, 50].

The work emphasizes electrostatic force of attraction as a substitute for chemical bonding via physisorption for interacting functionalized nanomaterials with metallic surfaces. In this process, counter ions are adsorbed onto positively charged metallic surfaces in aqueous electrolytes, leaving the surface with a negative charge. The negatively charged metallic site is drawn to the positively charged functionalized nanomaterial inhibitor group as a result of protonation, which causes physisorption. Chemisorption and physisorption are visually represented in Figure 17.3. A mixed mode, or physiochemisorption, is the most common adsorption mechanism for organic corrosion inhibitors. The potential of graphene oxide (GO) and its derivatives in the field of corrosion inhibition is further discussed in the paper [51].

Organic Inhibitors. A type of organic compounds called organic inhibitors can be employed to shield coatings from corrosion. These inhibitors function by adhering to the coating's surface and establishing a barrier that prevents corrosion. Compared to conventional inorganic inhibitors like chromates, which are poisonous and carcinogenic, organic inhibitors can be more ecologically benign. Benzotriazole, derivatives of the imidazoline, and fatty acid derivatives are among the frequently utilized organic inhibitors in coatings. While imidazoline derivatives are helpful for preventing corrosion on steel and iron substrates, benzotriazole is particularly good in preventing corrosion on copper and copper alloys. Oleic acid is one of the derivatives of fatty acids that may effectively preserve aluminum and its alloys from corrosion [9, 13, 52, 53].

By changing the surface characteristics, adhesion, and corrosion resistance of coatings, organic additives such polymers, graphene and carbon-based compounds, and organic inhibitors can improve their performance. The technique of integration, the choice, concentration, and dispersion of these additives are crucial elements that influence the coating's qualities. Coatings that perform better can be produced by optimizing these variables [54, 55].

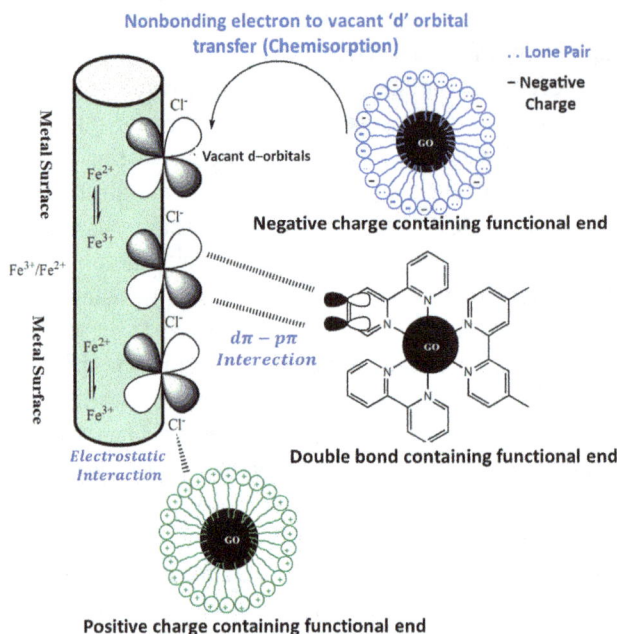

Figure 17.3: A collection of images shows how functionalized graphene oxide interacts with metal surfaces [51].

17.3.2 Effects on mechanical properties

Adhesion Strength. The term "adhesion strength" describes a coating's capacity to adhere to a surface. By altering the surface characteristics of the coating and the substrate, organic additives like polymers can strengthen coatings' adherence. Functional groups that enable chemical bonding between the coating and the substrate are introduced to achieve this. Additionally improving adhesion, polymers might expand the region at which the coating and substrate come into contact. By boosting the coating's surface area and creating powerful covalent connections with the substrate, graphene and other carbon-based compounds can likewise increase adhesion strength. These substances can also improve the coating's wettability, enabling it to distribute more uniformly across the substrate and improving adherence [13, 43, 47].

Hardness and Wear Resistance. Important mechanical characteristics of coatings that govern their capacity to withstand harm and keep their look over time include hardness and wear resistance. By raising the crosslink density of the coating, which increases its resistance to deformation and wear, organic additives like polymers can enhance the hardness and wear resistance of coatings. By producing a strong and long-lasting barrier on the surface of the coating, graphene and carbon-based compounds can also improve the hardness and wear resistance of coatings. By spreading the power of the scratching force over a broader region, these materials can also increase the coating's scratch resistance while limiting the coating's damage [2, 3, 55].

Elastic Modulus. The stiffness of a material, or its capacity to withstand deformation under stress, is referred to as elastic modulus. By adding hard segments to the polymer chain, which limits the mobility of the polymer molecules and stiffens the coating, organic additives like polymers can raise the elastic modulus of coatings. By creating a robust and stiff network of carbon atoms that fortifies the coating against deformation, graphene and other carbon-based compounds can also increase the elastic modulus of coatings. Additionally, by improving the coating's thermal stability, these materials can increase its resistance to high temperatures and thermal cycling [1, 52, 54].

The mechanical characteristics of coatings, including adhesion strength, hardness and wear resistance, and elastic modulus, can be significantly influenced by organic additions including polymers, graphene and carbon-based compounds, and organic inhibitors. To provide the coating the appropriate mechanical qualities, these additives must be carefully chosen, concentrated, and distributed [3, 43, 56]. Figure 17.4 shows multiple corrosion pits where the early "cauliflower" forms first developed. These pits are identified by yellow dotted lines. High carbon concentrations in these pits show that nickel (Ni) atoms are co-deposited with multiwalled carbon nanotubes (MWCNTs) and graphene oxide (GO) during electrodeposition brush plating (EBP). Layers of Ni grain clusters surround them. More MWCNTs are hidden by larger "cauliflower" formations, which suggests that MWCNTs have a propensity to aggregate in the coating.

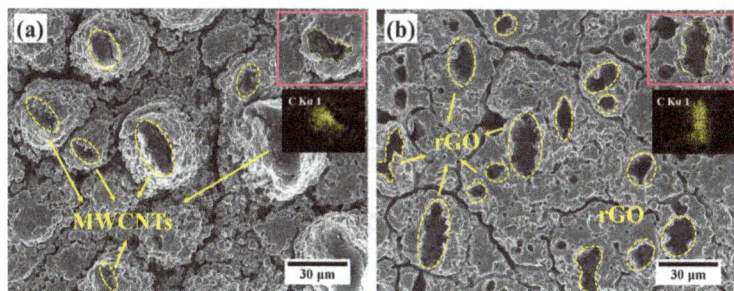

Figure 17.4: Surface corrosion morphologies of Coating 2 (a) and Coating 3 (b).

17.3.3 Effects on electrochemical properties

The electrochemical characteristics of coatings, which affect how well they can shield the underlying substrate from corrosion and other electrochemical processes, can also be influenced by organic additions. The electrochemical characteristics of coatings can be impacted by organic additions in one of the following three ways:

Polarization Resistance. A coating's capacity to resist corrosion is gauged by its polarization resistance. By creating a barrier on the coating's surface that prevents corrosive species (such oxygen, water, and ions) from diffusing to the substrate, organic additives can increase polarization resistance. Utilizing organic inhibitors, which adsorb onto the coating's surface and prevent the electrochemical processes that lead to corrosion, can further strengthen this protection [43, 47, 53].

EIS Stands for Electrochemical Impedance Spectroscopy. By measuring the impedance, or resistance to the passage of electric current, of coatings, EIS is a technique used to investigate their electrochemical behavior. Organic additions can alter a coating's electrical conductivity, permittivity, and dielectric characteristics, which in turn can change how impedant it is. The performance of conductive and antistatic coatings can be enhanced by enhancing the coating's electrical conductivity using polymers, graphene, and carbon-based compounds. Organic inhibitors can also change a coating's impedance by creating a barrier of protection on the surface, which makes electrochemical activities like ion transport and redox reactions more difficult to carry out. EIS may be used to examine how well organic inhibitors work to stop corrosion as well as to improve how they are applied and concentrated [2, 41].

Curves of Potentiodynamic Polarization. Potentiodynamic polarization curves are used to measure the current response as a function of applied potential in order to analyze the corrosion behavior of coatings. These curves' location and shape can be affected by organic additives, which can provide important details about the coating's corrosion protection system. For instance the addition of polymers, graphene, or compounds derived from carbon can modify the corrosion potential to values that are more negative or positive, suggesting a change in the coating's thermodynamic stability. Organic inhibitors can also alter the kinetics of the electrochemical processes that cause corrosion, lowering the corrosion current density or raising the passivation current density, to alter the polarization curves.

The polarization resistance, impedance, and potentiodynamic polarization curves of coatings can all be considerably impacted by the organic additions. Designing coatings with enhanced corrosion protection and specialized electrochemical performance can take advantage of these benefits. The development of high-performance coatings for diverse purposes will continue to advance with a better knowledge of these processes and further study on the function of organic compounds in coatings [1, 8, 54].

17.4 Synergistic effects of inorganic and organic additives

Combined Effects on Mechanical Properties. Beyond what any addition might do alone, the combination of inorganic and organic additives can have synergistic effects that improve the mechanical characteristics of coatings. These effects when combined can result in [57–59]:

1. A stronger contact between the coating and the substrate is possible thanks to the complimentary bonding processes of inorganic and organic compounds.
2. A coating that is more resilient and long-lasting thanks to the enhanced crosslink density and reinforcement supplied by both inorganic and organic compounds.
3. An improved coating's ability to endure deformation and stress due to an increased elastic modulus brought on by the combined stiffening effects of inorganic and organic additives.

Combined Effects on Electrochemical Properties. Additionally, inorganic and organic additives can cooperate to enhance the electrochemical characteristics of coatings, resulting in [60–62]:

1. Greater resistance to polarization due to the combined barrier effects of inorganic and organic additives, which can act as a more effective barrier against the penetration of corrosive species [63].
2. The capacity of the coating to withstand electrochemical operations is improved because to the complimentary electrical and dielectric contributions of the inorganic and organic additives.
3. More advantageous potentiodynamic polarization curves as a result of the additives' combined thermodynamic and kinetic impacts on the coating's corrosion behavior.

Optimal Formulations for Various Applications. In order to create the best formulations for a given application, it is possible to take use of the synergistic interactions between inorganic and organic additives. The following factors should be taken into account when choosing the right additive combination [64–66]:

1. The coating's desirable mechanical characteristics, including adhesion power, hardness, wear resistance, and elastic modulus, which are determined by the choice and quantity of inorganic and organic additives.
2. The coating's requirements for electrochemical performance, including impedance, electrical conductivity, and corrosion resistance, which can be impacted by the proportion and mix of inorganic and organic additives.
3. The application's working environment and environmental parameters, which may influence the choice of additives that offer the best defense against things like UV radiation, temperature extremes, and chemical exposure.

It is possible to build high-performance coatings with improved mechanical and electrochemical characteristics, customized to satisfy the unique requirements of diverse applications, by carefully choosing and optimizing the combination of inorganic and organic additives [67, 68].

17.5 Application examples

Aerospace Sector. The aerospace sector needs high-performance coatings to shield parts from corrosion and wear in intense temperatures and chemically aggressive conditions. Coating additives, both organic and inorganic, can enhance the mechanical and electrochemical characteristics, resulting in more durable and low-maintenance solutions. The following are some examples of coatings used in the aerospace industry [60, 64, 69]:
- Coatings for aircraft components, such as engines, landing gear, and fuel tanks, to prevent corrosion and increase wear resistance;
- Coatings for satellite components, such as solar panels and antennas, to withstand the harsh conditions of space [70, 71].

Offshore Structures. Offshore constructions, such as wind turbines and oil rigs, are subject to abrasive maritime conditions like seawater, waves, and wind. In order to maintain the strength and security of these buildings, corrosion prevention is essential. Coating additives, both organic and inorganic, can improve the mechanical and electrochemical characteristics, resulting in more durable and low-maintenance solutions. The following are some examples of coatings used on offshore structures [2, 40, 66]:
- Coatings for wind turbine blades and towers to increase resistance to erosion and fatigue;
- Coatings for oil and gas pipelines, platforms, and subsea structures to protect against corrosion and biofouling.

Electronics and Sensors. Coatings that shield against moisture, chemicals, and other external variables that might harm their performance and longevity are necessary for electronics and sensors. Coating additives, both organic and inorganic, can increase substrate adherence, electrical conductivity, and barrier qualities. Examples of coatings used in electronics and sensors include [3, 40, 42]:
- Printed circuit board coatings that increase electrical conductivity and guard against moisture;
- Coatings for sensors and MEMS devices that defend against environmental elements that might reduce their lifespan and performance.

Additionally developments in several sectors are being sparked by the continual improvements in the creation of functionalized thin film coatings for corrosion protection.

The creation of coatings with better qualities, such as greater durability, enhanced resistance to environmental conditions, and prolonged lives, can be aided by the incorporation of new technologies like nanotechnology and self-healing systems. Another intriguing area for future advances is the potential for 3D printing to make it possible to produce coatings with organic and inorganic additives efficiently and affordably [72, 73].

Aerospace, offshore constructions, electronics, and sensors are just a few of the areas where organic and inorganic coating additives are used. The mechanical and electrochemical characteristics of coatings can be greatly improved by these additions, resulting in longer-lasting and lower-maintenance solutions. For certain applications, the best solutions may be created by adjusting the additives' selection and concentration in accordance with the required performance criteria.

17.6 Challenges and future perspectives

To ensure sustainable and efficient solutions, a number of problems and future views must be taken into account as the demand for high-performance coatings with improved mechanical and electrochemical qualities continues to rise.

Environmental Concerns:

1. Environmental and health issues are brought up by the usage of some organic additives, such as hazardous substances and volatile organic compounds (VOCs). Therefore, it is necessary to create eco-friendly substitutes that deliver comparable or better performance without endangering the environment or people's health.

2. Given that these substances may leak into the environment and contaminate it, it is important to manage the disposal and recycling of coatings containing organic and inorganic additives.

3. Long-lasting and low-maintenance solutions can help to lessen the total environmental effect by utilizing coatings that are more resistant to environmental elements including UV radiation, temperature fluctuations, and chemical exposure.

Scalability and Cost-Effectiveness:

1. To address the expanding demand for high-performance coatings in diverse sectors, it is crucial to develop scalable production procedures for generating coatings with organic and inorganic additives.

2. The expense of raw resources, such as graphene and other carbon-based compounds, may prevent these additions from being widely used. This obstacle can be solved by creating techniques for generating these materials that are affordable and by maximizing the usage of additives in coating compositions.

3. The manufacture of coatings with organic and inorganic additives may be stream-lined, increasing efficiency and lowering costs. This is made possible by the integration of new manufacturing processes like nanotechnology and 3D printing.

New Materials and Techniques for Enhancing Corrosion Protection:
1. New approaches to improving the mechanical and electrochemical characteristics of coatings can be found in the discovery and development of new materials, such as improved polymers, bio-based additives, and nanomaterials.
2. Researching novel methods for coating application and curing, such as self-healing systems and stimuli-responsive coatings, can produce creative solutions with enhanced corrosion protection and longer lives.
3. By continuing to explore the underlying processes and interactions between organic and inorganic additives in coatings, it will be possible to improve their performance and create solutions that are specialized for certain applications.

High-performance coatings with organic and inorganic additives will be developed and used sustainably if these issues are addressed and future possibilities are taken into account. Numerous structures and goods will continue to be protected and preserved thanks to continued research and advancements in this area, which will also increase their durability and lower maintenance costs [74, 75].

17.7 Conclusion

Summary of the Main Results:
1. Both organic and inorganic additives can greatly enhance a coating's mechanical attributes, such as adhesion power, hardness, wear resistance, and elastic modulus.
2. The addition of organic and inorganic additives can also improve the electrochemical characteristics of coatings, such as polarization resistance, impedance, and potentiodynamic polarization curves.
3. Combining organic and inorganic additives can have synergistic effects that enhance the mechanical and electrochemical characteristics of coatings.
4. By modifying the selection and concentration of additives depending on the intended performance criteria, optimal formulations may be generated for particular applications.
5. To ensure sustainable and efficient solutions for corrosion prevention, issues relating to environmental concerns, scalability, cost-effectiveness, and the development of new materials and processes must be addressed.

Importance of Understanding the Influence of Additives. For a number of reasons, it is essential to understand how organic and inorganic additives affect coatings.

1. It makes it possible to create customized coatings that have improved mechanical and electrochemical characteristics to satisfy the demands of varied applications.
2. It helps to achieve the intended performance while reducing negative effects on the environment and manufacturing costs by optimizing the choice, concentration, and mix of additives.
3. It fosters breakthroughs and developments in the field of corrosion prevention by expanding our understanding of coating technologies and materials science.

Possibility of Additional Developments in Functionalized Thin Film Coatings for Corrosion Protection:

There is a lot of room for improvement in the field of functionalized thin film coatings for corrosion prevention, including:

1. The discovery and inclusion of innovative materials and additives, which will increase performance and provide environmentally acceptable solutions.
2. The combination of cutting-edge manufacturing processes with cutting-edge and emerging technologies, such as nanotechnology, 3D printing, and self-healing systems, to produce novel coatings with improved characteristics and longer lives.
3. The creation of application-specific solutions that answer the particular problems and demands of different markets and settings, thereby aiding in the preservation of priceless resources and infrastructure.

For the creation of high-performance coatings for corrosion protection, an understanding of how organic and inorganic additives affect the mechanical and electrochemical characteristics of coatings is essential. Defeating obstacles and embracing the future will assure sustainable and efficient solutions while advancing this crucial sector.

Reference

[1]   Rajkumar R, Vedhi C. Study of the corrosion protection efficiency of polypyrrole/metal oxide nanocomposites as additives in anticorrosion coating. Anti-Corros Methods Mater 2020;67 (3):305–12.
[2]   Izadi M, et al. The electrochemical behavior of nanocomposite organic coating based on clay nanotubes filled with green corrosion inhibitor through a vacuum-assisted procedure. Compos Part B: Eng 2019;171:96–110.
[3]   Kulyk B, et al. A critical review on the production and application of graphene and graphene-based materials in anti-corrosion coatings. Critl Rev Solid State Mater Sci 2022; 47(3):309–55.
[4]   Berdimurodov E, et al. Green β-cyclodextrin-based corrosion inhibitors: Recent developments, innovations and future opportunities. Carbohydr Polym 2022;119719.
[5]   Berdimurodov E, et al. MOFs-based corrosion inhibitors, in supramolecular chemistry in corrosion and biofouling protection. CRC Press, pp. 287–305; 2021.

[6]   Zhu M, et al. Insights into the newly synthesized N-doped carbon dots for Q235 steel corrosion retardation in acidizing media: A detailed multidimensional study. J Colloid Interface Sci 2022;608:2039–49.

[7]   Bahgat Radwan A, et al. Electrospun highly corrosion-resistant polystyrene–nickel oxide superhydrophobic nanocomposite coating. J Appl Electrochem 2021;51:1605–18.

[8]   Pandis PK, et al. Evaluation of Zn-and Fe-rich organic coatings for corrosion protection and condensation performance on waste heat recovery surfaces. Int J Thermofluid Sci Technol 2020;3:100025.

[9]   Voulgari E, et al. Effect of organic coating corrosion inhibitor on protection of reinforced mortar. J Mater Sci Chem Eng 2019; 7(01):20.

[10]   Berdimurodov E, et al. Inhibition properties of 4,5-dihydroxy-4,5-di-p-tolylimidazolidine-2-thione for use on carbon steel in an aggressive alkaline medium with chloride ions: Thermodynamic, electrochemical, surface and theoretical analyses. J Mol Liq 2021;327:114813.

[11]   Berdimurodov E, et al. Experimental and theoretical assessment of new and eco-friendly thioglycoluril derivative as an effective corrosion inhibitor of St2 steel in the aggressive hydrochloric acid with sulfate ions. J Mol Liq 2021;335:116168.

[12]   Berdimurodov E, et al. A gossypol derivative as an efficient corrosion inhibitor for St2 steel in 1 M HCl + 1 M KCl: An experimental and theoretical investigation. J Mol Liq 2021;328:115475.

[13]   Coelho LB, et al. Corrosion inhibition of AA6060 by silicate and phosphate in automotive organic additive technology coolants. Corros Sci 2022;199:110188.

[14]   Teijido R, et al. State of the art and current trends on layered inorganic-polymer nanocomposite coatings for anticorrosion and multi-functional applications. Prog Org Coat 2022;163:106684.

[15]   Strauss F, et al. Li2ZrO3-coated NCM622 for application in inorganic solid-state batteries: Role of surface carbonates in the cycling performance. ACS Appl Mater Interfaces 2020; 12(51):57146–54.

[16]   Suleiman RK, et al. Hybrid organosilicon-metal oxide composites and their corrosion protection performance for mild steel in 3.5% NaCl solution. Corros Scie 2020;169:108637.

[17]   Ashassi-Sorkhabi H, Kazempour A. Incorporation of organic/inorganic materials into polypyrrole matrix to reinforce its anticorrosive properties for the protection of steel alloys: a review. J Mol Liq 2020;309:113085.

[18]   Ong G, Kasi R, Subramaniam R. A review on plant extracts as natural additives in coating applications. Prog Org Coat 2021;151:106091.

[19]   Berdimurodov E, et al. Novel bromide–cucurbit[7]uril supramolecular ionic liquid as a green corrosion inhibitor for the oil and gas industry. J Electroanal Chem 2021;901:115794.

[20]   Yao T, et al. Effects of additive NaI on electrodeposition of Al coatings in AlCl 3-NaCl-KCl molten salts. Front Chem Sci Eng 2021;15:138–47.

[21]   Valero-Gómez A, et al. Microencapsulation of cerium and its application in sol–gel coatings for the corrosion protection of aluminum alloy AA2024. J Sol Gel Sci Techn 2020;93:36–51.

[22]   Becker M. Chromate-free chemical conversion coatings for aluminum alloys. Corros Rev 2019;37 (4):321–42.

[23]   Mohanty US, et al. Roles of organic and inorganic additives on the surface quality, morphology, and polarization behavior during nickel electrodeposition from various baths: a review. J Appl Electrochem 2019;49:847–70.

[24]   Bonin L, Vitry V, Delaunois F. Inorganic salts stabilizers effect in electroless nickel-boron plating: Stabilization mechanism and microstructure modification. Surf Coat Technol 2020;401:126276.

[25]   Jouyandeh M, et al. Synthesis, characterization, and high potential of 3D metal–organic framework (MOF) nanoparticles for curing with epoxy. J Alloys Compd 2020;829:154547.

[26]   Ulaeto SB, et al. Polymer-Based coating for steel protection, highlighting metal–organic framework as functional actives: a review. Corr Mat Degrad 2023; 4(2):284–316.

[27] Shin J, et al. Epitaxial metal–organic framework for stabilizing the formation of a solid electrolyte interphase on the sianode of a lithium-ion battery. ACS Sustainable Chem Eng 2022;10(32):10615–26.

[28] Seidi F, et al. Metal-organic framework (MOF)/epoxy coatings: A review. Materials 2020;13(12):2881.

[29] Berdimurodov E, et al. Novel cucurbit[6]uril-based [3]rotaxane supramolecular ionic liquid as a green and excellent corrosion inhibitor for the chemical industry. Colloids Surf A Physicochem Eng Aspect 2022;633:127837.

[30] Berdimurodov E, et al. Novel gossypol–indole modification as a green corrosion inhibitor for low–carbon steel in aggressive alkaline–saline solution. Colloids Surf A: Physicochem Eng Aspects 2022;637:128207.

[31] Rabiei H, et al. Antimicrobial activity and cytotoxicity of cotton-polyester fabric coated with a metal–organic framework and metal oxide nanoparticle. Appl Nanosci 2023;1–12.

[32] Nabipour H, et al. Metal-organic frameworks for flame retardant polymers application: A critical review. Compos Part A: Appl Sci Manuf 2020;139:106113.

[33] Haeri Z, Ramezanzadeh B, Ramezanzadeh M. Recent progress on the metal-organic frameworks decorated graphene oxide (MOFs-GO) nano-building application for epoxy coating mechanical-thermal/flame-retardant and anti-corrosion features improvement. Prog Org Coat 2022;163:106645.

[34] Qiu S, et al. Ultrathin metal-organic framework nanosheets prepared via surfactant-assisted method and exhibition of enhanced anticorrosion for composite coatings. Corros Sci 2021;178:109090.

[35] Berdimurodov E, et al. Novel glycoluril pharmaceutically active compound as a green corrosion inhibitor for the oil and gas industry. J Electroanal Chem 2022;907:116055.

[36] Berdimurodov E, et al. New and green corrosion inhibitor based on new imidazole derivate for carbon steel in 1 M Hcl medium: Experimental and theoretical analyses. Int J Eng Res Afr 2022;58:11–44.

[37] Berdimurodov E, et al. In: Guo, L, Verma, C, Zhang, DBT-E-FCI (eds) Chapter19 – Ionic liquids as green and sustainable corrosion inhibitors I. Elsevier; pp. 331–90; 2022.

[38] Berdimurodov E, et al. Thioglycoluril derivative as a new and effective corrosion inhibitor for low carbon steel in a 1 M HCl medium: Experimental and theoretical investigation. J Mol Struct 2021;1234:130165.

[39] Toorani M, et al. Superior corrosion protection and adhesion strength of epoxy coating applied on AZ31 magnesium alloy pre-treated by PEO/Silane with inorganic and organic corrosion inhibitors. Corros Sci 2021;178:109065.

[40] Kaseem M, Dikici B. Optimization of surface properties of plasma electrolytic oxidation coating by organic additives: a review. Coatings 2021;11(4):374.

[41] Gladkikh N, et al. Synergistic effect of silanes and azole for enhanced corrosion protection of carbon steel by polymeric coatings. Prog Org Coat 2020;138:105386.

[42] Kumar SSA, et al. New perspectives on Graphene/Graphene oxide based polymer nanocomposites for corrosion applications: The relevance of the Graphene/Polymer barrier coatings. Prog Org Coat 2021;154:106215.

[43] Bazli L, et al. Application of composite conducting polymers for improving the corrosion behavior of various substrates: A Review. J Compos Comp 2020; 2(5):228–40.

[44] Ubaid F, et al. Multifunctional self-healing polymeric nanocomposite coatings for corrosion inhibition of steel. Surface Coat Technol 2019;372:121–33.

[45] Yuan TH, et al. Corrosion protection of aluminum alloy by epoxy coatings containing polyaniline modified graphene additives. Mater Corros 2019; 70(7):1298–305.

[46] Wang X, et al. Comparative study of three carbon additives: Carbon nanotubes, graphene, and fullerene-c60, for synthesizing enhanced polymer nanocomposites. Nanomaterials 2020; 10(5):838.

[47] Babaei K, Fattah-alhosseini A, Molaei M. The effects of carbon-based additives on corrosion and wear properties of Plasma electrolytic oxidation (PEO) coatings applied on Aluminum and its alloys: A review. Surfaces and Interfaces 2020;21:100677.

[48] Berdimurodov E, et al. New anti-corrosion inhibitor (3ar,6ar)-3a,6a-di-p-tolyltetrahydroimidazo [4,5-d]imidazole-2,5(1 h,3h)-dithione for carbon steel in 1 M HCl medium: Gravimetric, electrochemical, surface and quantum chemical analyses. Arab J Chem 2020;13:7504–23.

[49] Anastasiia A, et al. Influence of the UV radiation on the corrosion resistance of the carbon-based coatings for the marine industry. OnePetro.

[50] Amin KM, Amin HMA. Carbon nanoallotropes-based anticorrosive coatings: Recent advances and future perspectives. corrosion protection of metals and alloys using graphene and biopolymer based. Nanocomposites 2021;81–98.

[51] Kumar A, et al. Functionalized and biomimicked carbon-based materials and their impact for improving surface coatings for protection and functionality: insights and technological trends. Coatings 2022;12. doi: 10.3390/coatings12111674.

[52] Xiong L, et al. Modified salicylaldehyde@ ZIF-8/graphene oxide for enhancing epoxy coating corrosion protection property on AA2024-T3. Prog Org Coat 2020;142:105562.

[53] Ashassi-Sorkhabi H, et al. Effect of amino acids and montmorillonite nanoparticles on improving the corrosion protection characteristics of hybrid sol-gel coating applied on AZ91 Mg alloy. Mater Chem Phys 2019;225:298–308.

[54] Liu X, et al. Unraveling the Formation Mechanism of a Hybrid Zr-Based Chemical Conversion Coating with Organic and Copper Compounds for Corrosion Inhibition. ACS Appl Mater Interfaces 2021; 13(4):5518–28.

[55] Kamaruzzaman W, et al. Assessment of corrosion efficiency and volatile organic compounds content for a green coating with novel additive of Leucaena leucocephala. J Sustain Sci Manag 2021; 16(4):37–52.

[56] Berdimurodov E, et al. 8–Hydroxyquinoline is key to the development of corrosion inhibitors: An advanced review. Inorg Chem Commun 2022;109839.

[57] Dyer SJ, Anderson CE, Graham GM. Thermal stability of amine methyl phosphonate scale inhibitors. J Pet Sci Eng 2004;43(3–4):259–70.

[58] Jafar Mazumder MA. A review of green scale inhibitors: Process, types, mechanism and properties. Coatings 2020;10(10):928.

[59] Bassioni G. Mechanistic aspects on the influence of inorganic anion adsorption on oilfield scale inhibition by citrate. J Pet Sci Eng 2010;70(3–4):298–301.

[60] Gryta M. Polyphosphates used for membrane scaling inhibition during water desalination by membrane distillation. Desalination 2012;285:170–76.

[61] Guo X, et al. Scale inhibitors for industrial circulating water systems: A review. J Water Chem Technol 2021; 43(6):517–25.

[62] Hasson D, Shemer H, Sher A. State of the art of friendly "green" scale control inhibitors: a review article. Ind Eng Chem Res 2011;50(12):7601–07.

[63] Berdimurodov E, et al. The recent development of carbon dots as powerful green corrosion inhibitors: A prospective review. J Mol Liq 2021;118124.

[64] Jordan MM, et al. The effect of clay minerals, pH, calcium and temperature on the adsorption of phosphonate scale inhibitor onto reservoir core and mineral separates. Houston, TX (United States): NACE International; 1994.

[65] Jordan MM, et al. Phosphonate scale inhibitor adsorption/desorption and the potential for formation damage in reconditioned field core. OnePetro.

[66] Ketrane R, et al. Efficiency of five scale inhibitors on calcium carbonate precipitation from hard water: Effect of temperature and concentration. Desalination 2009; 249(3):1397–404.

[67] Dagdag O, et al. Epoxy coating as effective anti-corrosive polymeric material for aluminum alloys: Formulation, electrochemical and computational approaches. J Mol Liq 2021;117886.

[68] Dewangan Y, et al. N-hydroxypyrazine-2-carboxamide as a new and green corrosion inhibitor for mild steel in acidic medium: experimental, surface morphological and theoretical approach. J Adhes Sci Technol 2022;1–21.

[69] Khormali A, Sharifov AR, Torba DI. Increasing efficiency of calcium sulfate scale prevention using a new mixture of phosphonate scale inhibitors during waterflooding. J Pet Sci Eng 2018;164:245–58.

[70] Dagdag O, et al. Nanomaterials for Corrosion Mitigation: Synthesis, Characterization & Applications. Functionalized nanomaterials for corrosion mitigation: Synthesis, characterization, and applications, ACS Publications. pp. 67–85, 2022.

[71] Kaur J, et al. Euphorbia prostrata as an eco-friendly corrosion inhibitor for steel: electrochemical and DFT studies. Chem Pap 2022;1–20.

[72] Dagdag O, et al. Graphene and graphene oxide as nanostructured corrosion inhibitors. Carbon Allotropes: Nanostruct Anti-Corrosive Mater 2022;133.

[73] Rbaa M, et al. Development Process for Eco-Friendly Corrosion Inhibitors. In: Eco-Friendly corrosion inhibitors, Elsevier. pp. 27–42; 2022.

[74] Verma DK, et al. N–hydroxybenzothioamide derivatives as green and efficient corrosion inhibitors for mild steel: Experimental, DFT and MC simulation approach. J Mole Struct 2021;1241:130648.

[75] Rbaa M, et al. Synthesis of new halogenated compounds based on 8-hydroxyquinoline derivatives for the inhibition of acid corrosion: Theoretical and experimental investigations. Mater Today Commun 2022;33:104654.

Shveta Sharma, Richika Ganjoo, Abhinay Thakur, Ashish Kumar

18 Role of surface functionalization on corrosion resistance and thermal stability of functionalized thin film coatings

Abstract: Corrosion is the process through which a metal deteriorates as a result of chemical reactions with its surroundings. Corrosion inhibitors are a highly efficient method of preventing corrosion on metal surfaces. New studies are focusing on how to make coatings more effective toward corrosion. One cutting-edge method involves chemically bonding the surfaces of modified fillers in order to increase the better dispersion of coating used and the surface of the material on which coating is applied. In this regard, the surfaces of the filler are activated and modified by grafting a variety of functional groups in order to make them active. Surfaces with coating material matrices that include modified compounds as fillers improved their adherence to metal and increased the various properties like thermal, anticorrosion, etc. of the various coating materials, which used to shield metals from corrosive environments. This was accomplished by improving the adhesion of the modified compounds to the metallic surfaces. Therefore, in this article anti-corrosion coatings are first described in detail, including their synthesis, characterization, and structure-property–performance relationship.

Keywords: Corrosion, Corrosion Inhibition, Coatings, Functionalization, Corrosion Analysis Techniques

18.1 Introduction

The process through which a metal deteriorates as a result of its reaction with its surroundings is known as corrosion. When it comes to safeguarding metal surfaces against corrosion, one of the most efficient strategies is to make use of corrosion inhibitors. The most common types of metal protection used today are protecting the metallic samples by using alloys, electroplating protection, saving the metals electrochemically, and by coating with the organic materials. Organic coating protection is a popular options due to its inexpensiveness, ease of construction, and high quality of performance in a variety of industrial fields [1, 2]. Along with organic coating, other materials like Graphene (which is two-dimensional nanomaterial of carbon) can also be coated on metallic surface to protect it from corrosion [3, 4]. Recent years have seen an explosion of interest in graphene nanosheets as a result of the extraordinary properties the material possesses. GNSs have been used extensively as nanofillers for the reinforcement of polymer–matrix composites due to their low density and large specific surface area [5–7]. It also depends upon the polymeric material used for coating for the protection of sample

https://doi.org/10.1515/9783111016160-018

[8]. One more benefit of coating is its cost effectiveness, because of which its usage is increasing day by day. Also, effectiveness of coating can be enhanced adding fillers, functionalization of the coating material. It is still difficult to choose the best intelligent coatings for corrosion protection of metal objects, especially when filler or dispersing materials are used to reinforce the polymer matrix [9]. In most cases, when filler is used as GNS, the production of polymer composites involves the incorporation of nano-fillers that are amenable to solution processing into polymer [10–12]. Nonetheless, the GNSs' robust aggregation susceptibility remains the fundamental obstacle to achieving homogenous distribution and efficient interfacial reactions using this approach. Covalently functionalizing the surface of the GNSs is an additional technique that can be utilized in the fabrication of GNS-based polymer frameworks. This is a wonderful approach to enhancing the contacts that exist between the GNSs and the polymeric frameworks. The graphite oxide (GO), which is the precursor of graphene oxide (GNS), has a large number of functional groups on it, such as hydroxyl and other similar groups. The reactive target sites for the covalent functionalization are afforded by these molecules. In addition, the chemical attached functional groups drastically changes the weak interactive forces in the nanofillers, which makes it simple to disperse them throughout the polymer matrix [13]. Because of this, surface treatment of GO is a good potential method for improving interoperability because it is a relatively simple task [14]. The authors Li et al. [15] investigated the characteristics of epoxy resin coatings with GO (with incorporated silane). Their findings demonstrated that the mechanical properties of epoxy resin were significantly enhanced by adding modified GO. Fillers can also be modified by s by cyclohexene oxide and its derivatives [14]. It is already mentioned in literature that various types of functionalization to fillers have drastically enhance the anti-corrosion properties of the composites.

18.2 Functionalized epoxy coatings for corrosion resistance

Zhang et al. [14] synthesized a functionalized silane coupling agent and by using which graphene oxide was modified and was utilized as filler in epoxy coating to make it more effective it against corrosion. For forming the functionalized GO, first the Go is formed and was mixed with deionized water and a suspension was formed and, in that suspension, 1.2 grams of 2-(3,4-Epoxycyclohexyl) ethyl triethoxysilane (ETEO) was added. The system underwent a reaction while in reflux for a period of twelve hours. Following the completion of the reaction, the suspension was passed through a filtration system, and any unreacted ETEO was washes off. At the end, the moisture from the sample was removed at an elevated temperature (60 °C) for one day (reaction is given in Figure 18.1).

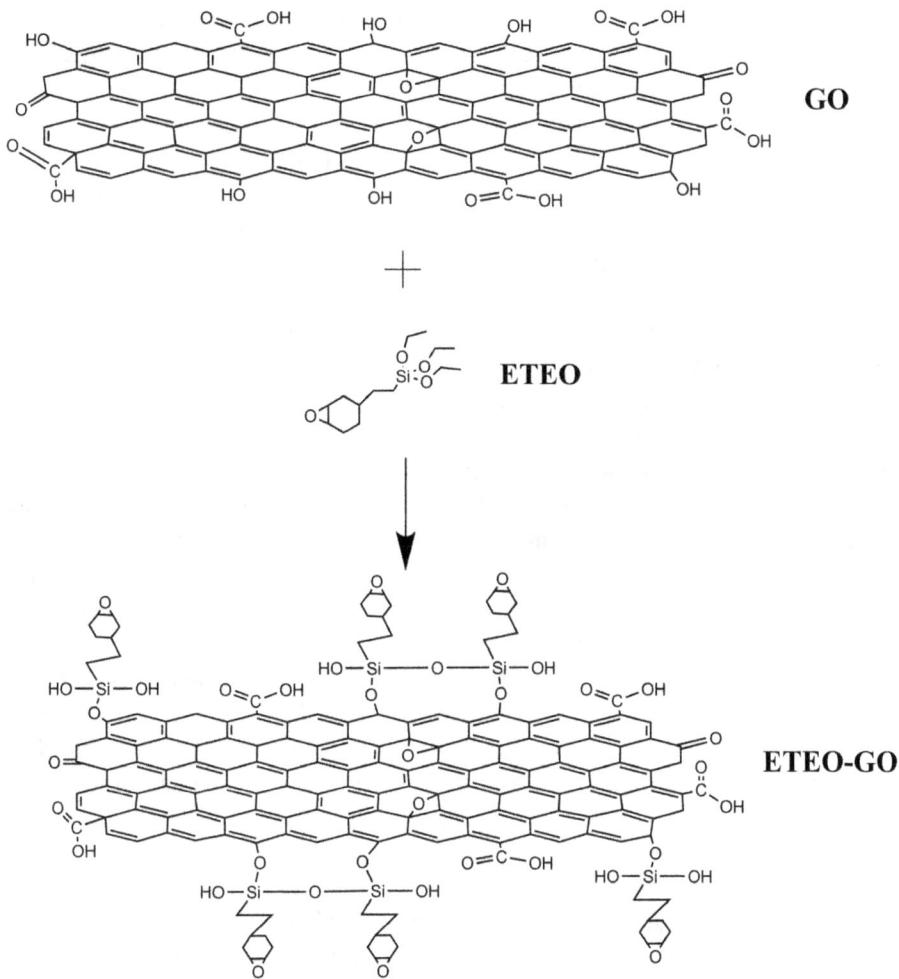

Figure 18.1: Preparation of ETEO–GO (source adapted from Open access (CCBY) [14]).

Coatings were then prepared by varying the percentage of filler, and coatings were then characterized with Fourier transform infrared spectroscopy (FTIR), XRD, X-ray photoelectron spectroscopy (XPS), SEM and TEM techniques. The coating's hardness and impact resistance were tested in order to get a better understanding of its mechanical properties. A pencil hardness tester was utilized in order to evaluate the surface's degree of abrasion in accordance with GB/T 6,739–2006 standard [16]. XPS was performed to find out the elements on the surface, And the spectra verified the C:O: Si ratio 60:40:0 and 59:33:8 in GO and ETEO-GO respectively along with functionalization of GO with ETEO. Further morphological investigation by SEM and TEM confirmed the lamellar structure in GO, in ETEO-GO, the surface became rough, fluffy and folded. It's possible that this is because the GO in question had a high surface energy, which led to a strong mutual attraction between the lamellae. Due to the presence of siloxane groups, the ETEO-GO surface energy was decreased after being modified by ETEO, which resulted in a reduction in the mutual attraction in lamellas. Further salt spray and EIS were also performed to check the anti-corrosion property. Salt spray test was performed after 400 h and it verified the presence of corrosion in metallic sample when coated with EP coating. On the other hand, a discernible nucleation of GO was seen in the GO coating. This was primarily the result of the imbalanced spreading of GO throughout the coating, which caused the outer layer of the GO coating to have varying densities at various locations. The number of pore spaces and cracking in the coating increased as a result of the GO agglomeration process. The formation of corrosion channels can be encouraged by the presence of micropores and microcracks. Therefore, a significant quantity of corrosive medium was able to reach the substrate by means of the corrosion channels. Subsequently, the substrate started to corrode, which resulted in the formation of corrosion products. The scuff section of the coatings that were amplified by ETEO-GO led to the creation of corrosion products with reduced levels. This was a successful outcome. Kumar and Gasem [17] investigated anti corrosion properties of coating material made up of polyaniline (PANI)-functionalized carbon nanotubes (f-CNTs) for metallic corrosion. First coating material was prepared electrochemically. Further coating was characterization was performed with ATR-IR, Raman spectroscopy, FE-SEM and contact angle techniques. Hardness was tested with dynamic hardness tester. Corrosion inhibiting tendency was checked and confirmed with electrochemical techniques and FE-SEM illustrated sample images coated with functionalized CNTs, PANI and PCNT. Microstructural analysis of PANI/f-CNTs, on the other hand, reveals unequivocally that the f-CNTs are firmly attached to and embedded within the PANI matrix [18]. Conventional load versus indentation depth curves were recorded for both bare and coated metallic samples. Depth sensing indentation is a highly effective experimental method for elucidating material mechanics. Analysis of curves reveals that coated substrates have shallower indentations than their uncoated counterparts. PANI films have already been shown to have the potency to exhibit hard-surfaced properties, which can be attributed to their aromatic ring having bonds present in conjugation, proneness to inter chain cross-linking, and dense surface nanostructures. This potential gas already been discussed in the scientific literature [19]. To fully understand how a coating will fare in an aggressive environment, it is common to conduct a contact angle analysis to establish a relationship between the ability of the coating to absorb water and how well it protects against corrosion. When compared to the contact angle of bare MS (98.42°), that of a PANI-coated substrate (100.63°), and this little high angle was because of the presence of a functional group (amine group) in the polymer matrix. In contrast to PANI-coated substrates, PCNT-coated ones showed a greater contact angle. This is because CNTs in PANI matrices are hydrophobic. As the percentage of f-CNTs in PANI coatings increased, it was also observed that the contact angle values rose. Based on our findings, PCNT3 coatings are more hydrophobic than other coatings because they have a larger contact angle (140.32°) [20, 21]. With the assistance of PDP, it was possible to see that the PCNT nanocomposite coating demonstrated a significantly better corrosion resistance property in comparison to the other two metallic samples (first on which PANI coating was used and other metallic sample was without any coating). It is common knowledge that a dense and stable PANI coating that has been supported with f-CNTs delivers an enhanced preventative barrier layer to reduce the deterioration of metal surfaces. While performing the PDP analysis, 3% NaCl was used as test solution and there was a clear change anodic region in the E_{corr} values of coated metallic sample as compare to uncoated metallic sample and corrosion current densities decreased. Liu et al. [22] used Friedel-Crafts acylation to produce the functionalized graphite by reacting it with 4-aminobenzoic acid. Ball milling was used as the preparation method for edge-functionalized graphite, which resulted in the production of two distinct types of edge-functionalized graphite. This approach was selected because it was published in the relevant literature. Further, acetone as solvent was utilized to dissolve the functionalized graphite and for that an ultrasonic bath for a period of twenty minutes and after that epoxy resin was mixed into it. Then sonication of the mixture was done once more for a total of twenty minutes so that the graphene could combine thoroughly with the epoxy resin. Characterization was performed with spectroscopic and gravimetric techniques along with other techniques like X-ray diffraction (XRD), Differential scanning calorimetry (DSC), SEM and Transmission electron microscopy (TEM). FT-IR revealed functionalization of graphite with 4-aminobenzoic acid. Quantization was checked with TGA. The unaltered graphite demonstrates almost no loss in weight. On the other hand, the weight of the graphite that has been functionalized gradually decreases. This decrease in mass is directly proportional to the number of 4-aminobenzoyl groups that have been grafted onto graphite using a covalent bond. Through the use of a mechanical shearing force, the graphite flakes will be broken up into flakes of a smaller size during the ball milling process. This will result in an increased number of corner spaces. Ball milling method makes the edges areas of the graphite very active, which is necessary in order to start the reactions that occur at graphite corner areas [23, 24]. Raman spectroscopic was performed to find out quality of the solid functionalized graphite and three bands D band, G band and 2D band were observed at 1,350 cm^{-1}, 1580 cm^{-1} and 2,700 cm^{-1} respectively. The Raman spectra of EFG and Ball-EFG are almost identical to those of pure graphite, with the exception of the D band. The fact that the D band has a relatively low intensity in both the EFG and the Ball-EFG suggests that the edge-functionalization process does not result in the destruction of basal planes. More so, the Ball-EFG 2D band becomes more distinct than the graphite one. Several research groups have found that the 2D peak's shape changes depending on the number of layers present in graphene flakes [25, 26]. El-Lateef et al. [27] investigated and generated Novel cerium oxide nanoparticles functionalized by gelatin. In this study Cerium oxide nanoparticles were loaded with gelatin for forming CeO_2-gelatin films. For this nanocomposite film CeO_2 and gelatin were taken in the ratio of 1: 20. Following preparation, CeO_2-gelatin nanocomposite films were obtained by desalting the film with deionized water and drying it for 72 h at 50° Celsius. The newly presented

CeO$_2$-gelatin was analyzed using techniques such field emission scanning electron microscopy, energy dispersive X-ray spectroscopy, transmission electron microscopy, chemical mapping, Fourier transform infrared spectroscopy, and thermogravimetric analysis. Using these characterization methods, we can conclude that a composite material consisting of CeO$_2$ and gelatin was successfully synthesized. In acidizing oil well media, the produced composite CeO$_2$-gelatin was employed as an eco-friendly coated film or X60 steel alloys. In addition, the role that CeO$_2$ percentage played in determining the film's make-up was studied. FE-SEM was used to categorize the corroded surfaces of covered and uncoated specimens both prior and after soaking the samples in the acidic media for 24 h for determining morphological characteristics of the samples. Figure 18.2A shows an example of a FE-SEM image of gelatin, which reveals a uniform and smooth surface devoid of sharp edges or isolated particles and thus consistent with the presence of solely organic matter. Surface morphology changes after CeO$_2$ integration (Figure 18.2B), and particles of varying sizes are visible, demonstrating the morphological success of the chemically designed heterogeneous gelatin/CeO$_2$ composite. These CeO$_2$-gelatin particles were confirmed by transmission electron microscopy (TEM) images at two different magnifications, as shown in Figure 18.2C and 18.2D. Both TEM images showed particles of the nanometer range. Figure 18.2E depicts the nanocrystals of CeO$_2$ with a very small diameter, and the investigation was performed with the selected area electron diffraction (SAED).

Further thermal analysis of the sample's gelatin, CeO$_2$, and CeO$_2$-gelatin was performed and illustrated in Figure 18.3. It can be inferred that CeO$_2$ is thermally stable up to the investigated temperature (7,000 C) because its weight loss is relatively low (approximately 3.48%). Weight loss at 700 °C was calculated to be 64.80% for gelatin and 66.45% for CeO$_2$-gelatin. Little shift is there in the CeO$_2$-gelatin shift and it confirms the strong interaction and stability in the composite.

In one more study, Chatterjee et al. [28] form and characterize amine functionalized graphene nanoplatelets (EGNPs) added into epoxy resins and investigated anti corrosion property of the composite. The EGNP/epoxy composites were mixed in the ratio 100:24 and kept in vacuum oven at 40 °C to remove any type of gas or moisture finally molded and cured at different temperature and time. Characterization was performed by SEM, TEM, and Raman spectroscopy. Raman spectroscopy revealed the increased D/G band intensity in EGNP samples treated with nitric acid, etc. This raised intensity of D band was due to defects caused by functionalization [29]. Minor shifts in the ID/IG ratio from 21.1% to 18.8% after amine treatment are consistent with a similar number of defects being present on the graphene surfaces and provide more evidence for the existence of functional groups on the walls of EGNPs [30]. FTIR spectra of the EGNPs after the different treatments showed that no band was there in pristine EGNPs confirming the absence of any type of functional group and new bands (1,735 cm^{-1} and 1,200 cm^{-1}) were there after the treatment with acid. The C@C stretching transition of the EGNP structure, at 1,585 cm^{-1}, also becomes more visible after the nitric acid treatment. After the reaction amine new bands 2,851 and 2,922 cm^{-1} were present [31, 32]. Wang et al. [33] prepared

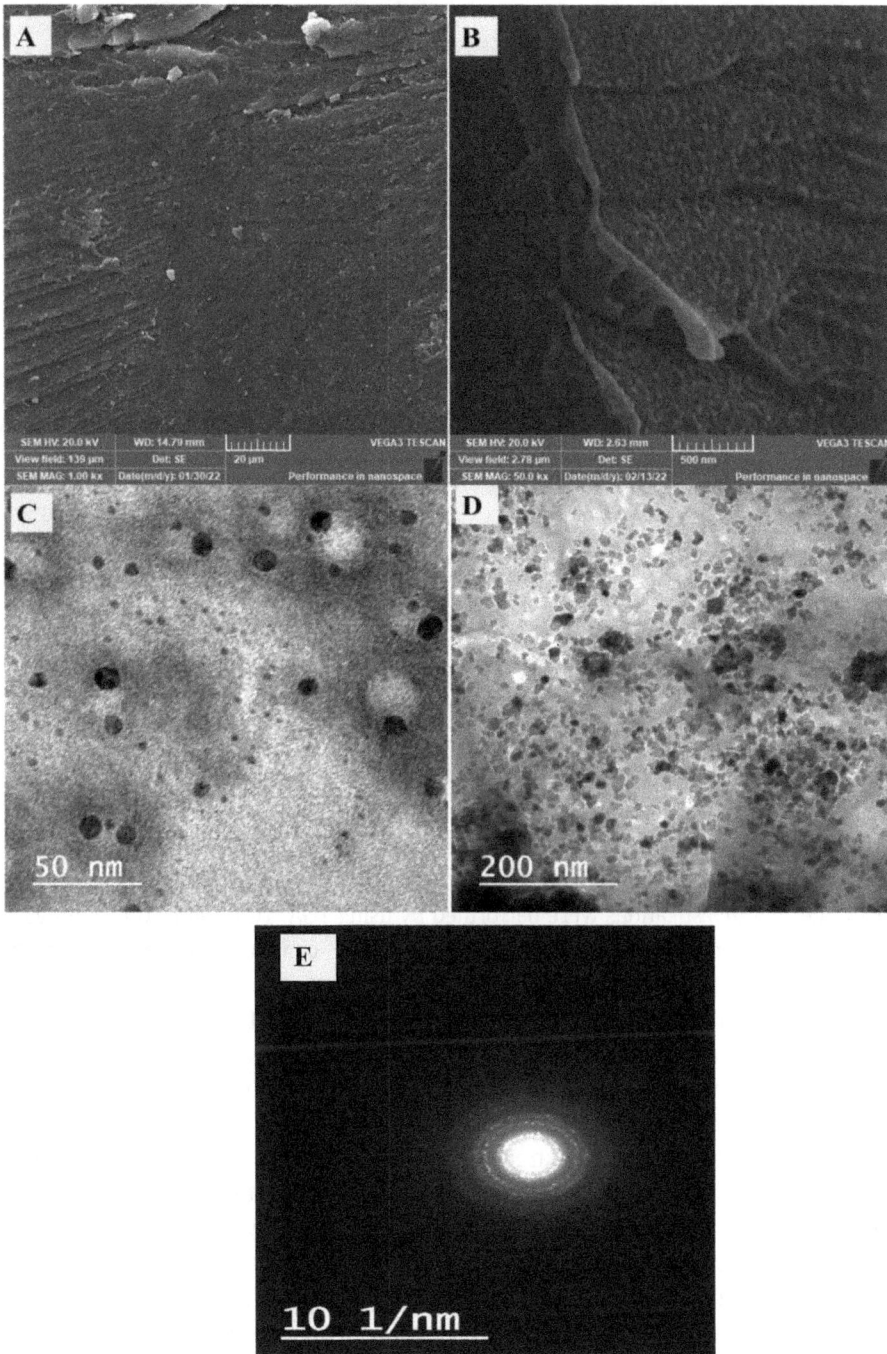

Figure 18.2: (A–E) FESEM of gelatin, and 5% CeO$_2$-gelatin composite, TEM of 5% sample at varied magnification and the SAED of resulted framework (source adapted from Open access (CCBY) [27]).

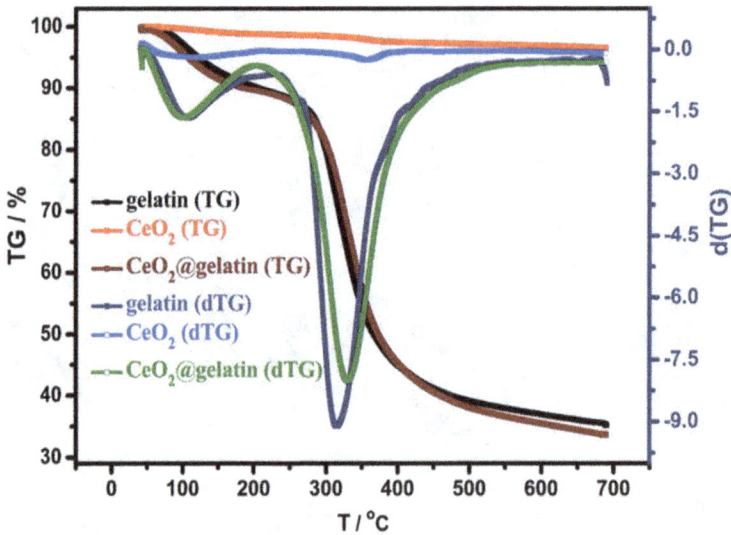

Figure 18.3: TGA analysis of gelatin, CeO$_2$, and CeO$_2$-gelatin samples (source adapted from Open access (CCBY) [27]).

Functionalized graphene nanosheets (f-GNSs), for that first Hummers' method was used to create graphite oxide (GO) powders from graphite, and a GNS suspension was made by chemically reducing a GO nanosheet dispersion in water with hydrazine. GNS by using hydroxyl functional groups reacted with 3- aminopropyltriethoxysilane (APTS) formed the APTS-functionalized graphene nanosheets [34, 35]. After that epoxy was used to form the composite and two types of composites (EP/GNS and EP/f-GNS) were formed. Characterization of the composite were performed by using SEM, TEM, AFM, XPS and FT-IR, thermal investigations were performed by using TGA. TGA and DTG profiles as a function of temperature for EP, GNS/EP, and f-GNS/EP composites were also studied. Below 100 °C the Pristine GNS gave a weight loss because of loss of water and removal of grafted oxygen when reacted with silane, resulted in stable functional. When heated to a temperature of 700 degrees Celsius, the f-GNS did not show any signs of significant mass loss. The inclusion of GNS into the GNS/EP composite led to the induction of thermal stabilization [36, 37]. TEM analysis depicted that some parts of the GNSs were difficult to make out because of the extremely thin nature of the GNS, whereas other parts of the GNSs were quite clear (dark areas), which are the direct result of flocculated or merged sections of the GNSs. The surface properties of f-GNSs were described as having the appearance of nanoplatelets that were wrinkled and measured several hundred nanometers in size. It was also clear in the TEM that the GNSs that were not functionalized were predominantly spread in the whole epoxy framework in the form of collected mass, where the f-GNSs were distributed throughout the matrix in a more uniform manner. After tensile tests, the additional study of the broken segments of all the three samples, i.e., EP, GNS/EP, and f-GNS/EP amalgams was performed by scanning electron microscopy (SEM). This was performed in order to obtain further knowledge regarding the interfacial reactions that occurs between epoxy resins and GNS. The pure unadulterated epoxy showed a fracture surface that was smooth and almost mirror-like, which is consistent with the embrittlement of a homogeneous material. It was very obvious that some untreated GNS was being pulled out of the epoxy matrix where it was embedded. In contrast, it was observed that the vast majority of the f-GNS were engrained in the epoxy framework, and it was also observed that none of the f-GNS could be seen to be drawn out. This suggests that the f-GNS has been chemically bonded with the epoxy matrix by forming the covalent interfacial bonds, which indicated that the breakage and deformation did not take place preferentially at the interface between the f-GNS and the EP. AFM observations were carried out in order to obtain a measurement of the GNS and f-GNS materials' respective thicknesses. From the analysis of the cross-section, we were able to determine that the height of the GNSs was approximately 1.015 nm, which corresponded very well with the reported thickness found in earlier literature [38, 39]. Chen et al. [40] investigated silane functionalized graphene oxide (PVSQ-GO) and the investigation demonstrated successfully interpreted the fact that there was good congruence between PVSQ-GO matrix and waterborne polyurethane and the reinforcement coatings showed very good anti-corrosion abilities. The surface morphology was interpreted by FTIR, Raman spectra, XRD, SEM, TEM, AFM, and for corrosion study electrochemical investigations and salt spray test (in accordance with ASTM B117 [41]) were considered were carried out. Open circuit potential, also known as OCP, was investigated as a possible point of reference for corrosion propensity to some extent. In this particular instance, specimens of coatings were submerged in 3.5% NaCl solutions for varying durations of time, and the OCP values were recorded throughout. When compared to any other coating, the specimen with a PVSQ-GO/WPU coating that contained 0.5 wt% showed the lowest rate of downward shift and exhibited a considerably large potentiality regardless of the immersion time. These results were regardless of the immersion time. Again, in EIS investigations it was found, that in the beginning of the immersion process, the coating containing 1.0 wt% of PVSQ-GO framework unveiled a larger impedance value in comparison to the coating containing 0.5 weight percent of PVSQ-GO/WPU. The greatest impedance value is maintained by the 0.5 wt% coating throughout all immersion times. Based on the Nyquist plots, it can be seen that all of the coatings displayed a decreased capacitance loop with the increased immersion time. This shows that the coatings' resistance to corrosion is decreasing [42]. The thermal stability of various materials was evaluated with the help of TGA analysis. The TGA investigations of both GO and PVSQ-GO were shown in the study. Because of the noticeable difference in weight loss, it was clear conclusion that after going through the modification process, GO got changed. Weight loss occurred in two steps in GO. First was weight loss of nearly 17.2 wt% at 180 °C. The evaporation of the trace amount of moisture that had become trapped between the graphene sheets was the root cause of this phenomenon. Secondly weight loss of 17.2 wt% at 800 °C, the final weight loss was roughly 57.6 wt %. This loss was attributed to the breakdown of functional groups specially which contained oxygen. PVSQ-GO showed weight loss stage only once. The PVSQ-GO composite was able to better preserve the strength of the internal structure because it featured with Si–O–C bonds. This was the primary factor that contributed to the composite's success. When silane was used to bring changes in the carbon materials, the incorporation of silane will definitely result in an improvement in the materials' thermal stability, which will allow for greater processing temperatures [43, 44]. While studying Raman spectra, GO displayed two distinct peaks, one of which was located at approximately 1,350 cm^{-1} (D band), and the other at 1,592 cm^{-1} (G band), which were related to oxygen-containing functional group structural flaws and sp2 carbon atom in-plane vibration, respectively [45]. Specifically, the PVSQ-GO composite was shown to have a more highly irregular structure. Enhanced relative intensity may be due to vinyltriethoxysilane-modified graphene oxide [35]. Atta et al. [46] studied the activation and modification of silica gel surfaces was accomplished by grafting 1,3-dihydrazide-2,4,6-triazine onto the hydroxyl groups of active silica surfaces. The investigation focused on the modified silica's chemical structure, as well as its thermal stability and surface morphologies. SEM investigation revealed that the roughness was repaired and reduced after the modification of activated silica gel with triazine hydrazide because of the hydrogen bonding [47]. In addition, both STHa and STHc epoxy frameworks exhibited stronger compatibility and performed exceptionally well in terms of their ability to guard against corrosion. Parhizkar et al. [48] investigated graphene oxide by modifying it with 3-aminopropyltriethoxysilane and then by adding it to epoxy, further effect of coating was studied for its anticorrosion effect. Result revealed that the corrosion rates on steel were reduced after being treated with a f-GO coating, which worked by blocking anodic and cathodic sites and enhancing the coating's barrier properties. The epoxy coating applied to the surface-treated substrate provided a higher level of corrosion protection, as measured by EIS. This is likely due to the coating's improved barrier performance, its enhanced adhesion properties, its reduced cathodic reaction rate. Further, phthalimide-functionalized benzoxazine monomers (pPP-BZ and oPP-BZ) were synthesized by Aly et al. [49] and were investigated for their anticorrosion properties for mild steel sample. Results revealed that poly(pPP-BZ) was more effectively saving the metallic sample from corrosion and as a conclusion, it would appear that adding imide structures to benzoxazine prepared the coating which was successfully mitigating the corrosion.

18.3 Conclusion

An in-depth investigation was conducted for having an insight into the coating for knowing about their bonding exist in them, thermal characteristics, and how the surface behaved in case of variety of coating materials that were bonded with functionalized fillers. Additionally, it was very clear in the literature that the addition of modified compounds as fillers to coating material networks improved the coating materials' adherence to metallic surfaces and improved their mechanical, thermal, and anticorrosion properties. Water, oxygen, and ions were effectively prevented from penetrating the metallic surface by the functionalized coating films. This protected metals from corrosive mediums. To better understand the mechanism of corrosion inhibitors and the relationship between structure and observed corrosion inhibition, more advanced characterization techniques and fundamental investigations need to be conducted. With this knowledge in hand, we may adjust the inhibitor's structural makeup to achieve the desired corrosion-inhibiting effects. In the case of coatings, suitable chemical alterations, can be endowed with improved anticorrosive properties.

References

[1]   Tallman DE, Spinks G, Dominis A, Wallace GG. Electroactive conducting polymers for corrosion control. J Solid State Electrochem 2002;6(2):73–84.
[2]   Nguyen TN, Hubbard JB, Mcfadden GB. A mathematical model for the cathodic blistering of organic coatings on steel immersed in electrolytes. J Coat Technol 1991;63(794):43–52.
[3]   Zhou F, Li Z, Shenoy GJ, Li L, Liu H. Enhanced room-temperature corrosion of copper in the presence of graphene. ACS Nano 2013;7(8):6939–47.
[4]   Sahu SC, Samantara AK, Seth M, Parwaiz S, Singh BP, Rath PC, Jena BK. A facile electrochemical approach for development of highly corrosion protective coatings using graphene nanosheets. Electrochem Commun 2013;32:22–26.
[5]   Liang J, Huang Y, Zhang L, Wang Y, Ma Y, Guo T, Chen Y. Molecular-level dispersion of graphene into poly (vinyl alcohol) and effective reinforcement of their nanocomposites. Adv Funct Mater 2009;*19*(14):2297–302.
[6]   Bai X, Wan C, Zhang Y, Zhai Y. Reinforcement of hydrogenated carboxylated nitrile–butadiene rubber with exfoliated graphene oxide. Carbon 2011;49(5):1608–13.
[7]   Kuila T, Bose S, Hong CE, Uddin ME, Khanra P, Kim NH, Lee JH. Preparation of functionalized graphene/linear low density polyethylene composites by a solution mixing method. Carbon 2011;49(3):1033–37.
[8]   Nawaz M, Yusuf N, Habib S, Shakoor RA, Ubaid F, Ahmad Z, Gao W. Development and properties of polymeric nanocomposite coatings. Polymers 2019;11(5):852.
[9]   Yang W, Feng W, Liao Z, Yang Y, Miao G, Yu B, Pei X. Protection of mild steel with molecular engineered epoxy nanocomposite coatings containing corrosion inhibitor functionalized nanoparticles. Surf Coat Technol 2021;406:126639.
[10]   Cao Y, Feng J, Wu P. Preparation of organically dispersible graphene nanosheet powders through a lyophilization method and their poly (lactic acid) composites. Carbon 2010;48(13):3834–39.
[11]   Xu Y, Hong W, Bai H, Li C, Shi G. Strong and ductile poly (vinyl alcohol)/graphene oxide composite films with a layered structure. Carbon 2009;47(15):3538–43.
[12]   Yang SY, Lin WN, Huang YL, Tien HW, Wang JY, Ma CCM, Wang YS. Synergetic effects of graphene platelets and carbon nanotubes on the mechanical and thermal properties of epoxy composites. Carbon 2011;49(3):793–803.
[13]   Fang M, Wang K, Lu H, Yang Y, Nutt S. Covalent polymer functionalization of graphene nanosheets and mechanical properties of composites. J Mater Chem 2009;19(38):7098–105.

[14] Zhang C, Dai X, Wang Y, Sun G, Li P, Qu L, Dou Y. Preparation and corrosion resistance of ETEO modified graphene oxide/epoxy resin coating. Coatings 2019;9(1):46.

[15] Li Z, Wang R, Young RJ, Deng L, Yang F, Hao L, Liu W. Control of the functionality of graphene oxide for its application in epoxy nanocomposites. Polymer 2013;54(23):6437–46.

[16] GB/T 6739-2006. Paints and varnishes-determination of film hardness by pencil test. 2006.

[17] Kumar AM, Gasem ZM. In situ electrochemical synthesis of polyaniline/f-mwcnt nanocomposite coatings on mild steel for corrosion protection in 3.5% nacl solution. Prog Org Coat 2015;78:387–94.

[18] Hinra H, Ebbesen TW, Tanigaki K. Raman studies of carbon nanotubes. J Chem B Phys Lett B 1993;202:509–12.

[19] Kang ET, Ma ZH, Tan KL, Tretinnikov ON, Uyama Y, Ikada Y. Surface hardness of pristine and modified polyaniline films. Langmuir 1999;15(16):5389–95.

[20] Boinovich LB, Gnedenkov SV, Alpysbaeva DA, Egorkin VS, Emelyanenko AM, Sinebryukhov SL, Zaretskaya AK. Corrosion resistance of composite coatings on low-carbon steel containing hydrophobic and superhydrophobic layers in combination with oxide sublayers. Corros Sci 2012;55:238–45.

[21] Sethi S, Dhinojwala A. Superhydrophobic conductive carbon nanotube coatings for steel. Langmuir 2009;25(8):4311–13.

[22] Liu K, Chen S, Luo Y, Jia D, Gao H, Hu G, Liu L. Edge-functionalized graphene as reinforcement of epoxy-based conductive composite for electrical interconnects. Compos Sci Technol 2013;88:84–91.

[23] Jeon IY, Choi HJ, Jung SM, Seo JM, Kim MJ, Dai L, Baek JB. Large-scale production of edge-selectively functionalized graphene nanoplatelets via ball milling and their use as metal-free electrocatalysts for oxygen reduction reaction. J Am Chem Soc 2013;135(4):1386–93.

[24] Lin T, Chen J, Bi H, Wan D, Huang F, Xie X, Jiang M. Facile and economical exfoliation of graphite for mass production of high-quality graphene sheets. J Mater Chem A 2013;1(3):500–04.

[25] Ferrari AC, Meyer JC, Scardaci V, Casiraghi C, Lazzeri M, Mauri F, Geim AK. Raman spectrum of graphene and graphene layers. Phys Rev Lett 2006;97(18):187401.

[26] Graf D, Molitor F, Ensslin K, Stampfer C, Jungen A, Hierold C, Wirtz L. Spatially resolved Raman spectroscopy of single-and few-layer graphene. Nano Lett 2007;7(2):238–42.

[27] El-Lateef HMA, Gouda M, Khalaf MM, Al-Shuaibi MA, Mohamed IM, Shalabi K, El-Shishtawy RM. Experimental and in-silico computational modeling of cerium oxide nanoparticles functionalized by gelatin as an eco-friendly anti-corrosion barrier on X60 steel alloys in acidic environments. Polymers 14(13):2544.

[28] Chatterjee S, Wang JW, Kuo WS, Tai NH, Salzmann C, Li WL, Chu BTT. Mechanical reinforcement and thermal conductivity in expanded graphene nanoplatelets reinforced epoxy composites. Chem Phys Lett 2012;531:6–10.

[29] Kudin KN, Ozbas B, Schniepp HC, Prud'Homme RK, Aksay IA, Car R. Raman spectra of graphite oxide and functionalized graphene sheets. Nano Lett 2008;8(1):36–41.

[30] Strano MS, Dyke CA, Usrey ML, Barone PW, Allen MJ, Shan H, Smalley RE. Electronic structure control of single-walled carbon nanotube functionalization. Science 2003;301(5639):1519–22.

[31] Geng Y, Li J, Wang SJ, Kim JK. Amino functionalization of graphite nanoplatelet. J Nanosci Nanotechnol 2008;8(12):6238–46.

[32] Hahn HT, Choi O, Wang Z. Development of nanoplatelet composites. California Univ Los Angeles; 2008.

[33] Wang X, Xing W, Zhang P, Song L, Yang H, Hu Y. Covalent functionalization of graphene with organosilane and its use as a reinforcement in epoxy composites. Compos Sci Technol 2012;72 (6):737–43.

[34] Hummers WS, Jr, Offeman RE. Preparation of graphitic oxide. J Am Chem Soc 1958;80(6):1339–1339.

[35] Stankovich S, Dikin DA, Piner RD, Kohlhaas KA, Kleinhammes A, Jia Y, Ruoff RS. Synthesis of graphene-based nanosheets via chemical reduction of exfoliated graphite oxide. Carbon 2007;45 (7):1558–65.

[36] Villar-Rodil S, Paredes JI, Martínez-Alonso A, Tascón JM. Preparation of graphene dispersions and graphene-polymer composites in organic media. J Mater Chem 2009;19(22):3591–93.

[37] Ramanathan T, Abdala AA, Stankovich S, Dikin DA, Herrera-Alonso M, Piner RD, Brinson LC. Functionalized graphene sheets for polymer nanocomposites. Nat Nanotechnol 2008;3(6):327–31.

[38] Kim H, Miura Y, Macosko CW. Graphene/polyurethane nanocomposites for improved gas barrier and electrical conductivity. Chem Mater 2010;22(11):3441–50.

[39] Yang H, Li F, Shan C, Han D, Zhang Q, Niu L, Ivaska A. Covalent functionalization of chemically converted graphene sheets via silane and its reinforcement. J Mater Chem 2009;19(26):4632–38.

[40] Chen C, Wei S, Xiang B, Wang B, Wang Y, Liang Y, Yuan Y. Synthesis of silane functionalized graphene oxide and its application in anti-corrosion waterborne polyurethane composite coatings. Coatings 2019;9(9):587.

[41] ASTM B117 standard practice for operating salt spray (fog) apparatus. West Conshohocken, PA, USA: ASTM International, 2018, Volume 03.02.

[42] Li J, Cui J, Yang J, Li Y, Qiu H, Yang J. Reinforcement of graphene and its derivatives on the anticorrosive properties of waterborne polyurethane coatings. Compos Sci Technol 2016;129:30–37.

[43] Ramezanzadeh B, Ahmadi A, Mahdavian M. Enhancement of the corrosion protection performance and cathodic delamination resistance of epoxy coating through treatment of steel substrate by a novel nanometric sol-gel based silane composite film filled with functionalized graphene oxide nanosheets. Corros Sci 2016;109:182–205.

[44] Yu Z, Lv L, Ma Y, Di H, He Y. Covalent modification of graphene oxide by metronidazole for reinforced anti-corrosion properties of epoxy coatings. RSC Adv 2016;6(22):18217–26.

[45] Das AK, Srivastav M, Layek RK, Uddin ME, Jung D, Kim NH, Lee JH. Iodide-mediated room temperature reduction of graphene oxide: a rapid chemical route for the synthesis of a bifunctional electrocatalyst. J Mat Chem A 2014;2(5):1332–40.

[46] Atta AM, Ahmed MA, Tawfek AM, El-Faham A. Functionalization of silica with triazine hydrazide to improve corrosion protection and interfacial adhesion properties of epoxy coating and steel substrate. Coatings 2020;10(4):351.

[47] Li J, Wang S, Liao L, Ma Q, Zhang Z, Fan G. Stabilization of an intramolecular hydrogen-bond block in an s-triazine insensitive high-energy material. New J Chem 2019;43(27):10675–79.

[48] Parhizkar N, Shahrabi T, Ramezanzadeh B. A new approach for enhancement of the corrosion protection properties and interfacial adhesion bonds between the epoxy coating and steel substrate through surface treatment by covalently modified amino functionalized graphene oxide film. Corros Sci 2017;123:55–75.

[49] Aly KI, Mahdy A, Hegazy MA, Al-Muaikel NS, Kuo SW, Gamal Mohamed M. Corrosion resistance of mild steel coated with phthalimide-functionalized polybenzoxazines. Coatings 2020;10(11):1114.

Harpreet Kaur, Abhinay Thakur, Ramesh Chand Thakur, Ashish Kumar

19 Challenges and future outlooks

Abstract: Functionalized thin film coatings have been extensively researched and developed to enhance the properties and performance of various materials, especially metals and their alloys. These coatings offer a range of benefits, including improved corrosion resistance, enhanced durability, and reduced friction, making them highly desirable for various industrial applications. This chapter discusses the challenges and future outlooks of functionalized thin film coatings on several metals and their alloys. It highlights the importance of selecting the appropriate deposition technique to achieve the desired coating properties and optimize the process to minimize defects. The chapter also outlines the challenges associated with the development of functionalized thin film coatings, including the need for improved coating adhesion, long-term durability, and resistance to environmental factors. It discusses the use of advanced materials, such as nanoparticles and graphene, to enhance the properties of coatings and mitigate some of the challenges associated with their development. In addition the chapter explores the future outlooks of functionalized thin film coatings on several metals and their alloys. It discusses the potential for these coatings to be applied in various industrial sectors, including aerospace, automotive, and electronics, and the opportunities presented by advances in nanotechnology and surface engineering.

Keywords: Functionalized thin film coatings, Metals and alloys, Corrosion resistance, Advanced materials, Challenges, Future outlooks

19.1 Introduction

Functionalized thin film coatings have attracted significant attention in recent years due to their potential to improve the properties and performance of various materials, especially metals and their alloys. These coatings offer a range of benefits, including improved corrosion resistance, enhanced durability, and reduced friction, making them highly desirable for various industrial applications. The development of functionalized thin film coatings on metals and their alloys is a highly interdisciplinary field that involves materials science, chemistry, physics, and engineering. The deposition of these coatings on metal surfaces is achieved through various techniques, including physical vapor deposition, chemical vapor deposition, electroplating, and others. The choice of deposition technique depends on the desired properties of the coating, such as thickness, adhesion, uniformity, and composition [1–7]. One of the primary challenges associated with the development of functionalized thin film coatings on metals and their alloys is the selection and optimization of the deposition tech-

https://doi.org/10.1515/9783111016160-019

nique. The chosen method should produce coatings with the desired properties and minimize defects such as cracks, delamination, and porosity. The optimization of the deposition process is crucial to achieving uniform coatings with the desired properties. Another challenge associated with functionalized thin film coatings is their durability and resistance to environmental factors. These coatings must be resistant to various environmental factors, such as temperature, humidity, and corrosive substances, to maintain their integrity and functionality over prolonged periods. The development of coatings that can withstand harsh environmental conditions is essential for their application in various industrial sectors. Despite these challenges, the future outlook for functionalized thin film coatings on metals and their alloys is promising. Advances in nanotechnology and surface engineering have enabled the development of novel materials and coatings with unique properties and functions. Moreover the increasing demand for high-performance coatings in various industries, including aerospace, automotive, and electronics, continues to drive research in the field. This chapter aims to provide an overview of the challenges and future outlooks of functionalized thin film coatings on several metals and their alloys [8–16]. The chapter begins by discussing the different types of coatings and their properties, including chemical, physical, and mechanical characteristics. It highlights the importance of selecting the appropriate deposition technique to achieve the desired coating properties and optimizing the process to minimize defects. The chapter then discusses the challenges associated with the development of functionalized thin film coatings, including the need for improved coating adhesion, long-term durability, and resistance to environmental factors. It explores the use of advanced materials, such as nanoparticles and graphene, to enhance the properties of coatings and mitigate some of the challenges associated with their development. In addition the chapter explores the future outlooks of functionalized thin film coatings on several metals and their alloys. It discusses the potential for these coatings to be applied in various industrial sectors, including aerospace, automotive, and electronics, and the opportunities presented by advances in nanotechnology and surface engineering [17–20]. Figure 19.1 shows the plasma-based synthesis of functionalized thin films [21].

19.1.1 Common types of functionalized thin film coatings

Functionalized thin film coatings are designed to improve the properties of various materials, particularly metals and their alloys. These coatings offer a range of benefits, including enhanced corrosion resistance, improved durability, and reduced friction. The following are some of the common types of functionalized thin film coatings:

– Corrosion-resistant coatings
Corrosion is a significant challenge faced by metals and their alloys, particularly when exposed to harsh environmental conditions such as humidity, temperature, and

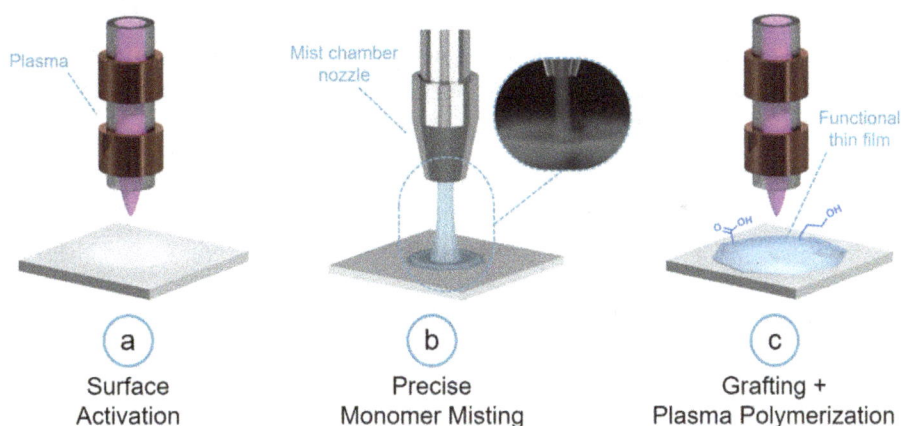

Figure 19.1: The process for synthesizing functionalized thin films using plasma-based methods from liquid monomers. Firstly, the substrate's surface is activated using plasma (a). Next, misted monomer is deposited onto the substrate to form a thin liquid layer (b), which is achieved using a mist chamber outlet with a focusing nozzle, an orifice of $\varnothing = 2$ mm, and a distance of 2 mm to the substrate. Finally, the deposited liquid monomer film is polymerized via plasma treatment (c). Adapted with permission from ref. [21]. MDPI [2021]. Distributed under CCBY 4.0.

corrosive substances. Corrosion-resistant coatings are designed to protect metal surfaces from corrosion by providing a barrier between the metal and the environment [17, 22–24]. These coatings can be applied to various metals, including aluminum, steel, and copper, to prevent corrosion and extend the service life of the metal.

– Wear-resistant coatings
Wear is a significant challenge faced by metal components, particularly in applications where they are subjected to high stress and friction. Wear-resistant coatings are designed to reduce friction and wear on metal surfaces, thereby improving their service life. These coatings can be applied to various metals, including steel, titanium, and aluminum, to increase their wear resistance and reduce the need for frequent replacement.

– Thermal barrier coatings
Thermal barrier coatings are designed to protect metal components from high temperatures and thermal shock. These coatings are typically applied to gas turbine components, such as blades and vanes, to protect them from the extreme heat generated by combustion. Thermal barrier coatings can be made of various materials, including ceramics and metallic alloys, to provide excellent thermal insulation and protect metal components from thermal damage.

– Anti-fouling coatings
Anti-fouling coatings are designed to prevent the buildup of biological and organic materials on metal surfaces. These coatings are typically applied to ship hulls and ma-

rine structures to prevent the growth of marine organisms, such as barnacles and algae, which can increase drag and fuel consumption. Anti-fouling coatings can be made of various materials, including polymers and silicone, to prevent the adhesion of biological materials to metal surfaces [25, 26].

– Self-cleaning coatings
Self-cleaning coatings are designed to repel dirt, dust, and other contaminants from metal surfaces. These coatings are typically applied to building facades, solar panels, and other structures exposed to the environment. Self-cleaning coatings can be made of various materials, including nanostructured coatings and hydrophobic materials, to prevent the adhesion of contaminants and make the surface easier to clean.

– Biocompatible coatings
Biocompatible coatings are designed to improve the biocompatibility of metal surfaces for medical applications. These coatings are typically applied to medical implants, such as orthopedic implants and dental implants, to improve their biocompatibility and reduce the risk of rejection by the body. Biocompatible coatings can be made of various materials, including biocompatible polymers and ceramics, to improve the compatibility of metal surfaces with living tissue.

19.2 Physical characteristics of functionalizing thin coatings

Thin film coatings are used in a wide range of applications, from electronic devices to medical implants. Functionalizing these coatings can further enhance their properties, allowing them to perform specific functions such as improved corrosion resistance, biocompatibility, or antimicrobial properties [27–32]. This section will provide an overview of different types of functionalized thin film coatings and their properties, including chemical, physical, and mechanical characteristics. Figure 19.2 shows the deposition of several distinctive units as a function of functionalized thin film coatings for their several useful application [33].

19.2.1 Chemical functionalization

Chemical functionalization is a process of modifying the surface of a thin film coating by adding chemical functional groups or reactive sites that can interact with the environment, modify the surface chemistry, and enhance the properties of the coating. Chemical functionalization can be used to improve the adhesion, biocompatibility, corrosion resistance, and other chemical properties of the coating. Some common

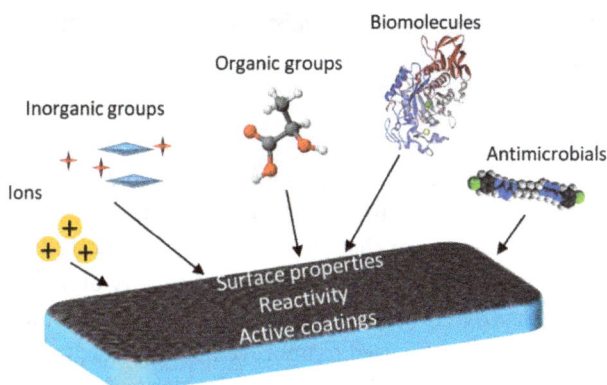

Figure 19.2: Depiction of the various functionalization groups of functionalized thin films. Adapted with permission from ref. [33]. MDPI [2021]. Distributed under CCBY 4.0.

methods of chemical functionalization of thin film coatings include self-assembled monolayers (SAMs), silane coupling agents, and plasma treatment.

Self-assembled monolayers (SAMs)

Self-assembled monolayers (SAMs) are a type of chemical functionalization method that involves the adsorption of a monolayer of molecules onto a surface. SAMs can be used to modify the surface chemistry of a thin film coating by introducing reactive functional groups that can bind to other molecules or surfaces. SAMs can be formed by the spontaneous adsorption of molecules onto a substrate from a solution or by covalent attachment of the molecules to the substrate surface. SAMs can be used to tailor the chemical properties of a thin film coating, such as its wettability, adhesion, and corrosion resistance. For example a SAM of carboxylic acid functionalized molecules can be used to anchor biomolecules onto a metal oxide surface, while a SAM of alkylsilanes can be used to improve the hydrophobicity of a surface.

Silane coupling agents

Silane coupling agents are another type of chemical functionalization method that involves the covalent bonding of a silane molecule to the surface of a substrate or a thin film coating. Silane coupling agents have reactive functional groups, such as alkoxy or amino groups, that can react with hydroxyl groups on the substrate surface, creating a covalent bond between the silane molecule and the surface. Silane coupling agents can be used to improve the adhesion of a thin film coating to a substrate by creating a chemical bond between the coating and the substrate surface. This can improve the durability and stability of the coating, as well as its resistance to wear and tear. Silane coupling agents can also be used to tailor the surface chemistry of the coating, such as its hydrophobicity or biocompatibility.

Plasma treatment

Plasma treatment is a type of chemical functionalization method that involves the exposure of a substrate or a thin film coating to a plasma. Plasma is a gas that has been ionized by an electric field, creating a mixture of charged particles, free radicals, and other reactive species. Plasma treatment can modify the surface chemistry of a substrate or a coating by introducing reactive functional groups that can bind to other molecules or surfaces. Plasma treatment can be used to improve the adhesion, wettability, and biocompatibility of a thin film coating. For example plasma treatment can be used to introduce carboxylic acid or amino groups onto the surface of a coating, which can then be used to anchor biomolecules onto the surface. Plasma treatment can also be used to create functionalized coatings that have specific chemical or physical properties, such as antimicrobial coatings or superhydrophobic coatings [32, 34–36]. One example of chemical functionalization is the use of self-assembled monolayers (SAMs) on metal oxide surfaces. SAMs are formed by chemisorption of molecules with reactive functional groups onto the surface. The chemical properties of the SAMs can be modified by changing the functional group of the molecules used. For example SAMs with carboxylic acid functional groups can be used to anchor biomolecules onto metal oxide surfaces. Another example of chemical functionalization is the use of silane coupling agents to modify the surface chemistry of thin film coatings. Silane coupling agents are molecules with reactive functional groups that can bond to the surface of the coating, creating a covalent bond between the coating and the silane molecule. This can improve the adhesion of the coating to the substrate, as well as modify its chemical properties.

19.2.2 Physical functionalization

Physical functionalization is a process of modifying the surface of a thin film coating by altering its physical properties, such as its roughness, porosity, or surface energy. Physical functionalization can be used to enhance the mechanical, optical, or thermal properties of the coating. Physical functionalization techniques can be classified into two main categories: top-down and bottom-up approaches.

Top-down approaches

Top-down approaches involve the modification of the surface of a thin film coating by removing or restructuring its outer layer or by creating surface patterns. These approaches are often used to improve the adhesion, friction, wear resistance, or optical

properties of the coating. Some common top-down approaches for physical functionalization of thin film coatings include:

Abrasive Techniques: Abrasive techniques involve the use of mechanical or chemical abrasion to remove or reshape the surface of a thin film coating. These techniques can be used to create surface patterns or to remove surface contaminants
that may affect the properties of the coating.

Ion Beam Techniques: Ion beam techniques involve the use of energetic ions to
bombard the surface of a thin film coating, creating surface modifications such as
nanocrystallization, densification, or amorphization. These techniques can improve the mechanical, optical, or thermal properties of the coating.

Laser Techniques: Laser techniques involve the use of laser radiation to modify
the surface of a thin film coating. Laser ablation can be used to create surface
patterns or to remove surface contaminants, while laser annealing can be used to
improve the crystallinity or the optical properties of the coating [37–39].

Bottom-up approaches

Bottom-up approaches involve the self-assembly or deposition of molecules or nanoparticles onto the surface of a thin film coating, creating a new layer or modifying
the existing layer. These approaches are often used to improve the surface energy,
biocompatibility, or electrical properties of the coating. Some common bottom-up
approaches for physical functionalization of thin film coatings include:

Layer-by-Layer Assembly: Layer-by-layer assembly involves the deposition of alternating layers of positively and negatively charged polyelectrolytes onto the
surface of a thin film coating, creating a multilayer film with controlled thickness
and composition. This technique can be used to improve the surface energy,
biocompatibility, or electrical properties of the coating.

Chemical Vapor Deposition: Chemical vapor deposition involves the deposition of
a thin film coating by reacting a precursor gas with the surface of a substrate [40,
41]. This technique can be used to deposit coatings with controlled composition,
thickness, and crystallinity, and can improve the mechanical, optical, or thermal
properties of the coating.

Sol-Gel Deposition: Sol-gel deposition involves the deposition of a thin film coating
by hydrolyzing and condensing a precursor sol onto the surface of a substrate.
This technique can be used to deposit coatings with controlled porosity, surface
area, or composition, and can improve the surface energy, biocompatibility, or
electrical properties of the coating.

19.2.3 Mechanical functionalization

Mechanical functionalization is the process of modifying the mechanical properties of a thin film coating, such as its hardness, toughness, or elasticity. These modifications can be achieved through various techniques, including the addition of reinforcing materials, the use of external stresses, or the modification of the coating's microstructure.

Reinforcement: The addition of reinforcing materials to a thin film coating can significantly improve its mechanical properties. Reinforcing materials can include nanoparticles, nanofibers, or even large-scale fibers. These materials are added to the coating during the deposition process, either by co-deposition or post-deposition methods, and can significantly improve the coating's hardness, stiffness, or strength.

External Stress: The application of external stress can also modify the mechanical properties of a thin film coating. For example compressive stress can improve a coating's hardness and strength, while tensile stress can improve its ductility and toughness. These stresses can be applied through various methods, including ion implantation, laser irradiation, or mechanical deformation.

Microstructure: The microstructure of a thin film coating can also play a significant role in its mechanical properties. By controlling the deposition parameters, such as the substrate temperature, deposition rate, or gas flow rate, the microstructure of the coating can be tailored to improve its mechanical properties [42–45]. For example increasing the substrate temperature during deposition can improve the crystallinity and hardness of the coating. In addition to these techniques, hybrid approaches that combine multiple techniques can also be used to achieve superior mechanical properties. For example the addition of reinforcing materials to a coating, followed by the application of compressive stress, can lead to a significant improvement in the coating's hardness and strength.

19.2.4 Applications of mechanical functionalization

Mechanical functionalization of thin film coatings has a wide range of applications across various industries. Some examples include:

Biomedical Applications: Mechanical functionalization can be used to improve the wear resistance and biocompatibility of coatings used in medical implants or devices. For example adding nanofibers to a polymer coating can significantly improve its mechanical properties, making it more suitable for use in orthopedic implants.

Protective Coatings: Mechanical functionalization can be used to improve the hardness and scratch resistance of coatings used for protective purposes. For example adding ceramic nanoparticles to a polymer coating can improve its hardness and scratch resistance, making it more suitable for use in scratch-resistant coatings for eyeglasses or mobile phone screens.

Aerospace Applications: Mechanical functionalization can be used to improve the toughness and fatigue resistance of coatings used in aerospace applications. For example adding carbon nanotubes to a metal coating can significantly improve its strength and toughness, making it more suitable for use in aircraft components.

19.3 Importance of selecting the appropriate deposition technique to achieve the desired coating properties and optimizing the process to minimize defects

The selection of an appropriate deposition technique is crucial in achieving the desired functionalized coating properties. Different deposition techniques have different advantages and limitations in terms of the properties of the resulting coating, such as its thickness, porosity, uniformity, and adhesion. Some common deposition techniques for thin film coatings include physical vapor deposition (PVD), chemical vapor deposition (CVD), electrochemical deposition, and spin coating.

19.3.1 Physical vapor deposition (PVD)

PVD is a popular technique for depositing thin films with high purity and uniformity, making it suitable for applications that require high precision and control over the coating properties. In PVD, a material is vaporized in a vacuum chamber, and the vapor condenses onto the substrate to form a thin film. PVD can be further classified into different techniques, such as sputtering, evaporation, and ion plating [46]. Sputtering is a PVD technique that involves the use of a plasma to generate ions that bombard a target material, causing it to eject atoms or molecules. The ejected material then condenses onto the substrate to form a thin film. Sputtering allows for precise control over the composition and thickness of the coating and can produce high-quality coatings with good adhesion and uniformity. Evaporation is another PVD technique that involves the heating of a material to its vaporization temperature, causing it to evaporate and condense onto the substrate. Evaporation can be carried out using various methods, such as resistive heating or electron beam heating. Evaporation is a simple and low-cost technique that can produce high-quality coatings with good uniformity and thickness control. Ion plating is a PVD technique that involves the use of an ionized gas, which bombards the substrate to improve adhesion and promote the formation of a dense and uniform coating. Ion plating can also be used to incorporate dopants into the coating to modify its properties.

19.3.2 Chemical vapor deposition (CVD)

CVD is a deposition technique that involves the use of a gas phase chemical reaction to deposit a solid material onto a substrate. In CVD, a reactant gas is introduced into a reactor chamber, where it reacts with a substrate surface to form a thin film coating. The reactant gas can be decomposed by thermal or plasma energy to produce a reactive species that reacts with the substrate surface. CVD can be further classified into different techniques, such as atmospheric pressure CVD and low-pressure CVD [47]. Atmospheric pressure CVD is carried out at atmospheric pressure, making it suitable for high-volume production. Low-pressure CVD, on the other hand, is carried out at reduced pressure, which allows for better control over the coating properties.

19.3.3 Electrochemical deposition

Electrochemical deposition is a deposition technique that involves the use of an electrolytic solution to deposit a metal or semiconductor coating onto a substrate. In electrochemical deposition, the substrate is placed in an electrolytic solution, and a current is applied to the substrate, causing a reaction to occur at the surface of the substrate. The reaction produces a coating that adheres to the substrate. Electrochemical deposition allows for the deposition of coatings with high adhesion and corrosion resistance. The properties of the coating can be controlled by adjusting the deposition parameters, such as the current density and solution composition.

19.3.4 Spin coating

Spin coating is a simple and low-cost technique that can be used to deposit thin films with good uniformity and thickness control. In spin coating, a solution containing the coating material is deposited onto a substrate, and the substrate is then spun at high speed to spread the solution evenly across the surface. The excess solution is then removed, leaving behind a thin film coating.

Optimizing the deposition process is critical to achieving a high-quality functionalized coating with minimal defects. Defects can arise during the deposition process due to a variety of factors, such as poor adhesion, poor uniformity, poor thickness control, or contamination. The following are some strategies that can be used to optimize the deposition process and minimize defects:

Substrate preparation: The substrate should be cleaned thoroughly to remove any impurities or contaminants that may interfere with the adhesion of the coating. The surface of the substrate should be smooth and free of any scratches or defects that may be amplified during the deposition process.

Deposition parameters: The deposition parameters, such as the temperature, pressure, and deposition rate, should be carefully controlled to ensure that the coating is deposited uniformly and with the desired thickness. Deposition parameters can also affect the microstructure of the coating, which can impact its mechanical and physical properties.

Gas flow rate and composition: In CVD, the gas flow rate and composition can impact the chemical reactions that occur during the deposition process, affecting the composition and properties of the coating. Careful control of the gas flow rate and composition can help minimize defects such as voids, cracks, or impurities.

Monitoring and control: During the deposition process, it is important to monitor the coating properties, such as thickness, uniformity, and composition. Real-time monitoring techniques, such as ellipsometry or spectroscopic ellipsometry, can be used to monitor the thickness and composition of the coating. The feedback from monitoring can be used to adjust the deposition parameters in real-time to ensure that the coating is deposited with the desired properties.

Post-deposition treatment: Post-deposition treatment, such as annealing, can be used to improve the properties of the coating, such as its adhesion, crystallinity, or density. Post-deposition treatment can also help to minimize defects, such as cracks, voids, or impurities, by healing or removing them.

Quality control: Once the deposition process is complete, it is important to perform quality control tests to ensure that the coating meets the desired specifications. Various techniques, such as X-ray diffraction, scanning electron microscopy, or atomic force microscopy, can be used to examine the microstructure and properties of the coating. In conclusion, selecting the appropriate deposition technique is crucial to achieving the desired functionalized coating properties. However, even with the right technique, defects can still arise during the deposition process [48–50]. Optimizing the deposition process by carefully controlling the deposition parameters, gas flow rate and composition, monitoring and control, post-deposition treatment, and quality control can help minimize defects and produce high-quality functionalized coatings. The use of advanced monitoring techniques and feedback control systems can help further improve the deposition process and ensure the consistent and reliable production of functionalized coatings.

19.4 Challenges associated with the development of functionalized thin film coatings

Functionalized thin film coatings are becoming increasingly important in a wide range of industries, including electronics, aerospace, biomedical, and energy. However, the development of these coatings presents several challenges, such as the need for improved coating adhesion, long-term durability, and resistance to environmental factors.

19.4.1 Improved coating adhesion

One of the main challenges associated with functionalized thin film coatings is achieving good adhesion to the substrate material. Poor adhesion can lead to premature failure of the coating, which can compromise the performance and longevity of the product. Adhesion can be affected by several factors, such as the surface roughness of the substrate material, the chemical compatibility between the coating and the substrate, and the deposition method used to apply the coating. To address these challenges, researchers are exploring new surface treatment methods to improve substrate roughness and enhance adhesion, as well as new deposition techniques that can deposit coatings with improved adhesion properties. For example plasma surface modification techniques, such as plasma cleaning and plasma activation, have been shown to improve the adhesion of functionalized coatings to a wide range of substrates [51–53]. Similarly, the use of layer-by-layer deposition techniques, such as electrostatic layer-by-layer deposition, can help improve adhesion by creating a multilayered structure that is strongly adhered to the substrate.

19.4.2 Long-term durability

Another challenge associated with functionalized thin film coatings is achieving long-term durability, particularly in harsh environments. Many applications require coatings that can withstand exposure to high temperatures, corrosive chemicals, and mechanical stress without degrading or delaminating. To address these challenges, researchers are exploring new materials and deposition methods that can improve the durability of functionalized coatings. For example the use of nanocomposite materials, such as graphene or carbon nanotubes, can improve the mechanical strength and thermal stability of coatings, making them more durable in harsh environments. Similarly, the use of atomic layer deposition (ALD) can help improve the durability of coatings by depositing thin layers of material with high purity and uniformity. ALD can also be used to deposit coatings with precise thickness control, which can help improve their resistance to wear and tear.

19.4.3 Resistance to environmental factors

Functionalized thin film coatings must also be resistant to environmental factors, such as moisture, heat, and UV radiation. These factors can degrade the properties of the coating over time, leading to a loss of functionality and reduced performance. To address these challenges, researchers are exploring new materials and deposition techniques that can improve the resistance of coatings to environmental factors. For example the use of polymeric materials, such as polyimides or silicones, can improve the resistance of coatings to moisture and UV radiation, making them more durable in outdoor or high-humidity environments. Similarly, the use of atmospheric pressure plasma deposition techniques can help improve the resistance of coatings to heat and chemical exposure. Plasma deposition can create coatings with high chemical and thermal stability, making them more resistant to environmental factors.

19.5 Use of advanced materials to enhance the properties of functionalized coatings

The use of advanced materials such as nanoparticles and graphene has shown great promise in enhancing the properties of coatings and mitigating some of the challenges associated with their development. These materials offer unique properties such as high surface area, thermal stability, and mechanical strength, which can be exploited to improve the performance of functionalized thin film coatings.

19.5.1 Nanoparticles

Nanoparticles are particles with sizes ranging from 1 to 100 nanometers. These materials offer a high surface area to volume ratio, which can be exploited to improve the properties of coatings. By adding nanoparticles to a coating, researchers can enhance properties such as hardness, wear resistance, and thermal stability. One example of a nanoparticle-enhanced coating is the use of silica nanoparticles to improve the properties of epoxy coatings. The addition of silica nanoparticles to epoxy coatings has been shown to improve the hardness, scratch resistance, and thermal stability of the coating. Similarly, the addition of alumina nanoparticles to coatings can improve the wear resistance and hardness of the coating. Nanoparticles can also be used to improve the adhesion of coatings to substrates. For example the addition of titanium dioxide nanoparticles to a coating can improve the adhesion of the coating to aluminum substrates. The high surface area of the nanoparticles provides more sites for bonding with the substrate, which enhances the adhesion of the coating [54–56].

19.5.2 Graphene

Graphene is a two-dimensional material composed of carbon atoms arranged in a hexagonal lattice. It is known for its exceptional mechanical, thermal, and electrical properties, making it a promising material for enhancing the properties of coatings. Graphene can be added to coatings to improve their mechanical properties, such as hardness and wear resistance. For example the addition of graphene to polyurethane coatings has been shown to improve their hardness and wear resistance. Similarly, the addition of graphene to epoxy coatings has been shown to improve their mechanical strength. Graphene can also be used to improve the thermal and electrical conductivity of coatings. For example the addition of graphene to polymeric coatings can improve their thermal conductivity, making them suitable for use in high-temperature applications. Graphene can also be used to improve the electrical conductivity of coatings, which is important in applications such as electronics and sensors.

19.5.3 Other advanced materials

In addition to nanoparticles and graphene, other advanced materials such as carbon nanotubes, metal-organic frameworks, and quantum dots have also been explored for their potential in enhancing the properties of coatings. Carbon nanotubes, like graphene, offer exceptional mechanical and electrical properties, making them suitable for use in coatings. For example the addition of carbon nanotubes to epoxy coatings has been shown to improve their mechanical strength and thermal stability. Metal-organic frameworks (MOFs) are porous materials composed of metal ions and organic ligands. MOFs offer high surface area and tunable porosity, which can be exploited to enhance the properties of coatings. For example the incorporation of MOFs into coatings can improve their gas sorption properties, making them suitable for use in gas separation applications. Quantum dots are semiconductor particles with sizes ranging from 1 to 10 nanometers. They offer unique optical properties, such as high fluorescence and tunable emission spectra, making them useful in applications such as sensors and optoelectronics. Quantum dots can be incorporated into coatings to impart these optical properties, enhancing their functionality in these applications.

19.6 Future outlooks of functionalized thin film coatings on several metals and their alloys

Functionalized thin film coatings have become increasingly important in the field of material science and engineering due to their ability to improve the properties of metals and their alloys. These coatings can provide enhanced corrosion resistance, wear resistance, and other desirable properties, making them useful in a wide range of applications. In this section, we will explore the future outlooks of functionalized thin film coatings on several metals and their alloys.

19.6.1 Stainless steel

Stainless steel is a popular material for a wide range of applications due to its excellent corrosion resistance and mechanical properties [57–61]. However, there are limitations to its use, particularly in environments with high chloride ion concentrations, which can lead to localized corrosion. To address this issue, researchers have explored the use of functionalized thin film coatings on stainless steel. One promising approach is the use of nanocrystalline coatings such as titanium nitride, which can provide enhanced wear and corrosion resistance. These coatings can be applied using techniques such as physical vapor deposition (PVD) or chemical vapor deposition (CVD). Another approach is the use of hybrid coatings, which combine a corrosion-resistant layer with a wear-resistant layer. These coatings can be applied using electrochemical deposition or spray coating techniques. In the future, functionalized thin film coatings on stainless steel are likely to become even more sophisticated, with a greater focus on the development of multi-layer coatings and the use of advanced materials such as graphene and carbon nanotubes.

19.6.2 Aluminum alloys

Aluminum alloys are widely used in the aerospace and automotive industries due to their excellent strength-to-weight ratio. However aluminum alloys are susceptible to corrosion, particularly in marine environments. To address this issue researchers have explored the use of functionalized thin film coatings on aluminum alloys. One promising approach is the use of anodizing, which involves the creation of a porous oxide layer on the surface of the aluminum alloy. This layer can be modified to improve its properties, such as by incorporating nanoparticles or adding a polymer coating. Another approach is the use of conversion coatings, which involve the conversion of the surface of the aluminum alloy to a more stable form, such as a phosphate or chromate coating. In the future, functionalized thin film coatings on aluminum alloys are likely to become even more specialized, with a greater focus on the development of coatings for specific applications such as marine environments or high-temperature applications. Additionally, there may be a greater focus on the use of advanced materials such as MOFs or quantum dots to enhance the properties of these coatings.

19.6.3 Copper alloys

Copper alloys are widely used in the electronics industry due to their excellent electrical conductivity. However, copper alloys are also susceptible to corrosion and wear, particularly in harsh environments. To address this issue, researchers have explored the use of functionalized thin film coatings on copper alloys. One promising approach is the use of diamond-like carbon (DLC) coatings, which can provide enhanced wear and corrosion resistance. These coatings can be applied using techniques such as PVD or chemical vapor deposition (CVD). Another approach is the use of polymer coatings, which can provide enhanced adhesion and corrosion resistance. These coatings can be applied using techniques such as electrochemical deposition or spray coating. In the future functionalized thin film coatings on copper alloys are likely to become even more specialized, with a greater focus on the development of coatings for specific applications such as high-frequency electronics or high-temperature environments. Additionally there may be a greater focus on the use of advanced materials such as graphene or carbon nanotubes to enhance the properties of these coatings.

19.6.4 Titanium alloys

Titanium alloys are widely used in the aerospace and biomedical industries due to their excellent mechanical properties and biocompatibility. However titanium alloys are also susceptible to wear and corrosion, particularly in harsh environments. To address this issue researchers have explored the use of functionalized thin film coatings on titanium alloys. One promising approach is the use of nitride coatings, which can provide enhanced wear and corrosion resistance. These coatings can be applied using techniques such as PVD or CVD. Another approach is the use of hydroxyapatite (HA) coatings, which can improve the biocompatibility of titanium alloys for biomedical applications. These coatings can be applied using techniques such as electrochemical deposition or spray coating. In the future functionalized thin film coatings on titanium alloys are likely to become even more specialized, with a greater focus on the development of coatings for specific applications such as aerospace or biomedical implants. Additionally, there may be a greater focus on the use of advanced materials such as nanofibers or nanoparticles to enhance the properties of these coatings.

19.6.5 Magnesium alloys

Magnesium alloys are lightweight materials that are increasingly being used in the aerospace and automotive industries. However magnesium alloys are also susceptible to corrosion and wear, particularly in harsh environments. To address this issue researchers have explored the use of functionalized thin film coatings on magnesium alloys. One promising approach is the use of anodizing, which involves the creation of a porous oxide layer on the surface of the magnesium alloy. This layer can be modified to improve its properties, such as by incorporating nanoparticles or adding a polymer coating. Another approach is the use of electroless nickel coatings, which can provide enhanced wear and corrosion resistance. In the future, functionalized thin film coatings on magnesium alloys are likely to become even more specialized, with a greater focus on the development of coatings for specific applications such as high-temperature environments or biomedical implants. Additionally there may be a greater focus on the use of advanced materials such as MOFs or carbon nanotubes to enhance the properties of these coatings.

Overall the future outlooks for functionalized thin film coatings on metals and their alloys are promising, with a greater focus on the development of specialized coatings for specific applications and the use of advanced materials to enhance their properties. As the demand for high-performance materials continues to grow, functionalized thin film coatings are likely to play an increasingly important role in the field of material science and engineering.

19.7 Potential of functionalized coatings in various industrial sectors

Functionalized thin film coatings have tremendous potential to be applied in various industrial sectors, including aerospace, automotive, and electronics. These coatings can provide a range of benefits, such as enhanced wear and corrosion resistance, improved thermal and electrical properties, and better biocompatibility. In this section we will explore the potential applications of functionalized coatings in these industries.

19.7.1 Aerospace

The aerospace industry is one of the primary users of functionalized thin film coatings. These coatings are applied to a range of components, such as engine parts, landing gear, and turbine blades, to improve their performance and durability in harsh environments. For example coatings such as titanium nitride or diamond-like carbon can provide enhanced wear resistance and reduced friction for engine components, while aluminum oxide or ceramic coatings can improve the thermal resistance of turbine blades. Another area where functionalized coatings have potential in the aerospace industry is in the development of lightweight materials. Lightweight materials are essential for reducing the weight of aircraft and improving fuel efficiency. Functionalized coatings can be applied to these materials to improve their properties, such as by providing enhanced corrosion resistance or improving their thermal stability.

19.7.2 Automotive

The automotive industry is another sector that can benefit greatly from the use of functionalized thin film coatings. These coatings can be applied to various components of a vehicle, such as engine parts, exhaust systems, and brake rotors, to improve their performance and durability. For example coatings such as chrome or nickel can improve the corrosion resistance of exhaust systems, while diamond-like carbon or tungsten carbide coatings can improve the wear resistance of engine components. Another area where functionalized coatings have potential in the automotive industry is in the development of electric vehicles. Electric vehicles require high-performance batteries, which can be enhanced with the use of functionalized coatings. For example coatings such as aluminum oxide or lithium phosphate can improve the conductivity and thermal stability of battery components, thereby improving the efficiency and reliability of electric vehicles [62–65].

19.7.3 Electronics

The electronics industry is another sector that can benefit greatly from the use of functionalized thin film coatings. These coatings can be applied to various components of electronic devices, such as circuit boards, displays, and sensors, to improve their performance and durability. For example coatings such as gold or silver can improve the conductivity of circuit boards, while diamond-like carbon or aluminum oxide coatings can improve the hardness and scratch resistance of displays. Another area where functionalized coatings have potential in the electronics industry is in the development of flexible and wearable devices. Flexible and wearable devices require materials that can withstand repeated bending and stretching without degrading. Functionalized coatings can be applied to these materials to improve their properties, such as by providing enhanced adhesion or improving their resistance to cracking.

19.7.4 Biomedical

The biomedical industry is another sector where functionalized thin film coatings have tremendous potential. These coatings can be applied to various biomedical devices, such as implants, prosthetics, and drug delivery systems, to improve their biocompatibility and performance. For example coatings such as hydroxyapatite or titanium nitride can improve the biocompatibility and osseointegration of dental and orthopedic implants, while diamond-like carbon or graphene coatings can improve the wear resistance and corrosion resistance of prosthetic joints. Another area where functionalized coatings have potential in the biomedical industry is in the development of drug delivery systems. Drug delivery systems require materials that can release drugs in a controlled manner over a prolonged period. Functionalized coatings can be applied to these materials to improve their properties, such as by providing enhanced drug release or improving their stability in the body. For example coatings such as titanium dioxide or zinc oxide can improve the efficiency of solar cells by increasing their absorption of sunlight, while graphene or carbon nanotube coatings can improve the conductivity and durability of fuel cell components. Another area where functionalized coatings have potential in the energy industry is in the development of energy storage systems. Energy storage systems are essential for managing fluctuations in energy supply and demand, and for enabling the integration of renewable energy sources into the grid. Functionalized coatings can be applied to energy storage materials, such as lithium-ion batteries, to improve their properties, such as by providing enhanced stability or improving their cycling performance.

19.7.5 Marine

The marine industry is another sector where functionalized thin film coatings have potential applications. These coatings can be applied to various marine components, such as ship hulls, propellers, and underwater structures, to improve their performance and durability in harsh marine environments. For example coatings such as antifouling coatings can prevent the growth of marine organisms on ship hulls, thereby reducing drag and improving fuel efficiency, while corrosion-resistant coatings can protect underwater structures from corrosion. Another area where functionalized coatings have potential in the marine industry is in the development of offshore wind turbines [66–68]. Offshore wind turbines require materials that can withstand the harsh marine environment and the effects of wind and waves. Functionalized coatings can be applied to these materials to improve their properties, such as by providing enhanced corrosion resistance or improving their durability in high winds.

19.7.6 Construction

The construction industry is another sector that can benefit from the use of functionalized thin film coatings. These coatings can be applied to various building components, such as windows, facades, and roofs, to improve their performance and durability. For example coatings such as low-emissivity coatings can improve the thermal efficiency of windows, while self-cleaning coatings can reduce maintenance costs and improve the appearance of building facades. Another area where functionalized coatings have potential in the construction industry is in the development of sustainable building materials. Sustainable building materials require materials that can reduce the environmental impact of buildings, such as by reducing energy consumption and carbon emissions. Functionalized coatings can be applied to these materials to improve their properties, such as by providing enhanced insulation or improving their durability in harsh environments.

19.8 Opportunities presented by advances in nanotechnology and surface engineering

Nanotechnology and surface engineering are two rapidly growing fields that have presented exciting opportunities for the development of functionalized coatings. By utilizing these advancements it is possible to create coatings with improved performance, durability, and functionality. In this section we will explore the potential opportunities presented by advances in nanotechnology and surface engineering in functionalized coatings.

19.8.1 Nanotechnology in functionalized coatings

Nanotechnology involves the study and manipulation of materials at the nanoscale, typically ranging from 1 to 100 nanometers. At this scale materials exhibit unique properties and behaviors that can be harnessed to create functionalized coatings with improved performance and functionality. The use of nanomaterials in coatings has already demonstrated significant improvements in areas such as mechanical, chemical, and thermal properties, as well as in biocompatibility and electrical conductivity.

Mechanical properties

Nanomaterials have the potential to significantly enhance the mechanical properties of coatings. For example the addition of nanoparticles to a coating can improve its hardness, scratch resistance, and wear resistance. This is because the small size of nanoparticles allows them to fill in the gaps between larger particles, creating a denser and more uniform coating structure. This results in coatings with improved mechanical properties, such as increased resistance to wear and abrasion.

Chemical properties

Functionalized coatings with improved chemical properties can be developed by incorporating nanomaterials into the coating formulation. For example nanomaterials such as silica or alumina can be used to create coatings with improved chemical resistance and adhesion. Additionally the use of nanomaterials can provide coatings with enhanced UV stability, corrosion resistance, and barrier properties.

Thermal properties

The incorporation of nanomaterials into coatings can also improve their thermal properties, such as thermal conductivity and insulation. For example coatings containing carbon nanotubes or graphene can improve thermal conductivity, while coatings containing aerogels can provide enhanced insulation. This has potential applications in various industries, such as in the development of thermal barrier coatings for aerospace and automotive applications.

Biocompatibility

Nanotechnology also has potential applications in the development of biocompatible coatings. By incorporating nanomaterials such as titanium dioxide or hydroxyapatite into coatings, it is possible to create coatings with improved biocompatibility, such as in medical implants. Additionally the use of nanomaterials can provide coatings with enhanced antimicrobial properties, which can be particularly useful in healthcare settings.

19.8.2 Surface engineering in functionalized coatings

Surface engineering involves the modification of a material's surface properties to improve its functionality and performance. By utilizing advanced surface engineering techniques, it is possible to create functionalized coatings with improved properties, such as improved adhesion, wear resistance, and corrosion resistance. The following are some potential opportunities presented by surface engineering in functionalized coatings.

Surface modification techniques

Surface modification techniques, such as plasma treatment or ion implantation, can be used to modify the surface properties of coatings. These techniques can be used to improve the adhesion of coatings to substrates, as well as to enhance the surface hardness and wear resistance of coatings. Additionally surface modification techniques can be used to create coatings with enhanced corrosion resistance, such as by introducing a corrosion-resistant layer to the surface of the coating.

Surface coatings

Surface coatings can be applied to a material's surface to improve its functionality and performance. For example coatings such as diamond-like carbon or nitride coatings can be applied to metal components to improve their wear resistance and hardness. Additionally surface coatings can be used to create coatings with improved biocompatibility, such as by applying a hydroxyapatite coating to medical implants [69–71].

Self-assembled monolayers

Self-assembled monolayers (SAMs) are a type of functionalized thin film coating that has shown great potential in various applications. However the development of SAMs has been limited by their relatively low stability and poor adhesion to substrates. Advances in nanotechnology and surface engineering have opened up new opportunities for improving the stability and adhesion of SAMs, as well as enhancing their functionality. One promising approach is the use of nanoparticles to enhance the properties of SAMs. Nanoparticles can be incorporated into the SAMs to increase their mechanical strength and durability, and to introduce additional functionalities. For example gold nanoparticles have been used to increase the stability and adhesion of SAMs on silicon substrates, while iron oxide nanoparticles have been used to enhance the magnetic properties of SAMs. Another approach is the use of surface modification techniques to improve the adhesion of SAMs to substrates. For example plasma treatment can be used to modify the surface properties of the substrate, creating a more favorable environment for SAM formation. Similarly, the use of interfacial layers, such as a thin layer of metal or metal oxide, can improve the adhesion of SAMs to the substrate.

In addition to SAMs advances in nanotechnology and surface engineering have also led to the development of other functionalized coatings with enhanced properties. One example is the use of graphene as a coating material. Graphene is a single-layer sheet of carbon atoms with unique mechanical, electrical, and thermal properties [72–76]. Graphene coatings have shown promise in various applications, including corrosion protection, energy storage, and sensor development. The use of nanocomposites is another promising approach for enhancing the properties of functionalized coatings. Nanocomposites are materials that consist of a matrix material and nanoparticles dispersed within it. By incorporating nanoparticles into the matrix material, the resulting nanocomposite can exhibit improved mechanical, electrical, and thermal properties. Nanocomposite coatings have been developed for various applications, including corrosion protection, wear resistance, and thermal barrier coatings. In conclusion it could be stated that the advances in nanotechnology and surface engineering have created new opportunities for developing functionalized coatings with enhanced properties. Nanoparticles, surface modification techniques, and nanocomposites can be used to improve the stability, adhesion, and functionality of functionalized coatings, making them suitable for a wide range of applications in various industries.

19.9 Conclusion

Functionalized thin film coatings offer numerous benefits, including improved corrosion resistance, wear resistance, and biocompatibility, among others. These coatings have the potential to significantly enhance the performance and durability of various metal alloys, making them ideal for use in various industries, including aerospace, automotive, and energy. However, the development of functionalized thin film coatings is not without challenges. One of the main challenges is achieving good adhesion between the coating and the substrate. Poor adhesion can lead to premature failure of the coating, rendering it ineffective. Another challenge is the need to optimize the deposition process to minimize defects and ensure consistent coating properties. Despite these challenges, the future outlook for functionalized thin film coatings is promising. Advances in nanotechnology and surface engineering have opened up new opportunities for enhancing the properties of coatings and improving their stability and adhesion. The use of advanced materials, such as nanoparticles and graphene, can significantly enhance the mechanical, electrical, and thermal properties of coatings, making them suitable for a wide range of applications. In addition the development of new deposition techniques, such as ALD and MLD, has allowed for precise control over the thickness and composition of coatings, making it possible to tailor their properties to specific applications. These techniques also offer the potential for high-throughput production, making them ideal for use in large-scale industrial applications. In conclusion, the development of functionalized thin film coatings presents a significant opportunity for enhancing the properties and performance of various metal alloys. While there are challenges associated with their development, advances in nanotechnology and surface engineering offer new opportunities for improving the stability, adhesion, and functionality of these coatings. As a result, functionalized thin film coatings are likely to play an increasingly important role in various industries in the coming years.

References

[1]   Bashir S, Thakur A, Lgaz H, Chung IM, Kumar A. Corrosion inhibition efficiency of bronopol on aluminium in 0.5 M HCl solution: Insights from experimental and quantum chemical studies. Surf Interfaces [Internet]. 2020;20(April):100542. Available from: https://doi.org/10.1016/j.surfin.2020.100542

[2]   Thakur A, Kaya S, Kumar A. Recent innovations in nano container-based self-healing coatings in the construction industry. Curr Nanosci. 2021;18(2):203–16.

[3]   Thakur A, Sharma S, Ganjoo R, Assad H, Kumar A. Anti-corrosive potential of the sustainable corrosion inhibitors based on biomass waste: A review on preceding and perspective research. J Phys Conf Ser. 2022;2267(1):012079.

[4]   Thakur A, Kaya S, Abousalem AS, Experimental KA. DFT and MC simulation analysis of Vicia Sativa weed aerial extract as sustainable and eco-benign corrosion inhibitor for mild steel in acidic environment. Sustain Chem Pharm [Internet]. 2022;29(July):100785. Available from: https://doi.org/10.1016/j.scp.2022.100785

[5]   Thakur A, Kaya S, Abousalem AS, Sharma S, Ganjoo R, Assad H, et al. Computational and experimental studies on the corrosion inhibition performance of an aerial extract of Cnicus Benedictus weed on the acidic corrosion of mild steel. Process Saf Environ Prot [Internet]. 2022;161:801–18. Available from: https://doi.org/10.1016/j.psep.2022.03.082

[6]   Thakur A, Kumar A, Sharma S, Ganjoo R, Assad H. Materials Today : Proceedings Computational and experimental studies on the efficiency of Sonchus arvensis as green corrosion inhibitor for mild steel in 0 . 5 M HCl solution. Mater Today Proc [Internet]. 2022;66:609–21. Available from: https://doi.org/10.1016/j.matpr.2022.06.479

[7]   Kumar A, Thakur A. Encapsulated nanoparticles in organic polymers for corrosion inhibition [Internet]. Corrosion protection at the nanoscale (p. 345–62). Elsevier Inc.; 2020. Available from: http://dx.doi.org/10.1016/B978-0-12-819359-4.00018-0

[8]   Thakur A, Kumar A. Recent advances on rapid detection and remediation of environmental pollutants utilizing nanomaterials-based (bio)sensors. Sci Total Environ [Internet]. 2022;834 (January):155219. Available from: https://doi.org/10.1016/j.scitotenv.2022.155219

[9]   Thakur A, Kumar A, Kaya S, Vo DVN, Sharma A. Suppressing inhibitory compounds by nanomaterials for highly efficient biofuel production: A review. Fuel [Internet]. 2022;312(September 2021):122934. Available from: https://doi.org/10.1016/j.fuel.2021.122934

[10] Bashir S, Thakur A, Lgaz H, Chung I-M, Kumar A. Computational and experimental studies on Phenylephrine as anti-corrosion substance of mild steel in acidic medium. J Mol Liq 2019;293:111539.

[11] Parveen G, Bashir S, Thakur A, Saha SK, Banerjee P, Kumar A. Experimental and computational studies of imidazolium based ionic liquid 1-methyl- 3-propylimidazolium iodide on mild steel corrosion in acidic solution experimental and computational studies of imidazolium based ionic liquid 1-methyl- 3-propylimidazolium. Mater Res Express 2020;7(1):016510.

[12] Thakur A, Savaş K, Kumar A. Recent trends in the characterization and application progress of nano-modified coatings in corrosion mitigation of metals and alloys. Appl Sci 2023;13:730.

[13] Thakur A, Kumar A, Kaya S, Marzouki R, Zhang F, Guo L. Recent advancements in surface modification, characterization and functionalization for enhancing the biocompatibility and corrosion resistance of biomedical implants. Coatings 2022;12:1459.

[14] Thakur A, Kumar A. Recent trends in nanostructured carbon-based electrochemical sensors for the detection and remediation of persistent toxic substances in real- time analysis. Mater Res Express. 2023;10:034001.

[15] Thakur A, Kumar A. Sustainable inhibitors for corrosion mitigation in aggressive corrosive media: A comprehensive study. Bio Tribo Corros [Internet]. 2021;7(2):1–48. Available from: https://doi.org/10.1007/s40735-021-00501-y

[16] Bashir S, Thakur A, Lgaz H, Chung I-M, Kumar A. Corrosion Inhibition performance of acarbose on mild steel corrosion in acidic medium: An experimental and computational study. Arab J Sci Eng [Internet]. 2020;45(6):4773–83. Available from: https://doi.org/10.1007/s13369-020-04514-6

[17] Darwish MSA, Mostafa MH, Al-Harbi LM. Polymeric Nanocomposites for Environmental and Industrial Applications. Int J Mol Sci 2022;23(3).

[18] Saif S, Tahir A, Asim T, Chen Y, Adil SF. Polymeric nanocomposites of iron–oxide nanoparticles (Ionps) synthesized using terminalia chebula leaf extract for enhanced adsorption of arsenic(v) from water. Colloids Interfaces 2019;3(1).

[19] Wang M, Wang K, Yang Y, Liu Y, Yu DG. Electrospun environment remediation nanofibers using unspinnable liquids as the sheath fluids: A review. Polymers (Basel) 2020;12(1).

[20] Qiao SJ, Xu XN, Qiu Y, Xiao HC, Zhu YF. Simultaneous reduction and functionalization of graphene oxide by 4-hydrazinobenzenesulfonic acid for polymer nanocomposites. Nanomaterials 2016;6(2).

[21] Barillas L, Makhneva E, An S, Fricke K. Functional thin films synthesized from liquid precursors by combining mist chambers and atmospheric-pressure plasma polymerization. Coatings 2021;11(11).

[22] Yadav VK, Yadav KK, Tirth V, Jangid A, Gnanamoorthy G, Choudhary N, et al. Recent advances in methods for recovery of cenospheres from fly ash and their emerging applications in ceramics, composites, polymers and environmental cleanup. Crystals 2021;11(9):1–20.

[23] Leonés A, Lieblich M, Benavente R, Gonzalez JL, Peponi L. Potential applications of magnesium-based polymeric nanocomposites obtained by electrospinning technique. Nanomaterials 2020;10 (8):1–33.

[24] Agboola O, Sunday O, Fayomi I, Ayodeji A, Ayeni AO, Alagbe EE, et al. A review on polymer nanocomposites and their effective applications in membranes and adsorbents for water treatment and gas separation. Membranes (Basel) 2021;11(139):25–33.

[25] Homaeigohar S, Elbahri M. Nanocomposite electrospun nanofiber membranes for environmental remediation. Materials (Basel) 2014;7(2):1017–45.

[26] Rahmati S, Doherty W, Amani Babadi A, Akmal Che Mansor MS, Julkapli NM, Hessel V, et al. Gold–carbon nanocomposites for environmental contaminant sensing. Micromachines 2021;12(6).

[27] Yadav S, Asthana A, Singh AK, Chakraborty R, Sree Vidya S, Singh A, et al. Methionine-functionalized graphene oxide/sodium alginate bio-polymer nanocomposite hydrogel beads: Synthesis, isotherm and kinetic studies for an adsorptive removal of fluoroquinolone antibiotics. Nanomaterials 2021;11 (3):1–25.

[28] Cova TF, Murtinho D, Aguado R, Aacc P, Valente AJM. Cyclodextrin polymers and cyclodextrin-containing polysaccharides for water remediation. Polysaccharides 2021;2(1):16–38.

[29] Sawicki R, Mercier L. Evaluation of mesoporous cyclodextrin-silica nanocomposites for the removal of pesticides from aqueous media. Environ Sci Technol 2006;40(6):1978–83.

[30] Kobkeatthawin T, Chaveanghong S, Trakulmututa J, Amornsakchai T, Kajitvichyanukul P, Smith SM. Photocatalytic activity of TiO2/g-C3N4 nanocomposites for removal of monochlorophenols from water. Nanomaterials 2022;12(16).

[31] Wanjeri VWO, Sheppard CJ, Prinsloo ARE, Ngila JC, Ndungu PG. Isotherm and kinetic investigations on the adsorption of organophosphorus pesticides on graphene oxide based silica coated magnetic nanoparticles functionalized with 2-phenylethylamine. J Environ Chem Eng [Internet]. 2018;6 (1):1333–46. Available from: http://dx.doi.org/10.1016/j.jece.2018.01.064

[32] Yuan C, Li R, Wu L, Hong X, He H, Yang G, et al. Optimization of a modified QuEChERS method by an n-octadecylamine-functionalized magnetic carbon nanotube porous nanocomposite for the quantification of pesticides. J Food Compos Anal [Internet]. 2021;102:103980. Available from: https://doi.org/10.1016/j.jfca.2021.103980

[33] Sartori B, Amenitsch H, Marmiroli B. Functionalized mesoporous thin films for biotechnology. Micromachines 2021;12(7).

[34] Katibi KK, Yunos KF, Man HC, Aris AZ, Nor MZM, Azis RS, et al. Contemporary techniques for remediating endocrine-disrupting compounds in various water sources: Advances in treatment methods and their limitations. Polymers (Basel). 2021;13(19):1–46.

[35] Youssef AM, El-Naggar ME, Malhat FM, El Sharkawi HM. Efficient removal of pesticides and heavy metals from wastewater and the antimicrobial activity of f-MWCNTs/PVA nanocomposite film. J Clean Prod. 2019;206:315–25.

[36] Amari A, Alzahrani FM, Katubi KM, Alsaiari NS, Tahoon MA, Rebah FB. Clay-polymer nanocomposites: Preparations and utilization for pollutants removal. Materials (Basel). 2021;14 (6):1–21.

[37] Saxena S, Saxena U. Development of bimetal oxide doped multifunctional polymer nanocomposite for water treatment. Int Nano Lett. 2016;6(4):223–34.

[38] Hu L, Yang Z, Wang Y, Li Y, Fan D, Wu D, et al. Facile preparation of water-soluble hyperbranched polyamine functionalized multiwalled carbon nanotubes for high-efficiency organic dye removal from aqueous solution. Sci Rep. 2017;7(1):1–13.

[39] Qutub N, Singh P, Sabir S, Umar K, Sagadevan S, Oh WC. Synthesis of polyaniline supported CdS/CdS-ZnS/CdS-TiO2 nanocomposite for efficient photocatalytic applications. Nanomaterials. 2022;12 (8):1–21.

[40] Saber-Samandari S, Saber-Samandari S, Joneidi-Yekta H, Mohseni M. Adsorption of anionic and cationic dyes from aqueous solution using gelatin-based magnetic nanocomposite beads comprising carboxylic acid functionalized carbon nanotube. Chem Eng J [Internet]. 2017;308:1133–44. Available from: http://dx.doi.org/10.1016/j.cej.2016.10.017

[41] Hosseini SA, Daneshvar E, Asl S, Vossoughi M, Simchi A, Sadrzadeh M. Green electrospun membranes based on chitosan/amino-functionalized nanoclay composite fibers for cationic dye removal: Synthesis and kinetic studies. ACS Omega 2021;6(16):10816–27.

[42] Wan Ibrahim WA, Nodeh HR, Aboul-Enein HY, Sanagi MM. Magnetic solid-phase extraction based on modified ferum oxides for enrichment, preconcentration, and isolation of pesticides and selected pollutants. Crit Rev Anal Chem 2015;45(3):270–87.

[43] Liu Y, Hou C, Jiao T, Song J, Zhang X, Xing R, et al. Self-assembled AgNP-containing nanocomposites constructed by electrospinning as efficient dye photocatalyst materials for wastewater treatment. Nanomaterials 2018;8(1).

[44] Kuhn R, Bryant IM, Jensch R, Böllmann J. Applications of environmental nanotechnologies in remediation, wastewater treatment, drinking water treatment, and agriculture. Appl Nano 2022;3 (1):54–90.

[45] Rodríguez C, Tapia C, Leiva-Aravena E, Leiva E. Graphene oxide–zno nanocomposites for removal of aluminum and copper ions from acid mine drainage wastewater. Int J Environ Res Public Health 2020;17(18):1–18.

[46] Khan ZU, Khan WU, Ullah B, Ali W, Ahmad B, Yap PS. Graphene oxide/PVC composite papers functionalized with p-Phenylenediamine as high-performance sorbent for the removal of heavy metal ions. J Environ Chem Eng [Internet]. 2021;9(5):105916. Available from: https://doi.org/10.1016/j.jece.2021.105916

[47] Deng S, Liu X, Liao J, Lin H, Liu F. PEI modified multiwalled carbon nanotube as a novel additive in PAN nanofiber membrane for enhanced removal of heavy metal ions. Chem Eng J [Internet]. 2019;375(May):122086. Available from: https://doi.org/10.1016/j.cej.2019.122086

[48] Mahmoud ME, Nabil GM, Zaki MM, Saleh MM. Starch functionalization of iron oxide by-product from steel industry as a sustainable low cost nanocomposite for removal of divalent toxic metal ions from water. Int J Biol Macromol [Internet]. 2019;137:455–68. Available from: https://doi.org/10.1016/j.ijbiomac.2019.06.170

[49] Martín DM, Ahmed MM, Rodríguez M, García MA, Faccini M. Aminated polyethylene terephthalate (PET) nanofibers for the selective removal of Pb(II) from polluted water. Materials (Basel). 2017;10 (12).

[50] Kabir A, Dunlop MJ, Acharya B, Bissessur R, Ahmed M. Polymeric composites with embedded nanocrystalline cellulose for the removal of iron(II) from contaminated water. Polymers (Basel). 2018;10(12):1–16.

[51] Anuma S, Mishra P, Bhat BR. Polypyrrole functionalized Cobalt oxide Graphene (COPYGO) nanocomposite for the efficient removal of dyes and heavy metal pollutants from aqueous effluents. J Hazard Mater [Internet]. 2021;416(November 2020):125929. Available from: https://doi.org/10.1016/j.jhazmat.2021.125929

[52] Ramalingam B, Parandhaman T, Choudhary P, Das SK. Biomaterial Functionalized Graphene-Magnetite Nanocomposite: A Novel Approach for Simultaneous Removal of Anionic Dyes and Heavy-Metal Ions. ACS Sustain Chem Eng. 2018;6(5):6328–41.

[53] Akhavan B, Jarvis K, Majewski P. Plasma polymer-functionalized silica particles for heavy metals removal. ACS Appl Mater Interfaces. 2015;7(7):4265–74.

[54] Elessawy NA, Gouda MH, Elnouby MS, Zahran HF, Hashim A, Abd El-Latif MM, et al. Novel sodium alginate/polyvinylpyrrolidone/tio2 nanocomposite for efficient removal of cationic dye from aqueous solution. Appl Sci. 2021;11(19).

[55] Badruddoza AZM, Shawon ZBZ, Tay WJD, Hidajat K, Uddin MS. Fe 3O 4/cyclodextrin polymer nanocomposites for selective heavy metals removal from industrial wastewater. Carbohydr Polym [Internet]. 2013;91(1):322–32. Available from: http://dx.doi.org/10.1016/j.carbpol.2012.08.030

[56] Abualnaja KM, Alprol AE, Abu-Saied MA, Mansour AT, Ashour M. Studying the adsorptive behavior of poly(Acrylonitrile-co-styrene) and carbon nanotubes (nanocomposites) impregnated with adsorbent materials towards methyl orange dye. Nanomaterials 2021;11(5).

[57] Trikkaliotis DG, Ainali NM, Tolkou AK, Mitropoulos AC, Lambropoulou DA, Bikiaris DN, et al. Removal of heavy metal ions from wastewaters by using Chitosan/Poly(Vinyl Alcohol) adsorbents: A review. Macromol 2022;2(3):403–25.

[58] Shi T, Xie Z, Mo X, Feng Y, Peng T, Song D. Highly efficient adsorption of heavy metals and cationic dyes by smart functionalized sodium alginate hydrogels. Gels 2022;8(6).

[59] Li Y, Huang L, He W, Chen Y, Lou B. Preparation of functionalized magnetic Fe3O4@Au@polydopamine nanocomposites and their application for copper(II) removal. Polymers (Basel) 2018;10(6).

[60] Li K, Miwornunyuie N, Chen L, Jingyu H, Amaniampong PS, Koomson DA, et al. Sustainable application of ZIF-8 for heavy-metal removal in aqueous solutions. Sustain 2021;13(2):1–11.

[61] Zhang J, Azam MS, Shi C, Huang J, Yan B, Liu Q, et al. Poly(acrylic acid) functionalized magnetic graphene oxide nanocomposite for removal of methylene blue. RSC Adv [Internet]. 2015;5 (41):32272–82. Available from: http://dx.doi.org/10.1039/C5RA01815C

[62] Lakouraj MM, Hasanzadeh F, Zare EN. Nanogel and super-paramagnetic nanocomposite of thiacalix [4]arene functionalized chitosan: synthesis, characterization and heavy metal sorption. Iran Polym J English Ed. 2014;23(12):933–45.

[63] Plohl O, Finšgar M, Gyergyek S, Ajdnik U, Ban I, Zemljič LF. Efficient copper removal from an aqueous anvironment using a novel and hybrid nanoadsorbent based on derived-polyethyleneimine linked to silica magnetic nanocomposites. Nanomaterials. 2019;9(2).

[64] Mishra S, Singh AK, Singh JK. Ferrous sulfide and carboxyl-functionalized ferroferric oxide incorporated PVDF-based nanocomposite membranes for simultaneous removal of highly toxic heavy-metal ions from industrial ground water. J Memb Sci [Internet]. 2020;593(May 2019):117422. Available from: https://doi.org/10.1016/j.memsci.2019.117422

[65] Musico YLF, Santos CM, Dalida MLP, Rodrigues DF. Improved removal of lead(ii) from water using a polymer-based graphene oxide nanocomposite. J Mater Chem A. 2013;1(11):3789–96.

[66] Liu X, Guan J, Lai G, Xu Q, Bai X, Wang Z, et al. Stimuli-responsive adsorption behavior toward heavy metal ions based on comb polymer functionalized magnetic nanoparticles. J Clean Prod [Internet]. 2020;253:119915. Available from: https://doi.org/10.1016/j.jclepro.2019.119915

[67] Singh S, Barick KC, Bahadur D. Functional oxide nanomaterials and nanocomposites for the removal of heavy metals and dyes. Nanomater Nanotechnol 2013;3(1):1–19.

[68] Dadashi Firouzjaei M, Akbari Afkhami F, Rabbani Esfahani M, Turner CH, Nejati S. Experimental and molecular dynamics study on dye removal from water by a graphene oxide-copper-metal organic framework nanocomposite. J Water Process Eng [Internet] 2020;34(January):101180. Available from: https://doi.org/10.1016/j.jwpe.2020.101180

[69] Gao Y, Truong YB, Cacioli P, Butler P, Kyratzis IL. Bioremediation of pesticide contaminated water using an organophosphate degrading enzyme immobilized on nonwoven polyester textiles. Enzyme Microb Technol [Internet] 2014;54(1):38–44. Available from: http://dx.doi.org/10.1016/j.enzmictec.2013.10.001

[70] Moustafa M, Abu-Saied MA, Taha T, Elnouby M, El-shafeey M, Alshehri AG, et al. Chitosan functionalized AgNPs for efficient removal of Imidacloprid pesticide through a pressure-free design. Int J Biol Macromol [Internet]. 2021;168:116–23. Available from: https://doi.org/10.1016/j.ijbiomac.2020.12.055

[71]   Namvari M, Namazi H. Synthesis of magnetic citric-acid-functionalized graphene oxide and its application in the removal of methylene blue from contaminated water. Polym Int 2014;63 (10):1881–88.

[72]   Yang S, Wang L, Zhang X, Yang W, Song G. Enhanced adsorption of Congo red dye by functionalized carbon nanotube/mixed metal oxides nanocomposites derived from layered double hydroxide precursor. Chem Eng J [Internet] 2015;275(January):315–21. Available from: http://dx.doi.org/10. 1016/j.cej.2015.04.049

[73]   Zare EN, Motahari A, Sillanpää M. Nanoadsorbents based on conducting polymer nanocomposites with main focus on polyaniline and its derivatives for removal of heavy metal ions/dyes: A review. Environ Res [Internet] 2018;162(January):173–95. Available from: https://doi.org/10.1016/j.envres. 2017.12.025

[74]   Zhang W, Ou J, Wang B, Wang H, He Q, Song J, et al. Efficient heavy metal removal from water by alginate-based porous nanocomposite hydrogels: The enhanced removal mechanism and influencing factor insight. J Hazard Mater 2021;418(March).

[75]   Mousavi SR, Asghari M, Mahmoodi NM. Chitosan-wrapped multiwalled carbon nanotube as filler within PEBA thin film nanocomposite (TFN) membrane to improve dye removal. Carbohydr Polym [Internet]. 2020;237(March):116128. Available from: https://doi.org/10.1016/j.carbpol.2020.116128

[76]   Anirudhan TS, Christa J, Shainy F. Magnetic titanium dioxide embedded molecularly imprinted polymer nanocomposite for the degradation of diuron under visible light. React Funct Polym [Internet] 2020;152(April):104597. Available from: https://doi.org/10.1016/j.reactfunctpolym.2020. 104597

Index

www.ingramcontent.com/pod-product-compliance
Lightning Source LLC
Chambersburg PA
CBHW080124220326
41598CB00032B/4951